ENERGY SYSTEMS

ENERGY SYSTEMS

A NEW APPROACH TO ENGINEERING THERMODYNAMICS

2nd Edition

RENAUD GICQUEL

CRC Press
Taylor & Francis Group
Boca Raton London New York Leiden

CRC Press is an imprint of the
Taylor & Francis Group, an **informa** business

A BALKEMA BOOK

CRC Press/Balkema is an imprint of the Taylor & Francis Group, an informa business

© 2022 Taylor & Francis Group, London, UK

Typeset by codeMantra

First edition © 2011 by Taylor & Francis Group, LLC

Library of Congress Cataloging-in-Publication Data
Names: Gicquel, Renaud, author.
Title: Energy systems : a new approach to engineering thermodynamics /
Renaud Gicquel, École des Mines de Paris, MINES ParisTech, France.
Description: 2nd edition. | Boca Raton : CRC Press, [2021] | Includes
bibliographical references and index.
Identifiers: LCCN 2020057636 (print) | LCCN 2020057637 (ebook)
Subjects: LCSH: Thermodynamics. | Energy facilities. | Energy transfer.
Classification: LCC TJ265 .G45 2021 (print) | LCC TJ265 (ebook) | DDC
621.042--dc23
LC record available at https://lccn.loc.gov/2020057636
LC ebook record available at https://lccn.loc.gov/2020057637

Published by: CRC Press/Balkema
                 Schipholweg 107C, 2316 XC Leiden, The Netherlands
                 e-mail: Pub.NL@taylorandfrancis.com
                 www.routledge.com – www.taylorandfrancis.com

ISBN: 978-0-367-72600-3 (Hbk)
ISBN: 978-1-032-00774-8 (Pbk)
ISBN: 978-1-003-17562-9 (eBook)

DOI: 10.1201/9781003175629

To my wife and companion,
To our three daughters
And to our grandchildren

Contents

Searching references in the thermoptim-unit portal (www.thermoptim.org)

The Thermoptim-Unit portal contains numerous references that can be used by the readers of this book.

There are several ways to search for references in this portal.

You can of course navigate through the different pages of the portal, which will allow you to discover all of its resources.

You may also use the portal's search engine: in the upper-right part of its screen, you will find a field that automatically searches for pages containing the words you enter.

For the references to the links cited in this book, we recommend that you search for them via a small website that brings them all together and allows direct access.

Its address is: www.s4e2.com/portal/web/index.html

Foreword to the first edition by John W. Mitchell

This is an ambitious book that presents a new approach to teaching energy systems. The author, Renaud Gicquel, has introduced two new paradigms to teaching the subject. The first is that the study of energy systems is initiated at the system level rather than the mechanism level. The second paradigm is the use of software that is integral to the study of systems and, ultimately, to learning the underlying thermodynamic principles. The success of these paradigms is demonstrated through the adoption of this approach by over 120 institutions of higher education in France and worldwide.

This book is divided into five main sections. The first section, titled *First Steps in Engineering Thermodynamics*, covers the entire subject of energy systems, with the first chapter describing the pedagogic approach used in the rest of this book. The second chapter focuses on the system level and illustrates the teaching philosophy. This chapter gives students exposure to those components that comprise complete energy systems such as a steam power plant, and what each component does. The emphasis is on understanding the concepts underlying the operation of each component, such as the change of phase from liquid to vapor in the boiler. Process diagrams are introduced to further understanding. Complete system models are created, and the performance is determined using the software Thermoptim. By starting at the system level, students early on become familiar with the vocabulary, components, and performance of a system before they become enmeshed in the thermodynamic relations used to compute properties and energy flows. Stressing conceptual understanding rather than calculation reinforces the important basic ideas.

The pedagogical approach is a significant departure from the typical methods used in teaching thermodynamics. Traditionally, one starts with basic conservation equations and property relations, with calculations geared toward determining values for a process rather than for the performance of a system as a whole. The emphasis is placed on a detailed understanding of the relations that govern thermodynamic behavior. Typically, applications to system follow late in the study after all the underpinning has been developed.

While traditional approaches have been used successfully for many years, educational research has established that alternate approaches can improve learning. Context is one key driver for student learning, and this book's approach wherein students start with models of actual systems commonly used for power and refrigeration establishes relevance to engineering practice. Ideas from problem-based learning, in which students are confronted with a problem to solve and then learn what is needed to obtain a solution, are also incorporated. Student motivation and conceptual understanding are important to mastering thermodynamic fundamentals that underlie performance.

The second paradigm is that the learning of thermodynamics is intertwined with the software. Thermoptim allows students to quickly create realistic system models and explore a number of "What-if" questions, such as what is the effect on power if the steam temperature in the boiler is raised a certain number of degrees. The component models are built on the fundamental relations, and the familiarity students gain working with models and observing behavior aids them in their later study of thermodynamics. Further, using simulation models from the start provides a bridge to engineering practice in which realistic situations require software for solution. It is essential that the student and practitioner understand the thermodynamic relations necessary to form a model, as is covered in later sections of this book, but it is not necessary to develop a model from basic relations each time. This intimately linking of software to content is in the vanguard of modern books on engineering topics.

The second section of this book is *Methodology, Thermodynamics Fundamentals, Thermoptim, Components*. With the background in systems, thermodynamics, and Thermoptim from the first section, the rigorous study of thermodynamics proceeds rapidly. The first law, second law and property relations are introduced, following traditional approaches to teaching thermodynamics, except that now the context of the material is clear, motivating the student. The coverage extends to a number of real property effects that are often not emphasized in traditional thermodynamic books; there are sections on the property relations for moist air, solutions and mixtures. The Thermoptim framework allows real property variations to be included in system analysis.

Models are constructed in Thermoptim analogous to the way they are in a real system, and a chapter is devoted to discussing model building. Conduits connect components through fluid transport, carrying information from one component to the next. In the steam power plant, for example, the pipe transporting steam from the turbine exhaust to the condenser carries with it all of the state information on the steam that the actual pipe would. Information is transported automatically, which reduces the need for the student to keep track of all properties at every point. Evaluation of the state of fluids throughout the system is accomplished analogous to measurement stations in a real system. The natural analog between the simulation and the real system facilitates learning.

The remaining chapters in this section cover a number of more advanced topics, such as different types of compression processes, complete and incomplete combustion, psychrometrics, heat exchangers and second law topics (exergy/availability). As with the earlier thermodynamic material, the presentation is rigorous and includes effects that are often not taught in first courses, but that are included in the Thermoptim components available for simulation. Worked examples are included so that students can explore the subject using the available models.

The last three sections of this book are devoted to different heating, cooling and power systems. The first of these is a strong and detailed exposition of conventional cycles. Gas turbine, internal combustion engine (spark ignition and diesel), steam power and cogeneration plants, mechanical and absorption refrigeration systems are discussed from both thermodynamic and practical points of view. The last two sections deal with innovative cycles, with an emphasis on low environmental impact. Such topics as the Kalina cycle, desalinization, cryogenics, fuel cells, solar power and sequestering carbon dioxide are discussed. These last three sections are beyond what would be covered in undergraduate courses, but the combination of references and Thermoptim components make these sections valuable for research projects and advanced study. As with the beginning chapters, the emphasis is always on conceptual understanding, with the implementation incorporated into Thermoptim.

This is a comprehensive book on energy systems with an almost encyclopedic coverage of the details of the equipment and systems involved in power production, refrigeration and air conditioning. The integration of technical content with advanced software allows a range of users from students who are beginning their study to those involved in research on promising cycles. From a teaching perspective, the initial focus on the system level combined with the simulation tool Thermoptim serves to quickly bring students up to speed on applications, and provides motivation for further study. This book promises to be one that engineers will keep on their desks for ready reference and study.

John W. Mitchell
Kaiser Chair Professor of Mechanical Engineering, Emeritus
University of Wisconsin-Madison
Madison, Wisconsin, USA

Foreword to the first edition by Alain Lambotte

By an innovative approach to thermodynamics and a progressive study of systems – ranging from the simplest to the most complex – *Energy Systems* comprises three books in one: a guide for beginners, a reference book for advanced learners and a support for development for professionals. I have been using the French version for several years personally, to develop training content and validate skill levels in the electricity utility company where I am engaged with.

My target audience consists of two groups: officials who have left university or the Haute École for some years and therefore have a certain "distance" towards mathematics, and of fresh graduates, for whom "thermodynamics is math", and who have little evocative image of the underlying physical reality.

This book, by its content and its character, is an encouraging and stylish manifesto of a new teaching practice of engineering thermodynamics. In contrast to the existing teaching methods on the matter, it spares the reader mathematical contingencies, the aggregation of knowledge and the immutable laws of thermodynamics in the first steps. Instead, learning by reflection has priority over the memorization of scientific knowledge that is not immediately essential. Mathematical relationships illustrate the point, but are not the heart of the matter. This is ideal for the technicians and engineers we train, who often have a much lower accurate mathematical level at their disposal than when they were still students.

The approach is fully system-oriented. It allows both basic learning and the development of innovative thought. After the author has explained the setup, quite naturally and visually, there is place for reflection and numerous applications. Technologies are presented simply at first, and subsequently with increasing detail. This makes the adequacy of the author's approach multiple: beginners, professionals and experts will all find answers to their questions.

In combination with the *www.thermoptim.org* portal and the possibilities this offers, *Energy Systems* is an appealing textbook and developing tool for students, teachers, researchers and practitioners. The exhaustive character, through the given systems and models and through the enormous amount of tools in the Thermoptim portal and simulator, allows the reader to easily self-develop new models to fulfil his or her needs. These can range from basic notions to the most advanced concepts in energy systems. All factors together make this set a very powerful reference that allows easier implementation into practice than any existing books on the subject.

Although I use *Energy Systems & Thermoptim*, and even overindulge in it, I am far from having been around the concept. Usage has changed my approach to thermodynamics, both in my engineering work and in preparing course content. The development of a much more accessible and user-friendly approach than encountered earlier made using it a pleasure, both personally and in training. Last but not least, it widely opens the doors to creativity, which is a major requirement for our energy future.

Alain Lambotte
Content Manager, Competence and Training Center
Electricity Utility, Belgium

About the author

Active as a full professor from 1986 to 2020, Dr. Renaud Gicquel (ORCID ID: 0000-0001-7960-087) has taught a wide variety of subjects, such as applied thermodynamics and global energy issues and energy system modeling. His recent research activities were focused on the optimization of complex thermodynamic plants (heat exchanger networks, cogeneration, combined cycles) and on the use of information and communication technologies for scientific instruction.

Renaud's special interest and passion is the combination of thermodynamics and energy-powered system education with modern information technology tools. To this end, he has developed various software packages to facilitate the teaching and learning of applied thermodynamics and the simulation of energy systems:

* *Thermoptim (Thermo-Calc)* professional software, 2000, CSTB, www.cstb.fr
* *Interactive Thermodynamic Charts*, 2000
* *Diapason* e-learning modules, 2004, www.thermoptim.org

The origin of this book comes from earlier works, *Introduction to Global Energy Problems* (1992, Economica in French) and the original *Energy Systems* (two volumes, 2001, Presses de l'École des Mines in French). These have formed the basis for refinement of the Thermoptim Software and a three-volume second edition of Energy Systems in French. The latter was very well received by more than a hundred higher education institutes in France and abroad, and the first English edition was prepared in 2011.

This second edition, while following the same global logic, does not go as much in the details. In order to simplify the content, sections that were intended for advanced readers have been removed and theoretical developments have been limited to the essential.

With a new approach to teaching and learning applied thermodynamics, including a large number of educational resources, Renaud Gicquel has made a major tour de force to significantly ease the learning of engineering thermodynamics. Together with the wealth of information, e-learning modules, guided explorations, examples and exercises of the Thermoptim portal, this work will provide any student and professional working on this subject with excellent tools to master the subject up to an advanced level.

Career

Renaud Gicquel is now retired. He was previously Professor at the École des Mines de Paris (Mines ParisTech). He was trained as a mining engineering at the École des Mines and got his doctorate in engineering from the Paris VI University. He started his career as a Special Assistant to the Secretary General of the United Nations Conference on New and Renewable Sources of Energy in 1980 in New York. He then became the Deputy Director in charge of Dwellings at the Energy Division of the Marcoussis Laboratories of the Compagnie Générale d'Electricité until January 1982, and the Head of multilateral issues in the International Affairs Service of the Ministry for Research and Technology in Paris in 1982. From 1983 to 1985, he was Adviser for International Issues of the National Center of Scientific Research (CNRS). In 1986, he founded with the late Michel Grenon Mediterranean Energy Observatory (OME), based in Sophia Antipolis. In 1990, Dr Gicquel created the ARTEMIS group, a research body for thermal energy research, together with the University of Nantes and ISITEM (now Polytech) in Nantes. He acted as a coordinator while fulfilling the position of Deputy Director at the Ecole des Mines de Nantes (EMN) during three years from 1991 as well. In 1987, he was named Head of the Centre of Energy Studies of the École des Mines de Paris.

General introduction

We have developed over the past thirty-five years a new way of teaching thermodynamics applied to energy conversion now used in more than 120 higher education institutions, at both undergraduate and graduate levels (engineering schools, universities) as well as in vocational training.

The change in the education paradigm we have introduced is based on a shift of knowledge acquired by students. The writing of equations describing changes undergone by fluids is drastically reduced, and the calculations are made by a simulator such as Thermoptim (www.thermoptim.org) without learners needing to know the details. They devote most of the time, on the one hand, learning technologies and, on the other hand, reflecting on the architecture of both conventional and innovative thermodynamic cycles, graphically building and setting models of various energy systems instead of theoretical cycles caricaturally simplified in order to be analytically calculable.

The new teaching method has the distinction of being at the same time much simpler than conventional approaches for the introduction to the discipline, and much more powerful for confirmed students who can go further in their studies thanks to a strong and open modeling environment. They can work on the real-world complex innovative cycles currently being studied in laboratories and companies.

We published in 2011 a book entitled *Energy Systems: A New Approach to Engineering Thermodynamics*, which had a double objective: on the one hand, to allow beginners to understand the design principles of energy systems and to have an overall vision of the different technologies usable for their realization, and, on the other hand, to provide its confirmed readers with advanced methods of analysis of these systems. It showed how a structured approach to studying energy technologies can be implemented using the Thermoptim software package.

This first book is approximately 1,100 pages and, despite the simplifications made by the proposed approach, it can be a bit difficult for a number of readers.

This book, while following the same global logic as the 2011 book, does not go as much in the details. In order to simplify the content, sections that were intended for advanced readers have been removed and theoretical developments have been limited to the essential.

As indicated by its title, this book deals with energy systems, i.e., energy conversion technologies (ECTs) considered as systems based on sets of elementary components coupled together. Although it mainly addresses thermodynamic energy conversion, it covers a very large field, including a variety of cycles: conventional as well as innovative power plants, gas turbines, reciprocating engines, Stirling engines, combined cycles, cogeneration, refrigeration cycles, new and renewable energy conversion, combustion, Generation I–IV nuclear energy conversion, evaporation, desalination, fuel cells, oxycombustion, air conditioning, etc.

Structure of this book

This book comprises the four main parts:

1. After a first chapter which presents the new pedagogical paradigm that we have developed over the past thirty years, it begins by introducing in Part 1 the essential concepts which must be understood for studying the cycles of three basic energy technologies: the steam power plant, the gas turbine and the refrigeration machine. The chosen approach follows a presentation called CFRP, which stands for Components, Functions and Reference Processes.

In the CFRP presentation, we start by describing the architectures of these various technologies and the technological solutions implemented. We then show that despite their diversity, the components only perform four main functions, themselves corresponding to three reference changes undergone by the fluids that pass through them.

This analysis, initially functional, leads to the essential notion of reference processes, models of fluid behavior in machines, and quite naturally leads to the study of the properties of these fluids.

The approach is as light as possible in order to allow students to learn to model, gradually and by example, heat conversion technologies. We show in particular that one can present the essential of the concepts without using a state function, which can be difficult to understand well, entropy.

After having introduced some essential notions of thermodynamics, energy exchanges of a system with its environment are analyzed and the first law is presented.

The representation of the cycles in the (h, ln(P)) chart makes it possible to visualize the physical phenomena at play and the few reference processes undergone by the working fluids.

The way in which these cycles can be modeled with the Thermoptim software package is the subject of guided explorations of pre-built models.

2. We are advocating for the lightest possible and progressive approach, in a spiral, where new concepts are only introduced at the precise moment when they are needed, even if it means going back to them later to give further details. It is light to limit the cognitive load of the learners, and progressive to maintain their motivation. Such an approach can be considered as minimalist.

This is why, once we have shown in Part 1 that learners can very simply model the basic thermodynamic cycles, it becomes possible in Part 2 to provide theoretical, technological and methodological supplements that will be used in more complex cycle studies of Parts 3 and 4.

To be perfectly consistent with what has just been said, it would be necessary to alternate the theoretical and methodological sections and the sections of application to technologies, which would risk giving this book a bushy appearance.

For the sake of clarity, we have therefore gathered in this second part most of the supplements presented.

Specific supplements will, however, be provided at the time they are used (e.g., moist mixtures).

Theoretical supplements presented in Part 2 relate to combustion, heat exchangers, and entropy. Methodological supplements deal with process integration (pinch method), exergy balances and productive structures. Technological supplements are related to steam systems components: boilers, steam generators, steam turbines and cooling towers.

3. In Part 3, we start from the cycles introduced initially, and we show how their efficiency can be improved, which allows learners to study main conventional power and refrigeration cycles.

The guiding principle behind these analyses is the reduction of irreversibilities; special attention is paid to those that arise from temperature differences with external sources and during internal regenerations. In addition, the value of staged compression and expansion is highlighted whenever possible.

In this context, exergy analyses are of great interest, because they make it possible to quantify irreversibilities. This is why, we present numerous exergy balances of the cycles studied, and the emphasis is on the interpretation of their results rather than on the underlying theory. As we explain in Chapter 6, obtaining them is greatly facilitated by the use of productive structures, so that the use of exergy balances is done without any difficulty, at least for Part 3.

Classical ECTs are reviewed and analyzed as systems implementing the components whose operation has been studied previously. The link is made between scientific knowledge and technological achievements, which are presented in more detail than in the previous chapters.

Chapters 12 and 13 deal with air conditioning and absorption cycles, now classic, which justifies their inclusion in Part 3.

4. Part 4 is devoted to the study of innovative cycles with low environmental impact: advanced gas turbine cycles, Stirling engines, future nuclear reactors, oxycombustion cycles, new and renewable energy thermodynamic cycles, evaporation, mechanical and thermal vapor compression, desalination, drying by hot gas and electrochemical converters.

Some examples of exergy balances are also provided. For certain cycles, such as those relating to evaporation or drying, their interpretation is a little more difficult than for those studied previously.

In this book, the learning of modeling is carried out, on the one hand, in a very progressive manner, the more complex concepts being introduced contextually, as and when required, and on the other hand, by realistic examples, the reader being invited to carry out guided explorations of forty-five models of about thirty different energy systems.

We speak of a light pedagogical presentation because we seek to limit as much as possible the baggage in mathematics and physics necessary for the understanding of these technologies; our objective is to make them accessible to readers unfamiliar with the language of specialists in thermodynamics.

During the first two-thirds of this book, that is to say up to Chapter 11, we have thus limited the use of theoretical developments and equations to the strict minimum. From these chapters to the end of this book, we have reintroduced them, considering that at this stage of advancement in this book, even readers most reluctant initially by the mathematical and physical aspects would find the motivation to delve into them if the topic really interests them. Many examples are provided, with the equations retained in the models, a number of which use the external classes of Thermoptim, where these equations are encapsulated. Java code for these classes is available.

In around 550 pages, readers are presented essential concepts which allow them to understand both the functioning of energy systems ensuring thermodynamic heat conversion and the way in which they can be modeled.

This book thus constitutes a much easier complement to the first edition of the book *Energy Systems*, which deals with the subject in a more in-depth and more exhaustive manner.

Three types of inserts are included in this book. They highlight key issues, such as self-assessment exercises, worked examples and guided explorations, and provide links to the associated numerical resources.

The whole book is illustrated by numerous real-life examples (about 125) of cycles modeled with Thermoptim, which provide the reader with models whose structure or settings he may customize as he wishes to perform various simulations. The files of these examples are available for teachers, but not for their students.

The list of these examples, of varying difficulty, is given in a portal called Thermoptim-UNIT (www.thermoptim.org) with some comments and suggestions on ways in which they can be used educationally, depending on the context and the objectives pursued by teachers.

More generally, many digital resources for teaching energy systems have been gradually collected into this portal whose content is freely accessible, with few exceptions, including solutions of some exercises and problems, which are presented in this book.

Among them, readers will find about forty-five self-assessment online activities; the main purpose of which is to allow them to check their understanding of the different concepts presented. Marks are not recorded or transmitted to anyone. They simply give an indication of the quality of the answers.

These self-assessment activities are of four types:

- Drag and drop onto image (ddi) exercises allow learners to check if they can find their way around a sketch or a chart. They operate by a simple drag and drop;
- Gap-fill exercises (gfe) with a contextual image require a little more concentration on the reader's part, but are very fruitful in ensuring that difficult concepts are well understood.

From a pedagogical point of view, it is an excellent exercise because they are asked to rephrase what has been presented in this book in order to construct sentences that make sense; the missing texts are offered in drop-down menus;

- Categorization exercises (cat) complement the previous two activities well: readers organize elements into categories and thus learn to distinguish their characteristics;
- Finally, single-choice questions and multiple-choice questions (quiz) allow them to test their knowledge in a fairly broad way, but they are not very user-friendly tools.

Do these exercises with attention and concentration: if you succeed, it means that you have mastered the underlying concepts. If you make mistakes, it means you need to rework them. Some of these notions are much more subtle than they seem at first glance, and it's only natural not to fully understand them right away, but with some practice you will.

Moreover, the goals we set ourselves have led us to identify three major transverse themes that are found throughout the presentations:

- Theoretical foundations (whose presentation has been simplified as much as possible);
- An original modeling approach;
- A detailed presentation of technologies, often absent from books on these subjects offered to learners.

Objectives of this book

For beginners, the major interest of Thermoptim is that it can help them to model complex energy systems simply, without having to write an equation or program. It discharges its users of many problems, including computational ones, and enables them to make analyses that they could not pursue otherwise, especially when starting out. It becomes possible, when learning the discipline, to focus on a qualitative approach; the calculations required for quantitative studies are made by the software. Note that the name of this tool has a double meaning: it does allow one to optimize thermodynamic systems, but above all, it allows one to learn this discipline with optimism…

Under these conditions, energy systems operation can be studied using a completely new pedagogy, where only a small number of elements must be considered: first those used to describe the systems studied, and second those used to set the simulator components.

If done this way, it is not necessary, at least initially, to take into account the details of the equations allowing calculation of the process involved: Thermoptim automatically does this. Learning the discipline is limited to these basic concepts and their implementation in the package. The memorization effort and cognitive load required for beginners are greatly reduced, allowing them to focus their attention on understanding the basic phenomenological concepts and their practical implementation.

Thus, Thermoptim permits, without writing a single line of code, to calculate energy systems from simple to complex.

In summary, we pursue three objectives:

- Allow learners first to understand the functioning of various components at stake in energy systems and how they are assembled, emphasizing the technological aspects;
- Show how it is possible through the use of Thermoptim to model and calculate them very simply but with great precision, and make the reader familiar with this working environment;
- Provide the readers with methodological guidance, simple or advanced, for analyzing the systems they study. In this spirit, a significant place is devoted to exergy methods that are increasingly regarded as among the best suited to perform optimization studies, as they can take into account both the amount of energy put into play and its quality.

A working tool on many levels

This book can be read and used as a working tool on several levels:

- For an introduction to the discipline with an illustration of its implementation in Thermoptim: you can then simply understand the different technologies, the underlying physical phenomena and the methodologies recommended, possibly using the package somewhat blindly;
- For a deepening of the field, it not only presents the calculation principles and the basic equations, but also explains how to build exergy balances, implement the pinch method, etc.;
- For software users, this book is a scientific complement to the documentation provided with the tool, allowing those interested to better understand how the calculations are made and even to personalize it. Note, however, that the first edition is much more detailed in this respect than this one.

It follows that this book is deliberately composed of a series of sections that are on different planes, some more theoretical, others more applied and technological, and others methodological, concerning in particular the use of Thermoptim.

In this way, readers wishing to do an overview of this book will be guided through the research of basic concepts needed, especially to use the software properly, while those who wish to invest further in the discipline will benefit from a consistent and progressive presentation.

Pedagogically, this book and Thermoptim can be used in various contexts, under both a traditional approach (neo-behaviorist or objectivist) and a more recent constructivist approach. We hope it will help colleagues to overcome difficulties that they may face in their practice.

For supporters of a conventional approach, Thermoptim allows them particularly to enrich the classical presentation by accurate simulations and to make students study multiple and realistic examples. In a constructivist approach, this book and the software are in addition tools allowing students to work independently to explore a wide field and analyze very open topics:

- For beginners, it is a structured environment that reduces the cognitive load while acquiring the vocabulary and basic concepts encapsulated in the screens. Once this vocabulary is learned, cooperative learning with peers and teachers is strengthened;
- Very quickly, it becomes possible to work on realistic problems and not caricatures (as conventionally internal combustion engines with perfect air as working fluid). In addition, during internships, students using Thermoptim are fully operational in the company where they work, which is very exciting for them and leads them to engage more thoroughly;
- For experienced users, Thermoptim allows them to study very complex systems (e.g., Areva Framatome used Thermoptim to optimize combined cycles and cogeneration cycles coupled with high-temperature nuclear reactors).

The reader will understand that we seek above all to make as accessible as possible the study of energy systems, demonstrating very concretely how realistic models can be developed to represent them. This bias has led us to voluntarily limit our discussion of scientific issues to key points, especially regarding the fundamentals of thermodynamics and heat transfer. As a result, this book loses in generality what it gains in ease of use: know-how is privileged compared to knowledge itself. Readers interested in further developments may if they wish refer to documents given in the bibliography.

Experienced engineers will find a coherent body of theory and practice through which they can become quickly operational, without having to get personally involved in solving equations or in the development of a computer modeling environment.

Mind maps

This book includes five mind maps whose role is to help readers easily find information on a given topic.

These knowledge maps are explanatory figures, which summarize in a graphical way the main issues pertaining to the topic, with the indication of the chapters of this book where they are dealt with.

After the general introduction, you will find a first map presenting the main features of the Thermoptim software, and four ones devoted to thermodynamic cycles:

- Rankine cycles for steam power plants;
- Refrigeration cycles;
- Gas turbine cycles;
- New and renewable energy cycles.

Thermoptim

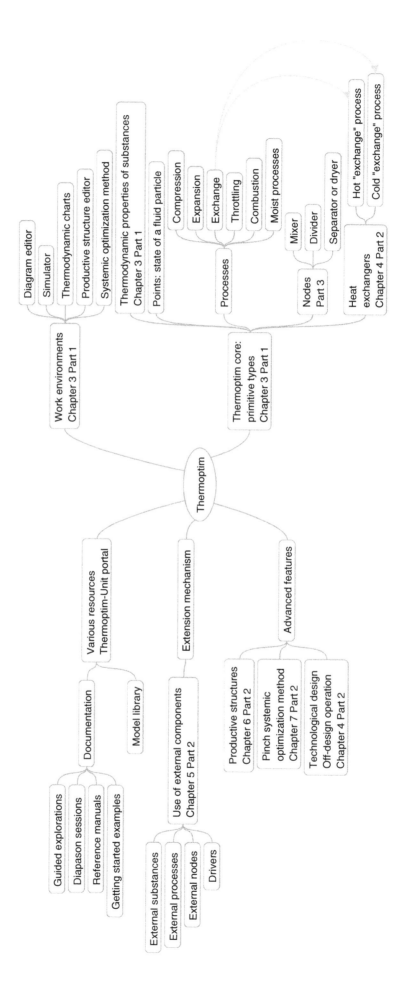

Rankine cycles for steam power plants

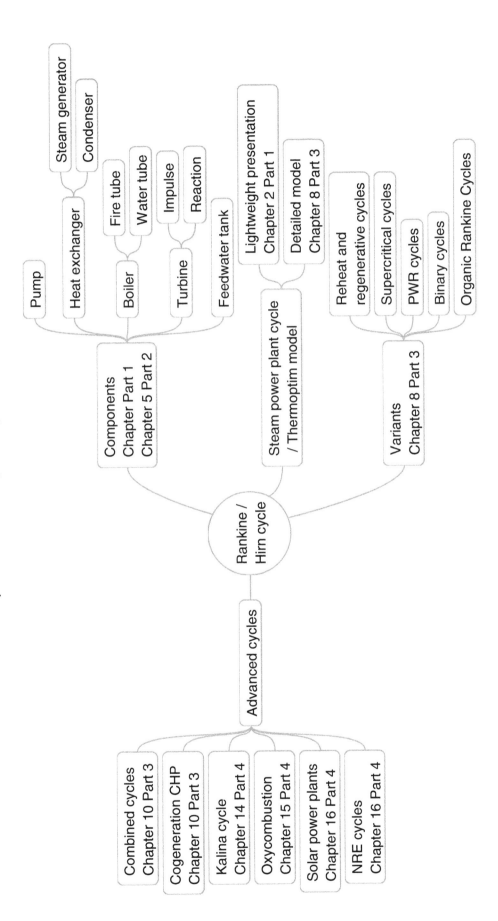

Rankine cycles for steam power plants

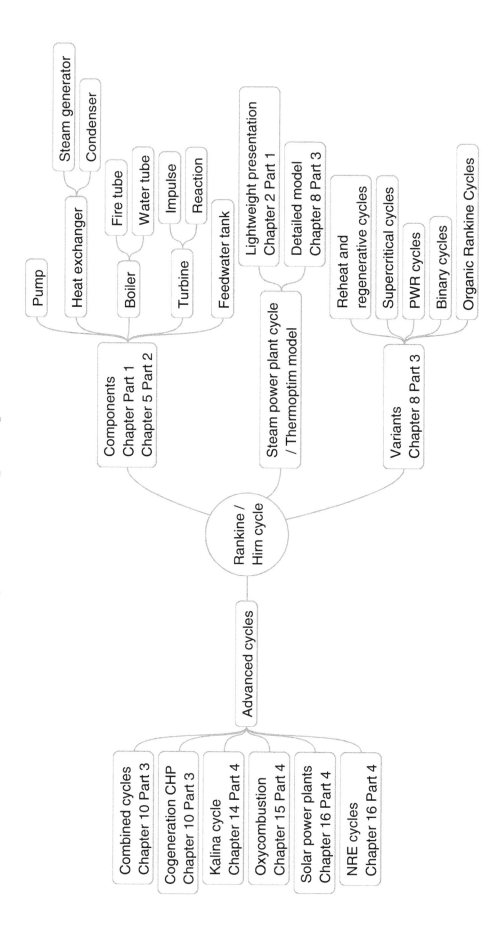

Gas turbine cycles

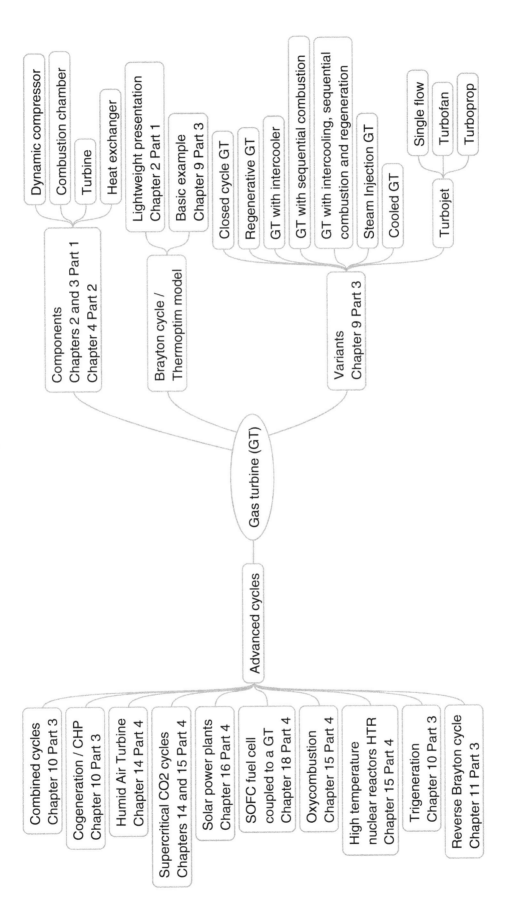

New and renewable energy cycles

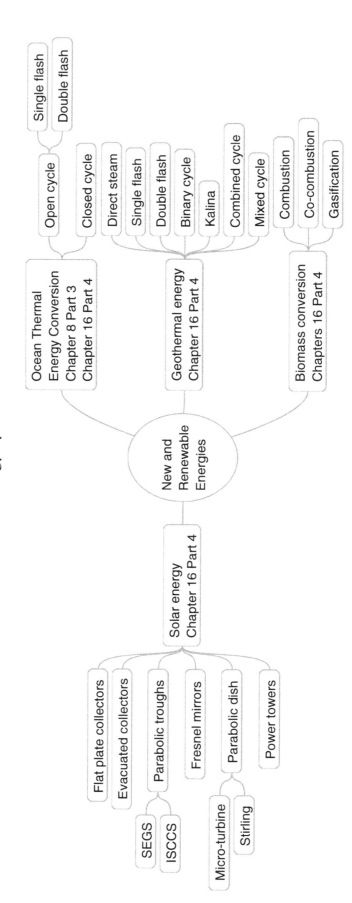

Credits list

French navy

These figures are reproduced with permission of the French Navy, Marine Nationale / S. Deschamps:

Figure 2.3, Figure 2.4, Figure 2.6, Figure 2.7, Figure 2.11, Figure 2.14, Figure 2.16, Figure 2.17, Figure 2.18, Figure 2.26, Figure 2.28, Figure 2.29, Figure 4.19, Figure 4.20, Figure 4.23

Siemens

These figures are reproduced with permission of Siemens Energy:

Figure 2.9, Figure 2.10, Figure 2.30, Figure 2.31, Figure 5.12, Figure 9.10

Techniques de l'ingénieur

All the figures below are reproduced with permission of the Éditions T.I., trademark Techniques de l'Ingénieur, Internet http://www.techniques-ingenieur.fr.:

Figure 4.17, Figure 4.21, Figure 4.22, Figure 4.24, Figure 12.10, Figure 12.18, Échangeurs de chaleur Description des échangeurs, B 2341

Auteur(s): André BONTEMPS, Alain GARRIGUE, Charles GOUBIER, Jacques HUETZ, Christophe MARVILLET, Pierre MERCIER et Roland VIDIL, 10 mai 1995

Figure 5.2, Génie énergétique, Différents types de chaudières industrielles, B1480

Auteur(s): Alain RIOU et Jean-Pierre DEPAUW, 10 févr. 1990

Figure 9.21, Génie Mécanique, Refroidissement des moteurs à combustion interne, B2830 v1

Auteur(s): Jean-Pierre MORANNE, 10 nov. 1986

Figure 9.42, Génie Mécanique
Suralimentation des moteurs de véhicules par turbocompresseur, BM 2 631
Auteur(s): Aimé PAROIS, 10 juil. 2001

Figure 9.53, Traité Mécanique et chaleur, Combustion dans les moteurs Diesel, B 2700

Auteur(s): Alain HAUPAIS, 10 févr. 1992

Figure 15.7, Génie Mécanique, Moteurs non conventionnels, BM 2 593

Auteur(s): Georges DESCOMBES et Jean-Louis MAGNET, 10 juil. 1998

Bosch

These figures are reproduced with permission of Bosch, Technical papers:

Figure 4.5, Figure 9.32, Figure 9.38, Figure 9.39, Figure 9.40, Figure 9.41, Figure 9.45, Figure 9.46, Figure 9.47, Figure 9.49, Figure 9.50

Framatome

These figures are reproduced with permission of Framatome:
Figure 8.18, Figure 8.19, Figure 8.20

Maywald

These figures are reproduced with permission of Mr. J. Maywald:
Figure 15.8, Figure 15.9, Figure 15.10

France évaporation

Figure 17.8 is reproduced with permission of France Évaporation

Cep mines-paristech

Figure 18.3 is reproduced with permission of CEP Mines-ParisTech

Symbols

A	Area	m^2
a	Hydrogen available for combustion	
C	Concentration	
C	Speed	m/s
C	Fluid velocity	m/s
C	Concentration factor	
C$_p$	Molar heat capacity at constant pressure	J/mol K
C$_r$	Volumetric efficiency	
c$_p$	Specific heat capacity at constant pressure	J/kg K
C$_v$	Molar heat capacity at constant volume	J/mol K
c$_v$	Specific heat at constant volume	J/kg K
D	Diameter	M
e	Excess air	
e	Thickness	M
f	Surface factor	
F	Thrust	N
F	Molar flow rate	Mol/s
H	Molar enthalpy	J/mol
h	Mass enthalpy	J/kg
h	Heat transfer coefficient	J/K1m2
h	Calorimetric coefficient	J/kg bar
h$_f$	Enthalpy of formation	J/mol
i	Octane number	
i	Intermittency factor	
i	van't Hoff factor of the solute	
I	Current density	A
k	Polytropic coefficient	
k	Dissociation rate	
k	Interaction coefficient	
k$_m$	Mixing entropy	J/mol K
K	Kinetic energy	J/kg
K$_{eb}$	Ebullioscopy constant	K kg/mol
Kp	Equilibrium constant	
l	Calorimetric coefficient	J/m3
L	Enthalpy of evaporation	J/kg
L	Length	m
Le	Lewis	Number
M	Molar mass	Kg/mol
Ma	Mach	Number
m	Mass	kg
mc	Specific fuel consumption	g/kWh

N	Number of moles	
nc	Number of cylinders	
N	Rotation speed	tr/mn
Nu	Nusselt number	
NTU	Number of transfer units	
P	Pressure	Pa or bar
Pr	Prandtl number	
p	Wetted perimeter	
P	Power	W
Q	Heat	J/kg
q'	Specific enthalpy (moist gas)	J/kg
R	Universal constant	8,314 J/mol K
R	Ratio of heat capacity rates of a heat exchanger	
r	Ideal gas constant	J/K
r	Compression ratio	
R	Richness	
R_e	Reynolds number	
S	Molar entropy	J/mol K
s	Mass entropy	J/kg K
S	Surface	m^2
Sb	Salt concentration in the feedwater tank	mg/l or °F
Sc	Salt concentration in the boiler	mg/l or °F
Sc	Schmidt number	
s	Volumetric efficiency	
T	Thermodynamic temperature	K
t	Temperature	°C or K
t	Dry bulb temperature	°C
t'	Wet bulb temperature	°C
t_r	Dew point	°C
u	Specific internal energy	J/kg
U	Molar internal energy	J/mol
U	Peripheral speed	m/s
U	Heat transfer coefficient	W/K m^2
v	Volume	m^3
v	Specific volume	m^3/kg
V	Volume	m^3
V	Molar volume	m^3/mol
V	Voltage	V
v_{spec}	Specific volume	m^3/kg
W	Specific mechanical energy	J/kg
w	Specific humidity	kg water/kg of dry air
W	Power density	W/kg
x	Length	m
x	Vapor quality	
x	Mole fraction	

x	Mole fraction in liquid phase (mixtures)	
x_h	Specific exergy	J/kg
x_q	Specific heat exergy	J/kg
y	Mass fraction	
y	Mole fraction in vapor phase (mixtures)	
Z	Global mole fraction (mixtures)	
D	Exact differential	
d	Differential form	
Ð	Variation	

Indices

A	Related to admission
C	Related to the critical point
C	Corrected
C	Related to compressor
C	Cold
Cc	Related to combined cycle
D	Related to expansion
dg	Related to the dry gas
F	Related to fuel
G	Related to the gas turbine
G	Related to the gas phase
H	Hot
I	Related to inlet
L	Related to liquid phase
mm	Related to the moist mixture
O	Related to the output
P	Related to polytropic
P	Related to products
R	Related to discharge
R	Related to reactants
S	Related to saturation
S	Related to isentropic
sat	Related to saturation
St	Related to the steam cycle
T	Total
T	Related to the turbine
Vap	Related to the vapor

Greek symbols

α	Recycling factor
α	Baumann coefficient

ε	Degree of reaction	
ε	Relative humidity	RH
ε	Effectiveness	
ε	Steam fraction	
ε	Volume ratio at the end and start of combustion	
	Initial expansion ratio	
φ	Heat flux	J/kg
γ	Ratio of thermal capacities	
η	Efficiency, effectiveness	
λ	Calorimetric coefficient	J/m^3
λ	Air factor	
λ	Thermal conductivity	W/K m
λ	Ratio of the combustion pressure to the compression pressure	
μ	Calorimetric coefficient	J/bar
μ	Viscosity	Kg/s m
μ_i	Molar chemical potential	J/mol
ω	Acentric factor	
ω	Rotation speed	rd/s
π	Heat created by irreversibilities	J/kg
π_i	Internal pressure	Ratio
π	Osmotic pressure	Pa or bar
θ	Carnot factor	
	Ratio of the turbine inlet temperature at the compressor inlet temperature	
ρ	Density	Kg/m
ρ	Volumetric compression ratio	
τ	Specific shaft work	J/kg
τ	Hydrogen utilization rate	
τ	Optical efficiency	
τ	Recirculation rate	
ξ	Mass fraction of fuel burned	
ξ	Temperature ratio in a Stirling engine	

Acronyms

AFC	Alkaline fuel cell
AFR	Air fuel ratio
AICVF	Association des Ingénieurs en Chauffage et Ventilation de France
API	American Petroleum Institute
ASHRAE	American Society of Heating Refrigeration and Air conditioning Engineers
AZEP	Advanced zero emission process
BDC	Bottom dead center
BWR	Boiling water reactor
CC	Combined cycle
CC	Composite curve
CCHP	Cooling, heating, and power generation
CF	Heat-power ratio
CFC	Chlorofluorocarbons
CFC	Cumulative frequency curves
CFDC	Carnot factor difference curve
CFR	Cooperative Fuel Research Committee
CFRP	Components, Functions and Reference Processes
CHAT	Cascaded humidified advanced turbine
CHP	Combined heat and power
CLC	Chemical looping combustion
COP	Coefficient of performance
CT	Combustion turbine
CTI	Cooling Technology Institute
DCS	Distillation and condensation system
deNOx	Destructive nitrogen oxides
ECT	Energy conversion technology
EPR	European pressurized reactor
ESC	energy specific consumption
GCC	Grand composite curve
GFR	Gas-cooled fast neutron reactor
GIF	Generation IV International Forum
GT	Gas turbine
GTMHR	gas turbine-modular helium-cooled reactor
GWP	Global warming potential
HAT	Humid air turbine
HCFC	Hydrochlorofluorocarbon
HFC	Hydrofluorocarbons
HHV	Higher heating value
HP	High pressure
HRSG	Heat recovery steam generator
HT	High temperature
HTR	High-temperature reactor
HX	Heat exchanger
ICE	Internal combustion engine
IHPTET	Integrated high-performance turbine engine technology
IHX	Intermediate heat exchanger
IP	Intermediate pressure
ISCCS	Integrated solar combined cycle system
LFR	Lead-cooled fast neutron reactor
LHV	Lower heating value

LMTD	Logarithmic mean temperature difference
LP	Low pressure
LPM	Locus of minimum pinches
LT	Low temperature
MCFC	Molten carbonate fuel cell
MED	Multieffect desalination
MEP	Mean effective pressure
MIEC	Mixed ionic-electronic conducting membrane
MIT	Massachusetts Institute of Technology
MON	Motor octane number
MP	Mean pressure
MSF	Multistage flash desalination
MSR	Moisture separator reheater
MSR	Molten salt reactor
MSWI	Municipal solid waste incinerator
MVC	Mechanical vapor compression
NCG	Non-condensable gases
NO_x	Nitrogen oxides
NRE	New and renewable energies
NTU	Number of transfer units
ODP	Ozone depletion potential
ORC	Organic Rankine cycle
OTEC	Ocean thermal energy conversion
OSC	operation specific consumption
PAFC	Phosphoric acid fuel cell
PBMR	Pebble bed modular reactor
PD	Parabolic dish
PDU	Productive or dissipative unit
PEMFC	Proton-exchange membrane fuel cell
PT	Parabolic trough
PT	Power towers
PURPA	Public Utility Regulatory Policies Act
PWR	Pressurized water reactor
R&D	Research and development
RH	Relative humidity
RO	Reverse osmosis
RON	Research octane number
SCWR	Supercritical water reactors
SEGS	Solar electric generating systems
SFR	Sodium-cooled fast neutron reactor
SG	Steam generator
SHF	Sensible heat factor
SHR	Sensible heat ratio
SOFC	Solid oxide fuel cell
SRBC	Steam recompression bottoming cycle
ST	Steam turbine
STIG	Steam injection gas turbine
TDC	Top dead center
TEWI	Total equivalent warming impact
TIT	Turbine inlet température
TVC	Thermal vapor compression
UHC	Unburned hydrocarbons
VHTR	Very high-temperature reactor
WFBC	Water flashing bottoming cycle
ZLD	Zero liquid discharge

Conversion factors

To Obtain	Multiply	By	Reverse Factor
Meter (m)	Inch (in)	2.540000E−02	3.937008E+01
Centimeter (cm)	Inch (in)	2.540000E+00	3.937008E−01
Meter (m)	Foot (ft)	3.048000E−01	3.280840E+00
Square meter (m²)	Square inch (in²)	6.451600E−04	1.550003E+03
Square centimeter (cm²)	Square inch (in²)	6.451600E+00	1.550003E−01
Cubic meter (m³)	Cubic foot (ft³)	2.831685E−02	3.531466E+01
Cubic meter (m³)	Cubic inch (in³)	1.638706E−05	6.102376E+04
Meter per second (m/s)	Foot per second (ft/s)	3.048000E−01	3.280840E+00
Meter per second (m/s)	Inch per second (in/s)	2.540000E−02	3.937008E+01
Meter per second (m/s)	Kilometer per hour (km/h)	2.777778E−01	3.600000E+00
Meter per second (m/s)	Mile per hour (mi/h)	4.470400E−01	2.236936E+00
Kilometer per hour (km/h)	Mile per hour (mi/h)	1.609344E+00	6.213712E−01
Radian per second (rad/s)	rpm (revolution per minute) (r/min)	1.047198E−01	9.549292E+00
Gram (g)	Ounce (oz)	2.834952E+01	3.527397E−02
Kilogram (kg)	Pound (lb)	4.535924E−01	2.204622E+00
Kilogram per cubic meter (kg/m³)	Pound per cubic foot (lb/ft³)	1.601846E+01	6.242797E−02
Kilogram per cubic meter (kg/m³)	Pound per cubic inch (lb/in³)	2.767990E+04	3.612730E−05
Pascal (Pa)	Pound-force per square foot (lbf/ft²)	4.788026E+01	2.088543E−02
Pascal (Pa)	Pound-force per square inch (psi) (lbf/in²)	6.894757E+03	1.450377E−04
Pascal (Pa)	Bar (bar)	1.000000E+05	1.000000E−05
Pascal (Pa)	Atmosphere, standard (atm)	1.013250E+05	9.869233E−06
Watt (W)	British thermal unit per hour (Btu/h)	2.930711E−01	3.412141E+00
Watt (W)	British thermal unit per second (Btu/s)	1.055056E+03	9.478170E−04
Watt (W)	Ton of refrigeration (12 000 Btu/h)	3.516853E+03	2.843451E−04
Joule per kilogram kelvin [J/(kg K)]	British thermal unit per pound degree Fahrenheit [Btu/(lb °F)]	4.186800E+03	2.388459E−04
Watt per square meter (W/m²)	British thermal unit per square foot hour [Btu/(ft²h)]	3.154591E+00	3.169983E−01
Watt per square meter (W/m²)	British thermal unit per square foot second [Btu/(ft²s)]	1.135653E+04	8.805507E−05
Joule per square meter (J/m²)	British thermal unit per square foot (Btu/ft²)	1.135653E+04	8.805507E−05
Joule per cubic meter (J/m³)	British thermal unit per cubic foot (Btu/ft³)	3.725895E+04	2.683919E−05
Joule per kilogram (J/kg)	British thermal unit per pound (Btu/lb)	2.326000E+03	4.299226E−04
Joule per kelvin (J/K)	British thermal unit per degree Fahrenheit (Btu/°F)	1.899101E+03	5.265649E−04
Watt per square meter kelvin [W/(m² K)]	British thermal unit per hour square foot degree Fahrenheit [Btu/(h ft² °F)]	5.678263E+00	1.761102E−01
Watt per square meter kelvin [W/(m² K)]	British thermal unit per second square foot degree Fahrenheit [Btu/(s ft² °F)]	2.044175E+04	4.891949E−05
Joule (J)	Watt hour (W h)	3.600000E+03	2.777778E−04
Joule (J)	British thermal unit (Btu)	1.055056E+03	9.478170E−04
Joule (J)	Foot pound-force (ft lbf)	1.355818E+00	7.375621E−01
Kelvin (K)	Degree Celsius (°C)	$T_K = t_{°C} + 273.15$	$t_{°C} = T - 273.15$

(Continued)

To Obtain	Multiply	By	Reverse Factor
Degree Celsius (°C)	Degree Fahrenheit (°F)	$t_{°C} = \dfrac{t_{°F} - 32}{1.8}$	$t_{°F} = 1.8t_{°F} + 32$
Kelvin (K)	Degree Fahrenheit (°F)	$T_K = \dfrac{t_{°F} + 459.67}{1.8}$	$T_{°F} = 1.8T_K - 459.67$
Kelvin (K)	Degree Rankine (°R)	$T_K = \dfrac{t_{°R}}{1.8}$	$T_{°R} = 1.8T_K$
Degree Celsius (°C)	Kelvin (K)	$t_{°C} = T_K - 273.15$	$T_K = t_{°C} + 273.15$

These values have been obtained from NIST Guide to the SI.

First steps in engineering thermodynamics

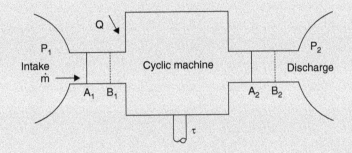

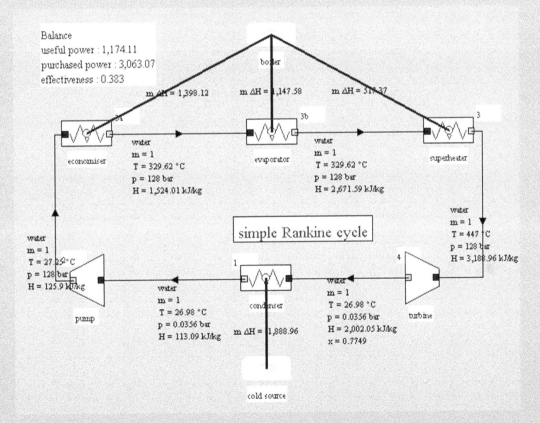

Part I, especially dedicated to beginners and fellow teachers, introduces in a very simple way basic concepts necessary to understand elementary thermodynamic cycles (steam power plants, gas turbines, refrigerators).

The approach is as light as possible in order to allow students to learn to model, gradually and by example, heat conversion technologies. We show in particular that one can present the essential concepts without using a state function which can be difficult to understand well, entropy.

DOI: 10.1201/9781003175629-1

A new educational paradigm

Introduction

This chapter presents the new educational paradigm that we have developed to teach thermodynamics applied to energy systems. This change is based on a shift of knowledge acquired by learners in initial or vocational training. The writing of equations describing processes undergone by fluids is drastically reduced, as the calculations are made by the simulator.

After having explained how pedagogy with Thermoptim differs from classical teaching, we show that energy systems involve only a small number of functions, which makes it possible to use a graphical environment to model them.

Once calculation problems have been resolved, most of the time spent in a class is devoted to presenting the technologies. More generally, the content of the course is determined using a model called RTM(E), which stands for Reality, Theory, Methods (and Examples).

We then explain how the course can be sequenced in three distinct steps, which are of variable duration, according to the pedagogical contexts.

The main pedagogic innovations brought by Thermoptim are listed, and a short comparison with other information and communication technology (ICT) tools with teaching potential is made.

General context

The context of the initial training of engineers has evolved considerably in recent years. Although their scientific and technical knowledge and the ability to mobilize it to solve concrete problems are still some of the features that continue to distinguish them most from other executives, like the latter they need more and more to pay attention to the nontechnical dimensions of their job, i.e., people management, project economics, product marketing, and environmental impact of technologies. In these circumstances, the time available to invest in technology and their motivation for doing so are now lesser than before. Moreover, the time devoted to technical subjects in the initial training programs of engineers is also declining gradually. In addition, practical work and projects have also often been found to be reduced as compared to studying the theory.

This evolution of the training specifications forces us to renew the pedagogies that we are implementing, but fortunately we also have new assets because virtual environments provide us with new opportunities to do so.

Although applied engineering thermodynamics can be considered as well-established science (its foundations were established more than a century ago), it continues to significantly evolve due to advances in the field of materials or those of control and command, to physical and geopolitical constraints on resources, and changes in regulations, all of which have

DOI: 10.1201/9781003175629-1

led to the development of devices that are more and more respectful of the environment. Considerable technological changes are still expected in the coming decades. They will call on strong skills in applied thermodynamics, in particular for the development of new integrated cycles with high efficiency and low environmental impact.

Our goal is to train our students as best as possible to meet these challenges.

Specific problems arise for learners in continuing education, or more broadly in vocational education, who are very different from students enrolled in initial education in the university system. Even though some went through higher education, they may have graduated years before and thus have forgotten many concepts that they have not used regularly, despite these concepts being implicit prerequisites to understanding the usual presentation of elementary cycles. Obviously, if their initial training is of a modest level, their bases in mathematics and physics are even more incomplete. As a result, the pedagogies conventionally used for students in their initial training are not tailored to their needs.

This observation, of course, does not only apply to the teaching of applied thermodynamics but also to a good number of training courses with a scientific content. It is much broader in scope and implies the need to develop specific digital resources if we want to cater for the needs of learners outside the traditional initial training system.

In college, students are used to changing from one topic to another by following their academic schedule and without questioning its overall logic (Mathematics from 8 a.m. to 10 a.m., English then…).

Trainees in vocational training ask that more attention be paid to showing them the purpose of the lessons followed, especially if they are theoretical. Their concern is to acquire one or more skills, and the link to the job must be clearly explained to them. They may even not be in a position to get involved in the training until they are explained the practical purpose of the curriculum.

Such a learner is therefore not at all part of the Cartesian deductive logic, which consists in starting by presenting the reminders of mathematics and physics before unfolding the theory, to end up with the practice, and which constitutes the global scenario of courses in initial training in higher education.

Even for students in initial training, this mode of presentation is generally not the most suitable. Nowadays, learners only become engaged if they perceive the meaning and the interest of the courses offered to them.

This is why we have been led to deeply modify the sequence of our course on energy systems by seeking to reduce as much as possible the recourse to mathematical formalism and by introducing it only contextually and when it is absolutely necessary.

Difficulties encountered in teaching applied thermodynamics

It is well known that thermodynamics is a difficult subject to teach. The problem has long been acknowledged, and many efforts have been made to remedy it, but until recently there was still a lack of solutions, despite the efforts made by the teachers and the developments in the curricula.

The theory/applications link, essential for understanding any discipline, is much less simple and intuitive in thermodynamics than it is in other fields of physics. In the classic approach of teaching experimental sciences, such as in electricity or mechanics, the theory and the applications to simple realizations are presented to the students at about the same time, with some practical work if possible. The relevance of simple models ($U = R\,I$, balance of forces) is then easily verified, and the link between theory and technology seems immediate; this is how we speak of the "laws" of elementary physics, when these are directly intelligible models which can explain the operation of many very useful technologies known to all (electric lamp, heating resistance, simple machines like the winch, the hoist, the inclined plane, the pendulum…).

For energy systems, it is unfortunately almost always impossible to find models that are both simple and precise. By barely caricaturing, one could say that classical approaches to the discipline are faced with a dilemma, the models to which they lead being either unrealistic or incalculable.

In these approaches, taking into account the difficulties that exist in determining with precision the properties of thermodynamic fluids, one is in fact generally led either to make somewhat oversimplifying hypotheses, or to adopt methods tedious to put into practice: for example, in almost all lessons taught in (under)graduate courses in the world, internal combustion engines are analyzed under the assumption that the technical fluid is air, which is itself assumed to behave as a perfect gas. How, under such assumptions, could we hope to motivate learners who today are concerned about the environmental impact of these technologies: an air engine has never polluted and will never pollute!

As for the refrigeration or steam cycle calculations, they are made using either digital tables requiring daunting interpolations or relatively imprecise paper charts.

This approach has two pitfalls that have the effect of demotivating learners:

* Since the assumptions are too simplistic, learners do not understand the practical relevance of the models they develop, which are very far from reality;
* Since precise calculation of cycles is tedious, learners are put off by the discipline.

If classical approaches have such limitations, it is in our view because they date from a time when the engineer had at his disposal only his slide rule and his logarithm table, and that they haven't changed for several decades...

Moreover, the time spent on the development of equations describing fluid properties and behavior of elementary components represents the bulk of the course, so that learners can ultimately only work on the basic discipline examples, without engaging with the study of innovative cycles, for which they are not equipped in terms of methodology.

Educational issues

Our approach in fact has its origins in the difficulties we encountered when we began teaching the discipline: we found ourselves in a situation of failure against targets that we set and which traditional teaching approaches could not achieve, namely, to make our learners capable, at the end of the course, to address the current energy challenges: reducing the environmental impact of technology, improving efficiencies in acceptable economic conditions...

The change in educational paradigm we have introduced is based on a shift in the knowledge to be acquired by learners. The writing of equations describing changes undergone by fluids is drastically reduced, since the calculations are made by a simulator like Thermoptim, without learners needing to know the details. They devote most of their time, on the one hand, to learning technologies and, on the other hand, to reflecting on the architecture of both conventional and innovative thermodynamic cycles, graphically building and setting models of various energy technologies.

The keystone of our educational change is undoubtedly the introduction of a simulator, the Thermoptim software package, which allows graphical modeling of energy systems. A brief presentation will better illustrate this change in educational paradigm.

Brief presentation of thermoptim

Let us take a look at the difference between a classic presentation of a gas turbine model and how it can be done with Thermoptim.

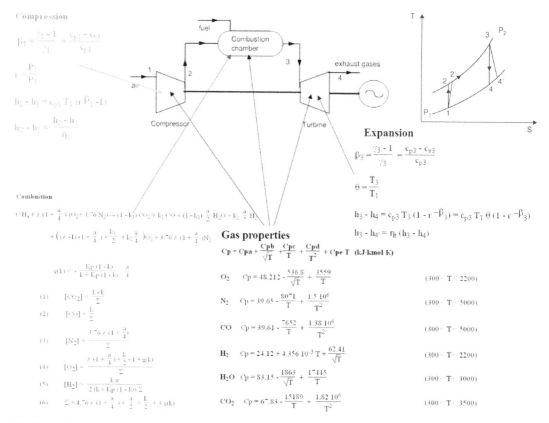

FIGURE 1.1

Conventional modeling of a gas turbine.

Figure 1.1 shows the twenty-two main equations that must be introduced in order to study, with reasonable accuracy, the performance of the simplest gas turbine in a conventional curriculum. It is of course possible to simplify, but the model becomes a caricature.

Equations describing the compression appear at the top left and those for expansion at the upper right. These equations are the simplest.

The more complicated ones are at the bottom left: they determine the composition of exhaust gases at the combustion chamber outlet.

And yet, for simplicity, we have firstly considered here a very simple fuel and secondly have omitted to give the equation for calculating the end of combustion temperature, which is an implicit equation, including an integral, one of the bounds of which is unknown.

In all cases, the gas properties require equations of the kind presented at the bottom of this figure. These equations are needed to calculate the energy properties of fluids.

All things considered, there are at least forty equations that must be taken into account to obtain a realistic model of the gas turbine.

Of course, even if each of the equations is known, the risks of error are numerous in the development of the model and its resolution, especially since they are interconnected and the resolution of some of them must be done iteratively.

Building a gas turbine model with Thermoptim is radically different from that which was traditionally implemented: we favor as we said a qualitative approach of phenomena, calculations required for quantitative studies are being conducted by software tools in a transparent manner for the learners; i.e., they have no need, at least initially, to know the details.

The use of equations is reduced to a minimum when introducing the discipline; the cognitive effort is mainly devoted to understanding concepts and technologies as well as their implementation, and it is only once the learner has acquired sufficient knowledge of the discipline that we consider the writing of equations becomes possible if relevant.

Specifically, as shown in Figure 1.2, the model is constructed by assembling icons placed on a diagram editor working plan; the machine architecture is very close to its physical sketch.

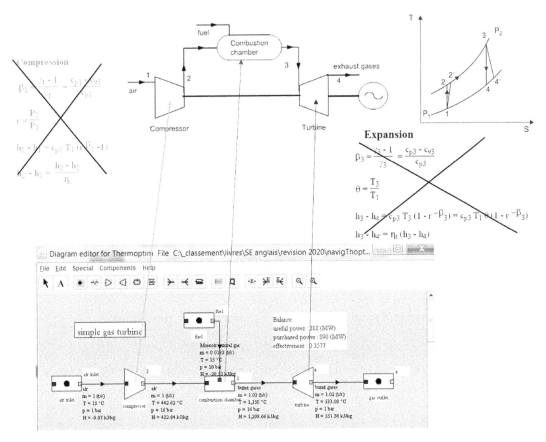

FIGURE I.2

Modeling with the new teaching method.

Each component is then set with a small number of characteristic values. As the thermodynamic properties of fluids are encapsulated in the software, computing performance presents no difficulty, thereby offering much greater precision than with the traditional approach.

Using a tool like Thermoptim, the time to calculate a thermodynamic cycle is massively reduced, and once the model is established, it is possible to perform sensitivity analysis and resolve within minutes problems which would take hours to resolve by conventional methods.

In addition, there is no risk of programming error or incorrect reading of properties. This results in considerable time savings on one aspect of things not essential at the educational level, namely, solving computational problems. Besides the time saved, the arduous nature of the work is greatly reduced, resulting in a significant gain in learner motivation, as they are no longer put off by cumbersome and tedious calculations.

Moreover, considerable gains in precision are obtained, and thus in the calculation likelihood, there is no need to resort to oversimplifying assumptions for the sole purpose of making possible the calculations.

Energy systems: a space of functions of reduced dimension

At the heart of Thermoptim, and this is fundamental, lies the following observation: despite their diversity, energy systems only implement a small number of functions. The functional analysis of these systems indeed shows that even if the technological solutions implemented can be very varied, their components fulfill only a limited number of different functions: the space of functions of energy systems is of reduced dimension.

For example, in all power cycles, a fluid is first compressed, then heated, expanded, and finally cooled, which corresponds to only three functions: exchange heat, compress the working fluid, and expand it by producing work. In all cycles running in reverse, the working fluid is first expanded, generally without producing work, then heated, compressed, and finally cooled, which corresponds to two of the same functions and a fourth.

While energy systems are very different technologically, their functional representations use only a small number of elements. It is therefore possible to design modeling environments that exploit this property. It is enough that they make available to users a small number of functions and make it possible to assemble them easily.

This is precisely what Thermoptim does, which is a modeling software package allowing the graphical construction of models of very many energy systems thanks to a small number of components corresponding to these elementary functions.

Educational innovations brought by thermoptim

Very quickly, it appeared that the pedagogical use of Thermoptim brought about a radical change in the attitude of the students toward the discipline: we hoped for an improvement, but without imagining that it would be so remarkable.

The main educational innovations brought by Thermoptim are the following:

* First of all, the tedious calculations are eliminated, and most of the quantitative aspects are taken care of by the computer; learners can concentrate more on the acquisition of the mental patterns of the discipline;
* The use of the simulator implicitly brings major benefits in the structuring of the learners' cognitive schemes: the screens of the software package make it possible to spatially structure the different concepts that the learners discover and manipulate. For instance, the main characteristics of a small fluid particle, such as its state functions and its calculation options, appear together in the screen of a point. By seeing them grouped each time they have a point screen in front of them the learners gradually get used to associating them, which helps them to structure their knowledge of the field. By appropriating the logical organization of the software, they structure their own mental patterns much better than if they were to follow only a classical course. In addition, they do it in a much more playful way;
* By using the software, beginners acquire the vocabulary and basic concepts, which are encapsulated in the screens presented and whose design has been made with care while ensuring that their content is as simple as possible. Once this vocabulary is learned, cooperative learning with peers and teachers is strengthened;
* The synoptic/diagram editor gathers in a synthetic way in a single screen almost all the relevant information on a thermodynamic cycle (its structure, the interconnections between its components, the values of state variables, the overall balance…);
* Thermoptim allows learners to solve realistic and not simplistic models as in the classical approach;
* Thermoptim is not only a good tutorial: its functionalities also make it a powerful professional simulator used by companies like EDF, CEA or Framatome;
* It places students' thinking at a higher conceptual and methodological level; the purely calculative aspects are outsourced to the computer. There is not only a reduction in cognitive load (Sweller, 1989) but also a significant increase in problem-solving ability;
* Finally, Thermoptim makes the students really operational, which is a major factor in their motivation and therefore in their engagement.

The possible disadvantage is of course the learning time of the tool, which must be as short as possible. The simulator is only a means at the service of applied thermodynamics teaching, and above all getting started with it must not take up too great a share of the time available.

As we will see later, guided explorations are an excellent way to reduce the difficulties and the time necessary to handle the software.

Second educational difficulty: the teaching of technological reality

An important point to note is that the existence of the simulator induces a displacement of educational difficulties. In our practice, the main initial hard point concerned thermodynamic calculations, and the use of Thermoptim made it possible to remove it. As a result, more than half of the course was subsequently devoted to explanations relating to technology, almost unknown to beginners. The idea was to present the different energy conversion technologies and make them as intelligible as possible. This type of knowledge is neither theory nor know-how. By deepening the reflection, we realized that the classic division between procedural and declarative knowledge needed to be refined, and we proposed a so-called RTM(E) model in which the knowledge to be transmitted is grouped into four main categories linked together, called Reality, Theory, Methods (and Examples).

The study of Reality (or even facts, nature, terrain, the world, technology…) by observation, analysis and experimentation allows one to develop or refine the Theory, that is to say an explanatory scheme highlighting the similarities of the different observations of Reality, and explaining them in a way that is both coherent and as simple and generic as possible. The Theory, on the one hand, thus constitutes a grid for reading Reality, and on the other hand serves as a guide for the development of Methods (and/or operational tools) for problem solving, calling if necessary on specific concepts.

This typology very fruitfully structures knowledge relating to a scientific discipline, especially if it is supplemented by the main application Examples, which illustrate very concretely how to solve (thanks to Methods and within the framework of a Theory) a class of problems relating to a particular aspect (of Reality).

More precisely, in the case of thermodynamics applied to engineering, we can distinguish the two main dimensions of Reality (first the properties of matter and then the presentation of technologies) and the two main dimensions of Methods (first the methodology for calculating components or thermodynamic systems, and then modeling in a software such as Thermoptim).

Learning a scientific discipline thus supposes the acquisition of both declarative knowledge for Reality and Theory, and procedural knowledge for Methods, which essentially correspond to know-how. When studying Examples, learners acquire knowledge of Reality, on the one hand, and the practice of Methods, on the other. It is on this occasion that they see how the theory can be applied to the real world, according to rules which, although possibly specific to an example, all fall within the same theoretical framework.

This is why, in addition to Thermoptim, well suited for teaching working Methods and Examples, but not for technological Reality or Theory, in 2004 we developed the e-learning modules called Diapason, which allow us to surmount these last difficulties in gathering, in a concise form and available at all times, all the information that students need to learn. A Diapason session presents a course or an exercise with a soundtrack. Using these sessions is very easy: a picture is loaded onto the screen, and when a soundtrack is provided, it is played so that the learner hears the explanations as if the teacher were present.

Today, videos are also available but only in French, and complement the existing Diapason sessions; this format is preferred for this type of course. As a sidenote, those videos are much more expensive to produce and more difficult to update than Diapason sessions.

A third tool has been more recently developed: guided explorations of pre-built models, which are presented in a particular browser capable of emulating Thermoptim.

The main interest of these tools is their excellent teaching efficiency:

* When they use them, learners are more active than in the classroom, in the sense that they regulate their own pace of work, but above all, they choose themselves the moments when they study, and are therefore available when they do; they learn better, all the more since they are free to go back or supplement the information presented to them by using other documents;

- As the audio tracks of the Diapason sessions have an average duration of less than a minute; learners' attention can be sustained when studying a step; and they move on to the next after a period of rest. The videos are a little longer, but the rhythm of reading can also very easily be personalized;
- When they work, students have all the educational resources they need; in case of doubt or if they have been absent, they can refer without any difficulty to the teacher's oral explanations.

The educational efficiencies of these tools are similar. Choosing among these tools depends on the learning styles and strategies preferred by the learners. Some will prefer the first, while others perhaps the second or third.

Sequencing the course

Once the content to be taught has been chosen, a point that we have not yet raised is also of prime importance. This is the sequencing of the course, that is to say the way of sequencing the different concepts so that the students can learn them as best as possible. In addition to the in-depth modification of the content taught which was mentioned above, the educational methods have therefore also evolved considerably in terms of form.

We have already mentioned the limits of the classical deductive approach whereby the theory is presented before the applications. It seems to us that an inductive and minimalist approach (Carroll, 1990; Köppe & Rodin, 1973), especially when discovering the discipline, much better meets the expectations of today's learners.

We think that learning is an iterative process, which lends itself well to a progressive pedagogy, going from the simple (but always realistic, this is fundamental) to the most complicated. For both cognitive and psychological reasons, it is better, and this is particularly true of applied thermodynamics, to start by showing students how the knowledge presented to them can be applied in practice, by limiting as much as possible conceptual difficulties. Remember that they must first of all become familiar with a new Reality, that they hardly know, and that this learning already results in a high cognitive load.

It seems preferable to initially show them that there are environments like Thermoptim with which they can learn thermodynamics in a painless way and obtain very precise results without writing a single equation. Once their initial reluctance is overcome and they have assimilated the vocabulary and basic concepts, it becomes possible to take a new step and if necessary introduce more equations.

The experience accumulated over the past ten years confirms that, once they realize that very effective methods for moving on to the application exist, learners who are initially very reluctant to engage with the Theory often ask for further study material: as soon as the psychological blockages induced by presentations of the discipline which are too axiomatic and of very little applicability are removed, the students become much more receptive with regard to the equations, undoubtedly because they no longer fear finding themselves incapable of putting them into practice.

From technology to theory: components, functions and reference processes

Concretely, to present the basic thermodynamic cycles, it is possible to adopt an original sequence, which we will call Components, Functions and Reference Processes (CFRP) presentation, where we start by describing the architectures of the various simple machines (Examples) and the many technological solutions implemented (Reality). We then show that despite their diversity, their components only perform four main functions, themselves corresponding to three reference changes undergone by the fluids that pass through the machines (Figure 1.3).

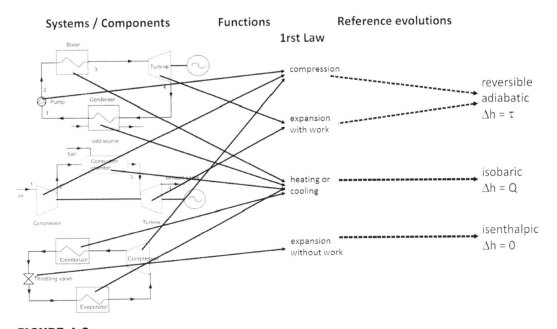

FIGURE 1.3

CFRP presentation, from components to reference processes.

After having introduced some essential notions of thermodynamics, energy exchanges of a system with its environment and the first law are presented (Theory), and we show that each reference process outlines a particular case of application of this first law.

Reference processes are behavior models of fluids in machines, which quite naturally lead one to examine the properties of these fluids, which is both Reality (study of matter) and Theory (for their representation). This leads to thermodynamic charts in which cycles can be visualized (Methods), and which constitute one of the working environments of the simulator and play a fundamental role in learning (Figure 1.4).

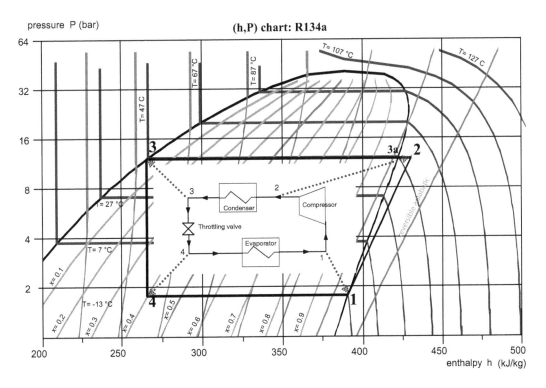

FIGURE 1.4

Simplified refrigeration machine cycle.

The study of the exchanges of energy taking place in the components of the machines leads moreover to introduce a minimum of notions of thermodynamics (Theory) by doing it in a contextual way and by justifying their necessity. This approach has the advantage that it avoids using the notion of entropy, which is always difficult for beginners to understand.

The use of the simulator (Methods) naturally finds its place in this context, because the components which it implements correspond precisely to the functions that have been identified previously. Cycle architectures are constructed by connecting these components in the graphic diagram editor; their setting is easily explained by comparison with reference processes. The cycles can then be viewed in the (enthalpy, pressure) chart, which is the easiest to understand (Figure 1.4).

The distinction between components and systems, which has been stressed several times, plays a fundamental role in the pedagogy that we recommend. As the number of functional thermodynamic components most commonly used is relatively small, the systems that can be imagined by assembling them are numerous and varied: at this level, there is still a considerable field of investigation for the decades to come.

Pre-built model guided exploration or model construction?

A tool like Thermoptim makes it possible to complete a classic teaching of thermodynamics with a great diversity of educational activities, which can be grouped into two main categories:

- Discovery and initiation, in particular by exploring pre-built models;
- Model construction, which concerns learners seeking to know how to model energy systems by themselves.

Depending on the objectives and especially the time available, the teacher can choose one or the other.

From 1998 to 2016, the main use made of this tool in higher education corresponded to the second category. It allows learners to get to the bottom of things and learn to build by themselves different thermodynamic cycles, which gives them great autonomy, a motivating factor, especially when they are on a work placement.

However, it assumes that their first steps in using the software are the subject of tutorial sessions benefiting from supervision by teachers who have a good command of the tool, and of certain manipulations requiring a little practice. In addition, this approach assumes that the learner has a license to use Thermoptim, which, although low in cost, is not free.

For other teaching contexts, the first category can also have many advantages. When learners do not build models by themselves, but explore and set pre-built models, the learning difficulties of the software are significantly reduced, as well as the need for coaching. As the guided explorations can all be studied using the free demo version of Thermoptim, it is not necessary in this case to purchase a license to use the software.

The scenario is presented in a specific Java browser capable of emulating Thermoptim, which offers different activities to learners, such as finding values in the simulator screens, setting them to perform sensitivity analyses, etc. Contextual explanations are offered gradually.

Guided explorations are defined in an HTML 5 file, which makes it possible to open and close Thermoptim files corresponding to the models studied, to plot their cycles in thermodynamic charts and to offer small quizzes to learners so that they can check their understanding of the methods presented.

This ensures that they do not waste time on handling errors that are not of educational interest, which is essential if we want to make the most of the time available. The risk of error decreases considerably, and if errors occur, learners only have to reset the browser by reloading the files they have.

In practice, we believe that these two modes of using Thermoptim are complementary and that, since the students have different learning strategies, one or the other is better suited to each case.

Three main steps

Based on feedback from students who have been trained with Thermoptim, it appeared desirable to structure the progression of a course on energy systems according to three main steps:

* The acquisition of concepts and tools, devoted to the refreshing of the basics of thermodynamics, the study of basic cycles, the discovery of the technologies used and the learning of Thermoptim;
* The consolidation of the concepts introduced during the first step, possibly with theoretical supplements on combustion, exchangers and exergy, the study of variants of basic cycles, combined cycles and cogeneration;
* Further development and personal application, giving rise to the study of innovative cycles and reflections on technological perspectives, on the occasion of mini-projects carried out alone or in groups.

First step

One of the great difficulties of learning a new discipline is that there is a minimum of fundamental background to know to be able to progress. You have to understand certain key concepts before you can go further and get to the heart of the matter.

During this first step, we generally try to ensure that learners:

* Acquire these key concepts;
* Master the tools that will allow them to deepen their knowledge later;
* Understand the basic examples (steam plant, gas turbine, refrigerator).

A classic deductive approach can be chosen for this, but inductive approaches of the type of CFRP presentation shown above also naturally find their place here, and underscore the recommendations of the supporters of the so-called minimalist approaches. They are particularly well suited for vocational training.

This way of proceeding also has a psychological dimension: to motivate the learners, it must be shown very quickly that they will be able to follow the course without much difficulty and even become operational. Once they are convinced, they will find in themselves the resources necessary to invest in training.

Second step

During the second step, the learners start by consolidating their knowledge by studying some theoretical and methodological supplements (notably on combustion, heat exchangers and exergy), and by analyzing variants of the basic cycles. Subsequently, they tackle combined cycles and cogeneration. This step can be an opportunity to think about how to improve basic cycles by reducing their irreversibilities.

As the studied cycles become more complex, the studied models require more work than the previous ones. Additional modeling with Thermoptim is therefore presented as and when required.

Third step

The first two steps offer a progression allowing learners, after around twenty hours of time investment, excluding reminders and supplements, to acquire a fairly complete overview of the basics of applied thermodynamics, but it is clear that they do not cover the whole subject.

The third step opens up a much wider range of possible activities, so that learners can define a work program that meets their personal aspirations, according to the objectives they are pursuing and the time available to them, and with the help of teachers who supervise them.

This third step allows them:

- On the one hand, to carry out further studies by tackling more complex subjects than those they have studied before, for example, the cycles involving gases whose humidity varies. Their knowledge of thermodynamics allows them to start putting their own models into equations. They discover the possibilities offered by the external class mechanism, which allows a user to extend the potential of Thermoptim by designing his own components in order to model technologies not available in the core of the software;
- On the other hand, to apply by themselves the knowledge acquired during the previous steps, by studying innovative cycles, by carrying out mini-projects alone or in a group, or even by carrying out reflections on technological perspectives.

Variable terms depending on the context

Note that the content of each of these steps may differ depending on the context, in particular on the pedagogical preferences of teachers and the learning styles of their students.

Those who are used to classical deductive approaches and who are in initial training may prefer a presentation with a few theoretical refreshers followed by explanations on technologies and cycles, while learners in vocational training may be more motivated by the inductive approach and minimalist CFRP presentation.

In the course on thermodynamic heat conversion that we offer in the form of two MOOCs, modeling and simulating corresponds to the first step and follows the CFRP approach, and classical and innovative cycles is limited to the second step. In both cases, the practical work with Thermoptim takes the form of guided explorations.

Only the first two steps could indeed be considered within the framework of MOOCs, as the third step requires a personalized monitoring of learners, which is impossible to provide in this type of course.

For these MOOCs, we also sought to lighten as much as possible the prerequisite mathematical background, which led us to avoid using the concept of entropy – the understanding of which can pose difficulties. The only activities using it are optional and have been included so that those who still want to study that topic can do so.

Among the digital educational resources that we have put online, all of the Diapason sessions and other guided explorations allow learners, on the one hand, to build models by themselves and, on the other hand, to train themselves on more difficult subjects such as thermal integration methods or the establishment of exergy balances thanks to productive structures.

Finally, we should point out that nothing prevents mixing up the modalities, starting for absolute beginners in the first step with the CFRP presentation, which is very easy to understand. In the second step, they can continue with a slightly more classic approach, whereby the different subjects are introduced sequentially. Once they have acquired the basics, learners are indeed more able to get along with a traditional disciplinary sequence.

In this book, the first part corresponds globally to the first step, while the second and third parts can provide elements for the second step, to be chosen according to the educational objectives pursued. For the third step, which must be personalized according to the learner's preference, the last three parts contain sections allowing many in-depth studies.

Comparison with other tools with teaching potential

To our knowledge, and apart from our own developments, the only currently available tools with teaching potential in this field are as follows, and only the first two have been developed with a view to changing educational paradigms in applied thermodynamics:

- Instructional software called CyclePad developed like an application of qualitative physics (Forbus & Whalley, 1994) to the teaching of thermodynamics, which has the advantage of being able to assist learners in their training in thermodynamic cycles, thanks to an inference engine written in Lisp that is able to guide modeling and to make a diagnosis of errors, but which has, on the other hand, significant limitations in its potential for application;
- Instructional software called TEST (The Expert System for Thermodynamics), which is a collection of Java applets developed by Bhattacharjee (2003), of the California State University at San Diego;
- Software called Pinch (Favrat & Staine, 1991), intended to illustrate thermal integration methods, developed at EPFL in Lausanne (Switzerland);
- An equation solver intended for energy engineering (Klein & Alvarado, 1993), named EES (Engineering Equation Solver). It supposes, however, that the user is able to describe all the equations of the system which is to be studied; i.e., users are assumed to be able to quantify their problems as well as qualify them, which is seldom the case for a beginning learner;
- Cycle-Tempo, a computer program developed by TU Delft (Delft University of Technology) as a modern tool for the thermodynamic analysis and optimization of systems for the production of electricity, heat and refrigeration (www.tudelft.nl).

By way of summary

The educational method that we recommend can be stated in various forms but rests on some main constants:

- Reduce the cognitive load of learners by limiting unnecessary theoretical developments as much as possible;
- Make learners operational thanks to the simulator that allows them to study real problems and not caricatures of reality because of overly simplifying assumptions;
- Shift the content of the teaching by reducing the equations and emphasizing the qualitative explanations relating to the physical phenomena that take place in the systems studied;
- Sequence the concepts presented using the functional approach and the RTM (E) model.

Historically, this method was first used in the form of directed work to build models either using or not using the Diapason sessions in classes supervised by teachers.

Since 2015, a complementary didactic effort has been made to prepare the guided exploration activities of pre-built models; the objective is to allow learners to focus on understanding the thermodynamic settings of models while minimizing the simulator learning time.

The training resources with Thermoptim were thus completed and diversified, and reached a large audience, at the national and international levels.

All of the digital resources relating to these new pedagogies have been brought together in the Thermoptim-UNIT portal (www.thermoptim.org); the main topics are as follows:

- Course pages;
- Summary thematic pages on components, systems and sectors;
- Diapason sessions;
- Videos (hosted in Canal U);
- Guided explorations;
- Written materials;
- Directed work guide sheets;
- Self-assessment activities;

- Methodological guides;
- Substance and component models allowing the Thermoptim core to be extended;
- Notes relating to the pedagogy of applied thermodynamics.

The rest of this first part, as we have said, corresponds more or less to the first step of the teaching method that we recommend. You will now be able to get to the heart of the matter and move on to practice.

We wish you an excellent learning of thermodynamics applied to energy systems.

Bibliography

S. Bhattacharjee, *The expert system for thermodynamics, a visual tour*, Pearson Education Inc, Upper Saddle River, NJ, 2003, ISBN 0-13-009235-5.

J. M. Carroll, *The Nurnberg funnel: Designing minimalist instruction for practical computer skill*, MIT Press, Cambridge, MA, 1990.

D. Favrat, F Staine, An interactive approach to the energy integration of thermal processes, CALISCE '91, EPFL, 1991.

K. Forbus, P. Whalley, Using qualitative physics to build articulate software for thermodynamics education. *Proceedings of AAAI-94*, 1994, pp. 1175–1182.

S. A. Klein, F. L. Alvarado, EES engineering equation solver, F-Chart Software, 4406 Fox Bluff Road, Middleton, Wisconsin 53562, 1993.

C. Köppe, R. Rodin, Guided exploration: An inductive minimalist approach for teaching tool-related concepts and techniques. Proceedings of the 3rd Computer Science Education Research Conference, CSERC'13, Arnhem Netherlands April, 2013, Publisher: Open Universiteit, Heerlen, P.O. Box 2960, Heerlen, Netherlands.

J. Sweller, Cognitive technology: Some procedures for facilitating learning and problem solving in mathematics and science, *Journal of Educational Psychology*, 81(4), 457–466, 1989.

Components, functions and reference processes

Introduction

In this chapter, we discuss the key concepts that should be presented to beginners in order to enable them to understand and study the cycles of three basic energy technologies: steam power plant, gas turbine and refrigerating machine. The chosen approach follows a presentation called CFRP for Components, Functions and Reference Processes.

In the CFRP presentation, we start by describing the architectures of the various technologies and the technological solutions implemented (boilers, turbines, pumps, condensers, turbocompressors, piston, scroll, screw, hermetic compressors, expansion or throttling valve, etc.). We then show that despite their diversity, the components only perform four main functions, themselves corresponding to three reference processes undergone by the fluids that pass through them.

After having introduced some essential notions of thermodynamics, energy exchanges of a system with its environment and the first law of thermodynamics are presented in open and closed systems, and we show that each reference process outlines a particular case of application of this first law. The importance of enthalpy is explained as it allows one to calculate work and heat exchange of a system with its surroundings.

Main functionalities associated with energy technologies

In this section, we will show that energy technologies only involve a very small number of functions. We will establish this by analyzing the components of three different technologies: the steam power plant, the gas turbine and the refrigeration machine. However, we will begin by defining two concepts essential for the whole book, that of working fluid and that of cycle.

Notions of working fluid and cycle

A steam power plant is essentially comprised of the following components:

- A boiler where the fuel (solid, liquid or gas) is burned for generating steam (usually superheated);
- A steam turbine in which this steam is then expanded. The turbine shaft provides the power output (Figure 2.1);
- A condenser in which the steam leaving the turbine is completely liquefied (water);
- A pump that restores the boiler pressure.

DOI: 10.1201/9781003175629-2

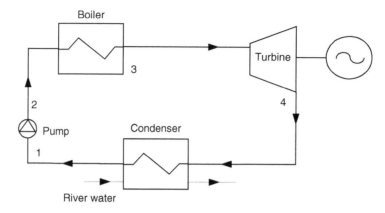

FIGURE 2.1

Sketch of a steam power plant.

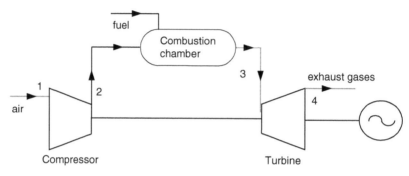

FIGURE 2.2

Sketch of a gas turbine.

In a gas turbine (Figure 2.2), air is drawn in on the left side of the machine by a compressor. A fuel burns with this compressed air in the combustion chamber, bringing the gases to high temperature.

These hot gases are expanded in a turbine, which is used first to drive the compressor and second to drive the alternator or an engine shaft. The gases are then released into the atmosphere on the right of the machine.

As in these two examples, all the thermal machines that convert heat into work or the reverse are **traversed by at least one fluid**, which undergoes various processes (such as heating, cooling, compression or expansion) and exchanges energy with the surroundings.

In a steam plant, it is water, in a gas turbine air and flue gas. As we will see later, in a refrigerator, it is a refrigerant.

Generally, this fluid is called **thermodynamic fluid** or **working fluid**.

In the steam plant, the working fluid passes successively through each of the components, to return to its starting state.

In many energy systems, as in this example, the fluid undergoes a series of processes which lead it to return to its initial state.

We then speak of a **cycle**. This is another fundamental notion that we will often use.

In the case of a gas turbine, as in all internal combustion engines, the exhaust gases are released into the atmosphere, so it is improper to speak of a cycle.

However, by extension, we come to speak of a cycle to qualify the representation of the succession of thermodynamic changes undergone by the fluids used in an energy technology.

Steam power plant

Steam power plants are used today mainly for centralized electricity generation. Most units are either conventional, i.e., boiler burning mainly coal or oil, or nuclear.

Technological aspects

Let us now briefly present the technologies implemented in the components of this steam plant.

The pump (Figure 2.3) is generally of the centrifugal type, multistage because it must achieve a high compression ratio (around 3,600 with our settings).

In conventional flame power plants, the boiler outlet conditions are around 560°C and 165 bar, leading to a thermodynamic efficiency close to 40%, but in certain so-called super-critical cycles, these values can reach 325 bar and 600°C. In the pressurized water reactor or PWR current nuclear power plants, the high pressure (HP) and the steam temperature are limited for safety reasons and hardly exceed 60 bar and 275°C.

A boiler (Figure 2.4) has three successive functions:

- Heat-pressurized feedwater (in the economizer) to the vaporization temperature corresponding to the HP;
- Vaporize steam in the evaporator;
- Finally superheat steam at the desired temperature.

FIGURE 2.3

Centrifugal pump. (Courtesy French Navy.)

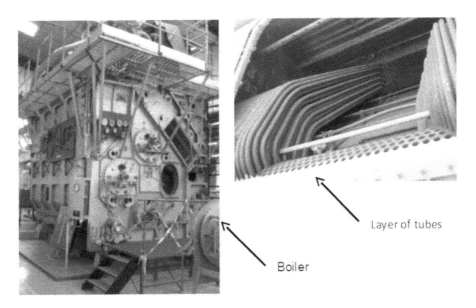

FIGURE 2.4

Boiler. (Courtesy French Navy.)

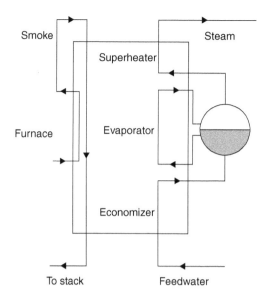

FIGURE 2.5
Boiler exchange configuration.

FIGURE 2.6
Steam turbine. (Courtesy French Navy.)

It behaves like a triple heat exchanger and may be represented in terms of heat exchange by the diagram in Figure 2.5.

Details on steam turbines will be given in Chapter 5. They are always multistage given the very important expansion ratio (Figure 2.6).

The condenser is a heat exchanger whose particularity is to work at a pressure lower than the atmosphere, given the low vapor pressure of water at ambient temperature (Figure 2.7).

Steam power plant cycle

The steam plant cycle, known as Rankine cycle or Hirn cycle, converts high-temperature heat into work on the turbine shaft. We are talking about a power cycle.

The setting values we give here (Figure 2.8) have been chosen so that the cycle will be easily plotted in the thermodynamic (h, P) chart, as we will see later.

At point 1, the water enters the pump in the liquid state and at very low pressure (LP) (about 1/30 bar).

It is compressed there and leaves it at point 2 at a pressure close to 128 bar, still in the liquid state.

In the boiler, a fuel (solid, liquid or gaseous) is burned, thus generating hot gases that are cooled by the working fluid, namely, water.

Condenser inside

Condenser (external view)

FIGURE 2.7

Condenser. (Courtesy French Navy.)

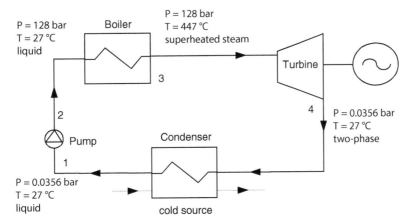

FIGURE 2.8

Steam plant cycle settings.

By staying at about the same pressure, the water goes from the liquid state at point 2 to that of superheated steam at point 3, where it is brought to the temperature of 447°C.

The superheated steam at point 3 is then expanded in a steam turbine, which allows mechanical power to be produced on its shaft.

It leaves at point 4 at the LP of the cycle, and its temperature again becomes equal to 27°C. Its state corresponds to a mixture of liquid and vapor, called two-phase state.

The steam exiting the turbine at point 4 is then completely liquefied at point 1 by cooling in the condenser. The pump then restores the water to the boiler pressure at point 2.

Note that the condenser is cooled by an external cold source, usually outside air or water from a river or sea.

The working fluid circuit has **two parts at different pressures: the HP at 128 bar** in the boiler and the **LP at 0.0356 bar** in the condenser.

This is a characteristic common to most of the cycles that we will study: they implement at least two pressure levels, namely, the pumps and compressors passing the working fluid from the LP to the HP, and the turbines and valves from the HP to the LP.

Functional analysis

The steam plant cycle can thus be broken down into three functions: compression, heat exchange and expansion.

1. The working fluid is compressed in the pump;
2. It exchanges heat in the boiler to bring it at high temperature;

3. The working fluid is then expanded in the turbine;
4. It exchanges heat in the condenser to cool it to the initial state.

In all power cycles, we find these three functions, succeeding each other in this order: the working fluid is compressed, then heated and finally expanded.

Gas turbines

Today, gas turbines are experiencing a very strong development in many applications: air transport, power generation, cogeneration and driving machines (compressors and pumps). Among the arguments in their favor are their small size, excellent power-to-weight ratio, their quick start, good performance and low emissions of pollutants.

Technological aspects

There are two main categories of gas turbines: industrial gas turbines – heavy and robust, but of average performance (Figure 2.9) – and gas turbines "derived from aviation" or "aeroderivative" (Figure 2.10) – much lighter and efficient, but also more expensive.

The industrial gas turbine shown in Figure 2.9 is characterized by silo combustion chambers (multifuel, emission control, radiative protection of turbine blades).

We speak of aeroderivative machines because these are variants of turbojets. To obtain high performance, their architecture can involve several sets of compressors and turbines rotating at different speeds.

The compressors, centrifugal or axial, are generally multistage.

The combustion chambers are normally constructed of refractory alloy.

FIGURE 2.9
SGT-400 industrial gas turbine. (Courtesy Siemens Energy.)

FIGURE 2.10
Compact SGT-A45 TR aeroderivative gas turbine. (Courtesy Siemens Energy.)

FIGURE 2.11

Flame tubes mounted in barrels. (Courtesy French Navy.)

They are often mounted in barrels around the axis of the turbine (Figure 2.11).

The turbines are also generally axial.

The main technological constraints are at the level of the first stages of the expansion turbine, which are subjected to the flow of exhaust gases at very high temperatures.

Gas turbine cycle

At point 1 (Figure 2.12), the air enters the compressor at atmospheric pressure (about 1 bar).

It is compressed there and leaves it at point 2 at 434°C and 16 bar.

In the combustion chamber, the working fluid is brought to the temperature of 1,065°C, always at the pressure of 16 bar. Given combustion, its composition varies: from air at point 2, it becomes burnt gases at point 3.

In the turbine, the working fluid is expanded to the pressure of 1 bar at point 4. Its temperature drops to 495°C.

The gases are then released into the atmosphere.

As with the steam power plant, this cycle, known as Brayton cycle, operates between two pressure levels: the LP at the intake and at the exhaust, and the HP in the combustion chamber, the compressor and the turbine passing the working fluid from one level to another.

Functional analysis

The gas turbine cycle can thus be broken down into the three successive functions that we identified earlier:

1. Compress the working fluid;
2. Heat it to high temperature;
3. Expand the working fluid.

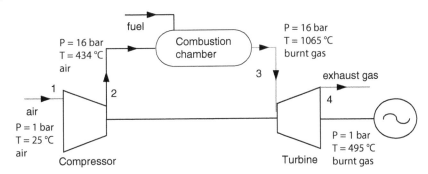

FIGURE 2.12

Gas turbine settings.

As we indicated during the presentation of steam power plants, we find these three functions in all power cycles, successive in this order: we compress, we heat and we expand.

Refrigeration machine

Principle of operation

In a vapor compression refrigeration installation, an attempt is made to maintain a cold enclosure at a temperature below ambient one.

To achieve this, a LP (and therefore low-temperature) refrigerant is evaporated in an exchanger placed in the cold enclosure. For this, the temperature T_{evap} of the refrigerant must be lower than that of the cold chamber T_{cc}.

The fluid is then compressed to a pressure such that its condensation temperature T_{cond} is higher than the ambient temperature T_a. It is thus possible to cool the fluid by heat exchange with the ambient air or with a cooling fluid, until it becomes liquid. The liquid is expanded at LP in a valve, without production of work, and directed into the evaporator. The cycle is thus closed.

Figure 2.13 shows that a refrigeration machine is comprised of four components:

* An evaporator;
* A compressor;
* A condenser;
* A throttling valve.

This cycle converts the work on the compressor shaft into production of cold at low temperature. We are talking about a cycle working in reverse or refrigeration cycle.

As for the steam power plant, we speak here rightly of a cycle because, successively traversing the four components of the machine, the working fluid undergoes a series of processes that lead it to return to its initial state.

Technological aspects

In a domestic refrigerator, the evaporator is generally formed of two corrugated flat plates welded one against the other, and the refrigerant circulating in the channels formed by corrugations.

It most often lines the freezer compartment of the refrigerator (it is on it that the layer of frost forms). The plate between the fluid passage channels serves as a fin to increase the thermal contact between the refrigerant and the cold compartment (Figure 2.14).

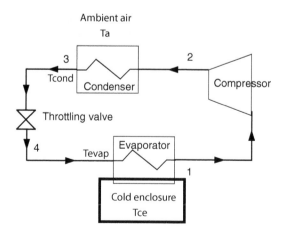

FIGURE 2.13
Sketch of a refrigeration machine.

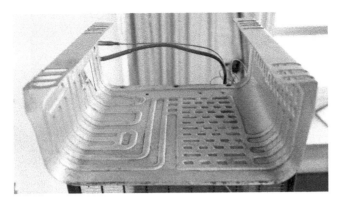

FIGURE 2.14

Evaporator. (Courtesy French Navy.)

FIGURE 2.15

Compressor.

This evaporator is connected to the rest of the machine by two pipes that pass through the insulating wall. One of them is connected to the outlet of the valve, while the other to the compressor intake.

Generally, the compressor is not directly visible because it is contained in a metal block mounted on rubber cushions, to avoid vibrations, and from where come an electric wire and two pipes for entering and leaving the fluid. It is a so-called hermetic piston or scroll compressor, which has the advantage that the engine is directly cooled and lubricated by the thermodynamic fluid (Figure 2.15).

Other types of compressors also exist, such as the so-called open or semi-open piston compressors (Figure 2.16), scroll compressors, screw compressors or centrifugal compressors.

The condenser (Figure 2.17) is the black grid located on the rear face of the refrigerator, made up of a pipe wound in serpentine and supported by metal tubes, which, on the one hand, increase the thermal exchanges with the air and, on the other hand, reinforce the mechanical rigidity. It is connected to the compressor outlet and to the valve inlet.

The throttling valve generally consists of a simple capillary tube, that is to say of very small diameter, and sometimes it is a thermostatic regulator (Figure 2.18).

Refrigeration machine cycle

Let us study how a refrigeration cycle works and what values can take its parameters (Figure 2.19).

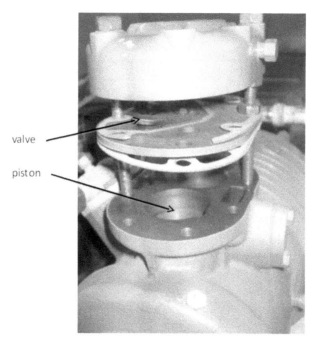

FIGURE 2.16
Open compressor. (Courtesy French Navy.)

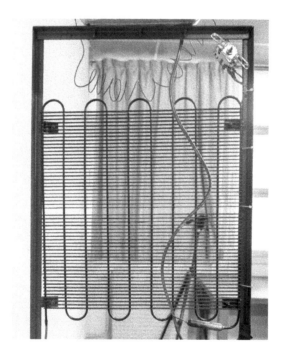

FIGURE 2.17
Condenser. (Courtesy French Navy.)

At point 1, a refrigerant, R134a, enters the compressor in the saturated vapor state: pressure of 1.78 bar and temperature T = –13°C.

It is compressed there and leaves it at point 2 at a pressure of 12 bar and at 53°C, in the vapor state.

The pressure of 1.78 bar was chosen so that the vaporization temperature of R134a at this pressure is lower than that of the cold enclosure, which is –8°C.

The pressure of 12 bar was chosen so that the vaporization temperature of R134a at this pressure is higher than that of ambient air, which is 35°C.

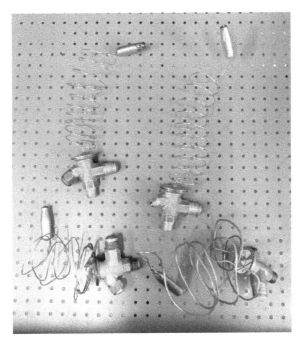

FIGURE 2.18

Throttling valves. (Courtesy French Navy.)

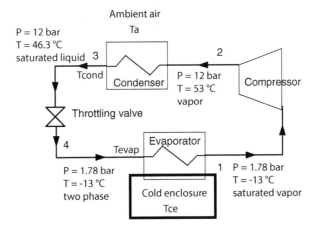

FIGURE 2.19

Refrigeration machine settings.

The cooling of the fluid in the condenser by exchange with the outside air comprises two stages: desuperheating in the vapor zone followed by condensation to point 3 in the saturated liquid state at 12 bar and 46.3°C.

The working fluid is then expanded without producing work to point 4 in the two-phase state at 1.78 bar and −13°C, before being directed to the evaporator.

As with the steam power plant and the gas turbine, the cycle operates between two pressure levels: the LP at the evaporator and the HP at the condenser, the compressor and the throttling valve passing the working fluid from one level to the other.

Functional analysis

This first overview of the operation of a refrigeration machine allowed us to see that this cycle calls upon two of the component types already encountered in the previous cycles:

1. A compressor;
2. Heat exchangers.

It also makes use of a component of a new type: the throttling valve.

Note that the refrigeration machine cycle can be broken down into four functions:

1. Compress the working fluid;
2. Exchange heat to cool it at HP;
3. Expand the working fluid without producing work;
4. Exchange heat to bring it to the initial state.

In all the refrigeration cycles, we find these three functions, succeeding each other in this order: we compress, we cool and we expand. Note that in some refrigeration cycles, expansion may take place in a turbine instead of a valve, with work production, but this is generally not the case.

Four basic functions

We have just seen that even if the technical solutions implemented are varied, working fluids are subjected to only **four different types of processes**:

* Compression;
* Expansion with work production;
* Expansion without work production;
* Temperature change (heating and cooling).

Therefore, **four functions** are sufficient to describe the operation of these machines:

* Compression can occur with the fluid being liquid or gaseous. In the first case, the component is a pump, while in the second, a compressor;
* Expansion with work production is generally made through turbines;
* Expansion without work production occurs in valves;
* Heating can be generated either in combustion chambers or boilers, or in heat exchangers. Cooling is usually done in heat exchangers.

This finding has a **very broad bearing**: in all engines, the working fluid is successively compressed, heated, expanded and cooled or released to the atmosphere, and, in all refrigeration cycles, it is compressed, cooled, expanded and heated or released to the atmosphere.

The different components of thermal machines can thus be grouped into a small number of functional categories. Furthermore, they can in certain conditions be calculated independently of each other when we know the corresponding thermodynamic changes.

Figure 2.20 summarizes the links between the components of the systems we have studied and these four functionalities.

From these functions, it is possible to represent a large number of energy technologies, from the simplest like those we have studied here, to large systems.

BOX 2.1
Key issues

Architecture and operation of basic cycles

In this chapter, we have introduced the cycles of the three simplest and most popular energy technologies.

It is important that you understand their architecture, how they work and know the names of their components.

These self-assessment activities will allow you to test your knowledge on these topics:

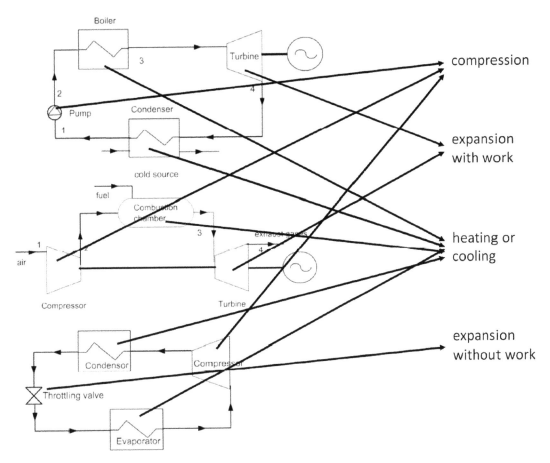

FIGURE 2.20

The four basic functions.

BOX 2.2
Key issues

Four basic functions

The first key concept is that only four functionalities are sufficient to describe the functioning of the machines we have studied:

- Compressions can be carried out with the fluid being liquid or gaseous. In the first case, the component is a pump, while in the second, a compressor;
- Expansion with work production is generally carried out in turbines;
- Expansion without work production (throttling) takes place in valves or filters;
- Heating can be carried out either in combustion chambers or boilers, or in heat exchangers. Cooling is generally done in heat exchangers.

We have also stated a very general observation: in all power cycles, the working fluid is successively compressed, heated, expanded and cooled or released in the atmosphere, and in all refrigeration machines, it is compressed, cooled, expanded and heated or released in the atmosphere.

The following self-assessment activities will allow you to check your understanding of these concepts:

Functions in a steam plant cycle, ddi
Functions in a gas turbine cycle, ddi
Functions in a refrigeration cycle, ddi

Energies brought into play in the processes

This second section introduces the thermodynamic concepts whose knowledge is essential to model energy systems.

After having started by a discussion on the performance of compressors and turbines used in a gas turbine, we will present the concepts of thermodynamic system and state.

The energy exchanges between a thermodynamic system and its surroundings will then be studied, which will make it possible to state the **first law of thermodynamics** in a closed and then open system.

We will show that the four main elementary functions identified in the previous section correspond to three reference processes undergone by the fluids that pass through the machines, each outlining a particular case of application of the first law of thermodynamics that we will present.

Operation of compressors and turbines

In a compressor, the ratio of the discharge pressure to the suction pressure is called the compression ratio, and in a turbine, its inverse is the expansion ratio.

Study of compression

When you compress a gas, it heats up, as everyone can see when inflating a bicycle tire. The compression of a gas is therefore accompanied by a simultaneous increase in its temperature and pressure. This unavoidable heating of the gas unfortunately results in additional compression work, more or less important depending on the quality of the compressor.

To fix ideas, let us take a look at the work required by the compressor of a gas turbine. Figure 2.21 shows, for a gas turbine of common characteristics crossed by a flow rate of

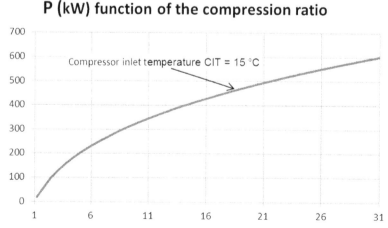

FIGURE 2.21

Compression power function of the compression ratio.

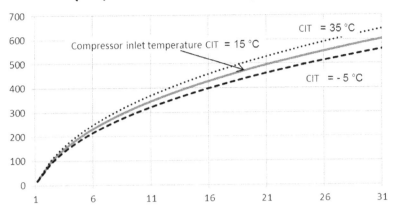

FIGURE 2.22

Influence of the outside temperature.

1 kg/s, the variation of the compression power as a function of the compression ratio, for a compressor inlet temperature (CIT) of 15°C.

For example, for a compression ratio of 6, the power required is almost 220 kW, and for a compression ratio of 21, it is worth 500 kW.

As can be seen, the higher the compression ratio, the more work required for the compressor.

The performance of a gas turbine compressor is very sensitive to the value of the suction temperature of the fluid, here the outside temperature. The latter evolving between winter and summer from –5°C to about 35°C, the compression power input varies by more or less 7% around its average value, for a compression ratio of 16, the minimum power to be supplied corresponding to winter, as shown in Figure 2.22.

The influence of the suction temperature on the compression power explains why it is always preferable to compress a gas at the lowest possible temperature, which leads to cooling it if it is technically possible.

It can be shown that the compression work of an ideal gas is, on the one hand, proportional to the temperature of the suction fluid (expressed in Kelvin) and, on the other hand, an increasing function of the compression ratio. If P_s and P_d designate the suction and discharge pressures of the compressor, and if CIT is the compressor inlet temperature, in °C:

$$\Gamma = (CIT + 273.15) f_c \left(P_d / P_s \right)$$

When the fluid cannot be assimilated to an ideal gas, this relationship remains valid as a first approximation.

Study of expansion

The work produced by a compressed gas being expanded in a turbine is all the more important as its temperature is high: it is given as a first approximation by a relation similar to that of a compression, if TIT is the turbine inlet temperature, in °C:

$$\Gamma = (TIT + 273.15) f_d \left(P_d / P_s \right)$$

Let us illustrate this point by studying the expansion that takes place in the turbine of a gas turbine. The higher the turbine inlet temperature (TIT) (equal to the temperature at the outlet of the combustion chamber), the higher the power output. The influence of this value is significant: for an expansion ratio of 16, a variation of 100°C in the temperature reached at the outlet of the combustion chamber causes the power output of the turbine to vary by more or less 8% around its average value (Figure 2.23).

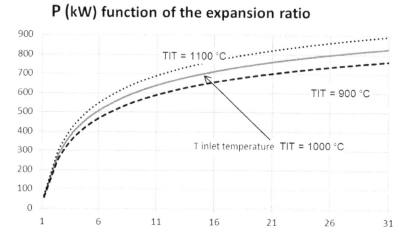

FIGURE 2.23

Power supplied by the turbine as a function of the expansion ratio.

For example, for an expansion ratio of 6, the power output is 500 kW, and for a compression ratio of 16, it is worth about 720 kW.

Study of the whole gas turbine

Now let us take a look at what is going on in the whole machine, the assembly of a compressor, a combustion chamber and a turbine.

In the combustion chamber, the chemical energy supplied to the cycle in the form of fuel is converted into thermal energy, which is used to bring the hot gases to the high temperature TIT.

Since the combustion chamber is roughly isobaric, the pressures upstream and downstream are the same. Since atmospheric pressure prevails at the compressor intake and the turbine exhaust, the compression and expansion ratios are equal. We will therefore only consider in the following the first of these parameters.

Figure 2.24 shows, for a gas turbine of common characteristics (CIT of 15°C and TIT of 1,000°C), crossed by a flow rate of 1 kg/s, the compression power (solid line), the power delivered by the turbine (dashed line) and the net power produced by the machine (dotted line), as a function of the compression ratio.

For example, for a compression ratio of 6, the compression power is around 220 kW, and the useful power output is 270 kW.

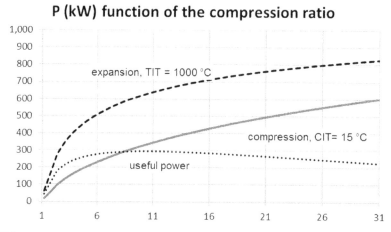

FIGURE 2.24

Power put in play.

If the compressed air was expanded at the temperature reached at the end of compression, without additional heat, the balance would be negative because of the losses that take place in the compressor and the turbine.

It is only because the constant pressure combustion reaction brings the gases to high temperature, that the work provided by the turbine is greater than that which is absorbed by the compressor.

The maximum net power is obtained here for a compression ratio of around 10, but the optimum is fairly flat.

The influence of the inlet temperatures of the compressor and the turbine is even more noticeable on the overall performance of the machine than on the components alone, because the useful power is equal to the difference between the power required by the compressor and that delivered by the turbine. For example, the useful power drops by 31.5% when the outside temperature rises from −20°C to + 20°C.

Conclusions

This study allowed us to highlight the influence of the main characteristics of the machine (compression ratio, inlet temperatures in the compressor and in the turbine) on the performance of a gas turbine.

It has shown that:

* it is preferable to cool a gas as much as possible before compressing it, and to heat it as much as possible before expanding it,
* and the energies put in play are proportional to the absolute temperature at the inlet of each component.

It also showed the influence of the compression or expansion ratio on the performance of the machine.

These are the points to which we will return later.

Note that a common mistake is to assume that the pressure should increase in the combustion chamber, as in those that constitute closed systems.

Indeed, the phenomenon is similar to that which takes place in a wall-mounted gas boiler, the combustion of which takes place at atmospheric pressure, the system is open, as well as in the devices that everyone knows, used for cooking or heating, whether it be gas stoves, fireplaces or others. They have no element varying with respect to the pressure, which therefore remains constant.

Technological aspects

Now that we have clarified these points, let us say a few words about the technologies implemented for compressions and expansions in energy systems.

Displacement compressors

A displacement compressor is characterized by the encapsulation, or imprisonment, of the fluid that passes through it in a closed volume, which is gradually reduced. A return of this fluid in the direction of decreasing pressures is prevented there by the presence of one or more mobile walls.

By design, positive displacement compressors are particularly suitable for treating relatively low fluid flow rates, possibly very variable, and under relatively HP ratios.

They are widely used in refrigerators and freezers.

Their operating principle is as follows (Figure 2.25): a fixed mass of gas at the suction pressure P1 is trapped in an enclosure of variable volume. To increase the pressure, this volume is gradually reduced, in a manner that differs depending on the technique used.

At the end of compression, the enclosure is placed in communication with the discharge circuit, so that the gas compressed at pressure P2 can exit.

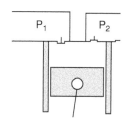

FIGURE 2.25
Piston compressor.

A new mass of gas at pressure P1 is then sucked into the upstream pipes, and so on, the operation of the machine is cyclical.

It is the same principle as the bicycle pump, well known to many amateur and professional cyclists.

Piston compressors are positive displacement compressors.

Automatic valves allow communication with the suction and delivery pipes.

In addition to reciprocating compressors, two other types of positive displacement compressors are widely used: screw compressors and scroll compressors (Figure 2.26), the latter being very widespread today because they have significant advantages in terms of maintenance. We will not speak here of screw compressors.

Scroll compressors are formed from two cylindrical spirals, one fixed and the other mobile, of identical shape.

The moving spiral is driven by an eccentric driving motion, which allows the two cylindrical spirals to roll by sliding over each other, enclosing pockets of gas of varying volume. The gas is sucked in at the circumference and discharged in the center. Compression is thus ensured.

The advantages of this device are the absence of valves, the simplicity of the mechanism and therefore its low cost and its silence, low mechanical losses, the possibility of turning at high speeds, the absence of vibrations, lightness, reliability and the low resistive torque.

We should also point out the rotary vane compressors, which are more widespread today than scrolls, which they compete with increasing available capacity.

Turbocompressors and turbines

Let us now come to the turbocompressors.

Unlike volumetric machines where the fluid is enclosed in a closed volume, a continuous flow of fluid is produced in a turbocompressor, to which energy is communicated by means of mobile blades driven by a rotor.

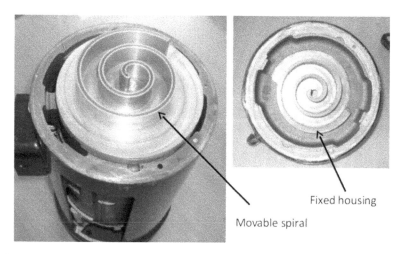

Fixed housing

Movable spiral

FIGURE 2.26
Scroll compressor. (Courtesy French Navy.)

A familiar example of a turbocompressor is a hair dryer or vacuum cleaner. In the first case, a small propeller rotating at high speed accelerates a flow of air which is then heated by electrical resistances. In the second case, the air drawn in by the turbomachine is discharged in the outside, creating a suction effect at the entrance to the device.

In general, a turbomachine consists of four elements in series (Figure 2.27):

* First of all, the inlet convergent C, or distributor, a fixed part whose function is to correctly orient the fluid veins entering the impeller, and to accelerate them slightly;
* The impeller I, or rotor, driven by a rotational movement around an axis. This wheel has vanes defining channels between which the fluid flow is distributed. It communicates to the fluid the mechanical energy of the blades in the form of kinetic, heat and pressure energy;
* The diffuser D is a fixed component that serves to transform in pressure part of the kinetic energy gained by the fluid when passing through the wheel. Depending on circumstances, the diffuser may or not involve blades. It is said to be partitioned or smooth;
* A volute V, also fixed, corrects the fluid lines on the periphery of the wheel, and guides them to the device outlet.

There are two main modes of fluid circulation with respect to the rotor of a turbomachine: the first mode of fluid circulation is radial circulation, widely used for land-based centrifugal turbocompressors, in particular for refrigeration or for supercharging engines (Figure 2.28).

The second mode of fluid circulation is axial circulation, almost always performed in aircraft turbocompressors (Figure 2.29).

A turbomachine frequently has several stages. This is called a multistage turbomachine.

In a turbocompressor, the guidance upstream of the wheel plays a secondary role in relation to the recovery of the kinetic energy at the outlet. We can therefore possibly do without the upstream distributor.

In a turbine, on the other hand, it is essential to have expansion and guide nozzles upstream of the impeller, while the diffuser only plays a secondary role and may possibly disappear.

As mentioned earlier, the main technological constraints in modern gas turbines are at the level of the first stages of the expansion turbine, which are subjected to the flow of exhaust gases at very high temperatures.

The most exposed parts are the rotor blades, which are very difficult to cool and are particularly sensitive to abrasion. The problem is even more difficult to resolve due to the fact that the shapes of the fixed nozzles and mobile blades of the turbines are very complex, especially in small units derived from aviation.

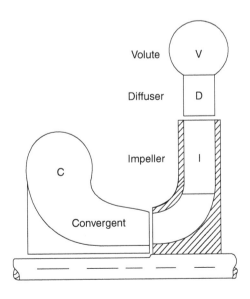

FIGURE 2.27

Turbomachine.

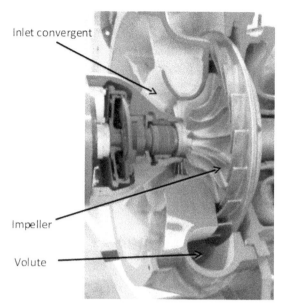

FIGURE 2.28
Centrifugal turbocompressor. (Courtesy French Navy.)

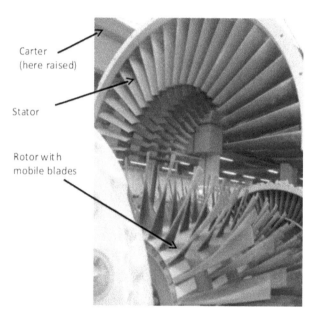

FIGURE 2.29
Multistage turbomachine. (Courtesy French Navy.)

It is therefore important to use a very clean fuel that does not contain particles or chemical components capable of forming acids. It is also necessary to limit the temperature at the end of combustion depending on the mechanical characteristics of the blades.

The materials used for turbine blades are refractory alloys based on nickel or cobalt, and manufacturers intend to make use of ceramics in the future.

As the cycle efficiency is itself an increasing function of temperature, the major technological developments have been devoted to the fabrication: first of efficient cooling systems of the blades and second of materials resistant at high temperatures. For half a century, there has been a gradual increase (about 20°C per year) of the TIT, now reaching 1,300°C–1,500°C.

To cool the turbine blades, air is taken at different levels of the compressor, depending on the desired pressure for reinjection into the turbine.

This air passes inside the blades, through a carefully designed set of baffles, and then is evacuated with the exhaust gases, either at the trailing edge or by allowing some porosity through the wall.

Figure 2.30 shows how the turbine blades are attached to the moving impeller using an assembly method called fire tree root.

Figure 2.31 shows the detail of a blade, with the cooling holes on the trailing edge. Manufacturers are constantly improving the performance of their blades, as well as manufacturing techniques. The blade in Figure 2.31, which was completely manufactured by a 3D printer, can withstand temperatures above 1,250°C at 13,000 revolutions per minute.

In aeroderivative gas turbines of the most powerful last generation, cooling of the blades is no more carried out by air circulation, but by water vaporization, which can benefit from the high heat transfer coefficients in a two-phase system.

The analytical and numerical models that will be presented in this book do not specifically take into account the cooling of the blades in the calculations of gas turbines.

FIGURE 2.30
Fire tree root blade disk assembly. (Courtesy Siemens Energy.)

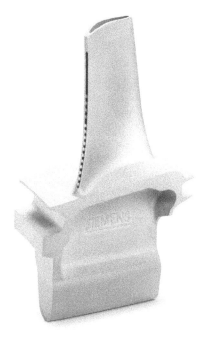

FIGURE 2.31
Modern turbine blade. (Courtesy Siemens Energy.)

Energies brought into play in the processes

To be able to model the energy systems which interest us, it is necessary to introduce a certain number of notions of thermodynamics.

Our aim is to adopt as simplified an approach as possible, and we will limit theoretical developments as much as we can.

Study these concepts well, even if they seem a little vague at first. Their meaning will become increasingly clear to you as you have the opportunity to put them into practice.

Notions of thermodynamic system and state

Let us begin by introducing the concept of **thermodynamic system**, which represents a quantity of matter isolated from what we call the **surroundings** by a real or fictitious **boundary**. This system concept is very general in physics and is found especially in mechanics.

These three concepts are fundamental and will be widely used later.

By way of illustration, Figure 2.32 shows that it is possible to define the boundary of the "steam power plant" system in several ways, represented by a rectangle drawn in thin lines.

The usual definition, identified by the letter A, is illustrated by the left part of the figure. The flows crossing the boundary are the hot gases coming from the hot source and the cooling fluid coming from the cold source. The mechanical power leaving the system is that which drives the electric generator. It implicitly assumes that the compression work of the pump is taken from that provided by the turbine.

The second definition of the system, identified by the letter B, is illustrated by the right part of the figure. It assumes that the compression work of the pump is not taken from the work provided by the turbine, but brought in the form of electricity by the surroundings. We will see below the implications of this change of boundary on the calculation of the energy balance of the system.

Remember that you must **always specify the boundary of the system** in question.

A second very important notion in practice is that of **state**. It allows you to characterize a system concisely.

The notion of **state of a system** represents "the minimum information necessary to determine its future behavior".

State variables and functions

This state is characterized by what is called a set of **state variables** making it possible to completely characterize a system at a given time.

In mechanics, position variables and speed are state variables.

For a **simple** thermodynamic system like a particle of matter in a pure fluid, there are several sets of state variables that meet this definition. The most used in the literature are the following pairs: (pressure, temperature), (pressure, volume), and (temperature, volume).

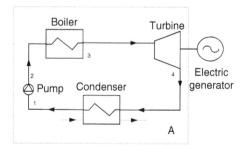
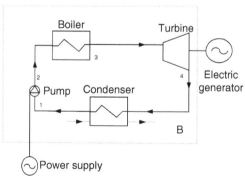

FIGURE 2.32

Steam power plant system boundaries.

For more complex systems, it may be necessary to add two other broad categories to these state variables:

- Chemical variables;
- Electrical variables.

A **state function** is a quantity whose value depends only on the state of the system, and not on its history.

We will note with a d an infinitesimal evolution of a state function: for example, dP for a small pressure variation.

However, during the evolution of a thermodynamic system, many quantities depend not only on the initial and final states of the system, but also on the way in which evolution takes place.

These quantities are often called **path functions**, to indicate this dependence.[1] This is the case of the work brought into play or the heat exchanged at the boundaries of the system. We will note with δ an infinitesimal evolution of a path function: for example, δW for a small variation of work.

In the general case, the calculation of these path functions is more complex than that of the state functions.

Open and closed systems

In thermodynamics, we are often led to distinguish two types of systems: **closed systems**, which do not exchange matter with the surroundings, and **open systems**, which exchange matter (Figure 2.33).

This distinction is important because thermodynamic properties are not expressed in the same way in closed systems and in open systems. Paradoxically even, they are generally easier to calculate for open systems, although they involve a transfer of matter.

At this stage of our presentation, it is important to note that the open or closed character of a system depends on the boundaries that we choose to define it, which can induce small semantic difficulties.

BOX 2.3
Key issues

Basic notions

These self-assessment activities will allow you to test your understanding of these concepts:

State and path functions, cat
Differences between closed system and open system, gfe

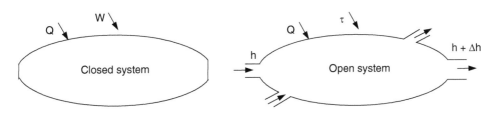

FIGURE 2.33
Open and closed systems.

---

[1] An analogy can be made with movements in the mountains: for a given altitude difference, the variation in potential energy of a mobile making the journey is always the same because it is a state function, while the distance traveled varies according to the path followed because it is a path function.

Energy exchange between a thermodynamic system and its surroundings

We will now study the energies that components of the energy systems exchange with their surroundings, and state the first law of thermodynamics, which is fundamental in practice.

In what follows, unless otherwise indicated, we will reason in a closed system on a quantity of material of 1 kg, and in an open system on a working fluid flow of 1 kg/s. This will allow us to lighten the notations by not showing the quantities or flow rates involved.

It is essential to note that thermodynamic systems associated with the components we are interested in, only exchange energy with the surroundings in **two** distinct forms:

* Heat, by heat exchange on the system boundaries. It is usually denoted by Q;
* Work, by the action of pressure forces on the boundaries; the work of gravity forces is neglected. This work is usually denoted W in closed and τ in open systems where it is often named shaft work.

We indicated previously that, to calculate these two forms of energy brought into play in a process, it is not enough to know the initial and final states of the system. It is also necessary to know the path followed during the process.

This is a classic problem: how can we calculate the global change of a variable during a process?

The solution, if it exists, concerns differential calculus. It consists in decomposing the process into what are called elementary processes, for which we can write the equations of physics by considering that the quantities remain constant. The variation in properties is then taken into account by "integrating" the differential equations thus written, which makes it possible to express the laws followed by the whole process. In the context of this simplified presentation, we limit ourselves to indicating how it can be done, but there is no need to go further (Figure 2.34).

One can also easily show that in an open system, for an infinitely small change, τ and Q are given by the following equations:

$$\delta\tau = v\,dP, \text{ or } dP = \rho\delta\tau$$

The useful work received by the movable walls of the system is equal to the product of the mass volume of the fluid by the elementary variation of the pressure within it. Applied to a compressor stage, this expression indicates that the compression ratio achieved is equal to the product of the density ρ of the fluid by the work communicated to the machine shaft.

The second equation is:

$$\delta Q = C_p dT - v\,dP$$

The latter expression simply expresses an experimental fact, which is an essential basis of compressible fluid thermodynamics: heat δQ exchanged with the surroundings causes a

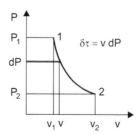

FIGURE 2.34

Differential calculus.

linear change in the thermodynamic state of the system, here characterized by two state variables, temperature T and pressure P.

If the pressure remains roughly constant, which is the case in most heat exchangers, dP = 0 and the variation in temperature of the fluid is proportional to the heat supplied to the system.

In the context of this presentation, we will not seek to solve these equations: we will determine the energies put into play using the well-chosen state functions.

Let us emphasize an important point: heat Q and temperature T should not be confused, even if relations exist between the two concepts:

- The temperature is a measure of the degree of agitation of the molecules of the working fluid: the more they are agitated, the higher its temperature;
- Heat is a transfer of thermal energy from one system to another when there is a temperature difference between them.

When a pure fluid condenses, its temperature remains constant, while it exchanges heat with the surroundings.

Energy conservation: first law of thermodynamics

Let us now come to the first law of thermodynamics.

We define the internal energy U of a system as the sum of the microscopic kinetic energies (comparable to the thermal agitation of particles) and the potential energies of microscopic interactions (chemical bonds, nuclear interactions) of the particles constituting this system. The internal energy corresponds to the intrinsic energy of the system, defined on the microscopic scale, excluding the kinetic or potential energy of interaction of the system with its surroundings, on the macroscopic scale, that which acquires a body because of its mass and speed.

The internal energy is defined with an arbitrary constant u_0, often chosen such that $U = 0$ for $T = 25°C$ and $P = 1$ atmosphere. Its unit in the International System (SI) is the Joule, but it is also frequently expressed in kWh, knowing that 1 Wh = 1 Watt for 1 h = 1 J/s × 3,600 s = 3,600 J.

The fundamental law that governs the behavior of thermodynamic systems is the conservation of energy, known as the **first law**.

It postulates that the variation of internal energy ΔU of a nonisolated system, interacting with its surroundings and undergoing an evolution, depends only on the initial state and the final state, which means that u is a state function.

The first law indicates that if the internal energy U of the system varies, it means that there is an exchange of energy with the surroundings in the form of work and/or heat, and that the sum of W and Q is equal to the variation of internal energy of the system. It can thus be written for a closed system in the form:

$$\Delta U = W + Q$$

What is remarkable is that U is a state function, while the two functions W and Q are only path functions.

Let us clarify a point important in practice: by convention, in this book, we will consider that the energy received by a system is counted positively (its internal energy increases), and that given away to the surroundings is counted negatively.

For a closed system, the first law can be stated as follows: the energy contained in a system which is isolated, or evolving in a closed cycle, remains constant whatever processes it undergoes. The various forms that the energy of a system can take – mechanical energy, heat energy, potential energy, kinetic energy – are all equivalent under the first law. Let us recall that, in our case, only heat and work are taken into account.

In this form, this law is very intuitive and readily accepted: it is a conservation law, which states that energy (like mass) is neither lost nor created.

Consider, for example, the thermal balance in winter of a house heated at a constant temperature of 18°C (Figure 2.35). To simplify the problem, we neglect the renewal of air and assume that it is heated by electricity, so that no fluid enters or leaves it.

The whole constitutes a closed system. The house has heat losses towards the surroundings through its walls, the environment being colder, as well as towards the ground. It may have gains due to sunshine.

The first law indicates that the heating power is equal to the thermal power due to these losses minus the gains due to the sun.

To be exact, one must take into account the effects due to the thermal inertia of the house, which means that at certain times energy is stored in its walls and at others, it is destocked and contributes to heating, but this reasoning is quite correct on average: the temperature of the house remains constant, its internal energy does not vary, and all the heating energy that enters the house is equal to that which comes out due to thermal losses.

For a system consisting of a phase of mass m, $U = m\,u$, u being the internal mass energy,

The unit of u is Joule/kg, but it is also frequently expressed in kJ/kg.

The internal energy, however, has meaning only if the system is closed, and needs to be generalized when considering a system in which matter enters and/or out of which it flows.

The principle of reasoning consists in following the evolution of a closed control volume in a periodic machine, and in calculating the work of external forces which is exerted during a cycle on all of its walls, by distinguishing the sections of passage of fluids (A1 and A2 at time t_0, becoming B1 and B2 at time $t_0 + \Delta t$ in Figure 2.36), the fixed walls, which obviously

FIGURE 2.35

House in winter.

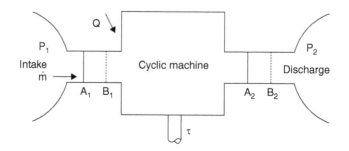

FIGURE 2.36

Machine in periodic regime.

do not produce or receive any work, and the movable walls, on which a certain work is exerted, which is called "**useful work**", that is to say seen by the user.

The work W exerted on a closed system can indeed be broken into two parts: one that is exerted on the mobile walls of the system if any, called shaft work τ, and one that is exerted on boundaries crossed by fluid coming out of the system and entering it. For open system components, which is the case for those that we are studying, this work, called flow work, equals $-\Delta(PV)$.

We therefore have: $W = \tau - \Delta(PV)$

Introducing a function called **enthalpy H**, such as $\Delta H = \Delta U + \Delta(PV)$, the **first law in open systems** is written:

$$\Delta H = \tau + Q$$

The first law is expressed as follows in open systems: the enthalpy change in an open system is equal to the sum of shaft work exerted on the moving walls and heat exchanged with the surroundings.

Enthalpy thus appears simply as a generalization in open systems of the internal energy in closed systems. In practical terms, just consider this state function as the **energy associated with the system under consideration**, neither more nor less.

As we will see later, this notion is fundamental in practice.

To illustrate the first law in an open system, let us take two examples.

To begin with, consider a flat solar thermal collector. Figure 2.37 shows the sectional view of such a collector. The absorber is composed of a metal plate on which are welded pipes in which the heat transfer fluid circulates. The heat losses towards the front of the collector are reduced by one or more glazings (2 in Figure 2.37), and those towards the rear by an insulation.

The incident solar flux is absorbed by the metal plate which is cooled, on the one hand, by the heat transfer fluid which heats up, and, on the other hand, by the thermal losses from the collector towards the surroundings.

The first law indicates that the solar energy received by the collector is equal to the sum of the heat losses of the collector and the enthalpy variation of the heat transfer fluid.

Consider as a second example an electric hair dryer.

It consists of a fan placed upstream of an electrical resistance. When the hair dryer is running, the fan draws in air which is slightly compressed and passes through the electrical resistance which heats it. The jet of hot air is then used to dry the hair.

The first law indicates that the electrical energy consumed by the device is equal to the change in enthalpy of the air passing through it.

For a system consisting of a phase of mass m, $H = mh$, h being mass enthalpy.

The unit of h is Joule/kg, but it is also frequently expressed in kJ/kg.

The passage from specific energies to power outputs is done by simply multiplying the first ones by the mass flow rate of the fluid that crosses the system considered.

The unit of power is Watt, but it is also frequently expressed in kW.

One Watt = 1 Joule/s

It is customary in physics to note a flow rate by a point above the quantity considered.

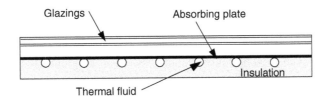

FIGURE 2.37

Flat solar thermal collector.

Thus, a mass flow is written $\dot{m}$, and an enthalpy power $\dot{H}$:

$$\dot{H} = \dot{m}h$$

As we said, for the sake of convenience and to simplify the notations, we will use, whenever possible, the unit of mass of the fluid considered.

Application to the four basic functions previously identified

We will show in this section that the enthalpy variation of the fluid flowing through them is sufficient to determine the energy put into play in the four elementary processes that were identified previously.

The results presented here correspond to industrial reality: for various technological reasons, we do not know how to manufacture, in most cases, components capable of both transferring heat and achieving effective compression or expansion. We therefore separate the different functions to be performed.

Compression and expansion with work

As we have seen, expansion can be made with and without work production. In the first case, the machine most commonly used is the turbine. In the second case, it is a simple valve or a filter.

Machines doing the compression or expansion of a fluid have a very compact design for reasons of weight, size and cost. For similar reasons, they rotate very fast (several thousand revolutions per minute). Each parcel of fluid remains there very briefly.

Moreover, fluids brought into play in compressors and turbines are often gas whose heat exchange coefficients have low values.

Short residence time, small areas of fluid–wall contact, and low exchange coefficients imply that the heat exchange is minimal and therefore usually negligible; this is referred to as **adiabatic**.

It follows that **the operation of these machines is nearly adiabatic**: $Q = 0$.

In a compression or expansion adiabatic machine, writing the first law in an open system shows that shaft work τ is thus equal to the change in enthalpy of the fluid Δh.

Expansion without work: valves, filters

There is a class of devices, such as the expansion valve of the refrigeration machine, where τ and Q are both zero, i.e., $\Delta h = 0$: they are static expanders such as valves and filters. The corresponding process is called an **isenthalpic throttling** or a **flash expansion**.

Heat exchange

Components that transfer heat from one fluid to another require large exchange areas, because heat fluxes transferred are proportional to these areas. Technical and economic considerations lead to the adoption of purely static devices. For example, large bundles of tubes in parallel are traversed internally by one fluid, while the other flows outside.

τ is zero because of the absence of movable walls.

In a heat exchanger, heat Q transferred or provided by one fluid to another is equal to its enthalpy change Δh.

Here, we re-introduce the flow in the equation since there is one for each of the fluids and it is important not to confuse them.

Note that if we seek to increase the heat exchange between the two fluids, we also try to limit the heat transfer with the surroundings of the exchanger, which we will assume to be null.

If we note Δh_h the variation in enthalpy of the hot fluid (which cools), and Δh_c that of the cold fluid (which heats up), the first law for the whole of the exchanger writes:

$$\dot{m}_h \Delta h_h + \dot{m}_c \Delta h_c = 0$$

$$\text{or} \quad \dot{m}_c \Delta h_c (>0) = -\dot{m}_h \Delta h_h (<0)$$

Combustion chambers, boilers

In a combustion chamber or boiler, there are no movable walls either, and $\tau = 0$.

Heat Q transferred to the fluid passing through is equal to the enthalpy change Δh.

Regarding the working fluid, a boiler can thus also be considered as a heat exchanger.

BOX 2.4
Key issues

First law, energies put into play in processes

Another key concept is a direct application of the first law of thermodynamics: for components that can be modeled in an open system, the variation in enthalpy of the fluid passing through them is sufficient to determine the energy put into play in the elementary processes corresponding to each of these functions.

Compression and expansion with work: $\tau = \Delta h$

Heat exchanges, combustion chambers, boilers: $Q = \Delta h$

Expansion without work: valves, filters: $\Delta h = 0$

This self-assessment activity will allow you to test your understanding of the first law:

First law in closed and open systems, gfe

Reference Processes

The previous section showed that the determination of the enthalpy change of the fluid flowing through them is enough to calculate the energy put into play in these four elementary processes.

But this information is not sufficient to characterize them completely. The physical analysis of their behavior allows us to identify the **reference processes** corresponding to the operation of components that would be ideal.

It is then possible to characterize the actual process by introducing an imperfection factor, often called effectiveness or efficiency, which expresses its performance referred to that of the reference process. This way makes it much easier to understand the processes undergone by the fluid.

The choice of reference processes is based on the analysis of physical phenomena that take place in the components: **it is an absolutely essential modeling choice.**

Lastly, as we shall see, the reference process is very useful when trying to plot the cycle studied on a thermodynamic chart.

Compression and expansion with work

We just saw that the compressors and turbines are machines in which heat exchanges with the surroundings are usually negligible, which is referred to as **adiabatic**.

Let us recall that the analyses of these components have shown that, in such a process, the temperature and the pressure of the fluid vary simultaneously.

In a perfect compressor, that is to say one where the working fluid would not be subject to friction or shock, the heating of the fluid and the work required to obtain a given compression ratio would be minimal. In a perfect turbine, the cooling of the fluid and the work produced for a given expansion ratio would be maximum.

The reference process for compression or expansion with work is thus the perfect or reversible[2] adiabatic. Its equation can be obtained by integrating the differential equation expressing that the heat exchanged is zero at any time, i.e., $0 = Cp \, dT - v \, dP$. For a perfect gas, for which $Pv = rT$, the corresponding curve is easily obtained. It is given by law $Pv^\gamma = Const$, with $\gamma = Cp/Cv$.

Expansion without work: valves, filters

For throttling or expansion without work conserving enthalpy ($\Delta h = 0$), the reference process is isenthalpic.

Heat exchange

Pressure drops correspond to the dissipation, by friction, of the mechanical energy of a moving fluid. They result in a drop in pressure and an internal heat supply, as will be explained in Chapter 6 in the section on entropy.

Any pressure drop in heat exchangers must be compensated for by an additional compression work and/or results in a lower expansion work. As the penalty induced by pressure drops is expressed in the form of mechanical work, we always try to minimize it.

If the pressure drops are low, the pressure remains more or less constant and the heat exchanges can as a first approximation be assumed to be isobaric. The reference process is therefore the isobar.

Combustion chambers, boilers

Similarly, combustion chambers and boilers can usually be regarded as isobaric.

Figure 2.38 summarizes the links existing between the components of the systems studied, the functions and the reference processes. It illustrates the genericity of the reference processes, which define the thermodynamic models corresponding to many technological components, and therefore their practical importance when one wishes to calculate the performances of the studied systems or to represent them in thermodynamic charts as we will see in Chapter 3.

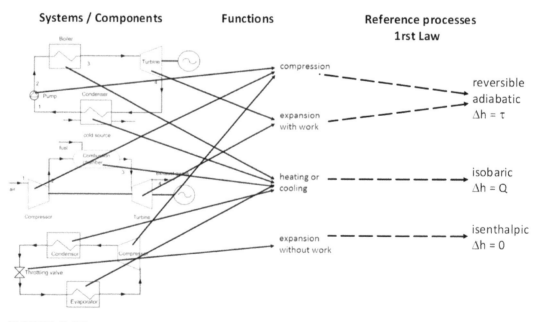

FIGURE 2.38

From components to reference processes.

---

[2] We will qualify as reversible a process without heat exchange with the surroundings and without friction losses.

BOX 2.5
Key issues

Reference processes, imperfection factors

Two other key concepts play a fundamental role: that of reference processes and imperfection factors.

The reference processes correspond to the functioning of components that would be perfect, for which a well-chosen variable or state function remains constant and to which we know how to associate a simple evolution equation.

It is then possible to characterize the real process by introducing an imperfection factor, often called effectiveness or efficiency, which expresses its performance compared to that of the reference process.

The pairs of reference process and imperfection factor that we have introduced are:

Compression and expansion with work: perfect or reversible adiabatic, isentropic efficiency

Heat exchanges, combustion chambers, boilers: isobaric, pressure drops

Expansion without work: valves, filters, isenthalpic (no imperfection factor)

This self-assessment activity will allow you to test your understanding of reference processes:

Reference processes, cat

Modeling of simple cycles in thermodynamic charts and thermoptim

Introduction

In this chapter, we start by studying the properties of pure substances and how the state of a fluid can be characterized. A deepening on water vapor is then carried out, due to its practical importance.

The ideal and perfect gas models are then briefly presented, and then a return to the concept of state is made, with a reflection on the choice of state variables to consider.

The thermodynamic charts in which the cycles can be visualized are analyzed, because they play a fundamental role in the learning of the discipline.

The cycles of simple energy systems are then plotted in the (h, ln(P)) charts, which makes it possible to visualize the physical phenomena at play and the few reference processes identified before.

Finally, we introduce the Thermoptim browser and show in guided explorations of pre-built models how these cycles can be modeled with the software package.

Properties and charts of pure substances

Properties of pure substances

Now that we have seen how to calculate the energies put into play in the different processes and what are the reference processes, some reminders on the properties of the working fluids are necessary. They will allow us to understand how these properties can be represented graphically in thermodynamic charts, and to learn how to use one of them, the (h, ln(P)) chart. We can then plot the cycles of the thermal machines that interest us.

Brief reminders on the properties of pure substances

To be able to understand how thermal machines work, it is essential to have a minimum of knowledge about the properties of the fluids that pass through them.

Let us recall that a pure substance can be in one or more of the three phases: solid, liquid or gaseous.

A phase is a continuous medium having the following three properties:

- It is homogeneous (which implies a uniform temperature);
- Speed in each point is zero in a suitable reference frame;
- It is subject to no external force at distance (uniform pressure).

DOI: 10.1201/9781003175629-3

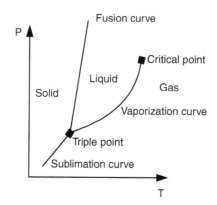

FIGURE 3.1

Phases of a substance.

When a given mass of a pure substance is present in a single phase, its state is defined by two variables, for example, its pressure and temperature. In the (P, T) plane, the three phases correspond to three areas, separated by three saturation curves (sublimation, vaporization, and fusion) joining at the triple point T (Figure 3.1).

When heating a solid at a well-chosen constant pressure, it turns into liquid, and we talk of fusion. This is how heated ice turns into liquid water.

If we continue to provide heat, the liquid turns to vapor, and we talk of vaporization.

So if you boil water in a pan, it evaporates and the amount of water left in the pan gradually decreases.

The phenomenon of vaporization and its opposite, condensation, play a fundamental role in the cycles of steam power plants and refrigeration installations. During these processes, the fluid is said to be in liquid–vapor equilibrium.

The temperature at which the liquid–vapor equilibrium takes place is an increasing function of the pressure exerted on the substance in question.

It remains constant as long as the vaporization is not complete. We call the law of saturation pressure the relation which gives the saturation pressure of a fluid as a function of temperature.

There is a pressure and temperature limit beyond which the liquid–vapor equilibrium zone disappears. We are talking about the critical pressure and temperature.

In the (P, T) plane, the critical point T represents the state where the pure vapor phase has the same properties as the pure liquid phase.

At higher temperatures and pressures, called supercritical, it is not possible to observe a separation between the liquid and gaseous phases: the border that separates the liquid and vapor phases disappears at the critical point.

Saturation pressure law

For the phase change to happen, it is necessary to provide or absorb energy, called **latent heat of change of state**. During the vaporization process, there are significant variations in the specific volume: vapor is about 600–1,000 times less dense than the liquid. This change in specific volume occurs at **constant pressure and temperature**.

For example, Figure 3.2 presents the law of saturation pressure P_{sat} (T) of water, below 100°C, which is the saturation temperature of water at 1 atm, i.e., say about 1 bar.

It shows that the saturation pressure of water at room temperature is very small. At 20°C, it is around 0.025 bar.

In the condenser of a steam plant, we try to condense the steam coming out of the turbine at the lowest possible temperature in order to increase the efficiency of the cycle. The condenser coolant is usually ambient air or river water, which are at low temperatures, close to 10°C in mid-season to fix ideas. The condensing temperature of the water in the condenser is therefore as low as possible, and it operates at a very low pressure, much lower than the atmospheric pressure.

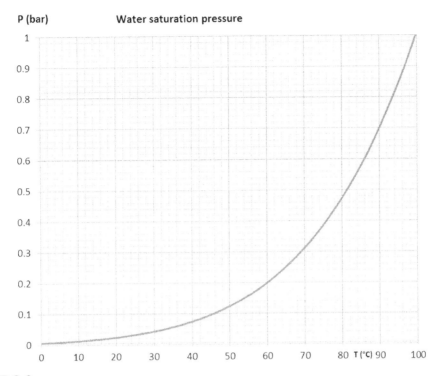

FIGURE 3.2

Water saturation pressure law.

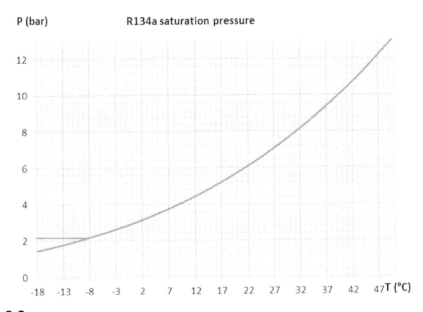

FIGURE 3.3

Condensation and evaporation pressures of a refrigeration cycle.

Figure 3.3 is an enlargement of the saturation pressure law of R134a, between −18°C and 50°C.

Let us examine its implications in terms of the value of the evaporation and condensation pressures of the working fluid of a vapor compression refrigeration installation where it is assumed that the cold enclosure is at the temperature $T_{ce} = -8°C$, and ambient air at $T_a = 35°C$.

The curve in Figure 3.3 shows that if this fluid is used as a refrigerant, the evaporation pressure must be less than approximately 2.1 bar for the evaporation to take place at a temperature T_{evap} below −8°C so that the cold enclosure can transfer heat to the fluid. At the condenser, the pressure must be greater than 8.8 bar so that the condensation temperature T_{cond} is greater than 35°C and that the cycle can be cooled by exchange with the ambient air.

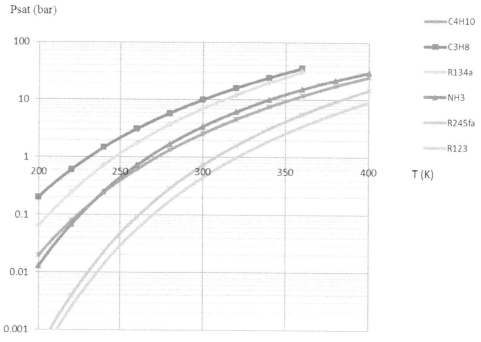

FIGURE 3.4

Saturation pressure law for several fluids.

A comparison of the condensation pressures of water and R134a, for example, at a temperature of 20°C, shows that the saturation pressures vary greatly depending on the thermodynamic fluid chosen, as illustrated by the curves in Figure 3.4.

For a given application, the saturation pressure law is one of the criteria for selecting the working fluid.

In practice, moreover, there must be a sufficient temperature difference between the working fluid and the sources with which it exchanges heat, so that the evaporation pressure is even lower and the condensation pressure is higher.

Characterization of the state of a fluid

A fluid can be present in one or more phases.

If it is single phase and we consider a small particle of fluid, its state can be defined by two quantities like its pressure P, its temperature T, its mass volume v, or its mass internal energy u or its mass enthalpy h.

These are the variables we will use most often.

Please note that pressures will generally be expressed in bar and not in Pascal, and temperatures in Kelvin by default and in °C otherwise.

At the liquid–vapor equilibrium, when the fluid is two-phase, the pressure and the temperature are no longer independent.

The state of the small particle of fluid can then be defined by a quantity like its pressure P or its temperature T, as well as:

* Either its vapor quality x;
* Or its mass volume v, its internal mass energy u or its mass enthalpy h.

Vapor quality

In the vapor–liquid equilibrium zone, the fluid is present in both liquid and vapor phases. In this central zone, isobars and isotherms are combined; the liquid–vapor change is taking place at constant temperature and pressure. The composition of the mixture is defined by its quality x, the ratio of the mass of vapor m_g to the total mass (m_g plus the mass of liquid m_l). Index g refers to the gas phase, and l to the liquid.

$$x = \frac{m_g}{m_g + m_l} \tag{3.1}$$

Enthalpy of vaporization

The length of the vaporization line gives the enthalpy (or heat) of vaporization L for the fluid state considered.

$$h_g - h_l = L$$

where L is a decreasing function of temperature, zero for T above the critical temperature. An approximate formula due to Clapeyron allows us to estimate L from the saturation pressure law:

$$L = T\left(v_g - v_l\right)\frac{dP_{sat}}{dT}$$

Calculation of pure substance two-phase properties

The properties of a two-phase pure substance are given by the law of phase mixture:

$$v = (1-x)v_l + xv_g$$
$$u = (1-x)u_l + xu_g$$
$$h = (1-x)h_l + xh_g = h_l + xL \tag{3.2}$$

Examples of practical use of phase changes

Let us give some examples illustrating either the practical use of phase changes or the constraints induced by the presence of a liquid–vapor equilibrium:

- When adding ice cubes to a lukewarm drink, they melt, which cools the drink. As the latent heat of fusion of ice is much larger than the heat capacity of the drink, we get the desired cooling effect without bringing too much water dilution;
- To transport methane over long distances by sea, it is liquefied at a temperature of –160°C, reducing its specific volume 600 times with respect to gas. It is thus possible to maintain atmospheric pressure in the tanks of the liquefied natural gas (LNG) ship. Although these tanks are very well insulated, you cannot avoid some heat exchange with the surroundings, which has the effect of vaporizing a small amount of gas that is used for propulsion;
- In contrast, butane or propane gas distributed for culinary purposes is confined in a liquid state at room temperature in thick metal cylinders, in order to resist the inside pressure of a few tens of bar (or hundreds of psi);
- All cooking done in boiling water takes place at 100°C if the pressure is equal to 1 atmosphere, and this is irrespective of the thermal heat supplied to the cooking. Thus, we can define the precise duration for cooking a recipe, for example, a boiled egg;
- The principle of the pressure cooker is to overcome this limit of 100°C by doing the cooking in a chamber at a pressure exceeding 1 atm. It can reach 110°C and 120°C, in order to cook food more quickly;
- An example of condensation is that which is deposited on cold surfaces in contact with moist air, like mist on a window, or the morning dew on leaves.

Deepening on water vapor

The interaction of water vapor with air is a complex phenomenon that deserves some explanation.

In this section, we propose to study water vapor in greater depth, as it is often overlooked. The choice of water is justified because it is undoubtedly the only condensable working fluid of which everyone has concrete knowledge.

Of the three phases of water, vapor is by far the least understood because it is invisible and odorless. Everyone now knows what ice or liquid water is, having had many experiences with it, but vapor is much less well perceived, to the point that some may be led to doubt its existence or at least to have a misconception of it, which will handicap their understanding of thermal machines with condensable fluid.

The difficulty is that the water vapor does not manifest itself in a way that is directly perceptible by our senses, except by touching it when we receive a jet of vapor on the skin. It is most often indirectly that we perceive water vapor, because of the manifestations of its interactions with the air and the cold surfaces around us.

A few explanations of these phenomena will allow you to get a more accurate idea of what water vapor is, and therefore the other vapors that we will discuss.

Water vapor is very present in our daily environment, under the short name of vapor.

Dieticians praise the virtues of steaming, which is carried out by placing in a saucepan closed by a lid a basket slightly raised from the bottom, in which the food to be cooked is placed, above a small quantity of water. During cooking, the water comes to a boil, and the resulting steam fills the available space in the pot, partly condenses on the pot and cooks food soaked in it. Both liquid and vapor phases are present in the pan, but since it is closed and the vapor is invisible, it cannot be perceived directly.

The steam is also manifested during cooking in a pressure cooker, by the jet that passes through the nozzle and that is partially visible. We will come back to this apparent paradox a little later.

Steam is commonly used for ironing, as the soles of modern irons have holes for steaming steam to moisten clothes to be hot-moistened, improving the treatment without risking leaving marks.

Do-it-yourselfers who have renovated homes may have had to use a stripper to remove the wallpaper from the walls, which greatly facilitates the task; the steam is diffused again through holes in the sole of the device.

Hammam enthusiasts appreciate baths taken in an atmosphere saturated with humidity brought by steam injected into the ambient air, which stimulates sweating.

Finally, in winter, the vapor present in the indoor air of homes causes condensation on the cold walls, which can result in disorders such as the development of molds.

Dry air is a mixture of oxygen (23% by mass), nitrogen (76% by mass) and certain gases (argon, CO_2) in small proportions (about 1% by mass), but dry air almost never exists: it almost always includes water, in gaseous, liquid or ice crystals form.

The thermodynamics of moist mixtures make it possible to study the properties of real air, which always includes a part of humidity. Further developments will be made in Chapter 12 on air conditioning.

When we are interested in a moist mixture, we are dealing with a mixture of gases, which do not condense, which is called dry gas, and water, which can condense. Everything happens in a way as if the moist mixture were a mixture of two substances: dry gas, the composition of which is unchanging, and water, capable of occurring in one or more phases.

Two quantities are commonly used to characterize the humidity of a moist mixture: the specific humidity w and the relative humidity ε, more commonly used by the general public.

We call specific or absolute humidity w the ratio of the mass of water contained in a given volume of moist mixture to the mass of dry gas contained in this same volume.

Note that Table 12.1 gives the equivalence between notations used in this book and Northern America for moist mixtures.

The curve in Figure 3.5 shows the value of the specific humidity of air saturated with water expressed in grams of water per kilogram of dry air, which varies considerably depending on the temperature: at 20°C, the saturated air contains about 15 g of water and at 60°C about ten times more.

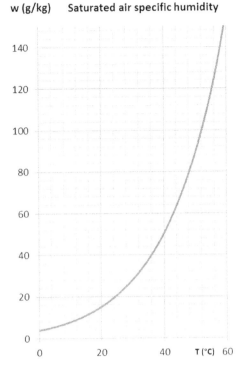

w (g/kg)   Saturated air specific humidity

FIGURE 3.5

Specific humidity of saturated air.

When, at a given pressure, the temperature drops below the water saturation temperature (given by the curve in Figure 3.2), the water vapor in excess of the specific humidity of saturated air begins to condense in the form of mist or on the cold wall, which delimits the system if it exists.

To be more precise, this is what happens on the physical plane: the (so-called partial) pressure of water in the air cannot exceed the saturation pressure at a given temperature.

Recall that the partial pressure P_i of a constituent is the pressure that this constituent would exert if it occupied alone the volume V of the mixture, and its temperature was equal to that of the mixture.

Mathematically, this is expressed by the relation $P_i = x_i\, P$, P being the total pressure and x_i the molar fraction of the constituent i. For water, we therefore have: $P_{H_2O} = x_{H_2O}\, P$.

The condensation condition is therefore written $P_{H2O} = x_{H2O}\, P \le P_{sat}(T)$, or:

$$x_{H_2O} \le \frac{P_{sat}(T)}{P}$$

The relative humidity ϵ briefly introduced above is precisely defined as the ratio of the partial pressure of water to its saturated vapor pressure:

$$\epsilon = \frac{P_{vap}}{P_{sat}}$$

Relative humidity is 1 (or 100%) when the vapor begins to condense. Otherwise, it is less than 1.

Figure 3.6 shows the so-called psychrometric diagram of humid air, on which the curves of relative iso-humidity appear, from 10% to 10%; the upper curve corresponds to saturation. In this diagram, the ambient temperature (called dry) is plotted on the abscissa, and the specific humidity w is plotted on the ordinate.

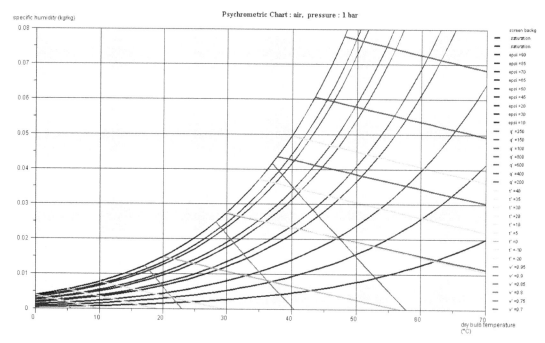

FIGURE 3.6

Psychrometric air chart.

The mass of vapor contained in the air results in a partial pressure of water vapor. If, at room temperature, this pressure exceeds that given by the saturation curve in Figure 3.2, there is condensation of the excess vapor fraction.

Let us take the example of the pressure cooker that we mentioned earlier. In normal operation, its interior temperature is close to 110°C.

The steam that escapes at high temperature from the security valve gradually cools in contact with the ambient air with which it mixes. Up to 1 or 2 cm above the nozzle, the jet contains only invisible vapor. Then, the water-saturated mixture begins to condense, forming a mist of perfectly visible microdroplets of water. This mist then evaporates when it mixes with dry air away from the pressure cooker.

The same phenomenon appears in the exhaust of the turbojets of planes flying at high altitude due to the fact that the gases burned by the combustion of kerosene contain water vapor. It explains the white streaks that signal their passage and disappear after a certain time; the condensed water vaporizes when the saturation conditions are no longer met.

The formation of clouds follows the same logic. It results from the cooling of a volume of air until the condensation of part of its water vapor. If the cooling process occurs on the ground (by contact with a cold surface, for example), fog is forming.

The condensation that appears in winter on the windows of a home or vehicle also comes from crossing the saturation threshold of humid air in the immediate vicinity of the cold surface. The morning dew on the plants reflects the same phenomenon.

Water in humid air can also condense directly to a solid state if the temperature is low enough, which is why snow and frost form in refrigerators. Snow cannons used by ski resorts use this phenomenon.

We hope this small section has convinced you of the existence and materiality of water vapor, although it is usually not directly noticeable. It is indeed fundamental that you have no doubts about it if you want to study thermodynamic cycles with condensable fluid under good conditions.

Model of ideal gases

An ideal gas is a gas whose internal energy and enthalpy only depend on temperature, which simplifies its modeling: its equation of state is $Pv = rT$.

From a microscopic description of the phenomena, the kinetic theory of gases makes it possible to find macroscopic properties such as volume, enthalpy, etc.

The simplest model is one where it is assumed that the size of the molecules and the interactions between them are negligible. The theory of ideal gases is based on the assumption that the molecules of the gas behave like hard spheres of negligible size, that is to say undeformable, which collide with each other and rebound elastically, without any interaction between them (Figure 3.7).

In an ideal gas, it is thus considered that the molecules do not have their own volume and only have kinetic energy due to their speed of movement. This model is mainly applied to gases under low pressure, far from their saturation curve.

This model leads to the ideal gas law: $Pv = rT$, in which T is expressed in Kelvin. It can be shown that, if C_p and C_v are constant (perfect gas):

$$\Delta u = C_v \left(T - T_0 \right)$$
$$\Delta h = C_p \left(T - T_0 \right)$$

where C_p and C_v are the mass thermal capacities (or specific heats) at constant pressure and volume, and T_0 is a reference temperature.

Many thermodynamic fluids in the vapor phase can be likened to ideal gases, in a wide range of temperatures and pressures. In particular, the temperature–pressure combination must deviate sufficiently from the possible condensation zone (that is to say that the pressure is not "too" high, nor the temperature "too" low). Such conditions are commonly achieved for the so-called "permanent" gases such as, at ambient pressure and temperature, hydrogen, oxygen, nitrogen, and the oxygen–nitrogen mixture that constitutes dry air.

The ideal gas model assumes that the heat capacities C_p and C_v are constants. This is only strictly verified for monoatomic gases (which do not have any mode of rotation or molecular vibration). This hypothesis is all the less satisfactory since the gas molecule has more atoms (and therefore possible vibrational modes).

An ideal gas differs from a perfect gas in that its thermal capacities are not constant, but only depend on the temperature.

Figure 3.8 gives the evolutions of the specific heat C_p for some typical mono-, bi-, and triatomic gases. The value of the molar C_p is given on the ordinate, and the temperatures appear on the abscissa. The dependence of C_p on the temperature is clearly visible.

An ideal gas still follows the law $Pv = rT$, and u and h are still independent of pressure, but their expression is more complicated than a linear function of temperature.

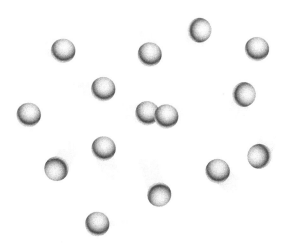

FIGURE 3.7

Ideal gas molecules.

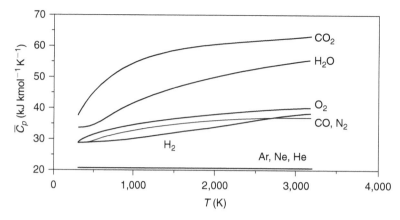

FIGURE 3.8
Molar specific heat C_p of some gases.

One of the remarkable characteristics of ideal gases is that a mixture of ideal gases behaves itself as an ideal gas. This is called the Dalton law. It states that the properties of the mixture are very easily calculated from those of the gases which compose it and their molar fractions x_i or their mass fractions y_i.

Recall that the molar fraction x_i of a constituent is by definition the ratio of the number of moles of this constituent to the total number of moles in the mixture, and that the mass fraction y_i of a constituent is by definition the ratio of the mass of this constituent to the total mass of the mixture.

$$x_i = \frac{n_i}{n} \quad \text{and} \sum x_i = 1$$

$$y_i = \frac{m_i}{m} \quad \text{and} \sum y_i = 1$$

Choice of state variables to consider

Now that we have introduced the first law, established its equation for open systems and showed just how it applies to the four basic changes experienced by fluids in machines that interest us, we can discuss the advantages and disadvantages of different state variables that one can consider using for displaying fluid properties in thermodynamic charts.

We have seen that several sets of state variables can be used to characterize a thermodynamic system. The most "natural" are temperature, pressure and volume, but there are others, like the enthalpy defined in the previous chapter, and second, they are not fully satisfactory for our goals, as we shall show.

Pressure P is essential, both because it directly determines the mechanical stress in components, and because, as we have seen, the reference process for fluid heating and cooling is isobaric.

Temperature T is also essential, but, unlike its predecessor, the isothermic process does not correspond to any process that concerns us.

Volume v intervenes very little in the analysis of interest, even in a closed system, because volume varies due to the existence of movable walls. In fact, its main practical interest is in the sizing of flow sections.

Enthalpy h is a fundamental variable too, because it is directly related to energy exchanges that take place in the machines. For a perfect gas, it is a linear function of temperature, very easily deduced, and in the liquid–vapor equilibrium zone, it provides additional information on the quality. Finally, remember that the isenthalpic process is the reference process for an expansion without work.

The analysis of processes undergone by fluids during compression and expansion with work showed that their reference process is reversible adiabatic. We indicated that for a perfect gas, it follows the law $Pv^\gamma = Const$.

Let us note, but incidentally given our focus for simplicity, that this law is that of the **isentrope**. **Entropy s** is also a state function widely used in thermodynamics, especially because the isentrope is the reference process for compression and expansion.

We will not use this concept in this first part, but it is better that you know what it corresponds to, if only because it often appears in reference documents.

BOX 3.1
Key issues

Fluid properties

The different states of fluids and gas modeling must be well understood before studying their representation in diagrams.
These self-assessment activities will allow you to test your knowledge on these topics:

Solid, liquid and gas phases, gfe
Perfect gas and ideal gas, gfe
Specific humidity and relative humidity, gfe.

Thermodynamic charts

Thermodynamic systems which we consider can be characterized by two state variables; they are called bivariant. This means that their thermodynamic properties can be plotted in a plane in the form of a thermodynamic chart.

By highlighting the reference processes of changes undergone by fluids and allowing us to calculate them, at least approximately, charts are among the basic tools of thermodynamics. Their interest is twofold:

* To help plot the cycles;
* To facilitate the estimation of the thermodynamic state of the various cycle points.

Because of the possibilities offered by software packages for calculating fluid properties that are increasingly common, the second interest tends to decline, while the former retains its relevance. The display on a chart of a cycle calculated using a computerized tool helps ensure that it does not contain an abnormal point due to an error when entering data.

A chart is presented in a graphical map with the plot of a number of remarkable curves, including families of state function isovalues.

Different types of charts

In practical terms, the main processes that take place in industrial processes involving pure fluids are, as we have seen, compression, expansion, heat exchange and throttling.

It is clear that the temperature T, pressure P and the steam quality x are state variables whose knowledge is necessary to study these processes and design equipment.

The above remarks show that the enthalpy h is also very important. Finally, knowledge of the specific volume v is necessary to size the pipes, since it allows us to convert mass flow rate into volume rate.

In conclusion, the most interesting quantities are in practice T, P, and h, and additional information on x and v may be necessary. So the abscissas and ordinates of the charts that we can consider should be chosen among them (recall that we exclude here the entropy).

The pair (T, h) is rarely retained because the isobaric and reversible adiabatic are represented by curves with inflection points that make their use difficult. In addition, variations of T and h are proportional when the fluid follows the ideal gas law.

Except for permanent gases, i.e., whose state is very far from their saturation conditions, the pair (T, P) is insufficient, because T and P are bound by the saturation law in the liquid–vapor equilibrium zone. It would, however, be quite suitable to represent gas turbine cycles.

For our purposes, as part of this lightweight educational presentation, these remarks are sufficient to allow us to conclude that the pair (**P, h**) is a set of state variables of particular interest. It is increasingly used, generally with a logarithmic scale for pressures. Its widespread use has been promoted by refrigeration engineers, and we will study it further.

Finally, the pair (P, v), Clapeyron chart, has a certain educational value, especially for the study of changes in closed systems. Its main drawback is its low visibility, the area of vapors being reduced and energy functions not appearing directly.

In conclusion, as this lightweight educational presentation does not include entropy, the pair (P, h) appears the best suited.

(h, ln(P)) chart

In the (h, ln(P)) chart (Figure 3.9), the enthalpy is the abscissa and pressure the ordinate, usually on a logarithmic scale.[1]

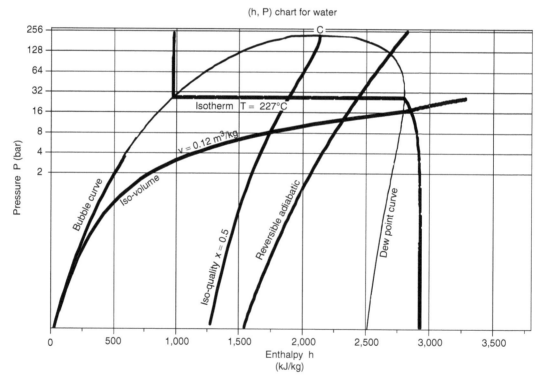

FIGURE 3.9

Isovalues in the (h, ln(P)) chart.

[1] The base logarithm b of a number is the power to which it is necessary to raise the base b to obtain this number: the logarithm of one hundred in base ten is 2, because $100 = 10^2$. Two large families of logarithms are commonly used: the decimal logarithm defined above and the natural logarithm, which is the inverse function of the exponential, and for which $b = \exp(1) = 2.718281828$. This is the one we will use, and we will note it ln for this reason.

The saturation curve separates the plan in several zones. Its summit is the critical point C, the left side, representing the saturated liquid (bubble curve), and its right side, the saturated vapor (dew point curve). Under this curve is the two-phase liquid–vapor equilibrium zone, and in the rest of the plan, that of simple fluid.

Remember, the critical point represents the state where the pure vapor phase has the same properties as the pure liquid phase.

For this chart to be used, it is equipped with reversible adiabatic curves, isotherms, isovolumes, and in the mixed zone, iso-quality curves.

In the "liquid" zone on the left of the chart, isotherms have a very strong negative slope: the compression of a liquid involves a very small work.

In the two-phase zone, pressure and temperature are related by the law of saturation pressure, and isotherms are horizontal. The enthalpy increases enormously, corresponding to the heat of vaporization that must be supplied to the fluid.

In the zone to the right of the saturation curve, the isotherms are curved downward, close to vertical for low pressure values. Indeed, the behavior of vapor then approaches that of an ideal gas, whose enthalpy depends only on temperature. Figure 3.10 shows a (h, ln(P)) chart of ideal gas.

Heating or cooling (isobars) is reflected in this chart by a horizontal segment, while an expansion without work (isenthalpic) by a vertical segment.

A reversible adiabatic is an upward curve with a slope equal to the inverse of specific volume. They are much less inclined in the vapor zone than in the liquid zone.

Iso-quality curves are contained inside the liquid–vapor equilibrium zone. They intersect at the critical point.

Isovolume curves converge in the liquid zone where they become independent of pressure.

In the chart in Figure 3.10 which corresponds to air, modeled as an ideal gas, the isotherms are vertical, the isovolumes and the reversible adiabatic, steeper curves being with downward concavity. If we plot the diagram in non-logarithmic coordinates (Figure 3.11), the isovolumes are straight lines, and the concavity of the reversible adiabatic is oriented upward.

Figure 3.12 shows the (h, ln(P)) chart of water. The zero of the enthalpies is in the liquid state on the saturation curve at 0°C.

Figure 3.13 shows the (h, ln(P)) chart of R134a refrigerant.

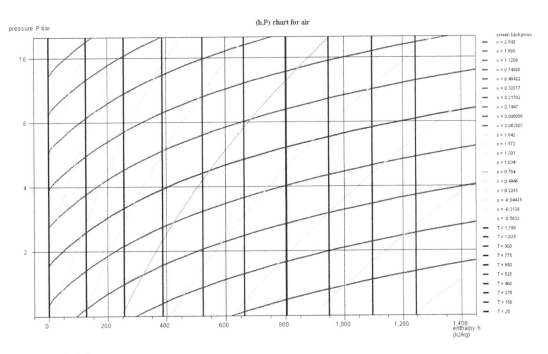

FIGURE 3.10

Air in the (h, ln(P)) chart.

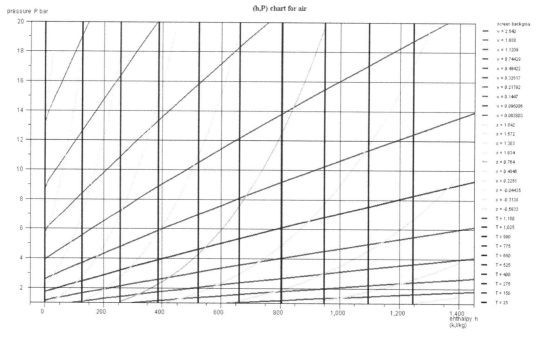

FIGURE 3.11

Air in the (h, P) chart.

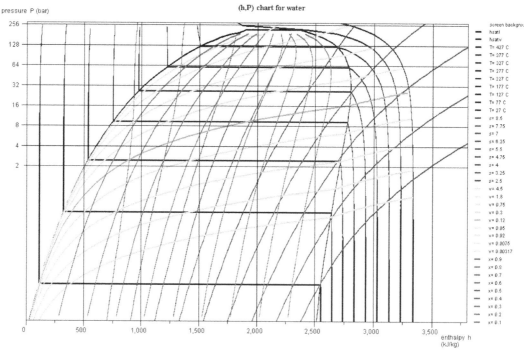

FIGURE 3.12

Water in the (h, ln(P)) chart.

Placing points in a diagram is not difficult, but requires attention and a little experience, since the known state variables are not the same depending on the case. It is therefore necessary to know how to find the appropriate isovalue curves and interpolate between them.

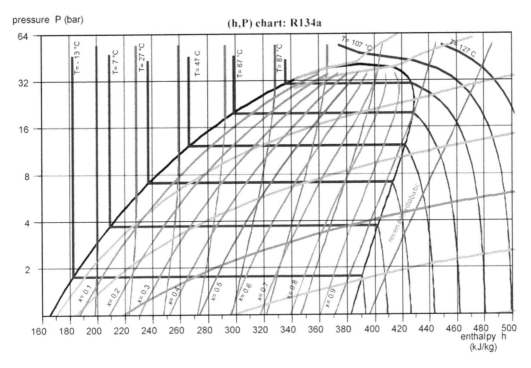

FIGURE 3.13

R134a in the (h, ln(P)) chart.

Plot of cycles in the (h, ln(p)) chart

Introduction

We now have almost all the elements to calculate the performance of the cycles of the steam power plant, the gas turbine and the refrigeration machine, by plotting them in the (h, ln(P)) charts, which will provide us with the values of the enthalpies put in play. We will do this after clarifying a few points, notably concerning the calculation of efficiencies.

We will only present a simplified version of the gas turbine cycle where combustion is replaced by heating, because the estimation of the thermodynamic properties of the combustion products requires special precautions. More realistic models taking combustion into account will be presented in the second part of this book.

In practical terms, the activities proposed in this section should be done by hand, with charts on paper. These can be obtained either commercially or in specialized books or printed directly from Thermoptim, but their accuracy is not very high. For the gas (h, ln(P)) charts, this will probably be the only solution, as this type of chart is seldom used.

Concept of isentropic efficiency

To determine the performance of the cycles, we will first assume that compressions and expansions are perfect, that is to say follow reversible adiabatic, then we will indicate how to take into account the irreversibilities in these machines.

Recall that we have indicated that it is possible to characterize a real process by introducing an imperfection factor, often called **effectiveness** or **efficiency**, which expresses its performance compared to that of the **reference process**.

The concept of isentropic efficiency makes it possible to characterize the performances of compressors and turbines, which, in reality, are not perfect, so that compression and expansion follow nonreversible adiabatic.

In the case of a compressor, the isentropic efficiency η is defined as the ratio of the work of the reversible compression to the actual work, and in the case of a turbine as the ratio of the actual work to the work of the reversible expansion. It should be noted that η is less than or equal to 1 for compression as well as for expansion (Figure 3.14).

To remember its definition, it suffices to remember that η is always equal to the smallest work divided by the largest work, and that irreversibilities have the effect of increasing the compression work and reducing the expansion work.

The values of these efficiencies, which depend very much on the technologies used, are typically between 0.75 and 0.9 in current applications.

Notion of cycle overall efficiency

It may be helpful at this stage to specify what is meant by the notion of efficiency or effectiveness, although we have already used it several times.

Let us in particular specify what is covered by these notions used to characterize the overall performance of a cycle, when establishing its energy balance. For a heat engine that converts heat into mechanical power, it is the ratio of the power output to the heat supplied to the machine:

$$\eta = \frac{\tau}{Q}$$

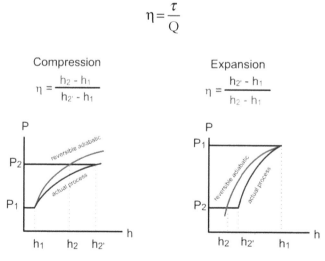

FIGURE 3.14
Isentropic efficiency definition.

In more general terms, when dealing with relatively complex cycles, one is led to adopt a broader definition of efficiency or effectiveness: it is the ratio of the useful energy effect to the purchased energy put in.

$$\eta = \frac{\text{useful energy}}{\text{purchased energy}}$$

- The purchased energy generally represents the sum of all the energies which one had to supply to the cycle coming from surroundings;
- The useful energy represents the net balance of the useful energies of the cycle, that is to say the absolute value of the algebraic sum of the energies produced and consumed within it participating in the useful energy effect.

Note that the result obtained depends directly on the way in which the boundaries of the cycle are chosen. If we refer to the two examples given above, the work supplied to the pump of a steam plant cycle will be counted as useful energy if we adopt the usual boundary identified by the letter A in Figure 2.30, and as purchased energy if we retain the boundary identified by the letter B.

This expanded definition of efficiency or effectiveness has the advantage of remaining valid for both power cycles and refrigeration cycles. In the latter case however, we no longer speak of efficiency because its value generally becomes greater than 1, but rather of COefficient of Performance (COP).

BOX 3.3
Key issues

Efficiencies

We will frequently use two concepts that must be well understood: that of isentropic efficiency and overall cycle efficiency.
These self-assessment activities will allow you to test your knowledge on these topics:

Concept of isentropic efficiency, gfe
Concept of cycle efficiency, gfe

Steam power plant

We will now explain how to draw in the (h, ln(P)) chart the steam power plant.

Perfect cycle

At the condenser outlet (point 1, Figure 2.8), water is in the liquid state at a temperature of about 27°C, under low pressure (0.0356 bar). In Figure 3.15, the point is easy to find at the intersection of the isotherm T = 27°C and the saturation curve.

The pump compresses water at about 128 bar, which represents a significant compression ratio (around 3,600).

Temperature T remains approximately constant during compression (1–2); point 2 is located at the intersection of the isotherm T = 27°C and the isobar P = 128 bar (ordinate 128 bar).

The pressurized water is then heated at high temperature in the boiler; heating comprises the following three steps:

- Liquid heating from 27°C to about 330°C, saturation temperature at 128 bar, just above isotherm T = 327°C: process (2–3a). Point 3a lies on the vaporization curve of the same isobaric curve;

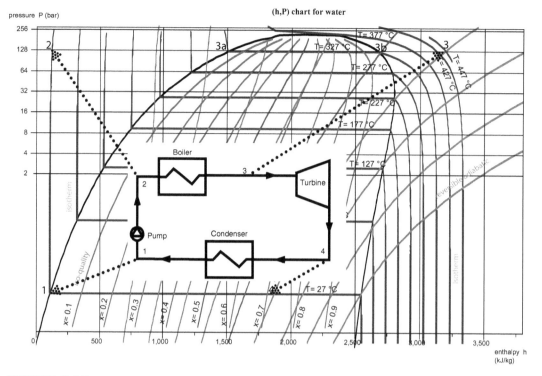

FIGURE 3.15

Construction of the steam power plant cycle.

- Vaporization at a constant temperature of 330°C: process (3a–3b). Vaporization is carried out at constant pressure and temperature; it results on the chart in a horizontal segment 3a–3b. Point 3b is therefore on the descending branch of the vaporization curve, or dew point curve, at its intersection with the horizontal line of pressure 128 bar;

- Superheating from 330°C to 447°C: process (3b–3). Point 3 is still assumed to be the same pressure, but at a temperature T_3 of 447°C. It is thus at the intersection of the horizontal P = 128 bar and the isotherm T = 447°C.

Point 3 is also on a concave curve inclined downward corresponding to a reversible adiabatic.

Process (3–4) is a reversible adiabatic expansion from 128 to 0.0356 bar. The point is in the mixed zone, e.g. on isotherm T = 27°C. Point 4 is at the intersection of the concave curve inclined downward and this isotherm. Quality x is between 0.7 and 0.8. Linear interpolation allows us to estimate its value, equal to 0.72.

The enthalpies of the points can be read directly by projecting these points on the x-axis, and the energies involved can be easily deduced. This allows us in particular to determine the cycle efficiency. Table 3.1 provides these values.

Note that reading the compression work in this chart is very imprecise, and it is best estimated from the integration of $\delta\tau = v\,dP$, very easy to make, with v being constant. It is therefore $\Delta h = v\,\Delta P$, with P being expressed in Pa, v in m³/kg and h in J.

Figure 3.16 shows the cycle in the chart, in which the points are connected.

The efficiency here is the ratio of the mechanical work produced (1,306 kW) to the heat supplied by the boiler ($h_3 - h_2 = 3,063$ kW).

Cycle taking into account deviations from reference processes

In reality, the turbine is not perfect, and expansion follows an irreversible adiabatic. It is customary to characterize the actual process by its isentropic efficiency η defined in the case of a turbine as the ratio of actual work to the reversible expansion work. Its value is typically about 0.85–0.9 in current applications.

TABLE 3.1

ENTHALPY BALANCE OF SIMPLE STEAM CYCLE

Point	Flow Rate (kg/s)	h (kJ/kg)	
1	1	113	
2	1	126	
3a	1	1,524	
3b	1	2,672	
3	1	3,189	
4'	1	1,870	

Process	τ (kW)	Q (kW)	Δh (kW)
(1–2)	13		13
(2–3a)		1,398	1,398
(3a–3b)		1,148	1,148
(3b–3)		517	517
(3–4)	−1,319		−1,319
(4–1)		−1,757	−1,757
Cycle	**−1,306**	**1,306**	
Energy efficiency		42.64%	

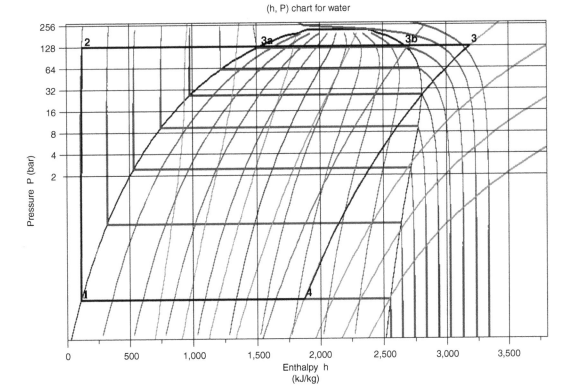

FIGURE 3.16

Steam power plant cycle in the (h, ln(P)) chart.

To find the actual point 4′, we first determine the value of the perfect machine work τ_s, then we multiply it by η, which gives the value of actual work τ.

The enthalpy of point 4 is equal to that of point 3 minus τ.

In this case, taking $\eta = 0.9$, we have:

$$\tau_s = h_3 - h_4 = 1,319\,\text{kJ/kg/K}$$

$$\tau = \eta\tau_s = 0.9 \times 1,319 = 1,187$$

$$h_4 = 3,199 - 1,187 = 2,002\,\text{kJ/kg/K}$$

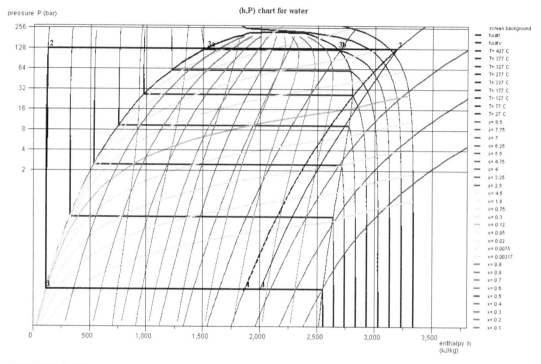

FIGURE 3.17

Steam power plant cycle with irreversible expansion.

Point 4′, which is still in the liquid–vapor equilibrium zone, is thus on isobar P = 0.0356 bar, i.e., on isotherm T = 27°C with abscissa h = 2,002 kJ/kg/K. Figure 3.17 shows the new look of the cycle (the previous cycle is plotted in dashed line).

The increase in the quality at the end of expansion is favorable, on the one hand, to the prolongation of the life of the turbine blades, for which the liquid droplets constitute formidable abrasives, and on the other hand, to the isentropic efficiency, which drops when the quality decreases, as this will be discussed in Chapter 5.

Here, we have not considered the pressure drops in heat exchangers. It would of course be possible to do so by operating on the same principle as described, and changing the pressure between the inlets and outlets of heat exchangers.

As shown in this example, the representation of the cycle in the (h, ln(P)) chart is very easy to understand: the heat exchanges, almost isobaric, correspond to horizontal segments, and the compression and expansion are reversible adiabatic, and less steep, as they are located far from the liquid zone.

Gas turbine

We are now going to construct the gas turbine cycle plot in the air (h, ln(P)) chart.

Perfect cycle

The gas turbine draws 1 kg/s of air at 25°C and 1 bar, and compresses it to 16 bar in a perfect compressor (Figure 2.12). This air is brought to the temperature of 1,065°C in the combustion chamber, and then expanded in a perfect turbine.

We assume initially that compressions and expansions are perfect, that is to say they follow the reversible adiabatic; then, we shall explain how to take into account the irreversibilities in these machines.

In the (h, ln(P)) chart in Figure 3.18, point 1 is on the x-axis, at the intersection with the isotherm T = 25°C (h = 0). Perfect compression along the reversible adiabatic leads to point 2,

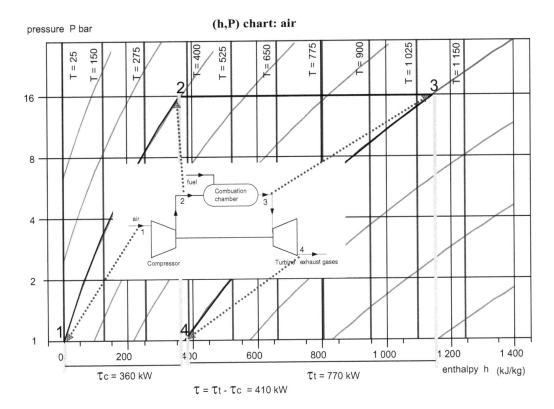

(h,P) chart: air

τc = 360 kW

τt = 770 kW

τ = τt - τc = 410 kW

FIGURE 3.18

Construction of the gas turbine cycle.

the intersection of this curve and the horizontal P = 16 bar. The enthalpy can be read on the abscissa and is h = 360 kJ/kg.

The heating in the combustion chamber leads to point 3, at the intersection of the isobar P = 16 bar and the isotherm T = 1,065°C (h = 1,140 kJ/kg).

Process (3–4) is a reversible adiabatic expansion from 16 to 1 bar. Point 4 is therefore at the intersection of the reversible adiabatic and the abscissa axis (h = 375 kJ/kg).

This chart helps to understand why the gas turbine can operate: the higher the temperature, the less steep the slopes of the reversible adiabatics, and therefore the greater the energy involved in the process, for a given pressure ratio: the enthalpy variation (3–4) increases much faster than the compression work (1–2). We therefore have an interest in working at P and T as large as possible.

For the selected setting, the compression work τ_c at 1 bar and 25°C is more than twice smaller than the expansion work τ_e at 16 bar and 1,065°C.

We also understand why the performance of the machine depends very much on the ambient air temperature: the lower it is, the steeper the slope of the reversible adiabatic and therefore the lower the compression work.

The useful energy is the absolute value of the algebraic sum of the work supplied to the compressor and produced by the turbine (410 kW), and the purchased energy is the heat supplied in the combustion chamber ($h_3-h_2 = 780$ kW). The efficiency is deduced from this, as shown in Table 3.2.

By assuming the turbomachines to be perfect, the efficiency is very high.

Cycle taking into account deviations from reference processes

In reality, the compressor and the turbine are not perfect, and the compression and expansion follow the nonreversible adiabatic. As we explained previously, the real process is characterized by an isentropic efficiency η, defined in the case of the compressor as the ratio of the reversible compression work to the actual work, and in the case of the turbine as the ratio of the actual work to the reversible expansion work.

TABLE 3.2

ENTHALPY BALANCE OF THE SIMPLE GAS TURBINE

Point	Flow Rate (kg/s)	h (kJ/kg)
1	1	0
2	1	360
3	1	1,140
4	1	370

Process	τ (kW)	Q (kW)	Δh (kW)
(1–2)	360		360
(2–3)		780	780
(3–4)	−770		−770
Cycle	**−410**	**780**	
Energy efficiency	52.60%		

To find the actual point $2'$, we therefore determine the value of the work τ_s required by the perfect compressor and we divide it by η.

As $\tau = h_{2'} - h_1$, the enthalpy of point $2'$ is equal to that of point 1 plus τ.

In the present case, taking $\eta = 0.85$, we have:

$$\tau_s = h_2 - h_1 = 360\,\text{kJ/kg}$$
$$\tau = \tau_s / \eta = 360/0.85 = 424$$
$$h_{2'} = 424\,\text{kJ/kg}$$

To find the actual point $4'$, we determine the value of the work τ_s delivered by the perfect turbine and we multiply it by η, which gives the value of the actual work τ.

As $\tau = h_3 - h_{4'}$, the enthalpy of point $4'$ is equal to that of point 3 minus τ.

In the present case, taking $\eta = 0,85$, we have:

$$\tau_s = h_3 - h_4 = 765\,\text{kJ/kg}$$
$$\tau = \eta \tau_s = 0.85 \times 765 = 654$$
$$h_{4'} = 1{,}140 - 654 = 486\,\text{kJ/kg}$$

Figure 3.19 shows the new appearance of the cycle (with dashes, the previous cycle being shown in solid lines). The efficiency drops considerably (it is 32% instead of 52%), because the net work equal to the difference in absolute value between the work of expansion and that of compression, is greatly reduced. The compression work increased by 65 kW, but above all the expansion work fell by 115 kW, the total loss being 180 kW.

Refrigeration machine

Let us now plot the refrigeration machine cycle in the R134a (h, ln(P)) chart.

Perfect cycle

In accordance with the rapid analysis that we made of the saturation pressure curve of R134a, the evaporation pressure of the R134a refrigeration compression cycle must be less than 2.1 bar, and its condensation pressure greater than 8.8 bar. We will retain 1.78 and 12 bar (Figure 3.3) so that the temperature difference with the hot and cold sources is sufficient.

At the evaporator outlet (Figure 2.19), a flow rate m = 1 g/s of fluid is fully vaporized, and thus, point 1 (see Figure 3.20) is located at the intersection of the saturation curve and the iso-bar P = 1.78 bar, or, equivalently, the isotherm T = −13°C.

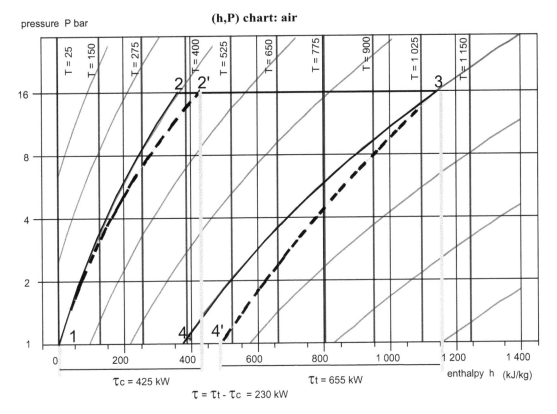

FIGURE 3.19

Simplified gas turbine cycles with and without irreversibilities.

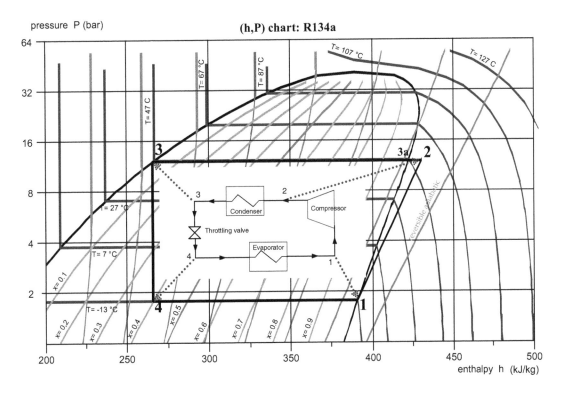

FIGURE 3.20

Simplified refrigeration machine cycle.

It is then compressed at 12 bar along a reversible adiabatic. Point 1 is located approximately one-third of the distance between two reversible adiabatics on the chart; it is possible, by linear interpolation, to determine point 2 on isobar P = 12 bar, between these two curves.

The refrigerant cooling in the condenser by exchange with outside air comprises two stages: de-superheating (2–3a) in the vapor zone followed by condensation along the horizontal line segment (3a–3). Points 3a and 3 lie at the intersection of the saturation curve and the isobar P = 12 bar, or, equivalently, the isotherm T = 47°C. Point 3a is located on the right, at the limit of the vapor zone, and point 3 on the left, at the limit of the liquid zone.

Expansion without work, and therefore an isenthalpic process, corresponds to the vertical segment (3–4); point 4 is located on the isobar P = 1.78 bar, or, equivalently, the isotherm T = –13°C, at abscissa $h = h_3$. Its quality reads directly from the corresponding iso-quality curve: it is x = 0.4.

The energies put into play can easily be determined by projecting these points on the x-axis. This allows us in particular to calculate the COP of the cycle, defined as the ratio of the useful effect (the heat extracted at the evaporator) to the purchased energy (here the compressor work).

The useful energy is the heat extracted from the evaporator (125 kW), and the purchased work energy is supplied to the compressor (40 kW). The ratio of the two, equal to 3.14, is greater than 1; the efficiency term is no longer suitable. We talk about the **COefficient of Performance (COP)** of the cycle. Table 3.3 provides these values.

Cycle taking into account deviations from reference processes

This cycle differs from that of a real machine on several points:

* The actual compression is not perfect, so that the compression work is higher than that which would lead to reversible adiabatic;
* To prevent aspiration of liquid into the compressor, which could deteriorate it as the liquid is incompressible, the gas is superheated by a few degrees (typically 5°C) above the saturation temperature before entering the compressor;
* Before entering the throttling valve, the liquid is subcooled by a few degrees (typically 5°C), as this firstly ensures that this device is not supplied with vapor, and secondly increases the refrigerator performance.

Here, we can also characterize the actual compression by an isentropic efficiency, defined as the ratio of the work of the reversible compression to the real work.

TABLE 3.3

SIMPLE REFRIGERATION CYCLE ENTHALPY BALANCE

Point	Flow Rate (kg/s)	h (kJ/kg)	
1	1	391	
2	1	430	
3a	1	422	
3	1	266	
4	1	266	
Process	**τ (kW)**	**Q (kW)**	**Δh (kW)**
(1–2)	40		40
(2–3a)		–8	–8
(3a–3)		–157	–157
(3–4)			
(4–1)		125	125
Cycle	**40**	–40	
Coefficient of performance		3.14	

To find the actual point 2′, we determine the value of work τ_s in the perfect machine, then divide it by η, which gives the value of actual work τ.

The enthalpy of point 2′ is equal to that of point 1 plus τ.

In this case, taking $\eta = 0.75$, we have:

$$\tau_s = h_2 - h_1 = 40\,\text{kJ/kg/K}$$
$$\tau = \tau_s/\eta = 40/0.75 = 53$$
$$h_{2'} = 391 + 53 = 444\,\text{kJ/kg/K}$$

The cycle changed to reflect superheating, subcooling and compressor irreversibilities is shown in Figure 3.21 (the previous cycle is plotted in dashed line).

The COP of the machine is of course changed: it drops to 2.57 because of the compressor irreversibilities.

The relevance of subcooling can easily be shown in the (h, ln(P)) chart because, for the same compression work, the COP increases as the magnitude of the subcooling grows. It is, however, limited by the need for a coolant.

We did not consider pressure drops in the exchangers. It would of course be possible to do so.

BOX 3.4
Key issues

Identification of basic cycles in (h, ln(P)) charts

We will frequently use two concepts which must be well understood, that of isentropic efficiency and overall cycle efficiency.

These self-assessment activities will allow you to test your knowledge on these topics:

Identification of a steam plant cycle in a (h, ln(P)) chart, ddi
Identification of a gas turbine cycle in a (h, ln(P)) chart, ddi
Identification of a refrigeration cycle in a (h, ln(P)) chart, ddi

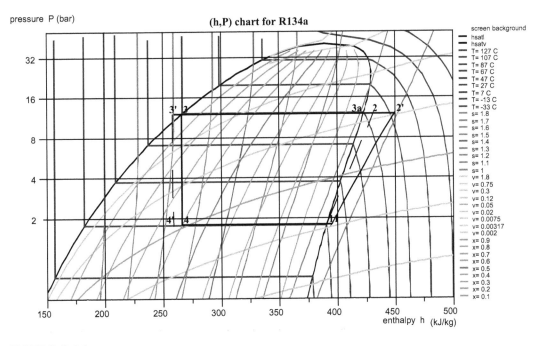

FIGURE 3.21
Refrigeration machine with superheating and subcooling.

Exploring models of simple cycles

This section is devoted to guided explorations of simple cycle models made with the Thermoptim simulator.

We will start by presenting the functionalities of the Thermoptim browser, then those of this simulator, which will allow you to follow educational scenarios putting into practice the knowledge exposed in the preceding sections.

Allow about forty-five minutes of work per model.

Each scenario has three main parts:

1. Discovery of the model, initial setting, link with what was seen previously;
2. Plot in the (h, ln(P)) chart;
3. Changes in the model settings.

Exploration of cycles modeled with thermoptim

At this stage, you have learned how to calculate simple thermal machines based on chart readings and have been presented all the concepts allowing you to move on to another stage: modeling with a simulator. You can then start to familiarize yourself with a tool like Thermoptim.

The great advantage of using a simulator is the ease with which you can vary all the settings of a model. In addition, the results you get are much more precise than the calculations performed on a chart, where you have to make approximate interpolations between the different lines of isovalues.

At first, it is not essential that you learn to build models by yourself. If you wish to do so, detailed online course sessions are also available in the Thermoptim-Unit portal.

We believe that before making your own models, you can learn a lot by analyzing the existing models through guided exploration activities.

Thermoptim browser

A browser capable of emulating Thermoptim allows you to carry out guided explorations of models of energy systems whose scenario is described in html files (Figure 3.22).

The scenario is presented in the browser which successively displays the different explanations and activities to be performed, emulating Thermoptim whenever necessary.

The browser thus offers you to find values in the simulator screens, to change the settings, to perform sensitivity analyses, to display cycles in thermodynamic charts…

This browser is written in Java and uses recent libraries. To download it, follow the instructions given in this page.[2]

If you are unable to install them on your machine, **the browsing files can be read by a standard Internet browser such as Internet Explorer, Firefox or Google Chrome**, but Thermoptim cannot be emulated automatically.

In this case, instead of simply clicking on the emulation button, you only need to follow the detailed instructions provided in the area below this button to be able to load in Thermoptim the files you need.

You can therefore perform the exercises even if the Thermoptim browser is not running on your computer.

You will not have to build by yourself the models of the cycles that you will study, so that you will not waste time on handling errors that are not of educational interest.

The risk of errors is greatly reduced, and if they do occur, simply reset the browser to reload the files you need.

The icon at the top left of the Welcome screen (Figure 3.23) gives you access to the different explorations available.

---

[2] https://www.s4e2.com/portal/browser/Thopt-browser.html.

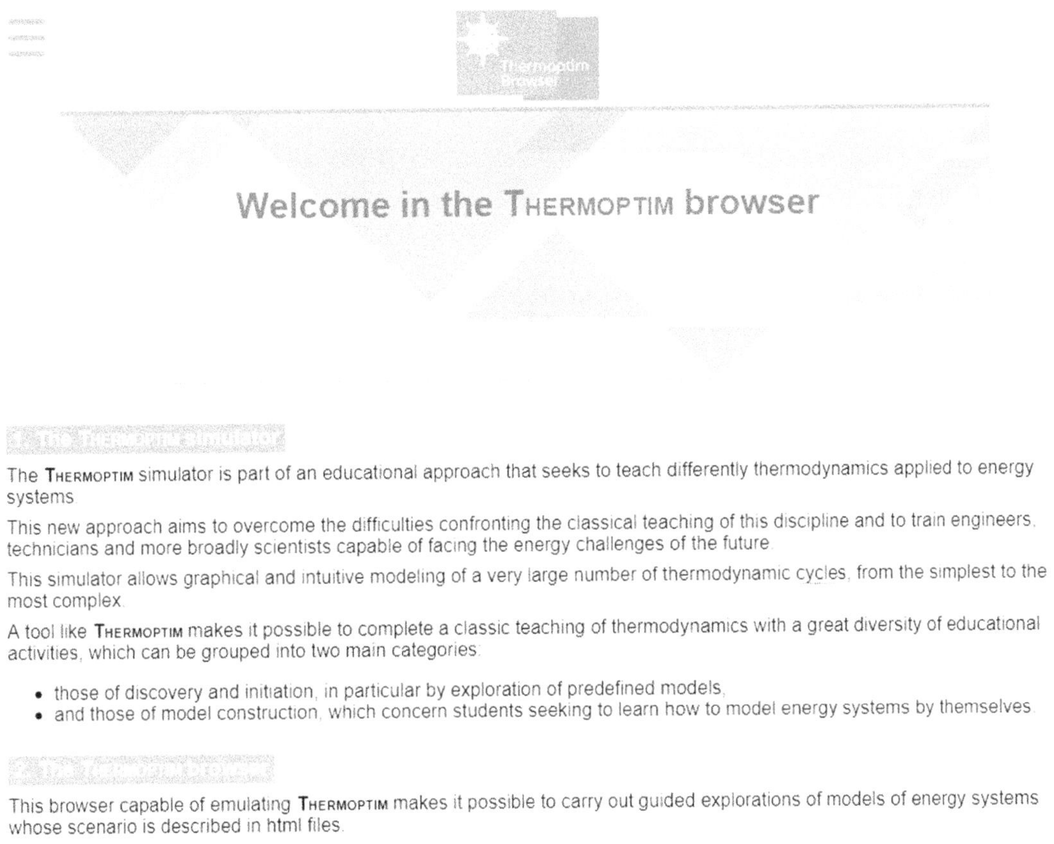

Welcome in the THERMOPTIM browser

The THERMOPTIM simulator is part of an educational approach that seeks to teach differently thermodynamics applied to energy systems

This new approach aims to overcome the difficulties confronting the classical teaching of this discipline and to train engineers, technicians and more broadly scientists capable of facing the energy challenges of the future.

This simulator allows graphical and intuitive modeling of a very large number of thermodynamic cycles, from the simplest to the most complex.

A tool like THERMOPTIM makes it possible to complete a classic teaching of thermodynamics with a great diversity of educational activities, which can be grouped into two main categories:

- those of discovery and initiation, in particular by exploration of predefined models,
- and those of model construction, which concern students seeking to learn how to model energy systems by themselves.

This browser capable of emulating THERMOPTIM makes it possible to carry out guided explorations of models of energy systems whose scenario is described in html files.

FIGURE 3.22

Thermoptim browser screen.

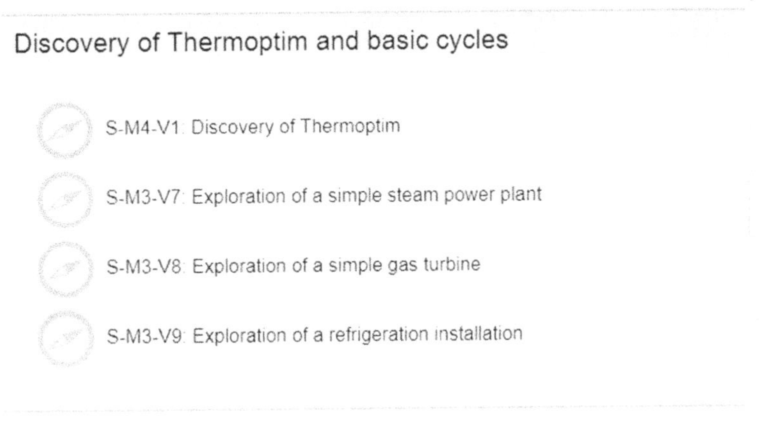

Discovery of Thermoptim and basic cycles

S-M4-V1: Discovery of Thermoptim

S-M3-V7: Exploration of a simple steam power plant

S-M3-V8: Exploration of a simple gas turbine

S-M3-V9: Exploration of a refrigeration installation

FIGURE 3.23

Browser menu.

The reference given corresponds to MOOCs which unfortunately are only available in French. They have been kept for consistency reasons.

Various activities can be offered depending on the case. They appear presented in a certain order, but can be studied in the order you choose.

In an activity, you have several choices of answers.

The browser can also ask you to enter a value, for example, read in a simulator screen, and it reacts to the entered value, depending on whether or not it is within the defined precision range.

Follow the instructions provided to download the browser and install it on your computer.

A discovery exploration of the software package (exploration S-M4-V1) allows you to familiarize yourself with Thermoptim, but it is preferable that you read the following section, which is much more complete.

Initiation to thermoptim

The Thermoptim software package provides a modeling environment integrating in a deeply interconnected manner a diagram editor/synoptic screen, interactive charts, simulation functions and an optimization method (Figure 3.24).

This tool was created to facilitate and secure the modeling of energy conversion technologies. You will find in the simulator screens many concepts that you have previously discovered:

* The components it implements correspond precisely to the functions that have been identified previously;
* Cycle architectures are built by connecting these components in the graphic editor;
* Their setting can be explained very easily by comparison with reference processes;
* The cycles can be viewed in the charts coupled to the simulator.

This section will initiate you to the main concepts that it is preferable to know in order to work with Thermoptim. However, to take full advantage of the possibilities of the software, we recommend that you complete it by consulting the reference manuals and the getting started examples available in the documentation of the tool.

Let us start by saying that Thermoptim is made up of:

* A core comprising the main elements, which already allows one to model many energy systems;

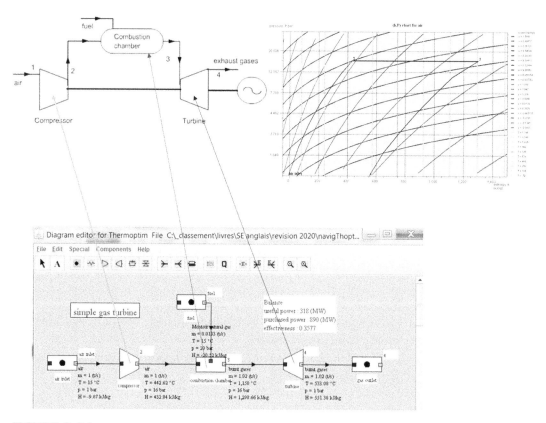

FIGURE 3.24

Modeling with Thermoptim.

- But it can also be extended to represent additional elements not available, making this environment highly customizable.

Three categories of extensions can be made:

- Substances, to add fluids not available in the core;
- Components representing specific energy technologies, such as solar collectors or fuel cells;
- Drivers or pilots, which are small programs that take control of Thermoptim and thus allow you to control the calculations it performs.

To refer to these extensions, we talk of external classes, a class representing an element of Java code, and the adjective external indicating that they are outside the core package.

To understand how Thermoptim works, three basic concepts should be borne in mind:

- The substances, which are characteristic of the various fluids put into play;
- Points, which represent an elementary particle of matter;
- Processes that are used to determine the changes undergone by fluids in the various components, such as compression, expansion or heat exchange.

Substances

To represent the various fluids that flow through the systems studied, Thermoptim uses four types of substances.

Thermoptim contains, on the one hand, about twenty pure gases, and on the other hand, as many compound gases as desired, defined by mixing the available pure gases.

The compound gases are subdivided into two categories: the protected gases whose composition cannot be modified by a user, and others, called unprotected. The reason for this distinction is simply to avoid that any modeling error may modify the composition of a known gas such as air or natural gas.

The properties of gases are based on classical models: the law $Pv = rT$ and a polynomial fit of the heat capacity C_p of the gas as a function of temperature for ideal gases; the perfect gas corresponds to the particular case where C_p is constant.

As shown in Figure 3.8, which gives the molar heat capacities of different gases, the perfect gas model is valid only for monoatomic gases, far from their saturation curve. For others, it is imperative to consider the ideal gas model.

Obviously, this type of model is only valid to represent the properties of the fluid away from the liquid–vapor equilibrium curve.

Condensable vapors

The third type of substance complements the previous two: about twenty condensable vapors. Note that only pure condensable vapors can be modeled as Thermoptim is not able in the general case, to calculate the properties of a vapor mixture.

Some substances, such as water, appear both as pure gases and as condensable vapor: they correspond to two different models, which are selected depending on the problem to solve. The names of pure gases include their chemical formula, which is mainly used in the combustion calculations, and those of vapors are generally common names. For instance, water is called H_2O as gas and water as steam.

External substances

The fourth type is external substances, which can either be simple, fully calculated in the Thermoptim external classes, or be mixtures, whose calculation is performed in specific software coupled to Thermoptim.

For a Thermoptim user, a substance is simply identified by its name.

FIGURE 3.25

Substance selection screen.

component name	molar fraction	mass fraction
CH4 ` methane	0.871	0.758966
C2H6 ` ethane	0.088	0.1437279
C3H8 ` propane	0.025	0.05987759
C4H10 ` n-butane	0.008	0.02525591
N2	0.008	0.01217253

FIGURE 3.26

Screen defining the composition of a compound gas.

Figure 3.25 shows the lists of available substances, where protected compound gases are deployed.

Figure 3.26 shows the screen that defines the composition of a gas, here natural gas at the LNG terminal of Montoir de Bretagne in France, with the names of its pure gas components and their molar and mass fractions.

Points

Thermoptim defines points, which represent a small amount of matter, whose thermodynamic state may be calculated when are known, for example, pressure and temperature.

They only provide access to intensive quantities; flows at stake are not specified at this stage.

Figure 3.27 shows the screen of a point, which can be calculated in the closed or open system, or allows you to determine the properties of moist mixtures which are modeled as a mixture of water and dry gas considered ideal (see Chapter 12 for additional explanations).

To set a point, you must start by giving it a name, here 1, then enter the name of the substance associated with it or choose from one of the lists presented above, which appear when you double-click in the field where "water" has been entered in this example.

You must then define a sufficient number of state variables, often the pressure and temperature, making sure that the method of calculation selected (by default, P and T known) is the correct one.

In the case of a condensable vapor, it is possible to set either the saturation temperature, knowing the pressure, or the saturation pressure, knowing the temperature. You must then specify the quality, here equal to 0.

The point can then be calculated by clicking the Calculate button.

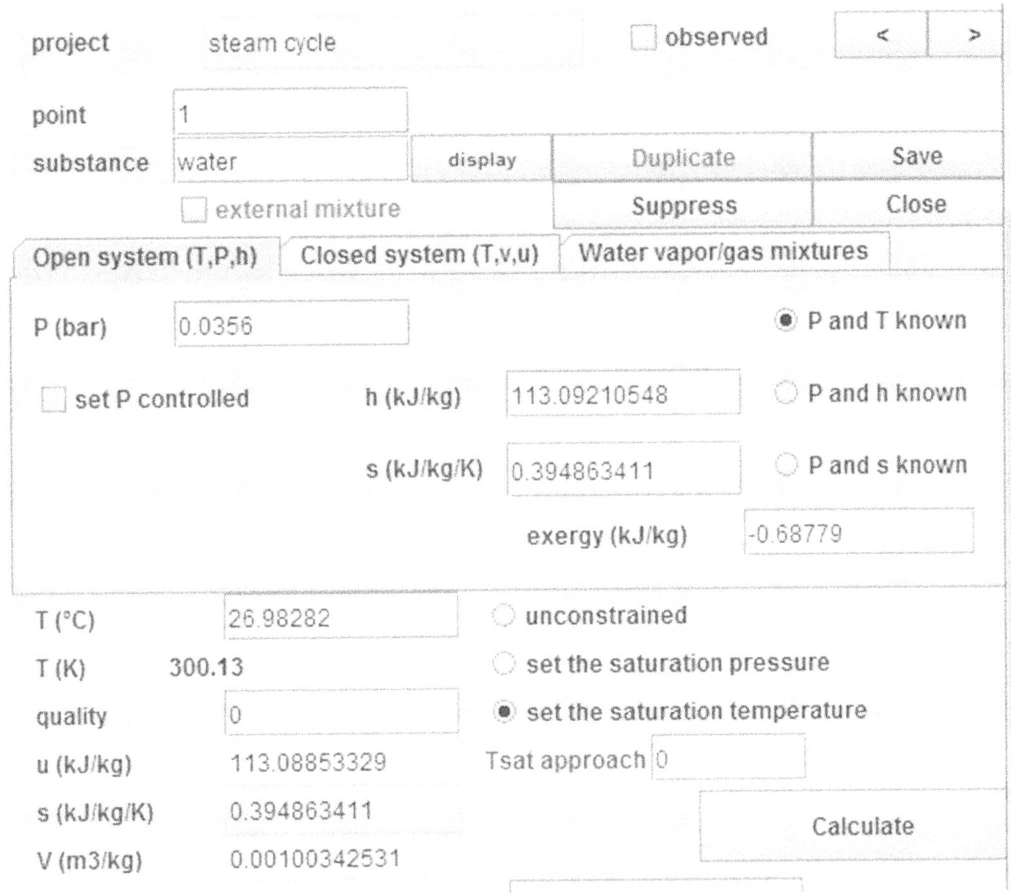

FIGURE 3.27

Point screen.

Processes

Processes correspond to thermodynamic changes undergone by a substance between two states. A process connects two points as defined above, an upstream or inlet point and a downstream or outlet point. In addition, it specifies the mass flow involved and, therefore, allows you to calculate the extensive state variables, including determining the variation of energy put in.

The most common processes were modeled and are directly accessible in the core. Knowing the condition of the fluid at the process inlet, Thermoptim may either solve the problem directly, or solve the reverse problem. In the first case, knowing the characteristics of the process, it computes the state at the process outlet and the energies put into play, and updates the downstream point. In the second case, it identifies the parameters of the process so that the selected change leads to the defined downstream point state.

Figure 3.28 shows an expansion process, which is a steam turbine. All process screens are structured similarly.

In the upper-left part appear the name of the process (here turbine), its type (here expansion), the type of energy that allows you to calculate the energy balance (here useful), and an option to set or not the flow through the process. If not set, the flow of the process located just upstream is automatically propagated.

On the left side appear summaries of the thermodynamic states of upstream and downstream points. Clicking the display button opens the screen of the corresponding point.

In the upper right appear the navigation buttons, allowing you to open, close and calculate the process.

In the bottom right are located the setting options specific to the process.

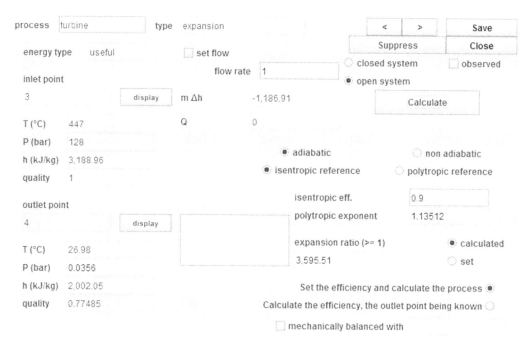

FIGURE 3.28

Process screen.

Compressions and expansions can be calculated in closed or open systems, taking as reference adiabatic or non-adiabatic and isentropic or polytropic[3] model. The compression ratio can be set or calculated.

The setting here is adiabatic expansion, and the isentropic efficiency is 0.85. The downstream point calculation is carried out by considering that the efficiency is known.

As a point does not allow one to specify the flow involved, it may be necessary to create special processes, known as process-points (Figure 3.29).

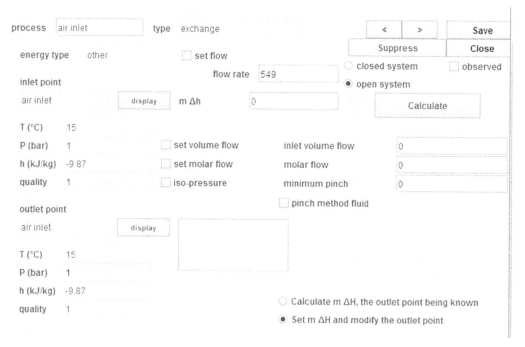

FIGURE 3.29

Process-point screen.

---

[3] See Chapter 8.

A process-point connects a point with itself and specifies the mass flow to be taken into account. It therefore technologically corresponds to a small pipe. Process-points allow you to define fluid inputs or outputs, to connect nodes, etc.

Other features

Fluids flow through machines, forming networks of varying complexity that have to be described. Processes allow you to represent part of these circuits. They are complemented by nodes which allow you to describe the network elements, which make up mixture, division or separation of fluids. In a node, several fluid branches are interconnected to form a single vein (Figure 3.30).

Three types of nodes exist in the core of Thermoptim: mixers, dividers and phase separators for two-phase fluids.

Figure 3.31 shows the screen of the Thermoptim simulator, which gives access to the list of points, at the top left and that of the processes, just below.

The list of nodes appears at the bottom left, but there is none in this example. Heat exchangers can also be defined, but we will not consider them for the moment.

Double-clicking in one of the lines displays the selected point or process screen.

Figure 3.32 shows the screen of the Thermoptim diagram editor with the icon palette.

Except for the left arrow allowing you to bring the mouse into its normal state, we can distinguish, from left to right, the icons that let you select different components and put them on the work plan:

* The A allows you to write a text;
* Then appear core processes, namely, process-point, heat exchange (heating or cooling of a fluid), compression, expansion with work, combustion chamber and isenthalpic throttling, which represents a filter or a valve creating a pressure drop.

The next block gives access to the three nodes: mixers, dividers and phase separators.

The next icon is the component giving the overall balance.

The Q is used to represent a heat source, intended primarily for educational purposes, to show clearly that the systems generally exchange heat with their surroundings.

Then, on the right of the palette are three icons used to represent the three types of external components that can be added to the core, which may be a process when there is a single fluid input and one output, a mixer when there are several inputs and one output, and a divider otherwise. Examples of these external components will be presented in the third and fourth parts of this book.

Figure 3.33 shows how the steam power plant cycle looks in the diagram editor.

FIGURE 3.30

Node screen.

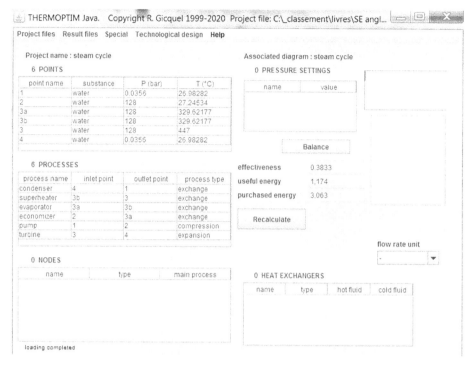

FIGURE 3.31

Simulator screen.

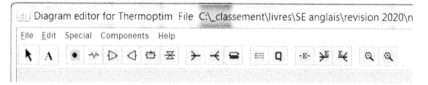

FIGURE 3.32

Screen of the Thermoptim diagram editor.

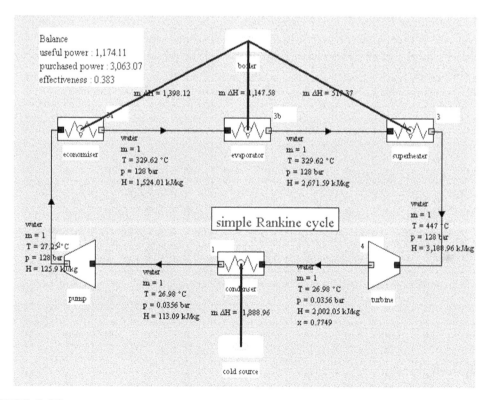

FIGURE 3.33

Synoptic view of the steam power plant.

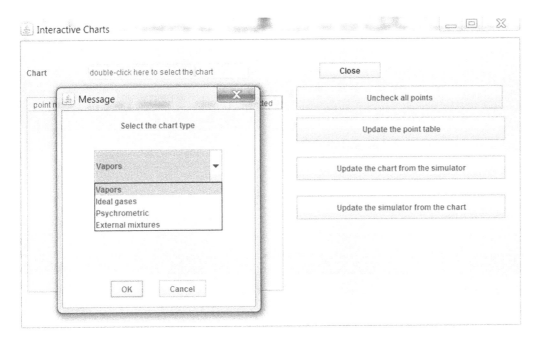

FIGURE 3.34
Display of thermodynamic charts.

The various components are connected by thin black lines oriented in the direction of flow of the fluid. The state of the fluid is indicated there.

Double-clicking on one of the processes will display its screen. Double-clicking on one of the oriented links opens the corresponding point.

The blue links here represent the connections to external sources, and the balance of the cycle appears in a small cartouche.

The thermodynamic charts are accessed via the Interactive charts menu line in the Special menu of the simulator screen (Figure 3.34).

When you double-click on the line "double-click here to choose the chart", the list of types of charts available is proposed.

You can choose the desired one.

Synoptic views of the simple cycles

The synoptic view of the steam power plant is given in Figure 3.33, and those of the gas turbine and the refrigeration cycles in Figures 3.35 and 3.36.

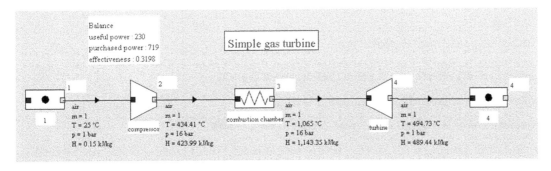

FIGURE 3.35
Synoptic view of the gas turbine.

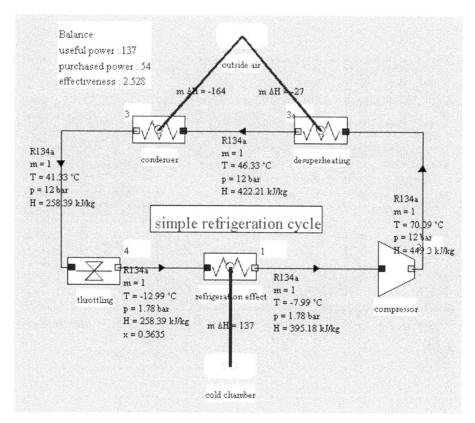

FIGURE 3.36

Synoptic view of the refrigeration machine with superheating and subcooling.

BOX 3.5

Guided educational exploration

Discovery of thermoptim (exploration S-M4-VI)

The objective of this exploration is to guide you in your first steps of using Thermoptim, by making you discover the main screens and functionalities associated with a simple refrigeration machine model.

You will discover the arrangement of the screens of the points and processes, the way they can be set and calculated, and the concepts of useful and purchased energies making it possible to draw up global energy balances and to determine the COP.

You will see the cycle plot in the (h, ln(P)) thermodynamic chart.

Three guided educational explorations were prepared for the cycles studied previously. They can be downloaded with the browser from the Thermoptim-UNIT portal.

BOX 3.6

Guided educational exploration

Exploration of a simple steam plant (Exploration S-M3-V7)

Exploration of a gas turbine (exploration S-M3-V8);

Exploration of a refrigeration machine (exploration S-M3-V9).

The objective of this exploration is to guide you in your first steps of using Thermoptim, by making you discover the main screens and functionalities associated with the cycle models you have studied so far.

You will discover how the screens of points and processes can be set and calculated, and the concepts of useful and purchased energies making it possible to draw up global energy balances.

You will visualize the cycles in the (h, ln(P)) thermodynamic chart, and you will carry out studies of sensitivity of the cycle vis-à-vis some parameters.

2
Complements for cycle studies

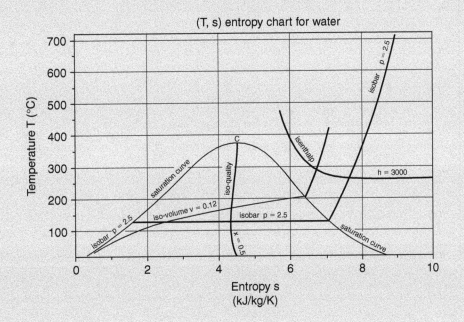

(T, s) entropy chart for water

DOI: 10.1201/9781003175629-2

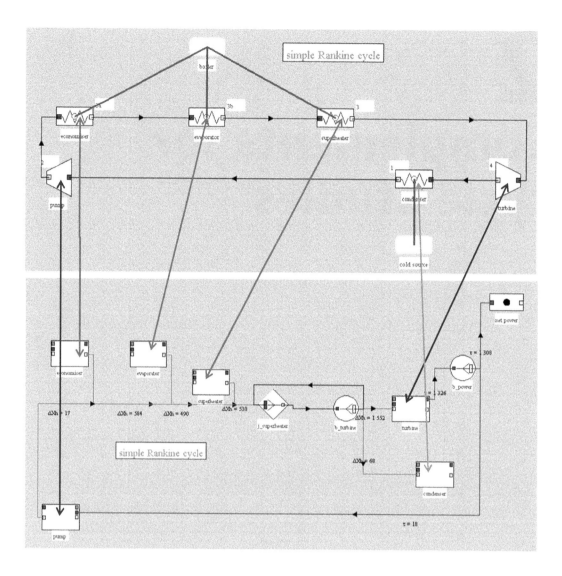

Part 2 provides theoretical, technological and methodological supplements that will be used in more complex cycle studies of Parts 3 and 4.

Theoretical supplements relate to combustion, heat exchangers and entropy. Methodological supplements deal with process integration (pinch method), exergy balances and productive structures. Technological supplements are related to steam system components: boilers, steam generators, steam turbines and cooling towers.

Combustion and heat exchangers

Introduction

The main objective of the second part is to provide readers with the supplements that will allow them to study the cycles presented in the third and fourth parts.

In the first part, we talked about combustion but, to simplify things, we did not specify how combustion can be calculated, and when we studied the model of the gas turbine, we replaced its combustion chamber by simple heating.

In addition, we have not given any information on how a heat exchanger like a steam plant condenser or a refrigerator evaporator can be calculated.

In this chapter, we provide supplements on these two topics widely used in energy systems:

* We first address combustion phenomena which we eluded so far in order to simplify things;
* We then study heat exchangers that are used to transfer heat from a hot fluid to a cold fluid.

Combustion

Combustion phenomena are of particular importance in the study of energy technologies, because they are the source of most heat and power production in the world: more than 90% of global consumption of primary commercial energy comes from burning coal, oil or natural gas.

The study of combustion is to determine the state and composition of combustion products, and consequently, their thermochemical properties, including the amount of energy, put into play in the reactions. Moreover, the combustion conditions largely determine the quantities of pollutants emitted by energy technologies.

The purpose of this section is to determine as precisely as possible the combustion characteristics that are important in terms of energy: the thermodynamic properties of the products and the energy released by combustion.

From these elements, it becomes possible to optimize the combustion, i.e., to obtain the best combustion efficiency by adjusting the air/fuel ratio or the excess air, which sets out the flame temperatures and losses by the fumes.

To achieve this goal, it is necessary to know how to characterize fuels and write the complete reactions, either stoichiometric or with excess or lack of air, taking into account the dissociation where it exists, which requires knowledge of general principles that govern equilibrium reactions (law of mass action), and finally calculate the energies released.

In the remainder of this book, we will always assume that the combustion reaction involves gaseous species and that the various components are treated as ideal gases. Dalton's law then tells us that the combustion products follow the ideal gas law.

DOI: 10.1201/9781003175629-4

Basic mechanisms

Combustion reactions are rather unusual chemical reactions, typically brutal, where the nonequilibrium is the rule and equilibrium is the exception. In these circumstances, it is still impossible to determine exactly during the reaction the evolution of thermodynamic functions characterizing the systems involved. They can only be calculated assuming that equilibrium conditions are met before and after the time of combustion.

Complete reactions, dissociation, law of mass action

Generally, a chemical reaction involving two reactants and two products can be written as follows:

$$v_1 A_1 + v_2 A_2 \leftrightarrow v_3 A_3 + v_4 A_4 \tag{4.1}$$

Experience shows that the unfolding of a combustion reaction is highly dependent on temperature.

At low temperature, the reactants generally do not react spontaneously. For the reaction to start, it is necessary to bring an activation energy, usually in the form of an electric spark. They say that the mixture is in false equilibrium. If the reaction is triggered artificially, it is brutal and total if the temperature remains low enough. The combustion is said to be complete, or without dissociation.

At high temperature, the reaction is usually incomplete. The composition of the reaction medium depends on the degree of progress of this reaction and the ratio of the number of moles of reactants that have been converted to the total number of moles of reactants. This degree of progress depends on the kinetics and the **law of mass action** at equilibrium.

This law is fundamental to studying the incomplete combustion reactions at equilibrium. It states that the rate of a chemical reaction at equilibrium is directly proportional to the product of the concentrations of the reactants.

In practice, for Equation (4.1), it writes, $K_p(T)$ being a function of the reaction and of the temperature:

$$K_p(T) = \frac{v_3 v_4}{v_1 v_2} \tag{4.2}$$

The reactants are not completely oxidized, and we talk of **dissociation**. If, starting from such a situation, we gradually decrease the temperature of reactants, the composition stabilizes and remains unchanged only below a certain threshold. It is said there is **quenching** of the reaction.

In the energy systems that we study, combustion will be either complete or incomplete, depending on cases. In this section, we give some explanations on how it can be calculated with some simplifying assumptions.

The general thermodynamic study of combustion reactions is part of the thermodynamics of chemical processes. In this chapter, we limit ourselves to only those aspects necessary for our purposes, i.e., mainly the study of variations of the thermodynamic functions of the reactive mixture caused by variations in chemical composition.

The study of the mechanism of the reaction kinetics (limit of false equilibria, means of triggering the reaction mixture in a false equilibrium, methods of propagation of the reaction, kinetics, etc.) is much more complex.

In what follows, as has been said in the introduction, we limit ourselves mainly to the study of combustion reactions in the gaseous phase. Even when fuel is introduced as a liquid (oil, diesel), the products are then present in gaseous form.

Combustion of gaseous fuel mixtures

A gaseous fuel mixture, in practice generally composed of air and fuel gas, is defined by its molar or mass composition.

It is customary to relate the composition of any mixture to that of the stoichiometric mixture, i.e., with exactly the amount of oxygen required for complete combustion. The richness R is defined as the ratio of the number of moles (or mass) of fuel in a predetermined amount of mixture to the number of moles (or mass) of fuel in the stoichiometric mixture.

R = 1 corresponds to stoichiometric mixture, R < 1 to excess air and R > 1 to excess fuel (lack of air). We will introduce below two more quantities directly related to richness, which are also widely used for the study of non-stoichiometric combustion: air factor λ and excess air e.

Conditions of self-ignition and propagation of combustion

A fuel mixture based on a given fuel is generally stable at room temperature. By heating, it is possible to trigger the combustion reaction from a certain temperature, called self-ignition. The H_2 mixtures burn spontaneously at 550°C, and those based on CO and CH_4 around 650° C. It sometimes happens that a self-ignition delay appears, which can reach 200°C for CH_4.

The auto-ignition temperature depends, of course, on the fuel and the composition of the mixture, but it is also a decreasing function of pressure, with some peculiarities (cold flames, long self-ignition delays), for which we refer to the literature.

The value of self-ignition temperature is that it defines an upper limit which must not be reached, whether for reasons of safety when handling the fuel, or for more specific applications, such as burning in gasoline engines.

When a fuel mixture is heated to a temperature above its self-ignition temperature, the reaction is triggered after some delay, the said auto-ignition delay, which becomes shorter at higher temperatures. This delay is related to basic mechanisms of combustion.

Moreover, in a given mixture of air and fuel, combustion, initiated locally by a spark, can only spread if the richness is between two limits: say lower inflammable R_l and higher inflammable R_h. If the richness is outside of this interval, the mixture is either too lean or too rich, and combustion cannot be maintained. These limits depend on the temperature and fuel type. At room temperature, limits R_l and R_h for most hydrocarbons are, respectively, close to 0.5 and 4. They diverge as the temperature increases.

Flame propagation

To avoid self-ignition of gaseous fuels, it is necessary that the mixture is ignited locally before reaching the self-ignition temperature one way or another. In practice, an electric spark is generally used for this purpose.

Once the mixture is ignited, combustion spreads from the ignition point, in a reaction zone called the flame front, corresponding to the surface which separates the burned gases from the mixture still intact. The case where the fuel and oxidizer form a homogeneous mixture before the start of the reaction is referred to as premixed flame. Experience shows that the propagation can take place in two very different modes: **deflagration** and **detonation**. The difference between them is the relative speed at which the flame front moves in the mixture.

In the deflagration, this speed is low, in the order of several cm/s to a few m/s, whereas in the detonation, it far exceeds the speed of sound in the mixture (several thousands of m/s).

In the deflagration, the combustion is mainly propagated by conduction: the layer of combustion gas heats the next layer of still intact mixture. When it is in turn brought to the auto-ignition temperature, it ignites, and so on. Since the gases are poor conductors, their propagation speed is low.

In detonation, the combustion is propagated by a shockwave that moves at a speed exceeding the speed of sound in the mixture, causing an almost instantaneous reaction at its passage, with sudden changes in velocity, pressure and temperature. Given this shockwave, detonation is a very dangerous combustion mode, which causes engines to deteriorate rapidly. Very abrupt changes in pressure induce, in any parts submitted to the thrust of the gases, very destructive steep front elastic waves. The joints are subjected to shocks, valves and seats leak, deformed by the wave. When it occurs in an engine, it is usually called **knocking combustion mode**. In addition, when the fuel is a hydrocarbon, the detonating combustion is always initiated by a decomposition of the molecule, releasing black smoke and thus making the engine dirty. We will study in detail in Chapter 9 what solutions are adopted to prevent detonation in internal combustion reciprocating engines.

Study of the deflagration

A mixture whose richness is between the flammability limits burns with a certain deflagration speed. This speed, often called laminar velocity, is the velocity of the flame relative to the reaction mixture that feeds it, in a laminar flow. Its value, which also depends strongly on the nature of the fuel, is a function of richness, temperature and pressure.

In general, for a given temperature, the maximum deflagration speed corresponds to a richness slightly above 1, which is often around 1.2. Deflagration speeds are also low, a few cm/s to a few m/s. For a given richness, they are an increasing function of temperature and a decreasing function of pressure.

Turbulence exerts a complex effect on the deflagration speed: it increases significantly if it is already noticeable at rest; it decreases near the flammability limit otherwise, and under these conditions, it can prevent the spread of combustion (blowing on a fire stirs it up if it is intense, blows it out otherwise).

Combustion of liquid fuels

The phenomena described above for gaseous fuels are similar in the case of liquid fuels, with some differences.

A droplet of a liquid fuel that is injected into a mass of air ignites spontaneously when the temperature exceeds a certain value, called fuel self-ignition temperature T_{si}. This temperature depends mainly on the fuel, and incidentally on air pressure, diameter and drop speed.

The self-ignition temperatures thus measured range from 350°C to 500°C depending on the fuel. Note that they are relatively low, much lower than those of gaseous fuel mixtures.

Ignition and combustion delays

Experience shows that when injecting a droplet of liquid fuel into an air mass whose temperature T exceeds the self-ignition temperature T_{si}, it does not catch fire instantly, but after some delay, the said inflammation delay, which depends mainly on the difference $(T - T_{si})$ and the type of fuel, and secondarily on the droplet diameter and velocity. The mechanisms involved are complex: a fraction of the liquid fuel generally evaporates to form, mixing with the oxidant, the reactive medium that self-ignites depending on its composition and pressure and temperature conditions.

In addition, complete combustion requires an additional delay, the said combustion, which, apart from the difference $(T - T_{si})$, depends essentially on the droplet diameter and velocity as well as air turbulence.

When T exceeds T_{si}, the ignition delay is never very large, but it rarely falls below 2 ms. The combustion delay is similar, provided that the fuel is finely pulverized (an average diameter of drops of a few microns) and injected at high speed, at a pressure exceeding 100 bar. Nevertheless, the overall ignition and combustion delay is of the order of several milliseconds.

Diffusion flames

Unlike premixed flames, where the fuel and oxidizer are mixed before the start of the reaction, the combustion of fuel introduced in the liquid state in the oxidizer is governed by the speed at which reactive chemical species interdiffuse. This is called the diffusion flame. The reaction zone of such a flame is thicker than that of a premixed flame. Richness varies from infinite value in the fuel to zero in the oxidizer. There are laminar diffusion flames (candles, oil lamps) and turbulent confined diffusion flames whose applications are numerous, either in industry or in conventional diesel engines.

BOX 4.1
Key issues

Main difference between diesel and gasoline engines

An important difference between a gasoline engine and a diesel engine is not in the mode of introducing fuel, which in modern gasoline engines is also injected, but when the fuel is introduced, which determines the nature of gas when the reaction starts and the type of combustion which takes place inside the engine, as will be explained in detail in Chapter 9.

In the gasoline engine, fuel is introduced well in advance so that the cylinder is full, when ignition occurs, of a substantially homogeneous gaseous mixture. In the diesel engine, the fuel is injected at the last moment and burned as fine liquid droplets as and when it is introduced (diffusion flame).

Quenching temperature

Thermodynamics of combustion at equilibrium allows one, from the law of mass action, to determine the degree of progress of reactions depending on the temperature and pressure. For this law to be applicable, it is necessary first that the reaction speeds are larger than the residence time of reactants in the combustion chamber, and second that cooling conditions for the fumes are known.

For most combustion reactions, the assumption of a medium at equilibrium can be chosen without introducing too much error. An exception should, however, be made for the oxidation reaction of nitrogen, which leads to the formation of NO_x. This reaction has a very slow kinetics, so that the degree of reaction is generally much less advanced than at equilibrium.

The study of combustion kinetics is also necessary if one wants to explain the quenching phenomena that appear when dissociation takes place (at temperatures above 1,800 K), followed by a sudden cooling of the fumes.

Due to dissociation, some combustion reactions are not complete. The speed of slow reactions may then not be sufficient in the case of a rapid decrease in temperature, so that molecular recombinations that would have taken place normally (that is to say at equilibrium, the reactions being reversible) cannot be attained.

In these conditions, unburned hydrocarbons may remain. It is said that there is quenching of the reaction. This explains the presence of CO in the flue gas, even with excess oxygen and for a homogeneous mixture. We call **the quenching temperature the temperature to be used in the law of mass action** to get the actual exhaust gas composition.

You should refer to the specialized literature for further developments on this subject.

Study of complete combustion

We will start by dealing with complete combustion, that is to say, such that all possible oxidations are carried out. After presenting the different ways of representing the oxidizer (usually air) and fuel, we will study successively the stoichiometric and non-stoichiometric combustion processes.

Air composition

A combustion reaction involves the oxidation of a fuel with oxygen or a substance containing oxygen, such as air. In most cases in practice, it is air, and it is accepted as a first approximation that all components other than oxygen are inert during the reaction.

The composition of normal air is given in Table 4.1. To simplify the study of combustion reactions, other inert gases are combined with nitrogen, by defining an atmospheric nitrogen, which allows us to consider the dry air as a mixture of two ideal gases, whose composition is indicated below. From these values, it follows that the atmospheric nitrogen/oxygen ratio is **3.76 in volume**, and 3.31 in mass.

In Thermoptim, calculations are made with greater precision, and the number of components is up to the user. Among the protected ideal gases, i.e., whose composition cannot be changed by the user, we, however, find the above-defined atmospheric air, in case you want to use it, for example, to check that the results obtained with it correspond well to those provided in this book.

Fuel composition

To study the quality of combustion, it is convenient to represent the fuel by a standardized formula reflecting its actual composition. For an ash-free fuel, the chemical elements to be retained are carbon, hydrogen, oxygen, nitrogen and sulfur.

The fuel can therefore be described by the formula: $C H_y O_x N_z S_u$.

The stoichiometric coefficients y, x, z and u can be easily determined:

- From mass fractions (C), (H), (O), (N) and (S), which gives:

$$y = \frac{12.01(H)}{1.008(C)}, x = \frac{12.01(O)}{16(C)}, Z = \frac{12.01(N)}{14.008(C)}, u = \frac{12.01(S)}{32.07(C)} \qquad (4.3)$$

- From the exact formula if it is a pure substance or a mixture of pure chemically defined substances, relating the chemical formula to the unit of carbon. For compound $C_n H_m O_p$, we get:

$$y = \frac{m}{n}, x = \frac{p}{n} \qquad (4.4)$$

- From volume or mole fractions if it is a gaseous fuel.

For a mixture of fractions $[H_2]$, $[CO]$, $\sum[C_n H_m]$, $[CO_2]$, $[N_2]$, $[H_2S]$, $[SO_2]$, we get:

$$y = \frac{2[H_2] + \sum m[C_n H_m] + 2[H_2S]}{[CO] + \sum n[C_n H_m] + [CO_2]}$$

$$x = \frac{[CO] + 2[CO_2] + 2[SO_2]}{[CO] + \sum n[C_n H_m] + [CO_2]}$$

TABLE 4.1
COMPOSITION OF NORMAL AIR

Component	Molar Fraction	Molar Mass (kg/kmol)	Mass Fraction
N_2	0.781	28	0.756
$Ar + CO_2$	0.009	−(40)	0.012
O_2	0.21	32	0.232
N_2 atmospheric	0.79	(28.15)	0.768

$$z = \frac{2[N_2]}{[CO] + \sum n[C_nH_m] + [CO_2]} \qquad (4.5)$$

$$u = \frac{[H_2S] + [SO_2]}{[CO] + \sum n[C_nH_m] + [CO_2]}$$

Stoichiometric combustion

We call a stoichiometric combustion a combustion without excess or lack of air, where all available oxygen is completely consumed.

Assuming that nitrogen reacts only in negligible proportions and is found after combustion in molecular form, the general equation of a stoichiometric combustion is:

$$CH_yO_xN_zS_u + \left(1 + \frac{y}{4} + u - \frac{x}{2}\right)O_2 \rightarrow CO_2 + \frac{y}{2}H_2O + uSO_2 + \frac{z}{2}N_2 \qquad (4.6)$$

Depending on circumstances, coefficient $(1 + y/4 + u - x/2)$ may be greater or lesser than unity, according to the existing availability of oxygen in the fuel. In most cases, $x \le y/2 + 2u$, but for some lean gases, the reverse situation may exist.

For most conventional fuels, the values of z and u are very low and often negligible, while the condition $x \le y/2$ is always verified. We can then simplify the symbolic formulation above by writing:

$$1 + \frac{y}{4} - \frac{x}{2} = 1 + \frac{a}{4}$$

which leads to the conclusion that formula CH_yO_x becomes:

$$CH_a + xH_2O$$

where a represents what is called hydrogen available for combustion referred to the complete oxidation of a carbon unit, and $a = y - 2x$.

With these conventions, the complete oxidation of fuels of formula $(CH_a + xH_2O)$ can be described by the following equation:

$$CH_a + \left(1 + \frac{a}{4}\right)(O_2 + 3.76N_2) \rightarrow CO_2 + \frac{a}{2}H_2O + 3.76\left(1 + \frac{a}{4}\right)N_2 \qquad (4.7)$$

while there is conservation of the constitutive water xH_2O.

This reaction is the fuel combustion **without excess or lack of air**.

If there is a lack of air, some of the carbon is oxidized into carbon monoxide (CO). To completely determine the reaction, we have seen above that it is necessary to study the conditions of chemical equilibrium of the dissociation reaction of carbon dioxide, which depend primarily on temperature and correspond to the following equation:

$$CO_2 \leftrightarrow CO + \frac{1}{2}O_2$$

Combustion quality

The combustion quality is primarily determined by two factors:

* The combustion efficiency, which characterizes the degree of fulfillment of the chemical reaction, or if one prefers, the absence of unburned hydrocarbons or incompletely oxidized compounds;

• The level reached by the combustion temperature, determined by the ratio of the amount of heat released by burning to the amount of reactants and products.

In practice, we can optimize combustion by playing on an excess of air. A large excess air ensures that at any point, enough oxygen is available for combustion and therefore reduces the risk of unburned hydrocarbons. However, it dilutes the fumes, whose temperature drops.

Non-stoichiometric combustion

When the combustion is not stoichiometric, it can be defined in several ways as we have seen, usually by its **excess air e**, or its lack of air (–e), or the **richness R**, or its inverse the **air factor λ**. The formulations vary according to the existing professional practice. For example, it is customary to speak of richness for gasoline engines, or excess air for boilers. In Thermoptim, we have selected the air factor, which is very useful in practice.

These quantities are related thanks to the following equations:

$$R = \frac{1}{1+e}$$

$$\text{and} \quad \lambda = 1 + e = \frac{1}{R} \tag{4.8}$$

Let us consider how a non-stoichiometric reaction is written as a function of air factor λ:

In case of excess air, λ is greater than 1, and there is too much oxygen. The reaction reads:

$$CH_a + \lambda\left(1+\frac{a}{4}\right)(O_2 + 3.76N_2) \rightarrow CO_2 + \frac{a}{2}H_2O + (\lambda-1)\left(1+\frac{a}{4}\right)O_2 + 3.76\lambda\left(1+\frac{a}{4}\right)N_2 \tag{4.9}$$

In case of lack of air, λ is less than 1, and while there is enough oxygen to oxidize all the carbon in carbon monoxide, and a fraction in carbon dioxide, the reaction becomes:

$$1 > \lambda > \frac{1+\frac{a}{2}}{2\left(1+\frac{a}{4}\right)}$$

(neglecting dissociation of CO_2 in H_2O)

$$CH_a + \lambda\left(1+\frac{a}{4}\right)(O_2 + 3.76N_2) \rightarrow \left[2\lambda\left(1+\frac{a}{4}\right) - \left(1+\frac{a}{2}\right)\right]CO_2$$

$$+ 2(1-\lambda)\left(1+\frac{a}{4}\right)CO + \frac{a}{2}H_2O + 3.76\lambda\left(1+\frac{a}{4}\right)N_2 \tag{4.10}$$

If the lack of air is more important, there is not even enough oxygen to form all the carbon monoxide.

Energy properties of combustion reactions

One of the main objectives of combustion studies is the determination of energy exchanges that occur when chemical reactions take place.

For this, it suffices to know the internal energies or enthalpies of reactants and products, because the energy exchange at stake does not depend on intermediate reactions, but only on the initial and final states of chemical species in reaction.

We define the combustion **heat of reaction** as the change in internal energy or enthalpy of the mixture, depending on whether the reaction takes place at constant volume or constant pressure.

Evidently, the calculation of heats of reaction implies knowledge of the internal energies U_r and U_p or enthalpies H_r and H_p of reactants and products. As mentioned previously, the principle of their calculation is based on the assumption that the different constituents are reacting in the gaseous state, and each of them is comparable to an ideal gas. Their mixture under these conditions is itself an ideal gas.

The internal energy and enthalpy of different substances, known as **heats of formation**, are determined in a reference state by methods that are outside the scope of this book; they are available in tables (e.g., JANAF Tables, Chase et al., 1985).

Heats of reaction

Once the enthalpies of formation of various chemical species are known, the calculation of heat of reaction is simple.

Let us use general equation (4.1) considered at the beginning of this chapter:

$$v_1 A_1 + v_2 A_2 \leftrightarrow v_3 A_3 + v_4 A_4$$

Let us note h_{fi} the enthalpies of formation of the A_i.
We obtain the heat of reaction by writing:

$$-\Delta H_r = v_1 h_{f1} + v_2 h_{f2} - v_3 h_{f3} - v_4 h_{f4} \tag{4.11}$$

In an equilibrium reaction, the heat of reaction ΔH_r is a function of v_i and thus x_i. Therefore, when the reaction is not complete or there is dissociation, the heat of reaction is different from that obtained under stoichiometric conditions.

Higher and Lower Heating Value

In most cases, the reaction products are found in the gaseous state at the end of combustion, but it is possible at low temperature that some of them are liquid or even solid, releasing a heat of condensation or solidification.

The problem arises especially during the combustion of hydrocarbons, since water is one of the products. The maximum energy release is obtained when the water contained in the fumes is liquefied. The complete value of the heat of reaction is named **higher heating value or HHV**. Where all the water produced remains in the vapor state, it is given the name of **lower heating value or LHV**. If only a fraction of the water condenses, the heat of reaction is an intermediate value between the HHV and the LHV.

Based on the above definitions, we identify four heating values which are particularly interesting: at constant pressure or volume, and lower or higher.

In fact, heating values at constant pressure and volume are substantially equal, but enthalpies of vaporization are far from negligible (for water, it is approximately 45 MJ/kmol at 0°C). So there are significant differences between the values of fuel HHV and LHV, and it is important to specify which one is used.

In conclusion, therefore, to determine the energies put into play in a complete combustion reaction, you should use the fuel heating value, specifying whether it is the HHV or the LHV.

The heating values of specific fuels commonly used in internal combustion engines are fairly close to each other. One can indeed show that hydrocarbon LHV range is from 48 to 44 MJ/kg when we consider products of distillation of increasing molar weight.

For gaseous fuels, heating values are generally expressed in kJ/m³N. They are obtained by dividing the heat of reaction in kJ/kmol by the volume occupied by a normal kilomol (0°C, 1 bar), i.e., 22.414 m³.

In Thermoptim, the LHV and HHV values of gases are displayed in the pure and compound gas screens, in kJ/kg.

Calculation of combustion in thermoptim

The combustion screen (Figure 4.1) is the most complex of those of processes, given the number of existing options. You should refer to the end of Chapter 3 for an overview of the process screens.

Declaration of fuel

In case of excess air, non-stoichiometric complete combustion of fuel CH_a with atmospheric air is given by Equation (4.9). Thermoptim uses a generalized equation of this type, where the oxidizer can be any compound gas, including oxygen, and fuel is either given as CH_a, or declared as a pure or compound gas. The definition of fuel is in the upper-right of the screen (Figure 4.2).

If the fuel is a pure or compound gas, it may contain one or more of the following reactants: CO, H_2S, $C_nH_mS_pO_q$, n, m, p and q being decimal numbers less than 100. Inert gases taken into account are Ar, CO_2, H_2O, N_2 and SO_2. Note that the fuel is always assumed to be gaseous, which may cause a slight error if it is in the liquid state.

The fuel and oxidizer can both contain reactive oxygen and inert gases, even if it is normal that the fuel contains no oxygen, nor the oxidizer fuel. It is thus possible to calculate complex combustion, such as that of a fuel mixture made before introduction into the combustion chamber (fresh charge of a gasoline engine), or that of a synthesis gas.

The reaction products are CO_2, H_2O, SO_2 as well as CO and H_2 if there is dissociation, and fuel if the reaction is not complete.

The software analyzes the chemical formulas of the fuel and oxidizer components, and deduces the reaction that takes place. Calculations can then be executed. Chemical formulas are obtained by decoding the names of the substances.

When the fuel is represented by a Thermoptim substance, the name that appears on the screen ("fuel" in Figure 4.2) should be that of a process (e.g., a process-point) to determine its

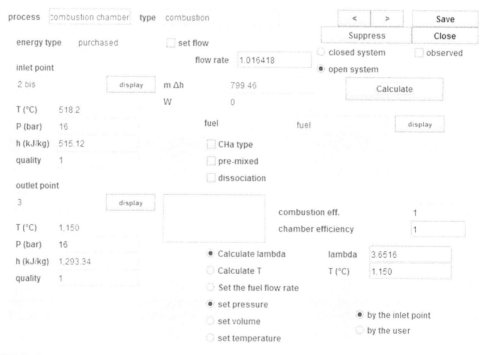

FIGURE 4.1

Combustion screen in Thermoptim.

FIGURE 4.2

Fuel process link.

component name	molar fraction	mass fraction
CH4 ` methane	0.871	0.758966
C2H6 ` ethane	0.088	0.1437279
C3H8 ` propane	0.025	0.05987759
C4H10 ` n-butane	0.008	0.02525591
N2	0.008	0.01217253

FIGURE 4.3

Fuel composition.

flow rate. The process is, of course, connected to the point specifying the substance name and settings.

The example below is taken from the Getting Started example[1] «Gas turbine». It relates to the natural gas available at the LNG terminal of Gaz de France at Montoir de Bretagne (France) and illustrates how to declare a fuel whose composition is given in Figure 4.3.

Each constituent, except for nitrogen, is a fuel, whose chemical formula appears in the first part of the name that sometimes is followed by a comment, separated from the formula by the character « ` ». By analyzing the chemical formula of each constituent and its molar fraction, Thermoptim completely characterizes the fuel.

This substance is associated with a point, allowing specification of its state, which is taken into account for the calculation of the outlet gas temperature. The flow rate is associated with a process-point, whose name serves as a parameter for the combustion.

To find the oxidizer flow rate, the software searches if there is a process whose outlet point is the inlet point of the combustion process, or if it is connected to a node. Messages inform the user if a problem occurs, either because there are several processes that lead to the inlet point, or because no process or node is connected to it.

When the fuel is given in the form of CH_a ("CH_a type" must be selected), it is necessary to enter the values of a, and of its enthalpy of formation h_{f0} (relative to the formulation CH_a). It is then impossible to specify the flow rate. You should, however, seldom use this setting.

Open systems and closed systems

It is necessary to specify if the combustion takes place in an open or a closed system.

For open systems, the pressure must be set to a certain value. It can be done either «by the user», which means that the value taken into account is that of the outlet point, or «by the inlet point», which corresponds to a combustion at constant pressure equal to that of the upstream oxidizer.

For closed systems, one can choose between three possibilities: set volume, set pressure or constant temperature combustion. For the first two cases, two different modes of setting, namely, the volume or the pressure, exist: «by the user» or «by the inlet point». In the third case, the combustion temperature is constant and equal to that of the upstream oxidizer.

In the last two cases (set pressure and constant temperature), the volume varies with combustion, so that a part of the energy liberated by the combustion transforms directly into mechanical energy due to the expansion of the volume.

---

[1] https://direns.mines-paristech.fr/Sites/Thopt/en/co/guide-prise-en-mains-tag.html

The first law of thermodynamics indicates indeed that internal energy variation (ΔU) is the algebraic sum of the heat ($-\Delta H_r\, \eta_{LHV}\eta_{th}$) and the work $\left(W = -\int P\,dv\right)$ received by the system.

Thermoptim calculates the value W and displays it just below the value of the energy liberated in the combustion. This value is subsequently taken into account as useful energy when the cycle balance is calculated. One can refer, for more details on this subject, to examples on diesel[2] and spark ignition[3] engines presented in the portal.

Characteristics of combustion

The dissociation of CO_2 to CO can be taken into account by checking the appropriate option. The dissociation reaction is given by equations:

$$CH_a + \lambda\left(1+\frac{a}{4}\right)\left(O_2 + 3.76N_2\right) \to \left(1-k_1\right)CO_2 + k_1 CO + \left(1-k_2\right)\frac{a}{2}H_2O + k_2\frac{a}{2}H_2$$

$$+\left\{\left(\lambda-1\right)\left(1+\frac{a}{4}+\frac{k_1}{2}+k_2\frac{a}{4}\right)O_2\right\} + 3.76\lambda\left(1+\frac{a}{4}\right)N_2$$

$$CO_2 + H_2 \leftrightarrow CO + H_2O \tag{4.12}$$

This equilibrium is independent of pressure, but is a function of temperature. With the assumption that the kinetics of combustion is sufficiently fast for the equilibrium to be reached, the law of mass action (4.2) can be written:

$$K_p = \frac{[CO][H_2O]}{[CO_2][H_2]} = \frac{k_1\left(1-k_2\right)}{k_2\left(1-k_1\right)} = f\left(T_q\right) \tag{4.13}$$

One can obtain an estimate of K_p from the approximate formula valid in limited temperature ranges:

$$\ln\left(K_p\right) = A - \frac{B}{T}$$

For reaction (4.12), in the range 1,000–2,200 K, covering much of the usual combustions, we take as values A = 3.387 and B = 3,753, T being expressed in Kelvin.

Thermoptim uses an approach of this type, but generalized.

If one chooses to take dissociation into account, a frame is displayed in which must be entered k_1, the dissociation rate of CO_2, and T_q, the quenching temperature which is used for calculating the constant K_p.

To account for any heat loss from the combustion chamber, not necessarily adiabatic, we introduce a thermal efficiency η_{th}, initialized to 1 by default (on the right at the center of the screen). This efficiency differs from the combustion efficiency η_{LHV}, calculated by the software based on the dissociation rate and the quenching temperature.

In the bottom left of the screen are located two fields: one corresponding to the air factor λ and the other to the temperature T_c at the end of combustion. It is possible to set either of these values, and calculate the other, or to calculate from both the fuel and oxidizer flows. The air factor may be higher or lower than 1. If it is below 1, the software considers this a lack of air combustion leading to the formation of carbon monoxide (CO). If the air ratio is too low for all available carbon to be oxidized to CO, a message warns the user.

---

[2] https://direns.mines-paristech.fr/Sites/Thopt/en/co/session-s38en-exercise.html.
[3] https://direns.mines-paristech.fr/Sites/Thopt/en/co/session-s39en-exercise.html.

Calculation options

The option «Calculate T» determines the outlet gas temperature T_c, from the value set for λ.

The mass flow rate of the combustion process is set equal to the sum of fuel and oxidizer flow rates, which means that the combustion process behaves, at the hydraulic level, as a flow rate mixer.

The option "«Calculate lambda»" determines λ, from the value of T_c set. Flow rate calculation rules are analogous to those of the preceding option. If the enthalpy released by the stoichiometric combustion does not allow it to reach the desired temperature, a message warns the user.

The option "Set the fuel flow rate" determines λ and T_c from the characteristics of the oxidizer and the fuel, which must be a gas. The mass flow of the combustion process becomes equal to the sum of the flows of fuel and oxidizer. If the fuel type is CH_a, nothing is done.

When combustion is calculated, the value of combustion efficiency η_{LHV} is determined.

Technological aspects

Technologically, energy production combustions take place in two general classes of devices: combustion chambers (reciprocating or continuous flow) and boilers. The first are used to produce high-pressure and temperature combustion gases that are then expanded, e.g., in a turbine. The latter realize simultaneously in the same chamber combustion and transfer to a fluid of the heat available in exhaust gases. We limit ourselves in what follows to a brief overview of these technologies.

Combustion chambers

For example, the combustion chamber of a gas turbine must satisfy severe constraints: ensure the complete combustion of fuel, minimize the pressure drop (which represents an increase in compression), ensure the good temperature stability at the turbine inlet and occupy a volume as small as possible while allowing proper cooling of the walls.

The chart in Figure 4.4 is a section of a flame tube-type combustion chamber, very commonly encountered in practice.

The compressed air exiting the compressor enters on the left side. It splits into two streams, one that provides wall cooling and the other that enters directly into the combustion chamber,

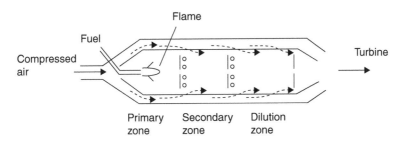

Cross-section of a gas turbine combustion chamber

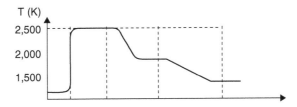

Temperatures in a gas turbine combustion chamber

FIGURE 4.4

Sketch of a combustion chamber.

where it serves as oxidizer for the fuel injected into the central part. Given the low excess air locally, the flame reaches a high temperature (up to 2,500 K) in the primary zone. Through holes at the periphery of the flame tube, the outside air mixes with exhaust gases in the transitional zone, where the temperature drops to around 2,000 K, and in the dilution zone, where one seeks to achieve a gas flow temperature as stable as possible to avoid the risk of local or momentary overheating.

In flame tube cylinder chambers, six to twelve tubes of this type are mounted in parallel around the axis of the gas turbine (Figure 2.11). They are interconnected in order to balance the pressures and enable the propagation of the ignition.

These flame tubes are very compact, and their dimensions amount to several tens of centimeters at most. Subjected to intense and high-temperature heat fluxes, the materials they contain are resistant steel sheets potentially coated with ceramic.

Figure 4.5 shows a diesel direct injection combustion chamber. The liquid fuel is atomized under high pressure and injected directly into the combustion chamber. The small drops of fuel ignite and burn in the air, which has been brought to high temperature during the compression phase. The detailed operation is explained in Chapter 9.

BOX 4.2
Key issues

Fundamental combustion notions

The fundamentals you should make sure to understand are the following: stoichiometry, air factor λ, CO_2 dissociation, quenching temperature and differences between HHV and LHV.

The following self-assessment activities will allow you to check your understanding of stoichiometric and non-stoichiometric combustion processes:

Stoichiometric combustion, gfe
Non-stoichiometric combustion, gfe

Boilers

Boilers are much larger than combustion chambers, because of the need to transfer fume heat to another fluid, which requires large exchange surfaces. In many applications, this fluid is pressurized water, which vaporizes inside the boiler, which then behaves like a triple heat exchanger as the water passes from liquid form (economizer), vaporizes (evaporator) and becomes steam (superheater).

A short presentation of boilers has been made at the beginning of Chapter 2 (Figures 2.4 and 2.5), and a specific section is devoted to them in Chapter 5.

BOX 4.3
Guided educational exploration

Exploration of a gas turbine model with combustion (exploration S-M3-V8-2)

The objective of this exploration is to guide you in your first steps of modeling a combustion in Thermoptim.

Heat exchangers

Heat exchangers are devices for transferring heat between two fluids at different temperatures. In most cases in energy conversion technologies, the two fluids are not in contact, and the transfer is through an exchange area.

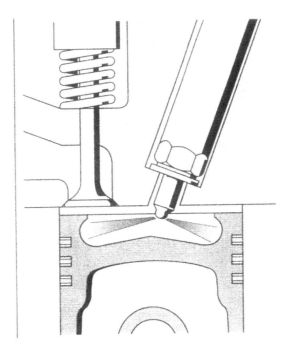

FIGURE 4.5
Diesel direct injection combustion chamber. (Courtesy Bosch.)

Within the dividing wall, the heat transfer mechanism is conduction, and on each of both surfaces in contact with fluids, convection almost always predominates. In many cases, fluids are single phase, whether gas or liquid.

However, there are three main types of heat exchangers in which phase changes occur: the evaporator where a liquid is vaporized, the condenser where a vapor is liquefied and evaporative condensers where both fluids change phase.

In practice, heat exchangers are of great importance in many energy systems in which heat is transferred.

The study of heat exchangers in an energy system can be done at very different levels of difficulty depending on the objectives that we pursue:

- The most basic approach, which is very simple to implement, is to balance the heat exchanger in terms of enthalpy by determining, for example, the outlet temperature of a fluid when the inlet temperatures and flow rates of the two fluids are known, as well as an outlet temperature;
- The problem can be posed in reverse, if the two outlet temperatures are known and only one inlet. The approach remains simple. One can indeed show that a heat exchanger has five degrees of freedom, and once five values are set among the inlet and outlet temperatures and flow rates, the sixth is directly deduced;
- It is also customary to characterize a heat exchanger by either an effectiveness or a minimum temperature difference between the two fluids (called pinch). Solving this problem is also fairly simple if four parameters are set. Note that, very often in the study of energy systems, heat exchanger calculations are not explored further. These three levels indeed already provide a sufficient information to conduct preliminary analysis of many cycles, especially at the functional level, apart from specific technological considerations. The theory shows that this approach allows us to characterize an exchanger by what we call the UA, the product of the internal heat exchange coefficient between the two fluids U by its exchange area A;
- To go further and determine the exchange surface required, calculations are much more complicated, requiring a thorough knowledge of the inner operation of heat exchangers, in order to estimate as precisely as possible the value of U. There is actually a very

important qualitative leap between the basic arithmetic of balancing the heat exchanger and what we call its technological design, which will only be addressed superficially in this book, especially in the context of guided explorations. A thorough discussion is, however, presented in Chapter 2 of Part 5 of the first edition.

In a heat exchanger, the fluid flows can be performed in multiple arrangements.

One can easily show that thermodynamically, the most efficient heat exchanger is the counter-flow heat exchanger (Figure 4.6), but other concerns than the thermodynamic effectiveness are taken into account when designing a heat exchanger: the maximum permissible temperatures in one fluid, or more often considerations of size, weight or cost.

It follows that the configurations of exchangers that are encountered in practice are relatively numerous.

However, we can gather these configurations in three main geometries:

* Counter-flow, in which fluids flow in parallel and in opposite directions;
* Parallel flow, in which fluids flow in parallel and in the same direction;
* Cross-flow, in which fluids flow in perpendicular directions.

In the following chapters, we will generally suppose that heat exchangers are counter-flow.

We will denote the hot fluid by index h, and the cold fluid by index c. Besides the geometric configuration, exchanger sizing or performance depends on many parameters and variables:

* Mass flows $\dot{m}_h$ and $\dot{m}_c$ passing through them;
* Inlet temperatures T_{hi} and T_{ci} and outlet temperatures T_{ho} and T_{co} of both fluids;
* Heat exchange coefficients U_h and U_c of each fluid;
* Thermal resistance of the wall;
* Surface A of the exchanger;
* Pressures of both fluids, almost constant;
* Thermophysical properties of fluids, which are used to determine coefficients U_h and U_c. The properties are essentially the heat capacity c_p, density ρ, thermal conductivity λ and viscosity μ.

In what follows, we will assume that the heat exchange coefficients U_h and U_c and thermophysical properties of fluids maintain a constant value at any time in the entire heat exchanger.

If this assumption is not verified, then in order to study the performance of the exchanger, it is necessary to divide it into small volume elements, in which these properties can be considered constant. The calculations are then much more cumbersome.

Finally, we always assume that the heat exchanger is globally adiabatic; that is to say, there is no heat exchange with the surroundings.

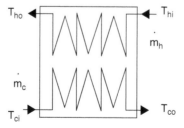

FIGURE 4.6
Heat exchanger sketch.

Heat flux exchanged, UA-LMTD method

It can be shown, where fluid flow is pure (parallel or counter-flow), that the heat flux exchanged between the two fluids is given by the formula:

$$\phi = UA\,\Delta T_{lm} \qquad\qquad (4.14)$$

$$\text{with} \quad \Delta T_{lm} = \frac{\Delta T_0 - \Delta T_L}{\ln\dfrac{\Delta T_0}{\Delta T_L}} \qquad\qquad (4.15)$$

A being the exchange area, U the overall heat exchange coefficient, ΔT_{lm} the **logarithmic mean temperature difference**, and ΔT_0 and ΔT_L the differences in fluid temperatures, respectively, at one end and at the other end of the exchanger, with the convention: $\Delta T_0 > \Delta T_L$.

Care must be taken that we speak of temperature differences between the two fluids at the entrance and exit of the exchanger (dimensions 0 and L), and not differences of inlet temperatures ($T_{hi} - T_{ci}$) and outlet temperatures ($T_{ho} - T_{co}$) of the fluids.

In the example in Figure 4.7, we have:

$$\Delta T_0 = T_{hi} - T_{co}$$

$$\Delta T_L = T_{ci} - T_{ho}$$

This formula shows that to transfer a given flux ϕ, if we wish to reduce the irreversibilities and therefore ΔT_{lm}, we need the product UA to be the largest possible, which can be done **either by increasing the surface**, but this directly affects the cost, **or by increasing the U value**, which is an aim sought by all heat exchanger designers.

In this chapter, the design of the heat exchanger will be mainly limited to determining the product UA of the exchanger surface A by the overall heat transfer coefficient U; the accurate estimation of the latter depends on the internal constructive details of the exchanger.

In general, the design of heat exchangers is a compromise between conflicting objectives, whose two main ones are:

- A large exchange area is desirable to increase the effectiveness of heat exchangers, but it results in high costs;
- Small fluid flow sections increase the values of the heat exchange coefficients U_h and U_c, and thus reduce the area, but they also increase pressure drops, which, as already said, generally induces additional compression work or reduced expansion work.

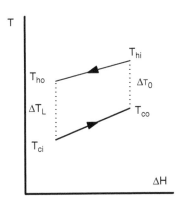

FIGURE 4.7

LMTD notations.

To size a heat exchanger, it is possible to use relation (4.14), but it assumes the logarithmic mean temperature difference is known, which is rarely the case.

Number of transfer units, effectiveness-NTU method

A method simpler to use, and especially for more general use, is the effectiveness-NTU, or the number of transfer units (NTUs), the method developed by Kays and London (1984). This method has the advantage of requiring the knowledge of fluid inlet temperatures, and not those at the outlet. In what follows, we will indicate with the index min, respectively, max, the minimum and, respectively, maximum values of the quantities relative to each of the two fluids.

We can show that ϕ_{max}, which would be obtained for a counter-flow heat exchanger of infinite length, is:

$$\Phi_{max} = \left(\dot{m}C_p\right)_{min} \Delta T_i$$

ΔT_i is the difference between inlet temperatures of both fluids.

By definition, NTU is defined as the ratio of the product UA of the heat exchange coefficient U by the area A to the minimum heat capacity rate.

$$NTU = \frac{UA}{\left(\dot{m}C_p\right)_{min}} \tag{4.16}$$

We call R the ratio (less than 1) of the heat capacity rates:

$$R = \frac{\left(\dot{m}C_p\right)_{min}}{\left(\dot{m}Cp\right)_{max}} < 1 \tag{4.17}$$

and ε the **effectiveness of the exchanger, defined as the ratio between the heat flux actually transferred and the maximum possible flux:**

$$\in = \frac{\Phi}{\Phi_{max}}$$

Figure 4.8 gives the appearance of the temperature profile in a heat exchanger. Note that here the abscissa used is the enthalpy exchanged, not the length of the exchanger, which explains, with the assumption that the heat capacities are constant, that this profile is linear, not exponential.

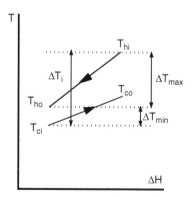

FIGURE 4.8

Effectiveness notations.

Note also that ΔT_{max} and ΔT_{min} are the temperature differences within each of the two fluids and not between them as in Figure 4.7.

As $\quad \Phi = \left(\dot{m}C_p\right)_{min} \Delta T_{max} = \left(\dot{m}C_p\right)_{max} \Delta T_{min}$ (4.18)

We have $\quad \epsilon = \dfrac{\Delta T_{max}}{\Delta T_i}$ (4.19)

Since $\phi = UA\,\Delta T_{lm}$

$$NTU = \frac{UA}{\left(\dot{m}C_p\right)_{min}} = \frac{\Delta T_{max}}{\Delta T_{lm}}$$ (4.20)

NTU measures the ability of the exchanger for changing the temperature of the fluid whose capacity rate is lower.

This is indeed an indicator of the quality of the equipment: the more efficient the heat exchanger, the more it allows for heating or cooling a fluid with a low temperature difference compared with the other.

For example, in the case of Figure 4.8, where we have:

$$\left(\dot{m}C_p\right)_{min} = \left(\dot{m}C_p\right)_h$$

$$\Delta T_i = T_{hi} - T_{ci}$$

$$\Delta T_{max} = T_{hi} - T_{ho} \quad \text{and}$$

$$\epsilon = \frac{T_{hi} - T_{ho}}{T_{hi} - T_{ci}}$$ (4.21)

With these definitions, it is possible to show that there is a general relation of type:

$$\epsilon = f(\text{NTU}, \text{R}, \text{flow pattern})$$ (4.22)

In practice, it suffices to have a set of relationships corresponding to flow patterns representative of exchangers studied, and the design of a heat exchanger is made on the basis, on the one hand, of the balance equations (4.18) and, on the other hand, of the internal equation (4.22).

If we know the flow rates of both fluids, their inlet temperatures and the heat flux transferred, the procedure is as follows:

* We start by determining the outlet temperatures of fluids from Equation (4.18);
* We deduce the fluid heat capacity rates $\dot{m}C_p$ and their ratio R;
* The effectiveness ε is calculated from Equation (4.19);
* The value of NTU is determined from the appropriate (NTU, ε) relationship;
* UA is calculated from Equation (4.16).

If you know UA, (4.16) gives the value of NTU, and you can determine ε from the appropriate (NTU, ε) relationship. Equation (4.18) allows you to calculate the heat transferred.

In both cases, the enthalpy balance provides the outlet temperatures. This method therefore connects the UA product to the exchanger performance, and the surface is deduced from an estimate of the value of the exchange coefficient U.

As shown in (4.21), ε measures the "temperature" effectiveness of the exchanger, i.e., the fraction of the maximum difference in temperature between the fluids which is communicated to that with the lower heat capacity rate.

The value of the method proposed by Kays and London is its generality because of the formal appearance of $\left(\dot{m}C_p\right)_{min}$ and $\left(\dot{m}C_p\right)_{max}$.

Phenomenological (ε, NTU) models provide only the exchanger UA product. As we said, the calculation of U depends on many factors they do not take into account.

Given its advantages, this method has been very successful and is now widely used. It is often even generalized to size heat exchangers other than those we have considered here, where it will be recalled we assumed that heat exchange coefficients U_h and U_c and fluid thermophysical properties keep a constant value.

Experience shows that, on a thermodynamic level, (ε, NTU) models generally represent well the behavior of many types of exchangers, even more complex than those we have studied so far.

Relationship between NTU and ε

Here, we give the relations between ε and NTU function of R for counter-flow heat exchangers, and their inverses, which can be expressed simply, as well as the corresponding graph in which curves represent ε function of NTU, for R varying from 0 to 1, top to bottom, by steps of 0.25.

For a counter-flow heat exchanger, analytical expressions are, for $R \neq 1$:

$$NTU = \frac{1}{1-R}\ln\left(\frac{1-\epsilon R}{1-\epsilon}\right) \tag{4.23}$$

$$\frac{\varepsilon = 1 - \exp\left(-NTU(1-R)\right)}{1 - R\exp\left(-NUT(1-R)\right)}$$

For $R = 1, \varepsilon = \dfrac{NTU}{1+NTU}$

The corresponding chart is given in Figure 4.9. Other correlations for parallel-flow, cross-flow and shell-and-tube heat exchangers can be easily found in the literature or in the first edition of this book (Sections 8.2.2.2 and the following).

Relationship with the LMTD method

The log-mean temperature difference method (LMTD) stems from Equation (4.14):

$$\phi = UAF\Delta T_{lm}$$

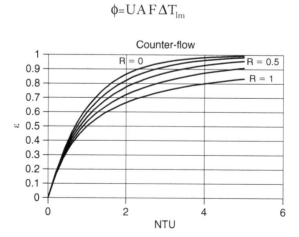

FIGURE 4.9

Counter-flow heat exchanger chart.

It is possible to switch from a calculation method to another using the relationship:

$$\Delta T_{lm} = \frac{\varepsilon \Delta T_i}{NTU}$$

BOX 4.4
Key issues

Effectiveness-NTU method

The following self-assessment activity will allow you to check your understanding of the effectiveness-NTU method:

Effectiveness-NTU method, gfe

Heat exchange coefficient U

The previous methods allow one to calculate the product UA of the heat exchange coefficient U by the heat exchanger surface A.

The precise determination of U and the study of pressure drops in heat exchangers are presented in Chapter 35 of Part 5 of the first edition. We will only present here the principles governing the calculation of U.

Thermal science teaches us that for a flat-plate heat exchanger, the overall heat transfer coefficient U is such that its inverse, called thermal resistance, is the sum of thermal resistances between the two fluids:

$$\frac{1}{U} = \frac{1}{h_h} + \frac{e}{\lambda} + \frac{1}{h_c} \tag{4.24}$$

with h_h being the heat transfer coefficient between the hot fluid and the wall, h_c the heat transfer coefficient between the cold fluid and the wall, e the wall thickness and λ the thermal conductivity of this wall.

If the wall is composed of several layers of different materials, or if some deposits have covered the wall, this formula becomes:

$$\frac{1}{U} = \frac{1}{h_h} + \sum_{k=1}^{n} \frac{e_k}{\lambda_k} + \frac{1}{h_c} \tag{4.25}$$

The overall thermal resistance is the sum of $(n + 2)$ thermal resistances – the largest of them is the one that slows down the more heat that is exchanged.

For example, in the case of a gas/liquid heat exchanger, the convection coefficient can be 10–100 times lower on the gaseous side than on the liquid side. The thermal resistance of the wall and the liquid side is generally negligible, and U is approximately equal to h gas side.

In this case, we seek to increase the gas side exchange surface by using fins. The previous formulas are complicated, and as the surfaces of cold and hot sides are not the same, we introduce a warm side global exchange coefficient U_h, such that its inverse is:

$$\frac{1}{U_h} = \frac{1}{\eta_{0,h} h_h} + \frac{e}{\frac{A_w}{A_h} \lambda} + \frac{1}{\frac{A_c}{A_h} \eta_{0,c} h_c} \tag{4.26}$$

with A_h and A_c being total exchange areas on hot and cold sides, A_w exchanger wall surface, and $\eta_{0,h}$ and $\eta_{0,c}$ overall fin effectiveness on hot and cold sides.

There is also a cold-side global exchange coefficient U_c, with of course:

$$U_c A_c = U_h A_h$$

Fin effectiveness

For a thin rectangular fin ($e \ll l$, Figure 4.10), we can show that the fin effectiveness, defined as the ratio of heat actually exchanged to the heat that would be exchanged if the fin was at the temperature of the base, is:

$$\eta_a = \frac{\tanh(ml)}{ml}$$

$$\text{with} \quad m = \sqrt{\frac{2h}{\lambda e}} \tag{4.27}$$

For fins of different shapes, the value of m changes. For example, for circular needle fins (Figure 4.11):

$$m = \sqrt{\frac{4h}{\lambda d}} \tag{4.28}$$

For other geometries, we refer to the literature (Incropera & Dewitt, 1996).

The total flux exchanged is equal to the sum of the fluxes exchanged, on the one hand, by the fins (area A_a) and, on the other hand, by the wall between the fins (area $A - A_a$). This defines an overall effectiveness η_0 equal to the ratio of the flux actually exchanged to the flux that would be exchanged if the entire surface was at the temperature of the base:

$$\eta_0 = \frac{hA_a \eta_a + h(A - A_a)}{hA} = 1 - \frac{A_a}{A}(1 - \eta_a) \tag{4.29}$$

Values of convection coefficients h

The values of convection coefficients h_h and h_c depend on fluid thermophysical properties and exchange configurations. Further explanations are given in Chapter 35 on technological design of the first edition of this book.

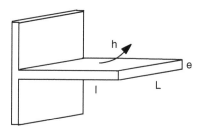

FIGURE 4.10
Rectangular fin.

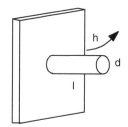

FIGURE 4.11
Needle fin.

These coefficients can be estimated from correlations giving the value of Nusselt number $Nu = \dfrac{hd_h}{\lambda}$, as a function of Reynolds $Re = \dfrac{\rho V d_h}{\mu}$ and Prandtl $Pr = \dfrac{\mu C_p}{\lambda}$ numbers, in which λ is the thermal conductivity and μ is the dynamic viscosity.

The hydraulic diameter d_h is equal to the ratio of four times the flow area S to the wetted perimeter p: $d_h = 4 \, S/p$ (if there are isolated walls, the perimeter of heat exchange must be considered).

Inside tubes, the formula most used is that of Mac Adams:

$$Nu = 0,023 Re^{0,8} Pr^{0,4} \tag{4.30}$$

For flow perpendicular to tubes, several options exist, such as Colburn:

$$Nu = 0,33 Re^{0,6} Pr^{0,33} \tag{4.31}$$

In general, the exponent of the Reynolds number is between 0.5 and 0.8, and that of the Prandtl number between 0.33 and 0.4.

Off-design calculation of heat exchangers

A behavioral model of a heat exchanger must answer the following question: how does the exchanger adapt when its input variables are changed? These are its inlet temperatures T_{hi} and T_{ci} and mass flows $\dot{m}_h$ and $\dot{m}_c$ passing through. A behavior model should determine the outlet temperatures T_{ho} and T_{co} of both fluids based on these four input variables. There are two more unknowns than for the sizing problem, but, as the heat exchanger is selected, we know its area A.

Chapter 35 of Part 5 of the first edition specifically addresses this issue. In this section, we limit ourselves to some elementary remarks.

(ε, NTU) models can be used as global models: for a given heat exchanger, if one knows the NTU (or UA) evolution law based on flow rates of both fluids passing through the exchanger, a (ε, NTU) model determines the evolution of ε when the operating conditions of the heat exchanger vary.

In Thermoptim, the "off-design" calculation mode allows one to calculate in a simplified manner off-design operation of the heat exchanger by the NTU method. In this case, the software considers that the four input variables of the exchanger are set. No correction is made on the exchange coefficients, but it is possible to modify by hand the UA value in the exchanger screen or to use a driver to do it automatically.

Guided exploration DTNN-1 explains how to proceed practically.

BOX 4.5
Guided educational exploration

Technological sizing of an air–water exchanger (exploration DTNN-1)

In this guided exploration, you will learn how the surface of a heat exchanger can be determined and how its behavior in off-design conditions can be calculated.

Heat exchanger pinch

A heat exchanger pinch is defined as the minimum temperature difference between both fluids. For a simple exchanger without phase change, it is:

$$\left(T_{ho} - T_{ci}\right) \quad \text{if} \quad \left(\dot{m}C_p\right)_h \leq \left(\dot{m}C_p\right)_c$$

$$\left(T_{hi} - T_{co}\right) \quad \text{if} \quad \left(\dot{m}C_p\right)_h \geq \left(\dot{m}C_p\right)_c$$

Consider the graphical representation of a double exchanger consisting of an economizer followed by an evaporator, in the $(T, \Delta H)$ graph, where the temperature is the ordinate and the enthalpy is the abscissa (Figure 4.12).

The fluid that evaporates is represented by the lower curve (1–2–3), which presents an angular point 2 corresponding to the beginning of the boiling. The hot fluid cools from 4 to 5, by decreasing its sensible heat.

On segments (1–2) and (4–5), we have $\Delta H = \dot{m}C_p\Delta T$,

and thus, $\Delta T = \dfrac{\Delta H}{\dot{m}C_p}$.

The slopes of segments (1–2) and (4–5) are equal to the inverse of the heat capacity rates of the corresponding fluids. For segment (2–3), we obviously have $\Delta T = 0$, because vaporization takes place at constant temperature for a pure component or azeotrope.

At point 2 on the diagram, P is the minimum temperature difference between the two fluids, called pinch. This point plays a fundamental role in the design of heat exchangers, as it represents the smallest difference in temperature in the device, and thus corresponds to the most constrained zone of the system.

The importance of the pinch is most evident in problems of complex heat exchanger network design, to the point that powerful optimization methods based on this concept have been developed. The pinch method implemented in Thermoptim is presented in Chapter 7.

The exchanger, sized according to the pinch, operates outside this point with temperature differences greater than the minimum pinch between the two fluids. The irreversibilities depend directly on the difference ΔT between the hot and cold fluids. They are even lower than the slopes of hot and cold fluids are of similar magnitude, that is to say that the ratio R of their heat capacity rates is close to unity. The heat exchanger that has the least irreversibility is the one whose heat capacity rates are equal and whose temperature difference is equal to the minimum allowable pinch.

Calculation of heat exchangers in thermoptim

In Thermoptim, a heat exchanger is not represented by a particular component, but by a connection established between two "exchange" processes representing the hot and the cold

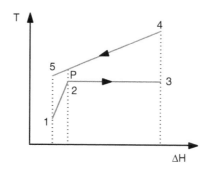

FIGURE 4.12

Pinch point.

fluids. This way of working has many advantages. In particular, it is possible to design fluid networks at the hydraulic level, to calculate their thermodynamic state and to thermally connect them only at a later time.

Creation of a heat exchanger in the diagram editor

To represent heat exchangers in the diagram editor, we use nonoriented links connecting two components of type "exchange".

An "exchange" component has in its center a small maskable exchanger connection port (Figure 4.13), which appears (as a small blue or red circle) only when the mouse is located above it, or when it is connected with another such component to represent a heat exchanger.

To make the connection, click on this port of one component and drag the mouse to the same port of another component and then release the button. During the connection, the name of the exchanger is asked for. Once the heat exchanger settings are entered, the port is blue for the cold fluid and red for the hot one.

A heat exchanger is created in the simulator when an exchanger link exists in the diagram and the two "exchange" components that it connects correspond to two "exchange" processes in the simulator which are sufficiently well defined (it is in particular necessary that their inlet and outlet temperatures are different, that they have been calculated so that the value of ΔH is not zero and that one is warming while the other cools). By default, the exchanger is initialized with a "counter-flow" type.

Heat exchanger screen

The connection is done in two stages: one for the hot part and one for the cold part.

The screen (Figure 4.14) contains information about the hot fluid in the middle left part, while that for the cold fluid is in the right part. To connect a fluid, if not done in the diagram editor, you must double-click the name field, then choose the process from the list. Besides the values of temperatures, flow rates, heat capacities and enthalpies put into play, constraints exist on the temperatures and flow rates used to manage the heat exchanger calculations, allowing one to distinguish among the variables of the problem, those set and those to be calculated.

Possible types of exchangers are counter-flow, parallel, cross-flow, mixed or not, and multi-pass shell and tube (p-n), but we will limit ourselves to counter-flow, as indicated before.

In the bottom left, there are three buttons to optionally specify the absence or presence of implicit constraints on temperatures (see below). In the lower right are placed options for defining the calculation mode ("design" or "off-design").

The heat exchanger calculation results are shown at the bottom of the central part: UA, R, NTU and LMTD.

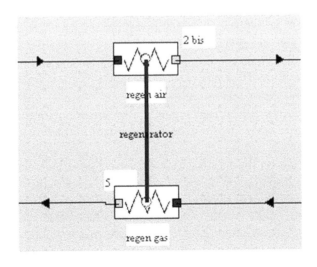

FIGURE 4.13
Exchanger link.

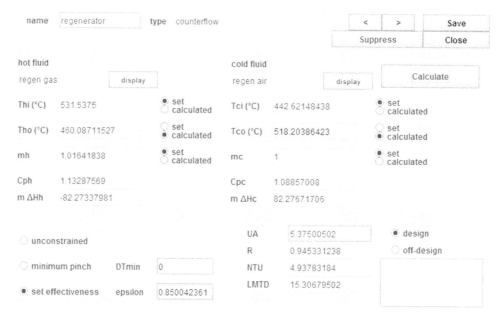

FIGURE 4.14

Heat exchanger screen.

Simple heat exchanger design

A heat exchanger connects two « exchange » processes. One of them is a hot fluid which corresponds to a substance that gets colder, while the other is a cold fluid that gets hotter. Once the coupling of the processes is made, the design problem is the following: it is necessary, on the one hand, to conserve enthalpy in the heat exchanger and, on the other hand, to respect some temperature constraints.

Given that there are four temperatures (two for each fluid) and two flow rates, **the problem has five degrees of freedom once the enthalpy is conserved**. One can furthermore show that **at least one of the two flow rates has to be set**; otherwise, the problem is indeterminable.

For temperatures, one can either set explicit constraints (for example, the fluid inlet temperatures) or set implicit constraints (the value of the heat exchanger effectiveness, or the pinch minimal value). To set an effectiveness value, enter it in the ε field, and select option «set effectiveness». To set a «minimal pinch», enter its value in the corresponding field, and select this option.

As explained above, for the problem to have a solution, it is necessary to set a total of five constraints, among which there is a flow rate. If one of them is implicit (effectiveness or minimal pinch), four have to be explicit (three temperatures and one flow rate, or two temperatures and two flow rates), or otherwise, all five have to be explicit (a single flow rate or a single temperature is free). These conditions are necessary, but not sufficient. This is why the software analyzes all the constraints proposed. If there is a solution, it is found. Otherwise, a message tells the user that the calculation is impossible.

Remember that the design of heat exchangers is always made with the implicit hypothesis that thermophysical properties of the fluids remain constant in the heat exchanger. However, this hypothesis is not made during the calculation of the processes. Therefore, when one recalculates a temperature on the basis of heat exchanger equations, slight differences may exist between the values displayed on the heat exchanger screen and that of the corresponding processes, or between the values of the enthalpies put into play in both hot and cold fluids. If a high accuracy is required, one may have to make first a heat exchanger design calculation, then iterate. Convergence is obtained quickly. Also, note that if one of the two flow rates is not set in the exchanger, it is recalculated, even if the corresponding process is "set flow".

When a heat exchanger is designed, if one of the flow rates is calculated, that of the exchange process is recalculated, as well as those of the processes located upstream as long as they are directly connected. The propagation stops when a node or a combustion process is encountered.

Even if there are many configuration options available on the exchanger's screen, those that appear in practice fall into only a few categories.

Most generally, the inlets of the exchanger are known, that is to say the temperatures and the flow rates of the two fluids T_{ci}, m_c, T_{hi}, m_h, that is to say four constraints. It therefore remains to set the fifth one.

It can be depending on the case:

- An outlet temperature (for example, in the study of the steam generator of a steam power plant cycle, the state of each point of the steam cycle is known, including the outlet temperatures of the three elements of the exchanger, namely, the economizer, the evaporator and the superheater, and the design consists in determining the value of the temperature of the hot fluid at the outlet of each of these elements;
- The fifth constraint can also be an effectiveness, as, for example, in the study of a gas turbine regenerator;
- Or the pinch value, i.e., the minimum temperature difference between the two fluids.

Thermocouplers

Thermocouplers complete heat exchangers by allowing components other than exchange processes to connect to one or more exchange processes to represent a thermal coupling. This mechanism does not encompass the exchanger mechanism: two exchange processes cannot be connected by a thermocoupler.

This mechanism has a number of benefits, because it can be used to represent many thermal couplings that do not constitute a heat exchange in the traditional sense, like, for example, cooling the walls of the combustion chamber of a reciprocating engine, cooled compression, and above all supply or removal of heat from multifunctional external components.

Figure 4.15 is an illustration of this: an absorption refrigeration cycle, whose absorption–desorption system is defined and integrated in an external process, is fed with a vapor

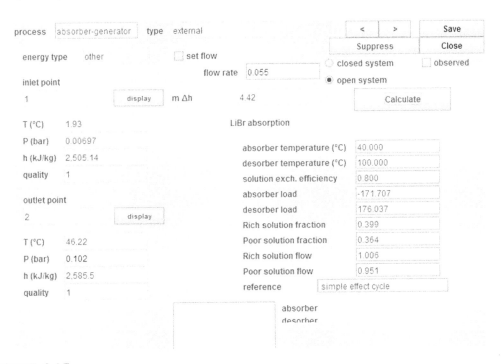

FIGURE 4.15

Absorber-generator screen.

that exits the evaporator and then enters the condenser. This cycle makes use of mixture LiBr-H$_2$O presented in Chapter 13, whose properties are modeled either directly in the external process or in an external substance, and requires high-temperature heat supply to the desorber and medium temperature heat removal from the absorber. The representation of these heat exchanges is possible thanks to the thermocoupler mechanism: the external process calculates the thermal energies that must be exchanged, and the thermocoupler recalculates the corresponding "exchange" process, which updates its downstream point.

The types of thermocouplers used by an external component appear in the lower right-hand corner of the screen. Double-click on one of the types to open the screen of the corresponding thermocoupler.

Given that thermocouplers are a type of heat exchanger, it is valuable as stated before to define them by values such as effectiveness ε, UA, NTU or LMTD, which can be calculated using similar equations. The component sends to each of its thermocouplers the equivalent values for flow rates, inlet and outlet temperatures, and thermal energy transferred, which they must take into account in their calculations. Specific methods are provided in the external class code and are not user-modifiable.

However, there are limits to the similarities with exchangers: for example, temperature crossovers unacceptable in an exchanger may occur in a thermocoupler, leading to absurd values.

So it is best to transmit values that are unlikely to lead to this type of situation. One possible solution is to assume that the thermocoupler is isothermal for calculations of characteristics that are similar to exchanger characteristics. For example, a combustion chamber may be assumed to be at mean temperature between its upstream and downstream points when calculating the cooling. This assumption is not absurd and may prevent a temperature crossover between the cooling fluid and the gases that cross the component.

In the case of the absorption machine presented above, we assumed that the absorber and the desorber were isothermal.

The absorber thermocoupler screen is given in Figure 4.16. If we had not taken the temperature of the absorber as a reference for the exchange calculations, keeping the temperatures of the steam entering and exiting the external process, we would have ended up with a temperature crossover.

For external processes and nodes that accept several thermocouplers, the potential complexity of the calculations prevents the exchange process from driving the thermocoupler. Its

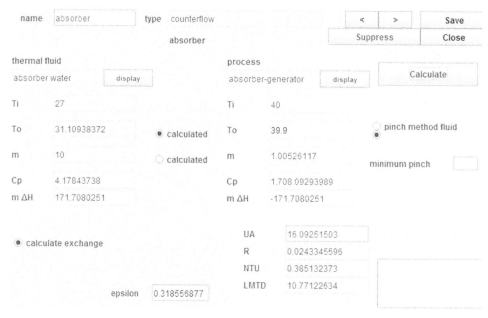

FIGURE 4.16

Thermocoupler screen.

thermal load is always set by the external component. This is why there are fewer options for calculating a thermocoupler than for a heat exchanger: the user can only choose between calculating the outlet temperature of the exchange process (at a given flow rate) and the flow rate, when the temperature is set.

Note that on the thermocoupler screen, the external component fluid can be selected as a pinch fluid and a minimum pinch value can be entered (see optimization method, Chapter 7).

Finally, note that the off-design calculation mode is not available for thermocouplers, as their behavior can deviate significantly from that of a conventional heat exchanger.

Technological aspects

There are a wide variety of heat exchangers developed to solve different problems. Capacities range from a few watts (electronics) to several hundred megawatts (condensing power plants). Uses relate as well to industry as to the residential and tertiary sectors, agriculture and transport. The temperature range varies from a few Kelvin for cryogenic applications to over 1,000°C for some applications. The materials are usually metal, but also plastic or ceramic. Among the common configurations, the following can be distinguished.

Tube exchangers

These are cheaper and thus more widespread. Among them, two categories are particularly used for energy applications: the shell-and-tube exchangers and air coils.

Shell-and-tube exchangers

Being very robust and economical, they are particularly used for liquid–liquid or liquid–vapor exchange (Figure 4.17). In the energy field, applications of this type of exchanger are numerous (hot gas heat recovery, evaporator and condenser of refrigeration machines etc.) (Figures 4.18 and 4.19).

A bundle of parallel tubes is welded or bolted at both ends to the thick perforated plates and traversed by one of the two fluids, while the other flows outside the tubes, in different modes depending on the type of exchanger. At both ends, boxes distribute or collect the fluid that passes inside the tubes, while a shell, usually cylindrical, ensures the confinement of the other fluid. The tubes may be arranged in a triangular or square pitch, and the second is more compact and better in terms of heat exchange, but more difficult to clean. These exchangers

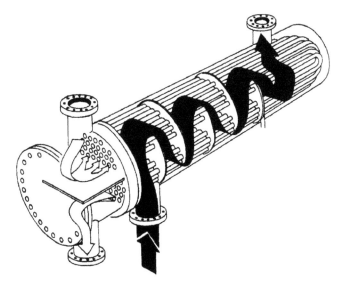

FIGURE 4.17
Shell-and-tube heat exchanger. (Courtesy Techniques de l'Ingénieur.)

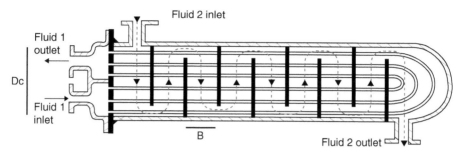

FIGURE 4.18
Cutaway of a shell-and-tube heat exchanger.

FIGURE 4.19
View of a shell-and-tube heat exchanger. (Courtesy French Navy.)

are called (p, n), with figure p representing the number of passes of the fluid outside the tubes and figure n being the number of passes of the fluid inside the tubes.

Technologically, the disadvantages of these exchangers are first that they represent a non-evolutive geometry, second that they can be difficult to clean, and finally that they are unsuitable if the temperature difference between the two fluids exceeds 50°C, as the difference in the dilatation of the constituent metals results in high stresses.

Air coils

Finned heat exchangers, called air coils, or just coils, are used when a fluid is a gas and the other a liquid or a fluid which is vaporized or condensed. An example of a cooling coil is given in Figure 4.20.

In these exchangers, the gas-side convective coefficients are much smaller than others, and the addition of fins increases the exchange surface and thus thermally balances the heat exchanger (Figure 4.21).

Plate heat exchangers

They are composed of embossed plates of various profiles, in which the two fluids exchanging heat pass alternately. They have a variable but highly developed exchange surface, which enables a very compact design (Figures 4.22 and 4.23).

More expensive than tubular heat exchangers, they offer better exchange coefficients and can (for those which are not welded but gasketed) be modified by changing the number of plates, making them scalable.

Plate-fin heat exchangers are a variant with extended surfaces primarily used for gas–gas heat transfer. Corrugated fins are placed between plates in order to enhance heat transfer and give stiffness.

FIGURE 4.20

Cooling coil. (Courtesy French.)

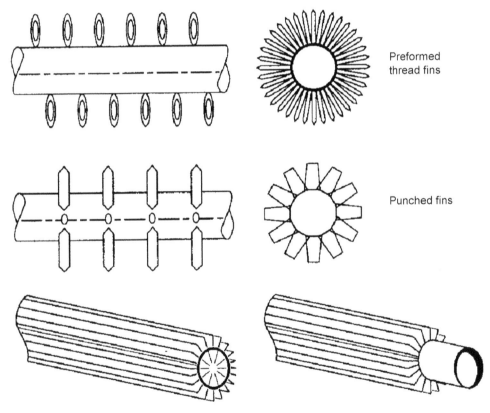

Preformed thread fins

Punched fins

FIGURE 4.21

Examples of fins. (Courtesy Techniques de l'Ingénieur.)

Other types of heat exchangers

There are many other types of heat exchangers, but they are little used for energy conversion, so we will not detail them. Let us include as a reminder heat pipe exchangers (Swanson, 2000), (Figure 4.24), often used between two gas flows but also very expensive, or regenerator rotary or static exchangers, where the two fluids exchange heat through a solid matrix.

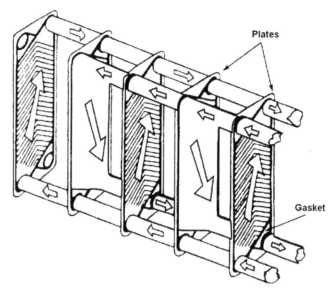

FIGURE 4.22

Flows in a plate heat exchanger. (Courtesy Techniques de l'Ingénieur.)

FIGURE 4.23

Plate heat exchanger. (Courtesy French Navy.)

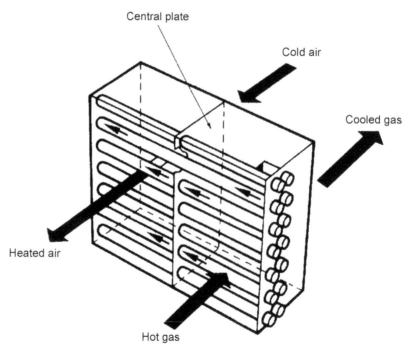

FIGURE 4.24

Heat pipe. (Courtesy Techniques de l'Ingénieur.)

BOX 4.6

Key issues

Fundamental heat exchanger notions

A heat exchanger can be modeled by the NTU method. While it is perfectly appropriate for studying the insertion of a heat exchanger in a thermodynamic cycle, such a phenomenological model, however, only gives access to the heat exchanger UA product, while the assessment of U can be particularly complex, as indicated. The success of this model is such that it is often used as a behavior model for a particular heat exchanger by adding a UA evolution law, depending, for example, of fluid flows through the exchanger.

The concepts you should make sure you understand are UA, effectiveness ε and NTU, as well as pinch.

BOX 4.7

Guided educational exploration

Exploration of a simple steam plant condenser (exploration S-M3-V7-2)

The objective of this exploration is to guide you in your first steps of setting a heat exchanger in Thermoptim.

Bibliography

M. W. Chase et al., Janaf thermochemical tables, *Journal of Physical and Chemical Reference Data*, 14 (Suppl. 1), 1985.

I. Glassman, *Combustion*, 3rd Edition, Academic Press, Inc, Dec 1996.

E. M. Goodger, *Combustion calculations, theory, worked examples and problems*, The MacMillan Press, London, 1977.

F. P. Incropera, D. P. Dewitt, *Fundamentals of heat and mass transfer*, 4th Edition, John Wiley and Sons, New York, 1996.

S. Kakac, H. Liu, *Heat exchangers, Selection, rating, and thermal design*, 2nd edition, CRC Press, Boca Raton, 2002, ISBN 0-8493-0902-6.

W. M. Kays, A. L. London, *Compact Heat Exchangers*, Mac Graw Hill, New York, 1984.

K.K. Kuo, *Principles of combustion*, John Wiley & Sons, Oct. 2000.

D. Spalding, *Combustion and mass transfer*, Pergamon Press, Oxford, 1979.

L. W. Swanson, *Heat pipes, The CRC handbook of thermal engineering*, (Edited by F. KREITH), CRC Press, Boca Raton, 2000, ISBN 0-8493-9581-X.

Steam systems components

Introduction

In this chapter, we provide supplements on several types of components commonly used in steam systems:

* We first discuss boilers and steam generators, with a special attention to the design constraints of heat recovery steam generators (HRSGs);
* The second section gives additional information on turbines, specifically steam turbines, presenting the different types of turbines and their characteristics (back pressure, condensing, extracting, impulse, reaction…);
* The major part of this chapter deals with the behavior of cooling towers, which are a particular type of heat exchanger that allows heat to be rejected into the surrounding air in the form of both sensible heat and latent heat due to the increase in its humidity. Cooling towers are widely used in practice, which justifies these developments. Various models used by manufacturers and design engineering offices are presented;
* Lastly, we explain the mechanism of external classes which allows one to extend the modeling possibilities of Thermoptim, illustrating it by the example of direct cooling towers.

Boiler and steam generators

Boilers

There are two main types of boilers, known from the fluid that circulates inside the tubes: fire-tube boilers and water-tube boilers.

* In the first type (Figure 5.1), the flame develops in a corrugated tube, then flue gases pass inside tubes, in one or more passes, with water being on the outside;
* Within the second type (Figure 5.2), water circulates by natural or forced convection between two drums placed one above the other, through a network of tubes (except in supercritical boilers which will be studied in Chapter 8). The flame develops in a furnace lined with tubes that absorb the radiation. A second tube bundle receives its heat from the flue gases by convection. The water rises in the tubes subjected to radiation and falls by the convection assembly.

Fire-tube boilers can achieve flue gas temperatures lower (220°C–250°C) than water-tube boilers (300°C) without an economizer, which gives them a slightly better effectiveness.

However, the former are limited to capacities lower than the latter, for reasons of mechanical strength and safety (very large volume of water under pressure).

DOI: 10.1201/9781003175629-5

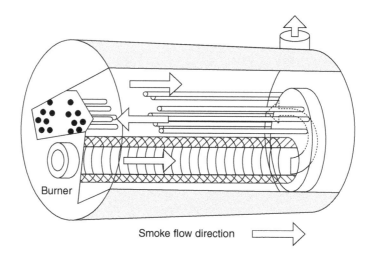

FIGURE 5.1
Fire-tube boiler.

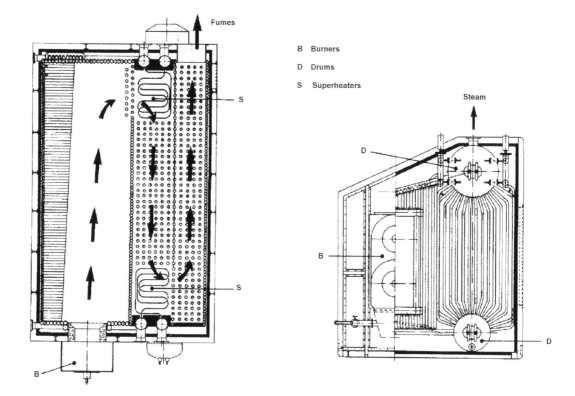

FIGURE 5.2
Water-tube boiler. (Courtesy Techniques de l'Ingénieur.)

Their main domain of use is the supply of saturated steam at low pressure (LP) (<15 bar), which represents over 60% of the French fleet of boilers, against 20%–25% for water-tube boilers, which are well suited for the supply of superheated steam at medium and high pressure (HP).

A boiler has three successive functions:

- Heat-pressurized feedwater (in the economizer) to the vaporization temperature corresponding to the pressure;
- Vaporize steam in the evaporator;
- Finally superheat steam at the desired temperature.

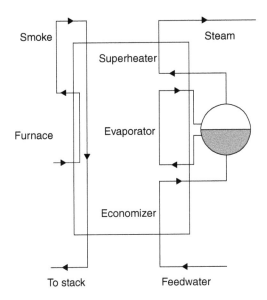

FIGURE 5.3
Boiler exchange configuration.

It behaves like a triple heat exchanger, and may be represented in terms of heat exchange by the diagram in Figure 5.3.

In a conventional high-temperature boiler, flame temperature reaches very high values (1,400°C–1,500°C). Heat is transmitted to steam mainly in the form of radiation, with very high flux densities (100–200 kW/m^2). Steel tubes are, for economic reasons, limited to a maximum temperature of 650°C, which requires them to be very well cooled from the inside. Given the low gas exchange coefficients, this is only possible if the fluid that passes through the tubes is liquid, or better still, two-phase.

For this reason, the first exchanger is the furnace evaporator. Upon exiting the furnace, gas temperature has dropped significantly (800°C–900°C) and convection takes over from radiation. The second series of gas–gas exchangers corresponds to superheaters. At the outlet, fumes are cooled at around 600°C, and residual enthalpy is then used in economizers, gas–liquid heat exchangers, which ensure the heating of water at its boiling temperature at the pressure considered. Where appropriate, a heater can then be used (in large boilers) to preheat combustion air.

The exhaust temperature of flue gases in the atmosphere should be as low as possible to maximize the effectiveness of the boiler. The need to avoid condensation of smoke prevents, however, cooling as much as desirable, if it contains sulfur oxides that can form corrosive acids.

BOX 5.1
Key issues

Architecture and configuration of a boiler

The following self-assessment activities will allow you to check your understanding of boilers:

Architecture of a boiler, ddi
Configuration of a boiler, gfe

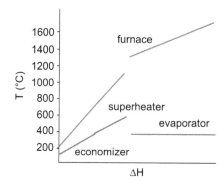

FIGURE 5.4

Temperatures in a boiler.

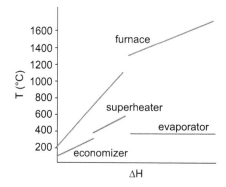

FIGURE 5.5

Temperatures in a boiler with evaporator approach.

The graph in Figure 5.4 illustrates in an enthalpy diagram heat exchange in a classic steam plant boiler. The upper curves correspond to furnace and smoke cooling, while the lower ones to the heating of the water. The discontinuity between the two segments of the upper curve is due to the furnace radiative contribution.

As shown, heat exchange is done with very large temperature differences, always greater than 180°C. At the end of the economizer, they are above 250°C. If the boiler control is not perfect, there are in these conditions significant risks that the vaporization starts in economizer, which is not designed for that. To prevent this malfunction, liquid heating is stopped before reaching the boiling point, maintaining an approach temperature difference of the order of a few tens of degrees. The missing enthalpy is then supplied by the evaporator. The graph in Figure 5.4 is amended as shown in Figure 5.5.

Steam generators

According to what has been stated above, we call steam generators (SG) devices where heat is not provided by combustion. Among them, we will focus here particularly on devices for recovering heat in thermal effluents, including gas (the heat recovery steam generators(HRSGs) are sometimes called recovery boilers). Nuclear power plant steam generators are presented in Chapter 8.

Most HRSGs are variations of water-tube boilers. The main differences come from the fact, on the one hand, that the temperature levels of effluents are much lower than those achieved in a boiler, and on the other hand, that heat exchange takes place only by convection. Nothing in these conditions prevents the superheater being positioned upstream of the evaporator. The HRSG operating sketch and enthalpy diagram are shown in Figure 5.6.

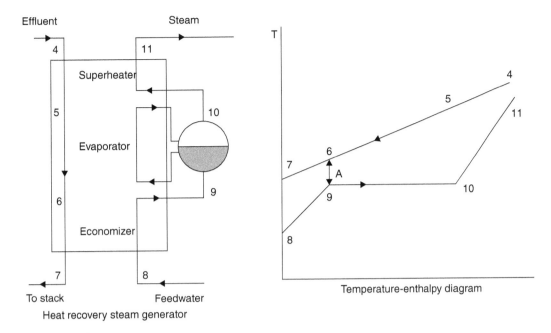

FIGURE 5.6

Steam generator exchange configuration.

At point 6 on the diagram, you can see a minimum A in the temperature difference between the two fluids, called the pinch (see Chapter 4). This point plays a fundamental role in the design of HRSGs, as it represents the smallest difference in temperature in the facility. To reduce the cost of equipment, it is preferable that the pinch is not too low. However, the thermodynamic optimization of the complete system requires it to be as small as possible.

Particular attention should be paid to the HRSG design at this pinch, which must exceed a minimum value of about 10°C–15°C; otherwise, vaporization cannot be done at the desired temperature.

Moreover, exhaust gas temperature T_7 (stack temperature) must be high enough to avoid condensation of sulfur oxides that may be present. This temperature differs depending on the fuel used. The lowest values (<90°C) are obtained with natural gas. With light oil, T_7 is recommended to be greater than 120°C–130°C and 150°C–170°C with heavy oil.

But a second requirement must be met: one must avoid, at the beginning of the economizer, any condensation on heat exchanger tubes at a temperature just above T_8, steam cycle condensation temperature. For this, it is sometimes necessary to provide a partial economizer recirculation.

In the same manner as in classic steam plant boilers, it may be necessary to restrict liquid reheat to avoid boiling in the economizer. As the differences in temperature between the water system and the effluent are much smaller than in a boiler, the temperature difference may be lower.

Boiler operation

The evaporator consists of two drums superimposed and connected by two bundles of tubes called hot leg and cold leg (Figure 5.7). In the upper tank, the emulsion produced by the vaporization in the hot leg crosses water separators and steam dryers. Separated water goes back down through the cold leg and mixes with water from the feedwater tank.

In the evaporator passes a total flow greater than that of steam produced. We call recirculation rate τ, which is defined as follows:

$$\tau = \frac{total\,flow\,rate}{steam\,flow\,rate}$$

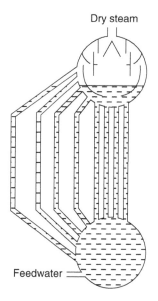

FIGURE 5.7
Water-tube drums.

The recirculation flow is obviously equal to $(\tau-1)$ times the steam flow.

According to boilers, water flows either naturally or through a pump. The current values of τ are around 8–10.

Optimization of pressure level

In the case of a boiler, due to the very high temperatures that are reached in the furnace, as shown in Figure 5.5, there is no pinch: the boiler effectiveness is independent of the steam system pressure level, which can be chosen solely based on use constraints.

In the case of energy recovery from effluents, this is not necessarily the case: the recovery effectiveness depends on both the effluent temperature and the pressure level of the steam system.

As shown in the graph in Figure 5.8, for a given steam pressure (and hence a given pinch value), and for a set maximum enthalpy, we find that when T_{max} varies, it is as if the effluent line pivoted around the pinch.

Because the effluent cools from T_{max} (the maximum temperature) to T_{rej} (the exhaust temperature), T_{rej} should be as low as possible so that energy recovery is maximized.

Calling T_{rejlim} the limit exhaust temperature compatible with the corrosive condensation constraints, we can qualify the maximum energy recovery by an effectiveness, which depends only on the nature of the effluent:

$$\epsilon_{ef} = \frac{T_{max} - T_{rej}}{T_{max} - T_{rejlim}}$$

As shown in Figure 5.8, at given T_{max}, T_{rej} increases (and therefore ϵ_{ef} decreases) when the network pressure level increases. Optimization of steam system can no longer be done independently of the HRSG. We shall see in Chapter 10 the implications this has for the design of combined cycles.

Steam turbines

The general layout of a turbine is given in Figure 5.9. A stator followed by a rotor are crossed in series by steam at HP and temperature. The stator is made of guide nozzles which

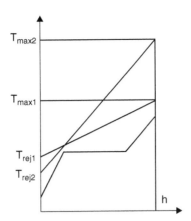

FIGURE 5.8

Effluent line.

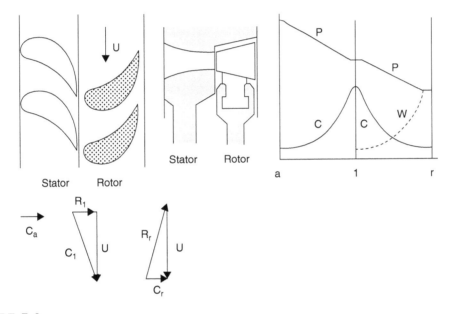

FIGURE 5.9

Velocity and pressure profile in an axial turbine stage.

accelerate the fluid, while the wheel (rotor) converts into mechanical energy at least part of the enthalpy available at its inlet. U is the tangential velocity of the rotor.

In a turbine, the evolution of the fluid is an expansion. The compressible flow equations which we will not develop here indicate that for a subsonic regime, the section of the vein should decrease and the velocity increase. This evolution takes place in two steps (Figure 5.9): in the stator, the absolute velocity C increases, while in the impeller the relative velocity R increases, and the absolute velocity C decreases.

For a given rotational speed N, the performance map of a turbine has the appearance of Figure 5.10, where a group of variables called corrected mass flow proportional to the Mach number in the flow appears on the ordinate, and the inverse of the expansion ratio on the abscissa. The index 0 relates to the turbine inlet conditions.

As can be seen, there is a corrected mass flow rate limit corresponding to the establishment of a sonic flow at the nozzle throat: the flow is said to be choked.

This point is very important, because it means that the flow rate which passes through a turbine cannot exceed a certain limit when the expansion ratio exceeds the value of approximately 2.

We will not develop in this book the study of the compressible flow equations and the relations of similitude, because they are not useful for our purpose. The interested reader may

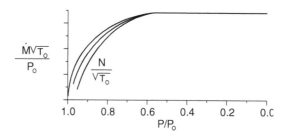

FIGURE 5.10
Characteristics of a turbine.

refer either to the specialized literature or to Section 7.3.3 (Similarity and Performance of Turbomachines) of the first edition of this book.

Different types of steam turbines

Depending on their use, there are four broad categories of turbines (Figure 5.11):

* Condensing turbines, where steam is completely expanded at a pressure of about 0.02–0.04 bar, and then liquefied in a condenser cooled by ambient air or by water. This type of turbine is mainly used in the power production facilities;
* Back-pressure turbines, in which vapor pressure is expanded from HP (>40 bar) to LP (about 4 bar). This type of turbine allows the production of mechanical power or electricity thanks to the high temperature and pressure that can be obtained in a boiler, while using the residual enthalpy for various processes;
* Extracting and condensing turbines, in which vapor undergoes a partial expansion at an intermediate pressure (about 20 bar) in a high-pressure section. One part is directed to a user network, while the rest of the steam is expanded in a low-pressure section, as in a condensing turbine. This type of turbine finds an important field of application in cogeneration plants whose requirements for heat are likely to vary considerably over time;
* Extracting and back-pressure turbines, in which steam escapes at LP in a LP network instead of being condensed.

Degree of reaction of a stage

Most steam turbines are multistage axial turbines. The study of the energy balance of a stage shows that the enthalpy change takes place partly in the nozzle and partly in the rotor. We call the degree of reaction ε the fraction of the enthalpy change that takes place in the rotor.
By definition, ε is between 0 and 1.
Steam turbines can be grouped into two broad classes:

* **Impulse turbines,** in which the degree of reaction ε is 0: any expansion of the fluid is then carried out in fixed blades or nozzles, upstream of the wheel, and pressures upstream and downstream of the rotor are equal;
* **Reaction turbines,** where $\varepsilon = 0.5$: expansion is then evenly distributed between the nozzle and the wheel.

Each of these two turbine types has advantages and disadvantages of its own: the impulse turbines are generally used for multistage turbine head stages or for small-capacity units, while reaction turbines turn out to be well adapted for low-pressure turbines.

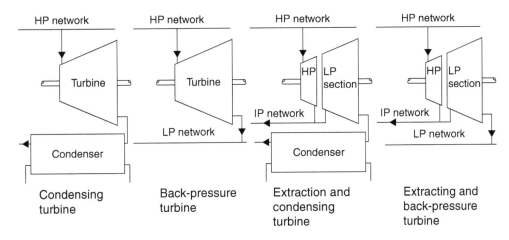

FIGURE 5.11

Different types of steam turbines.

FIGURE 5.12

Steam turbine. (Courtesy Siemens Energy.)

Figure 5.12 shows a cutaway of a Siemens compact steam turbine SST-111, which generates around 8 MW of electricity.

This image has the particularity of clearly highlighting the reduction gear, which converts the very high rotational speed of the turbines into a much lower rotational speed on the motor axis connected to the large-diameter gear wheel.

Diametrically opposed, one distinguishes, in the background the high-pressure part of the steam turbine, composed of action stages of very small diameter, and, towards the front on the left, the low-pressure block composed of reaction wheels of much larger diameter.

Degradation of expansion isentropic efficiency function of steam quality

When the expansion of a fluid (usually steam) extends beyond the dew point curve, vapor begins to condense and droplets appear. The phenomenon is not immediate, because the equilibrium conditions are not met, but it occurs gradually, more and more as the steam quality increases. The presence of these droplets leads to:

* Mechanical damage of the front edges of turbine blades;
* Reduction of the expansion isentropic efficiency.

For these two reasons, the permissible quality limit at the end of expansion should not be below about 0.8. To account for expansion isentropic efficiency degradation, we can use relation (5.1), with α being the Baumann coefficient, close to 1, and η_{hum} and η_{dry} being, respectively, the isentropic efficiencies for the humid zone and dry steam.

$$\frac{\eta_{hum}}{\eta_{dry}} = 1 - \alpha(1 - x) \tag{5.1}$$

This relationship shows that an increase of 1 point of the quality reduces the isentropic efficiency by about 1 point. It is not currently taken into account in the wet expansion calculated by Thermoptim. The reader can refer to the book by de Vlaminck and Wauters (1988) for detailed explanations on this subject.

Temperature control by desuperheating

As we shall see, the optimization of a cycle generally leads to work with superheat temperatures technologically as high as possible. To regulate the temperature around this value, it is common to use steam desuperheating, an operation that involves injecting into superheated steamsome pressurized cold water from the feedwater tank. The injection, which is usually done between two elements of the superheater, is controlled by a valve that adjusts the flow injected according to the superheater outlet temperature.

The industrial cogeneration facility studied in Chapter 10 specifically implements such a desuperheating.

Note that this setting mode introduces a thermodynamic irreversibility because it mixes two flows at very different temperatures. In cycles where the conventional heat source temperature is very high, this irreversibility does not affect the overall efficiency, but this is no longer necessarily the case when the source is average temperature hot effluent, as in a cogeneration facility or combined cycle.

Cooling towers

The performance of steam systems operating in a closed circuit is very sensitive to conditions in which steam may be condensed. Indeed, a condenser is a heat exchanger of a particular type, phase change, whose thermal equilibrium determines the condensation temperature of steam and thus its condensation pressure.

The cycle LP, which directly defines the power that the turbine can produce, is thus set by the condenser.

Heat power transferred to the condenser is typically removed by cooling water flowing inside the tubes, which must be renewed (open-loop cooling) or cooled by contact with ambient air (air cooling).

For systems that operate in an open loop, cooling water is drawn and released directly in the sea or a river, after passing through the condenser, while in others, condenser cooling water circulates in the closed circuit cooled with air, usually in a cooling tower.

In both cases, the amount of water required is very important, which is a burden, especially in the context of global warming and depletion of water resources. Innovative binary cycles presented in Chapter 8 are today proposed to reduce this constraint.

As heat exchangers have been studied in Chapter 4, we will not develop more water-cooled condensers. In this section, we present cooling towers, previously not addressed.

Principle of operation of cooling towers

A cooling tower is a heat exchanger of a particular type that discharges heat in the surrounding air in the form of both sensible heat and latent heat due to the increase of its moisture.

By working this way, it is possible to cool a fluid at a temperature a few degrees above the ambient air wet bulb temperature (and possibly below its dry bulb temperature), at the cost of a water consumption of about 5% of that which water cooling would require. Both economically and environmentally, cooling towers are very interesting systems, in particular in hot and dry climates.

Note that moist gas equations will be presented in Chapter 12 on air conditioning.

There are two main types of cooling towers, called direct contact or open cycle, and indirect contact or closed cycle.

In a direct contact tower (Figures 5.13 and 5.14), hot water is cooled in **contact** with ambient air either by spraying fine droplets or by runoff along flow surfaces. Both fluids are in contact, heat is exchanged by convection, and part of the water vaporizes, thereby increasing the humidity. If it is not saturated, it starts to cool in an almost adiabatic process, before warming up along the saturation curve. Therefore, water may come out at a temperature lower than ambient air.

An indirect contact tower comprises two circuits known as external and internal. In the latter, the cooling fluid, which can be arbitrary, remains confined in a tube bundle around

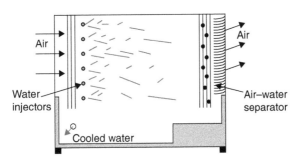

FIGURE 5.13

Direct contact cooling tower. (Courtesy Techniques de l'Ingénieur.)

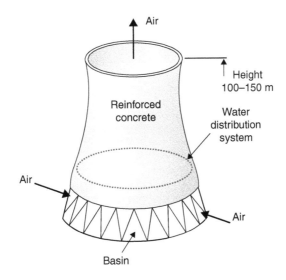

FIGURE 5.14

High-capacity power plant cooling tower. (Courtesy Techniques de l'Ingénieur.)

which the external cooling circuit water runs. It warms on contact, then is cooled by exchange with ambient air by the same mechanism as in a direct contact tower.

Phenomenological model

The theoretical modeling of cooling towers is quite complex given the multiplicity of transfers that take place. Note that we are content here to provide overall results and that the moist gas equations will only be given in Chapter 12. The 2000 ASHRAE Handbook (Chapter 36) presents the theory proposed in 1961 by Baker and Shyrock, which considers (Figure 5.15) three media exchanging mass and energy:

* The flow of water that cools;
* The air flow which is heated and whose moisture increases;
* An interstitial film.

The assumptions are:

* Air leaving the tower is nearly saturated;
* The interstitial film is saturated moist air, at the water temperature;
* The Lewis number is equal to 1;
* Thermal resistance on the liquid side is negligible compared to that on the air side (for indirect contact towers, it is assumed that the air–liquid film temperature gradient is equal to that of air–water cooling).

The energy balance of a small tower element of surface ΔA writes with our usual notations:

$$\Delta Q_{water} = \dot{m} C_{p\,water} \Delta T_{water}$$

This energy is used first to heat air (ΔQ_s) and second to evaporate water absorbed by air (ΔQ_L).

$$\Delta Q_s = k_s \Delta A \left(t'' - t_a \right)$$

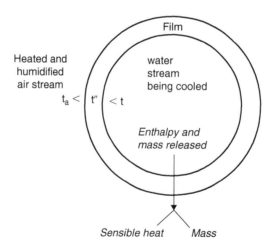

FIGURE 5.15

Sketch of transfers.

with k_s being the heat exchange coefficient in the presence of mass transfer, t'' being the film temperature (assumed equal to that of water) and t_a that of air.

$$\Delta Q_L = k' \Delta A \, L_{water} \left(w''_{sat} - w_{asat} \right)$$

with k' being the conductance referred to mass transfer, w''_{sat} the film saturation moisture and w_{asat} that of air.

We have thus:

$$\Delta Q_{water} = k_s \Delta A \left(t'' - t_a \right) + k' \Delta A \, L_{water} \left(w''_{sat} - w_{asat} \right)$$

In turbulent regime, with Sc and Pr being the Schmidt and Prandtl numbers and Le the Lewis number, it can be shown that k_s and k' are related by the following equation:

$$\frac{k_s}{k' C_{p\,air}} = \left[\frac{Sc}{Pr} \right]^{2/3} = Le^{2/3} \tag{5.2}$$

It turns out that for water–air mixture, $Le \approx 1$ as long as $t_{water} < 50°C$, as shown by Merkel in 1925 when he proposed to consider that for cooling towers, $Le = 1$.

With this assumption, the above equation is simplified, showing the specific enthalpy q':

$$\Delta Q_{water} = \dot{m} C_{p\,water} \Delta T_{water} = k' \Delta A \left[C_{p\,air} \left(t'' - t_a \right) + L_{water} \left(w''_{sat} - w_{asat} \right) \right]$$

$$\Delta Q_{water} = k' \Delta A \left[q'_{sat} \left(t'' \right) - q'_{sat} \, t_a \right]$$

This equation can be modified and allows us to write, per unit of water mass flow:

$$\frac{k' \Delta A}{\dot{m}_{water}} = \frac{C_{p\,water} \Delta T_{water}}{q'sat(t'') - q'sat(t_a)} \tag{5.3}$$

This equation can be numerically integrated along the cooling tower, leading to results in Figure 5.16, which shows the evolution of temperatures in the tower as a function of reduced distance between its inlet and its outlet. The water temperature curve is, of course, located above that of air.

This gives a performance indicator of the cooling tower, which appears in good agreement with experimental data.

Let us now recall a characteristic of heat exchangers, the number of transfer units (NTU) introduced in Chapter 4.

Equation (4.20) gives the value of NTU:

$$NTU = \frac{\Delta T_{max}}{\Delta T_{ml}}$$

NTU can be physically interpreted as the ratio of the maximum variation in temperature in one of the fluids to the log-mean temperature difference between them. This is indeed an indicator of the quality of the device: the more efficient the heat exchanger, the more it allows for heating or cooling a fluid with a low temperature difference with the other.

The concept of NTU can be generalized by considering not the ratio of the temperature variations, but that of the enthalpy differences. One may in particular put in the denominator the average logarithmic difference of specific enthalpy of saturated air taken at the temperatures of the two fluids, and in the numerator the variation of enthalpy Δh of one of the two fluids, the two being equal.

$$NTU = \frac{\Delta h}{\Delta h_{ml}}$$

or

$$NTU = \frac{\Delta h}{\Delta q'_{sat\,ml}}$$

In these conditions, Equation (5.3) of the tower elementary heat balance that we integrated is like an elementary NTU and integration provides the NTU of the tower. To differentiate it from normal NTU in exchangers, we will note it with index m (Merkel).

$$\Delta NTU_m = \frac{k'\Delta A}{\dot{m}_{water}} = \frac{C_{P\,water}\,\Delta T_{water}}{q'sat(t'') - q'sat(t_a)}$$

$$NTU_m = \int \frac{C_{P\,water}\,\Delta T_{water}}{q'sat(t'') - q'sat(t_a)}$$

It is indeed an indicator of how difficult it is to realize the heat transfer in the cooling tower. Incorporating complex coupled phenomena of energy and mass exchange and dependent on technology implementation, it is not possible to give a simple expression.

We call the "approach" the pinch of the tower, i.e., the difference between the outlet temperature of the water and the wet bulb temperature of ambient air (4°C in the example in Figure 5.16) and the "range" the value of ΔT_{water} (8°C in this example).

NTU_m is usually given as a function of four variables: $\dfrac{\dot{m}_{water}}{\dot{m}_{air}}$ and T'_{air}, the wet bulb air inlet temperature, as well as either the two water temperatures T_{wi} (inlet) and T_{wo} (outlet), or the values of range $(T_{wi} - T_{wo})$ and approach $\left(T_{wo} - T'_{air}\right)$.

If they operate in the same environment with the same water inlet temperature and fluid mass flow of the same ratio, two different cooling towers will cool more or less water depending on the quality of their design. The most efficient results in the lower value of T_{wo}, and thus a higher value of NTU_m.

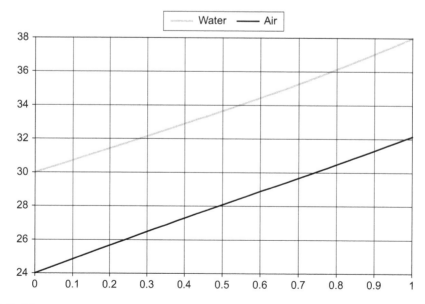

FIGURE 5.16

Profile of temperature in a direct contact cooling tower.

Behavior models

Many authors have investigated the modeling of cooling towers and sought expressions of NTU_m that are both easy to calculate and in agreement with experimental results. Since this is primarily an indicator of performance, it is not in itself unacceptable that there is a degree of arbitrariness in its definition, to the extent, however, that conventions are well specified.

Merkel model

Experience shows that when we modify the four input variables of a given cooling tower, NTU_m takes different values. Its behavior can be represented by equations (5.4) or (5.5), with n and m having negative values between −1.1 and −0.35, and C being between about 1 and 2.5.

C, m and n are called the tower coefficients.

$$NTU_m = C\left[\frac{\dot{m}_{water}}{\dot{m}_{air}}\right]^n \qquad (5.4)$$

or

$$NTU_m = C\left[\dot{m}_{water}\right]^n\left[\dot{m}_{air}\right]^{-m} \qquad (5.5)$$

Note that one of the difficulties encountered when attempting to identify parameters C, n and m is that available data are often incomplete. Manufacturers provide the values of inlet air wet bulb temperature, range and approach, as well as water flow, but not the value of the air inlet humidity. They are often content to give the power consumed by the air fan, or at best its volume flow, so that the identification of the thermodynamic state of air in the tower is inaccurate. When conducting tests, it is also often very difficult to measure the actual airflow that passes through the machine.

Figure 5.17 shows very good correspondence between the values of NTU_m calculated for the same model of cooling tower (Marley F1221) with four different fan engines (40, 50,

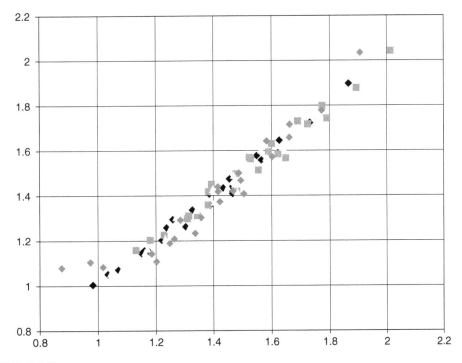

FIGURE 5.17
Mapping of manufacturer data and NTU_m correlation.

TABLE 5.1
VALUES OF COEFFICIENTS

	FI221 40 hp	FI221 50 hp	FI221 60 hp	FI221 70 hp
C	2.19	1.87	1.93	2.06
n Merkel	−0.860	−0.968	−1.009	−1.171
r^2	0.970	0.982	0.963	0.853

60 and 70 hp). This figure was obtained from values provided in the manufacturer's catalog and volume flow rates communicated to us. A linear regression allows us to identify, for correlation (5.5), the values of the coefficients given in Table 5.1, with correlation coefficients r^2.

Relation (5.4) also shows that for a given water flow, the higher the airflow, the smaller the tower can be. However, pressure drop and thus fan consumption are increasing functions of the airflow (air velocities are generally between 1.5 and 4 m/s).

For indirect contact towers, the fact that the fluid to be cooled is contained within the internal circuitry increases thermal resistance. We obtained good correlations by increasing slightly the approach, by values between 0.7°C and 1.6°C. This is easily explained when one considers that this device leads to lower NTU_m value, which is consistent with what should happen physically.

Cooling technology institute calculator

The Cooling Technology Institute has developed a tool called Toolkit (www.cti.org), which integrates the differential equation giving NTU_m. A comparison with our results yields a very small difference, less than 2%, presumably due to a difference in fluid property modeling.

Using the mass flow of water being noted L and that of air G, the expression of NTU_m in American literature is generally:

$$\Delta NTU_m = \frac{K_a V}{L} = \int \frac{C_{p\,water} \Delta T_{water}}{q'sat(t'') - q'sat(t_a)}$$

In English units, $C_{p\,water}$ is equal to 1, and may disappear from the expression.

The cooling tower characteristics obtained have the appearance given in Figure 5.18, in bi-logarithmic coordinates. Abscissa is the ratio of the mass flow of water to that of air L/G, and the ordinate is NTU_m. The design value is displayed in orange, as well as a curve representing the behavior model law (5.4). Yellow curves are iso-approach lines. By selecting a rectangle on the chart, zoom is shown (double-click restores the original scale).

$$\frac{L}{G} = \frac{\dot{m}_{water}}{\dot{m}_{air}}$$

This chart can be used to characterize a given tower:

*   The value of NTU_m for design conditions is calculated;
*   The line of slope n and constant C is plotted and adjusted on the design point;
*   For other operating conditions, determined by the fluid flows, the abacus allows us to find, at abscissa L/G and on the tower characteristic line, the new values of NTU_m and approach, which determines the outlet water temperature and therefore the capacity. The air outlet temperature is deduced. For another cooling or other air wet bulb temperature, it is possible to recalculate another abacus. Knowing the values of variables (NTU_m, L/G) from the original chart, we get the new approach.

For example, the design conditions of Figure 5.18 correspond to an approach of 3.55°C (obtained by right-clicking the point in question).

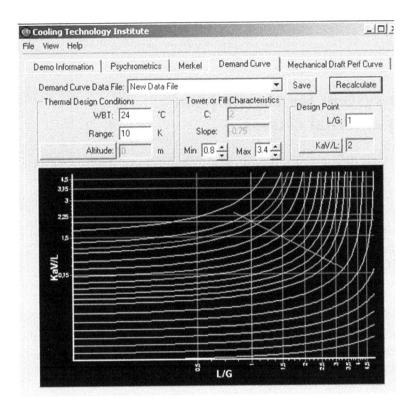

FIGURE 5.18
Cooling tower characteristics.

Analogies with heat exchangers

Since it is the change in enthalpy of saturated moist air which best characterizes the air thermodynamic behavior, a number of authors consider that it is as if water did exchange with a fictitious fluid whose specific heat at constant pressure is equal to the average heat capacity of moist saturated air $C_{p_a}^{sat}$, but it is also possible to express it in terms of specific heat of dry air.

We must therefore beware of the conventions used to avoid mistakes in calculations. The key is to be perfectly uniform throughout the approach: we must ensure that conventions used for identifying the values of C and n from experimental values are the same as those used to calculate ε and NTU.

The interested reader should refer to the referenced documents for more details.

One can, for example, calculate as average heat capacity that of moist saturated air between the minimum and maximum states of water:

$$C_{p_a}^{sat} = \frac{h_a^{sat}\left(T_a^{min}\right) - h_a^{sat}\left(T_a^{max}\right)}{T_a^{min} - T_a^{max}} \tag{5.6}$$

The flux transmitted to the air can be written as: $\phi = \varepsilon \left(q'sat(T_{wi}) - q'(T_a)\right)$.

It is ε times the maximum flux, equal to the product of the mass flow of dry air (index da) by the difference between the specific enthalpy of the air at water inlet temperature and its specific enthalpy at the entrance of the tower. Furthermore:

$$R = \frac{C_{p_{water}}}{C_{p_a}^{sat}}$$

and

$$NTU_m = C\left[\frac{\dot{m}_{water}}{\dot{m}_{air}}\right]^n$$

The values of C and n are identified from experimental data and the appropriate (ε, NTU) relation (see Chapter 4). Values of these parameters are proposed in the literature (Mills 1999).

Extension system for thermoptim by adding external classes

At this stage of our reflections, it is necessary to evoke a subject of which we have not yet spoken, that of the extension of Thermoptim by external classes.

It is important to do it now because we will use in the following a whole series of such external classes.

All documentation on this mechanism is available in the Thermoptim-UNIT portal, which we recommend you refer to.

Through version 1.4, only the components available in the Thermoptim core could be assembled in models, which limited the potential of the software. A number of users wished to be able to define their own elements and/or their own substances.

Thermoptim interface with external classes (Java code elements) provides the solution and facilitates the interoperability of the software with the outside world, especially with other applications developed using Java. Java programs are called classes, and we call these extensions of Thermoptim **external classes**.

The benefits are twofold:

* Create Thermoptim extensions from common primitive type set, by adding external modules that define elements that automatically appear on the screens in a seamless fashion. Thus, users can add their own substances or components not available in the basic set;
* Drive Thermoptim from another application, either to guide a user (smart tutorial) or to check the code (driver or controller, access to thermodynamic libraries).

Figures 5.19 and 5.20 show how the external substances are added to Thermoptim list of substances, and then replace an internal substance on the point screen. They are just as easy to use as if they were part of the software.

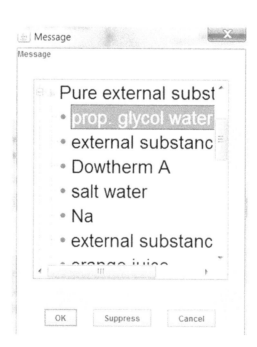

FIGURE 5.19

List of substances.

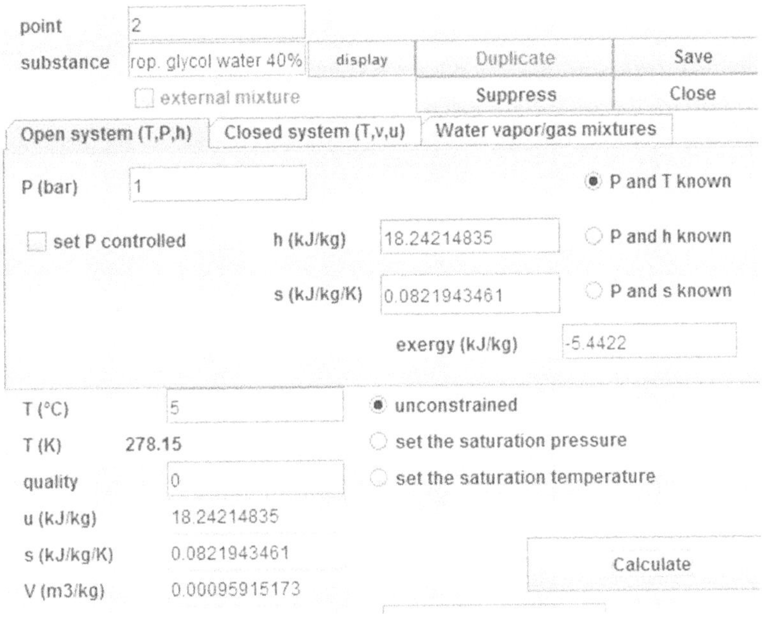

FIGURE 5.20
Substance screen.

Let us point out now that we will use in the rest of this book the CTP Lib library, to model various fluids not present in the Thermoptim core. This library can be downloaded from the Thermoptim-UNIT portal (www.thermoptim.org). It allows in particular one to model many fluids using a Peng Robinson equation of state (see Chapter 13), and also has very precise specific equations for CO_2 and for the pair (NH_3-H_2O), which will be used for the study of absorption machines (Chapter 13) and the Kalina cycle (Chapter 14).

Figure 5.21 shows how an external component representing a solar collector appears in the diagram editor, and Figure 5.22 shows the screen of the corresponding process, composed partly of Thermoptim internal code and partly of external code (lower third on the right).

Using external components

Using existing external classes is very simple and poses no difficulty as you will see for yourself.

Representation of an external component in the diagram editor
Specific icons were added to represent the external components (for processes, for mixers, and for dividers). The external component is then selected when the simulator is updated from the diagram as indicated below.

Loading an external class
To load an external process (for an external node, the operation mode is the same), you can either:

* From the simulator screen click on the column header of the process array, then choose External and finally select the type of external process you want from the list;
* Or, from the diagram editor, build the external component graphically, then update the simulator from the diagram. In the case of an external process, by default it is a "heat source/sink" type.

Once this default process is created, double-click on the label "source/sink" to access the list of all external processes available. Choose the one you want, and it will be loaded.

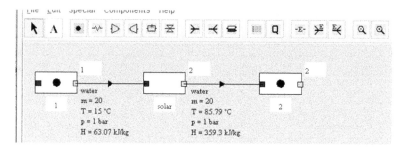

FIGURE 5.21

External process.

| process | solar | | type | external | | | < | > | Save |
| | | | | | | Suppress | | Close |

energy type   other   □ set flow   ○ closed system   □ observed

inlet point   flow rate   20   ● open system

1   display   m Δh   5,924.57   Calculate

T (°C)	15	solar collector	
P (bar)	1		
h (kJ/kg)	63.07	glass transmittivity	0.8
quality	0	thermal loss coefficient	8.0
		solar flux (W/m2)	1200.0
outlet point		collector area (m2)	10000.0
2	display	outdoor temperature (°C)	10.0
T (°C)	85.79	solar collector efficiency	0.4938
P (bar)	1	reference	reference solar collector
h (kJ/kg)	359.3		
quality	0		

FIGURE 5.22

External process screen.

Let us clarify two important points:

- First, all available external classes are accessible from the model library of the Thermoptim-UNIT portal;
- Then, the code for these classes is distributed in open source and can be customized at will by users who wish to do so. They just need to know how to program in Java.

Modeling a direct contact cooling tower in thermoptim

As an example, let us look at how a cooling tower can be modeled using external classes.

A direct contact cooling tower is crossed by two separate streams: air and water, which exchange matter and energy through an interface. It behaves like a quadrupole with two fluid streams as inputs, and two others as outputs.

This poses a slight difficulty in building a Thermoptim model, since the only components available are either processes or nodes. The solution is to form the quadrupole by combining an inlet mixer and an outlet divider; the two are connected by a process-point playing a passive role.

For the model to be coherent, the calculations made by both nodes are synchronized. More precisely, the divider takes control of the mixer, whose role is limited to perform an update of the coupling variables associated with the input streams. The model structure is given Figure 5.23.

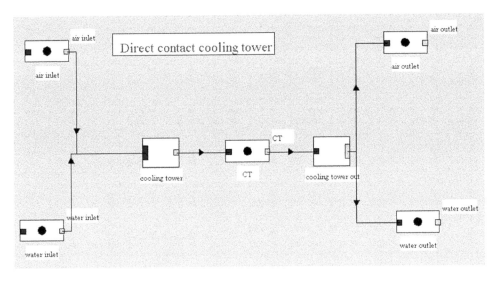

FIGURE 5.23
Diagram of the direct contact cooling tower component.

In these conditions, we can reason on the overall enthalpy level knowing the inlet and outlet air side conditions and the inlet ones on the water side: model results are consistent with experimental values and those supplied by manufacturers.

Flow rates of both inlet streams are set by conditions upstream of the component and not recalculated. If water flow is insufficient for its cooling (up to moist bulb temperature of incoming air) to provide air with the enthalpy required, a message warns the user.

Recall that, as usual in calculations of moist functions, values are referred to dry gas, whose composition is invariant, while other calculations in Thermoptim are relative to the actual gas composition, i.e., values are referred to the moist gas. It is therefore necessary to make the corresponding conversions.

The model that we can choose, when we know the outlet air temperature, is the following:

1. We begin by calculating inlet moist air properties and determining the dry gas mass flow from that of moist gas; the inlet relative humidity ε is displayed on the screen;
2. The outlet relative humidity is set equal to 1, and the outlet moist air properties are calculated, which gives the specific and total enthalpy to be brought to air;
3. The flow of water carried by air is determined, and the outlet moist air composition is changed;
4. The water enthalpy balance provides its outlet temperature, which must exceed the moist bulb temperature of air entering;
5. Values downstream of the node are updated;
6. NTU is calculated by integrating the Merkel equation (NTU integrated) established above and on the basis of the mean logarithmic enthalpies; effectiveness can be deduced from the classical relationship for counter-flow heat exchangers.

The tower is as we said represented by an external mixer connected to an external divider, and calculations are made by the latter. Classes are called DirectCoolingTowerInlet and DirectCoolingTower.

The cooling tower component screen is given in Figure 5.24. We wish here to cool 1 kg/s of water from 30°C to 25°C. With a rate of 0.9 kg/s of air at 18°C and relative humidity equal to 0.5, the exhaust air temperature is 20.1°C and 7.5 g/s water is evaporated. The tower capacity is 20.9 kW.

On this screen, we see the two values of NTU. That which is in the editable field is calculated by the approximate formula based on the logarithmic mean differences of enthalpies of saturated air between the tower inlet and the tower outlet, while the one displayed at the

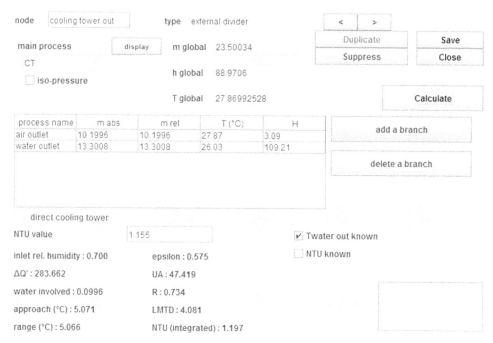

FIGURE 5.24
Cooling tower component screen.

bottom is the result of the integration of the Merkel differential equation for a counter-flow tower. It is in principle more accurate, but we can see that their values are close.

Example: external class **directcoolingtower**

BOX 5.2

Worked example

Refrigeration machine condenser with cooling tower

This example corresponds to an R134a refrigeration machine ensuring the production of 200 kW of cooling at −12°C, whose condenser is cooled by air at 25°C.

The objective is to compare the performance of the machine depending on whether one uses an air exchanger with a pinch of 16°C or a cooling tower; the minimum pinch between water and the refrigerant is below 12°C. The result is a coefficient of performance (COP) increase of 16%–19% when the cooling tower is used.

It is presented in the portal guidance pages (https://direns.mines-paristech.fr/Sites/Thopt/en/co/fiche-guide-fg6.html).

Without cooling tower (Figure 5.25), with a maximum airflow of 30 kg/s and a pinch of 16°C, simulation leads to a COP of 2.15, condenser outlet air temperature is equal to 34.7°C and R134a condensation temperature is equal to 50.4°C, i.e., condensing pressure of 13.3 bar.

The tower settings can in turn (Figure 5.26) be done as follows: we set air inlet T and ε (here 25°C and 70% RH) and inlet water T (here 31.1°C) and flow (13.4 kg/s). Balancing the condenser and desuperheater gives the power to extract ΔH and the water outlet temperature (26.03°C). We choose option "Twater out known", and the calculation of the tower provides Tair = 27.89°C.

COP with the tower is equal to 2.57 instead of 2.15, representing an increase of 16%. And yet we have neglected the fan power, air flow passing from 30 to 10 kg/s, or the change in isentropic efficiency, due to the variation in the compression ratio which dropped from 8.87 to 7.33.

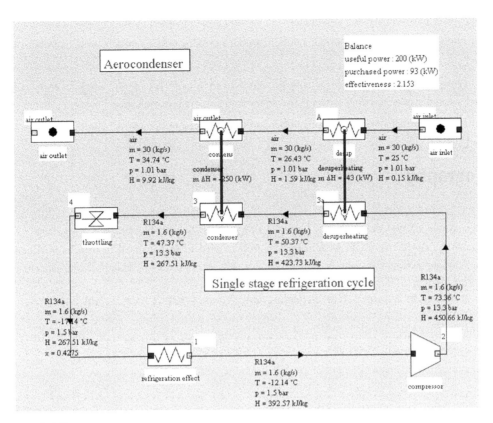

FIGURE 5.25

Synoptic view of the system with air exchanger, $\eta_{is} = 0.8$.

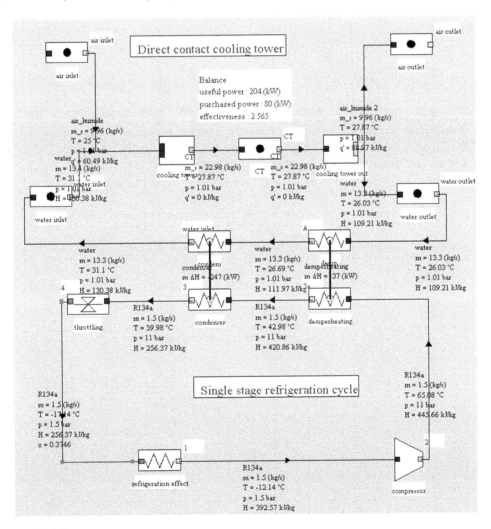

FIGURE 5.26

Synoptic view of the facility, $\eta_{is} = 0.8$.

This example shows the interest of constructing external classes allowing to represent in Thermoptim components absent from the core of the software package.

We will use many of them in the following chapters.

Bibliography

D. R. Baker, H. A. Shryock, A comprehensive approach to the analysis of cooling tower performance, *ASME Transactions, Journal of Heat Transfer*, 83(3): 339–349, Aug. 1961.

J. E. Braun, S. A. Klein, J. W. Mitchell, Effectiveness models for cooling towers and cooling coils, ASHRAE Transactions, Vol. 92, Part 2, 1989, pp. 164–174, ASHRAE Handbook 2000, "Cooling Towers", Chapter 36.

M. De Vlaminck, P. Wauters, *Thermodynamique et Turbines*, CIACO, Louvain-la-Neuve, 1988.

F. Merkel, "Verdunstungskuhlung," VDI Forschungsarbeiten, No. 275, Berlin, 1925.

G. Riollet, *Théorie des turbines à vapeur et à fluide compressible*, Techniques de l'Ingénieur, Traité Mécanique et chaleur, B 330-2.

L. Vivier, *Turbines à vapeur et à gaz*, Ed. Albin Michel, Paris, 1965.

Further reading

A. Bricard, L. Tadrist, *Echangeurs de chaleur à contact direct*, Techniques de l'Ingénieur, Traité Génie énergétique, BE 9 565.

P. Lemoine, *Refroidissement des eaux*, Techniques de l'Ingénieur, Traité Génie énergétique, BE 2 480.

P. Lemoine, *Réfrigérants atmosphériques*, Techniques de l'Ingénieur, Traité Génie énergétique, BE 2 481.

B. Manas, *Aéroréfrigérants secs*, Techniques de l'Ingénieur, Traité Génie énergétique, BE 2 482.

G. Manfrida, S. Stecco, *Le Turbomacchine*, Pitagora Editrice, Bologne, 1990.

A. F. Mills, *Cooling towers, The CRC handbook of thermal engineering*, Ed. F. Kreith, CRC Press, Boca Raton, FL, 2000, ISBN 0-8493-9581-X.

M. Pluviose, *Turbomachines hydrauliques et thermiques*, Exercices commentés, Eyrolles, Paris, 1988.

6

Second law, entropy and exergy

Introduction

In the first part of this book, we introduced the basic concepts of thermodynamics without resorting to the notion of entropy, which can be difficult to master for some learners. In this chapter, we introduce it as simply as possible and discuss the main implications of the second law of thermodynamics, in particular the Carnot cycle and the irreversibilities that take place in real processes.

The (T, s) entropy chart is presented, and the cycles of the simple energy systems are plotted in it, with a discussion about their difference with the Carnot cycle.

Finally, a significant place is devoted to exergy methods that are increasingly regarded as among the best suited to perform optimization studies, as they can take into account both the amount of energy put into play and its quality. To simplify the construction of exergy balances, we use the productive structures, network of productive and dissipative units (PDUs).

Entropy

Heat in thermodynamic systems

In the first part, we indicated that the heat exchanged by a system with the surroundings is, for an infinitesimal process, given by the following expression:

$$\delta Q = C_p dT - vdP$$

It expresses an experimental fact, which is the essential basis of the thermodynamics of compressible fluids: heat δQ exchanged with the surroundings is a linear function of the thermodynamic state of the system.

This equation is, however, valid only if there is no irreversibility inside or at the boundary of the fluid mass. If any, the relationship becomes:

$$\delta Q < C_p dT - vdP$$

We can then write:

$$\delta Q = C_p dT - vdP - \delta\pi$$

$\delta\pi$, essentially a positive term, has a very simple physical meaning: it is the heat generated by mechanical friction within the fluid. A straightforward interpretation for isobaric heating is that the fluid temperature rise dT is greater than $\delta Q/C_p$ because of irreversibilities.

Although it differs profoundly from heat received from the surroundings, it changes the thermodynamic state of the system in the same way.

DOI: 10.1201/9781003175629-6

δQ is the **heat exchanged with the surroundings**, counted positively if it is received by the system and negatively otherwise, and $\delta\pi$ the **heat dissipated by internal friction and shocks** if any. It is always positive or zero.

In practice, it is important to distinguish these two forms of heat; otherwise, serious errors of reasoning can be made. In particular, processes without heat exchange with the surroundings, called adiabatic, are such that $\delta Q = 0$, whether the seat of irreversibility or not, that is to say whether $\delta\pi$ is zero or not.

Introduction of entropy

In the first part, we stressed how important in practice are **reference processes** corresponding to the processes that fluids would undergo in perfect machines.

We have also shown that, for compressors and turbines, the reference process is the reversible adiabatic, and the corresponding law for a perfect gas is given by $Pv^\gamma = \text{Const}$ with $\gamma = C_p/C_v$.

It is obtained by solving the differential equation $\delta Q = 0 = C_p\, dT - v\, dP$, which is quite simple replacing v by rT/P.

Given the importance of this law, it is interesting to try to generalize it to find a formulation that is valid, on the one hand, for all fluids and not only perfect gases, and, on the other hand, for all the processes, whether or not they bring into play irreversibilities.

Let us come back to the equation:

$$\delta Q = C_p dT - v dP - \delta\pi$$

which is also written:

$$\delta Q + \delta\pi = C_p dT - v dP$$

This equation is an inexact differential and not an exact differential. We call an integrating factor an expression by which one multiplies an inexact differential in order to turn it into an exact differential.

We can show that in the general case as for a perfect gas, $1/T$ is one of the simplest integrating factors for this equation. For the perfect gas, this integral expression is written:

$$s = s_0 + C_p \ln\frac{T}{T_0} - r\ln\frac{P}{P_0}$$

As can be seen, function s is a state function formally very close to the calorimetric equation providing heat exchanged by a system with its surroundings, as it can be deduced by using as integrating factor $I = 1/T$. It is called the **entropy of the system**.

In the $(h, \ln(P))$ charts that we presented in the first part, the isovalue curves of the reversible adiabatic are those for which $ds = 0$. They are called **isentropes**.

The concept of entropy is thus introduced naturally, and entropy is the state function closest to heat exchanged with the surroundings. There is no evidence in this approach, however, that this concept remains valid for nonperfect gases. With the more axiomatic definition we will give by introducing the second law, a perfect gas becomes a special case of the general theory.

Second law of thermodynamics

Limits of the first law of thermodynamics

A major limitation of the first law of thermodynamics is the failure to take into account the quality of energy: indeed, different forms of energy expressed in kWh are equivalent, but the possibilities of converting one form of energy into another are not.

Thus, work can still be fully converted into heat, but the converse is not true. Work is one form of energy whose quality is among the best, and which can therefore be taken as a reference.

We can rephrase this by saying that a possible indicator of the energy quality is its **ability to be converted into work**.

The first law postulates the equivalence of different forms of energy, but it does not take into account an essential experimental fact, which is that when a system interacts with its surroundings, the energy processes that it undergoes can only take place in a privileged sense, which cannot be reversed without a qualitative change in the system.

Concept of irreversibility

One can convert electricity into work using an electric motor of efficiency over 98%, or conversely convert mechanical work into electricity through a generator of equivalent efficiency, which means that these two forms of energy are about the same quality.

In the example of power conversion work we have just given, we stated that the machines used had excellent efficiencies, close to but slightly lower than 1: experience indeed shows that, whatever precautions are taken, some energy is degraded. The first law teaches us that the total energy is conserved, but some of its quality declines and ultimately ends up as heat, because of friction, Joule losses etc.

These losses are called irreversibilities, because the process work → electricity → work is not completely reversible: part of the initial work is converted into heat.

Irreversibilities encountered in energy facilities that interest us, except those taking place in combustion reactions, can be grouped into two broad classes that we discuss briefly a little later:

* Irreversibilities stemming from temperature heterogeneity;
* Mechanical irreversibilities due to viscosity.

Heat transfer inside an isolated system, conversion of heat into work

In the particular case of heat, it is always experimentally verified without exception that the transfer of heat between two media at different temperatures is from the warmer body (the one whose temperature is higher) to the colder (whose temperature is lower).

Also, a heterogeneous system composed of two media at different temperatures not isolated from each other always evolves towards a homogeneous state at intermediate temperature.

In addition, when attempting to convert heat into work, experience proves that it is first necessary to have two heat sources: one at high temperature and the other at low temperature (Figure 6.1).

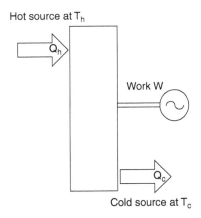

FIGURE 6.1
Conversion of heat into work.

Moreover, the larger the temperature difference between the sources, the greater the amount of work that can be converted from 1 kWh of heat. Temperature thus also appears as a possible indicator of the heat quality.

A further remark can be made at this stage: it is the existence of a temperature difference between two bodies that allows for work production.

Statement of the second law

The second law complements the first by introducing a function called entropy, which is used to quantitatively characterize the effects of irreversibilities taking place in a system and explain the phenomena we have just discussed.

A rigorous and comprehensive presentation of the second law requires significant developments due to the precautions that must be taken first in writing assumptions and explanations of their connection with experience, and second in demonstrations which must be made.

Given our objectives, this is not justified here, especially as for applied thermodynamic calculations of practical interest, the main advantage of this second law can be summarized in two points:

* First, entropy is, as we have shown, the state function most closely related to the heat Q exchanged with the surroundings. It thus intervenes implicitly or explicitly in many equations governing the operation of energy components;
* Second, the generation of entropy alone allows us to quantify all the irreversibilities taking place in these components and at their boundaries, which is fundamental.

The second law states that entropy s has the following properties:

* s is a function of the system state variables;
* In any elementary process involving heat exchange δQ with the surroundings, we have $\delta Q \leq T ds$; equality is satisfied if and only if the process is perfect (not irreversible).

Writing: $ds = \dfrac{\delta Q}{T} + d_i s$

$d_i s$, positive or zero, is called "entropy generation".

This relationship can also be written as we mentioned above:

$$\delta Q + \delta \pi = T ds$$

$\delta \pi$ is called "uncompensated work" or "uncompensated heat". It is positive for an irreversible process, and zero otherwise.

Carnot effectiveness of heat engines

Power thermal machines are designed to transform heat into mechanical energy. Let us assume that heat to transform is provided by an external source or reservoir, said heat source, whose temperature T_1 is set. A second source or reservoir, known as cold source, is required to evacuate part of the heat. Its temperature T_2 is necessarily lower than T_1.

The second law is due to S. Carnot, who in 1824 showed that the efficiency of an ideal heat engine cycle, is given by: $\eta = 1 - \dfrac{T_2}{T_1}$

Carnot has shown that this efficiency does not depend on the nature of the machine and fluids used, but only T_1 and T_2.

In a (T, s) chart where s is the x-axis and T the y-axis, called entropy chart which we will present later in more detail, this cycle is represented by the rectangle ABCD, described clockwise (Figure 6.2).

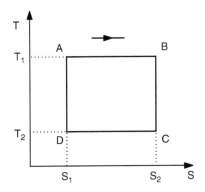

FIGURE 6.2

Carnot cycle.

- AB is a segment of isotherm T_1 described from left to right. The fluid receives heat (isothermal expansion in the case of a perfect gas);
- CD is a segment of isotherm T_2 described from right to left. The fluid releases heat (isothermal compression in the case of a perfect gas);
- BC is a segment of isentrope S_2 described from top to bottom (reversible adiabatic expansion in all cases);
- DA is a segment of isentrope S_1 described from bottom to top (reversible adiabatic compression in all cases).

This relation is probably rightly one of the best known in thermodynamics. It is of great practical importance, given its implications:

- First, with the assumptions made (reversible cycle without friction and heat transfer without temperature differences), it is easy to show that this effectiveness is the highest that can be achieved by a simple fluid heat engine operating between the two sources at T_1 and T_2. It is therefore a maximum limit, and real cycles generally have much lower efficiencies;
- Second, the Carnot effectiveness of all reversible machines operating between the two sources at T_1 and T_2 is the same and only depends on their temperatures, not on the working fluid used;
- For these reasons, we have preferred to speak of effectiveness; some authors consider that the efficiency term generally used to describe this limit cycle is not the most appropriate, and it is better to keep it to describe the performance of a real machine compared to that of an ideal machine;
- In the foregoing, we were interested in power heat engines. A second class of heat engines is used in reverse cycles to transfer heat from a cold source to a hot one, thanks to a mechanical power input. This is called a refrigeration cycle, or heat pump;
- The above reasoning can be transposed to the case of these reverse cycles, and it can be shown that the maximum effectiveness of these cycles (their coefficient of performance COP) is equal to that of the reverse cycle, given by Equation (6.1) or (6.2), depending on whether one looks at the heat exchanged with the cold source (refrigeration cycle) or with the hot one (heat pumps):

$$\eta = \frac{Q_2}{\tau} = \frac{T_2}{T_1 - T_2} \tag{6.1}$$

$$\eta = \frac{Q_1}{\tau} = \frac{T_1}{T_1 - T_2} \tag{6.2}$$

To convert heat into mechanical energy, in almost all the cycles used, the thermodynamic fluid is successively compressed, then heated and finally expanded, as we saw in the first part. If the cycle is open, the fluid is then discharged into the external environment; if the cycle is closed, it is cooled and then compressed again.

The existence of a wide variety of technological solutions is mainly explained by the multiplicity of existing heat sources, as cold sources that can be used in practice are relatively few in number, generally ambient air, or even a river.

Recall that the first law of thermodynamics indicates that the energy of the system is conserved, while Carnot's theorem says that only the fraction η of the heat supplied by the hot source is converted into mechanical energy.

As these two proposals may seem contradictory, some additional explanations are preferable.

Consider a hot source providing a machine with a quantity of heat Q_h at temperature T_h.

The conversion of this heat into a perfect machine exchanging with a cold source at the temperature T_c produces (Figure 6.3):

- On the one hand, a work W equals $\eta\, Q_h$, with η being equal to the Carnot effectiveness;
- On the other hand, a heat Q_c at low temperature equals to $(1-\eta)\, Q_h$.

The first law is duly respected, since Q_h is equal to $W + Q_c$, and Carnot's theorem has made it possible to determine η, which represents the fraction of the heat at high temperature which is converted into work.

Table 6.1 and Figure 6.3 show you how Carnot's effectiveness varies when the temperature of the hot source varies from 100°C to 1,300°C; the temperature of the cold source is 15°C.

The values reached for different types of cycles converting heat into mechanical power or into electricity are indicated.

In practice, the actual efficiency is much lower than that of the Carnot cycle, on the one hand, because the hypotheses which make it possible to establish this formula are almost

TABLE 6.1

CARNOT EFFICIENCY, $T_0 = 15°C$

	T_1 (°C)	Carnot Eff. (%)
High-temperature gas turbine	1,300	82
Flame steam power plant	560	65
PWR nuclear power plant	300	50
Vacuum tube collector solar thermal power plant	100	23

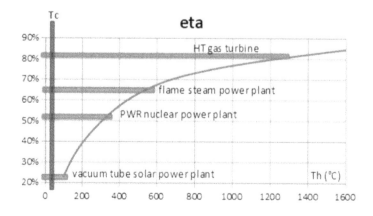

FIGURE 6.3

Carnot's effectiveness of various power plants.

never met, and on the other hand, due to the imperfections of the machines used, which have the effect of lowering their performance because they generate what we call irreversibilities.

The differences of a real cycle with the Carnot cycle come from the following points, among other things:

- First of all, in practice there must be a certain temperature difference between the machine and the hot and cold sources, which is a primary cause of irreversibility;
- The heat exchange with the hot and cold sources is isobaric and not isothermal, because, for technological reasons already mentioned, we generally do not know how to achieve isothermal expansion and compression;
- Then, it is exceptional that the hot and cold sources can themselves be considered as isotherms: most often, it is a fluid that exchanges heat between two temperature levels;
- Finally, when compression and expansion are adiabatic, they are not reversible due to mechanical irreversibilities.

The actual engine cycles therefore deviate significantly from the Carnot cycle.

BOX 6.1
Key issues

Converting heat to work, carnot cycle

The following self-assessment activities will allow you to check your understanding of the Carnot cycle:
Converting heat to work, gfe
Differences between actual motor cycles and the Carnot cycle, gfe

Irreversibilities in industrial processes

Heat exchangers

In a heat exchanger, heat can only be transferred between two fluids if a certain temperature difference exists between them. Indeed, for both technical and economic reasons, the exchange surfaces between these fluids are necessarily finite. It is then possible to show that the heat exchange is accompanied by an increase in entropy at the boundary between the two streams (at temperatures T_1 and T_2), given by:

$$d_i s = \frac{T_1 - T_2}{T_1 T_2} \delta Q$$

with δQ being the heat absorbed by fluid 2. These irreversibilities are often described as external, because they take place at the boundary of the system.

The larger the temperature difference between the fluids, the greater the entropy creation. In a heat exchanger, the energy (enthalpy) transferred by the hot fluid to the cold fluid is conserved, but because the transfer is taking place with temperature decrease, entropy increases.

In energy systems, the temperature difference that must exist between two fluids that exchange heat generally constitutes a significant source of irreversibility.

Compressors and turbines

We stated above that the compression and expansion devices are generally adiabatic ($\delta Q = 0$).

In an adiabatic reversible compressor or turbine, there is no creation of entropy. We say that evolution is **isentropic**, a very important concept in practice, because it characterizes the reference process against which actual changes are characterized.

In real machines, as we saw in the first part of this presentation, we can take into account irreversibilities by introducing what we called an isentropic efficiency, equal to the ratio of isentropic work to real work for compressors, and the reverse for turbines, i.e., the ratio of actual work to isentropic work.

(T, s) entropy chart

In the first part, we excluded entropy from variables we have considered to represent the fluid properties. Now that we have introduced this state function, it is natural to question its relevance in this regard.

Experience shows that the entropy and Mollier charts, which use as coordinates pairs (T, s) and (h, s), are particularly interesting, especially the first, to which we limit our analysis here.

In the entropy chart (Figure 6.4), entropy is the abscissa and temperature the ordinate. The vaporization curve again separates the plane into two areas, defining the two-phase zone and the zone of simple fluid. The critical point C is still at its maximum.

One of the advantages of such a chart (see Figure 6.5) is that any perfect cyclic process is reflected in the plane (T, s) by a contour (Γ), whose area A measures, depending on the sign, either the amount of heat Q brought into play, or the work provided or received τ.

Indeed, $Q = \int T \, ds$ by definition.

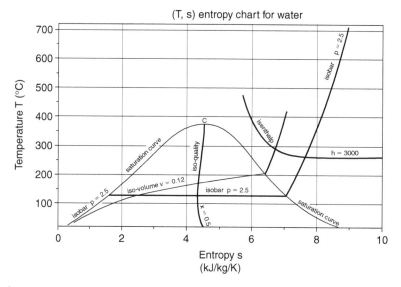

FIGURE 6.4

Water (T, s) entropy chart.

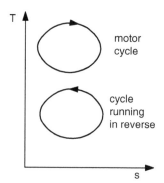

FIGURE 6.5

Types of cycles in the entropy chart.

As the cycle is closed, Q + τ = 0, and |τ| = A.

The rule of signs is as follows:

- If the cycle is described clockwise, work is negative, thus transferred by the fluid to the surroundings: we talk of a power cycle or of a cycle running forward;
- If the cycle is described counterclockwise, it is the opposite: we talk of a refrigeration cycle, or of a cycle running in reverse.

Form of isobars

To the left of the saturation curve, liquid isobars are curved upward. Liquid isentropic compression has almost no effect on the temperature; liquid isobars are virtually merged with the ascending branch of the vaporization curve. The chart is very imprecise in this area, and it is preferable to use a table or a code giving the thermodynamic properties along the vaporization curve.

Inside the two-phase zone, temperature and pressure are bound by the saturated pressure law, and isobars are horizontal.

Right of the vaporization curve, isobars are rising and, for an ideal gas, become exponential. They can be deduced from each other by horizontal translation.

If the pressure exceeds the critical pressure, isobars are strictly ascending curves, which do not intersect the vaporization curve.

Form of isenthalps

In the vicinity of the saturation curve, isenthalps are curved downward, with a strong negative slope.

Gradually as the distance from the vaporization curve increases, the gas approaches the corresponding ideal gas, and isenthalps become horizontal, as enthalpy is a sole function of temperature.

Of course, the plot of an isentropic process is very simple in this chart: it is a vertical segment.

BOX 6.2
Key issues

Entropy chart

The following self-assessment activities will allow you to check your understanding of the entropy chart:

Placement of points in a (T, s) entropy ideal gas chart, ddi

Placement of points in a (T, s) entropy condensable vapor chart, ddi

Plot of cycles in the entropy chart, qualitative comparison with the Carnot cycle

The Carnot cycle is the one that leads to the best effectiveness; it is almost always interesting to compare actual cycles to it. In this section, we perform such a comparison for each of the cycles that we studied in the first part.

Steam power plant

In the water entropy chart (Figure 6.6), to increase readability, we have not shown isovolumes. Points 1 and 2 showing compression in the liquid state are almost superimposed, and

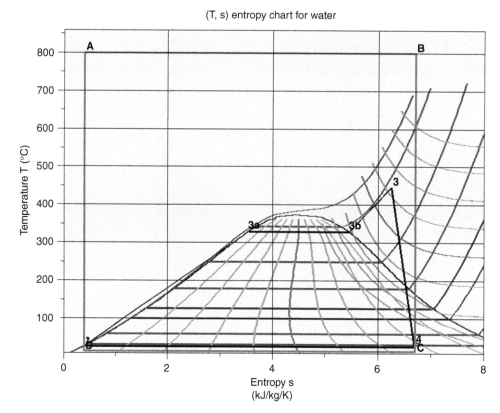

FIGURE 6.6
Steam power plant and Carnot cycles in the entropy chart.

heating in the liquid state almost coincides with the liquid saturation curve. Vaporization is done in a horizontal line segment.

The isobaric superheating (3) is the maximum peak of the cycle, and the irreversible expansion is reflected by an increase of entropy; the point 4 is located within the vapor–liquid equilibrium zone (quality equal to 0.775).

BOX 6.3
Guided exploration

Exploration of a steam plant in the entropy Chart (C-M4-V8)

The objective of this guided exploration is to make you discover the cycle of a steam power plant in the entropy (T, s) thermodynamic chart.

It completes exploration (S-M3-V7), where the cycle was presented, with explanations on its settings and its representation in the (h, ln(P)) chart.

In order to plot the Carnot cycle (A–B–C–D), we have assumed that $T_1 = 800°C$ and $T_2 = 15°C$ (in practice, T_1 is higher). The comparison of both cycles leads to the following comments:

1. The condenser (process 4–1) is a heat exchanger of finite size; condensing water cannot be at the same temperature as the cold source, which is a first difference with the Carnot cycle. However, we can consider that the heat exchange is nearly isothermal and irreversibilities are small;

2. Compression (process 1–2) in the pump may at first approximation be assumed isentropic; its irreversibilities are low. However, it deviates significantly from the Carnot cycle because the compression end temperature is about the same as that of the cold source, rather than that of the hot one;

3. We have seen that the isobaric pressurized water heating is done in three steps:
   - Liquid heating in the economizer up to the saturation temperature at the pressure considered (process 2–3a);
   - Vaporization at constant temperature in the evaporator (process 3a–3b); Superheating up to the maximum temperature in the boiler cycle (process 3b–3). Given the shape of isobars on the entropy chart, this isobaric heating process deviates much from the Carnot cycle, which states that the heat engine exchanges heat at constant temperature with the hot source. In particular, the temperature difference with the hot source is maximum in the economizer: heat at high temperature is used to warm water below 100°C.
4. Turbine (process 3–4) has an isentropic efficiency close to 0.85. Again, the difference with the Carnot cycle is significant.

As illustrated in the figure, the shape of the steam power plant differs appreciably from that of the Carnot cycle: it looks rather like a triangle than a rectangle, and its surface is much smaller. There is room for improvement as we shall see in Chapter 8.

Gas turbine

In the air entropy chart (Figure 6.7), irreversible compression and expansion result in an increase of entropy (air inlet-2 and 3–4). Heating in the combustion chamber is isobaric. However, as the working fluid composition changes during the combustion, the properties of the burnt gases are not exactly the same as those of air, and we should not in principle plot points 3 and 4 on the same chart as points 1 and 2. There is indeed a change in entropy due to the combustion, which explains why points 2 and 3 and points air inlet and 4 seem not to be on the same isobar.

In order to plot the Carnot cycle (A–B–C–D), we have assumed that $T_1 = 1,150°C$ and $T_2 = 15°C$. The comparison of both cycles leads to the following comments:

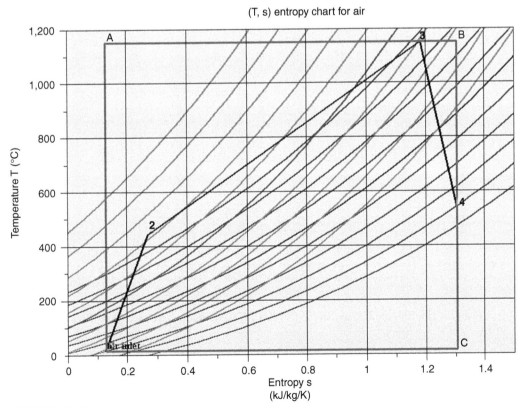

FIGURE 6.7

Gas turbine and Carnot cycles in the entropy chart.

BOX 6.4
Guided exploration

Exploration of a gas turbine in the entropy chart (C-M4-V9)

The objective of this guided exploration is to make you discover the cycle of a gas turbine in the entropy (T, s) thermodynamic chart.
It completes exploration (S-M3-V8), where the cycle was presented, with explanations on its settings and its representation in the (h, ln(P)) chart.

1. Compression (process 1–2) cannot be assumed to be isentropic, due to irreversibilities taking place in the compressor. It departs from the Carnot cycle;
2. Combustion (process 2–3) takes place at constant pressure. Given the shape of the isobars in an entropy chart (similar to exponential), the difference is important between this process and the Carnot cycle, which states that the heat engine exchanges heat at constant temperature with the hot source;
3. Turbine (process 3–4) has an isentropic efficiency close to 0.9. Again, the difference with the Carnot cycle is significant;
4. Hot gases leaving the turbine are then discharged directly into the atmosphere, which is an important form of irreversibility, since their heat is lost.

As illustrated in the figure, the shape of the gas turbine cycle differs appreciably from that of the Carnot cycle: it looks rather like a diamond than a rectangle, and its surface is much smaller. Improvements can be obtained by conducting cooled staged compression, staged expansion with reheats, regeneration and combined cycle arrangement.

Refrigeration machine

In the R134a entropy chart (Figure 6.8), to increase readability, we have not shown the iso-volumes. Point 1 (partly masked by point C), slightly superheated, is placed on isobar 1.8 bar right of the saturated vapor curve. Irreversible compression results in an increase of entropy. Cooling with outside air has three stages: desuperheating (2–3a) in the vapor phase, condensation along the horizontal segment (3a–3b) and a slight subcooling (3b–3), which almost coincides with the liquid saturation curve.

Isenthalpic throttling (3–4) leads to an increase of entropy; point 4 is located within the zone of vapor–liquid equilibrium (quality equal to 0.363).

BOX 6.5
Guided exploration

Exploration of a refrigeration installation in the entropy chart (C-M4-V10)

The objective of this guided exploration is to make you discover the cycle of a refrigeration installation in the entropy (T, s) thermodynamic chart.
It completes exploration (S-M3-V9), where the cycle was presented, with explanations on its settings and its representation in the (h, ln(P)) chart.

In order to plot the reverse Carnot cycle (A–B–C–D), we have assumed that $T_1 = 40°C$ and $T_2 = -8°C$.

Unlike previous power cycles, the reverse Carnot cycle appears inside the actual refrigeration cycle. The main difference in the shape stems from the compression curve (1–2–3a) due to the shape of isobars in the vapor zone. The surface of the actual cycle is larger than that of the Carnot cycle.

The comparison of both cycles leads to the following comments:

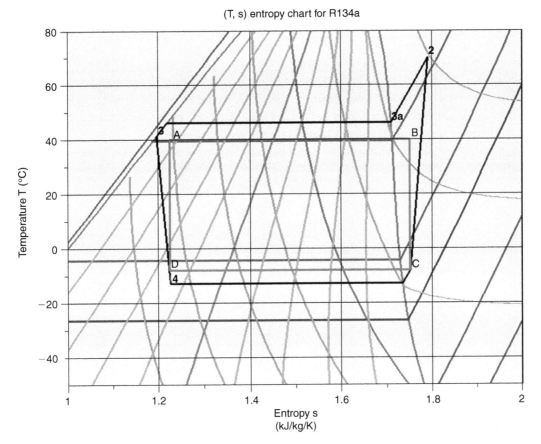

FIGURE 6.8

Refrigeration machine and reverse Carnot cycles in the entropy chart.

1. The evaporator is a heat exchanger of finite size; the refrigerant cannot be at the same temperature as the cold chamber, which is a first difference with the reverse Carnot cycle. However, we can consider that the heat exchange is nearly isothermal;

2. Compression cannot be assumed to be isentropic, due to irreversibilities taking place in the compressor, which induces a new gap with the reverse Carnot cycle;

3. Cooling and condensation of the refrigerant by heat exchange with ambient air generally cannot be isothermal, given the thermodynamic properties of refrigerants: you must first desuperheat the vapor, then condense it. Given the shape of isobars on an entropy chart, the difference with the reverse Carnot cycle is very important here;

4. The expansion of the condensed refrigerant could theoretically be close to isentropic, but the technological reality is different, and this is for three reasons: first, the expansion of a two-phase mixture is done, unless special precautions are taken, with low isentropic efficiency; second, the work at stake is very small; and finally, especially for low-capacity refrigerators, there is no suitable and cheap expansion device. Therefore, in practice, one resorts to expansion valves or even simple capillary static devices that perform isenthalpic throttling. Again, the gap with the reverse Carnot cycle is significant.

Exergy

The theory of exergy aims to develop an integrated analysis method that includes the first two laws of thermodynamics, and thus allows us to take into account both the amount of energy put into play and its quality, which the first law cannot do. Its value is that it provides a quite rigorous thermodynamic framework to quantify the quality of any system, open or closed, in steady state or not.

The first to introduce this concept was G. Gouy, who late in the nineteenth century, defined the concept of usable energy, now renamed exergy by many authors. The basic idea is to consider that a thermodynamic system interacts with what is called its environment, which behaves like an infinite reservoir at constant temperature and pressure and fixed composition, meaning that the system studied is small enough not to disturb this environment.

The environment will serve, for example, as a cold source for a power cycle, or hot source for a refrigeration cycle. Since the environment state determines the performance of the system studied, the theory of exergy can take it into account implicitly.

In the framework of this book, we content ourselves with a brief and steady-state (and therefore not rigorous) presentation of exergy analysis. Readers wishing to engage further in this direction will find especially in the work of Borel and Favrat (2005) or (Moran, 1989) a comprehensive theory of the subject.

Presentation of exergy for a monothermal open system in steady state

Let us consider a steady-state open system exchanging energy with its environment assumed to be at uniform temperature T_0 (Figure 6.9).

The energy equation is given by the first law:

$$\Delta h + \Delta K = \tau + Q \qquad (6.3)$$

and the entropy production by the second law:

$$Q + \pi = T_0 \Delta s \quad 0 \leq \pi$$

$$\Delta s - Q/T_0 \geq 0$$

If we assume that all the heat is provided by the environment, and T_0 is the temperature at the boundary of the open system, Q can be eliminated by combining these two equations, giving:

$$\tau = \Delta h + \Delta K - Q \geq \Delta h + \Delta K - T_0 \Delta s$$

$$-\tau \leq -\left(\Delta h + \Delta K - T_0 \Delta s\right)$$

In most cases, we can neglect the kinetic energy changes ΔK.

For an open system, function $x_h = h - T_0 s$ is usually called **exergy**.

The maximum work that the system can provide is equal to the reduction of its exergy.

In practice, the actual work provided by the system may be less than this value. We define the exergy loss Δx_{hi} as being equal to the difference between the maximum work possible and the actual work provided:

$$\Delta x_{hi} = \left(-\tau_{max}\right) - \left(-\tau_{actual}\right) = \left(-\Delta x_h\right) - \left(-\tau_{actual}\right)$$

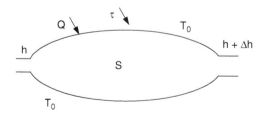

FIGURE 6.9

Monothermal system.

$$\Delta x_{hi} = T_0 \Delta s - \Delta h - \left(-\tau_{actual}\right)$$

If the process is furthermore adiabatic:

$$\left(-\tau_{actual}\right) = -\Delta h \text{ and } \Delta x_{hi} = -\Delta x_h + \Delta h = T_0 \Delta s$$

Note that this description is not entirely satisfactory because it is clear that for a closed system, the definition of exergy leads us to consider function $u - T_0 s$, so that there is some ambiguity.

Given the success of the concept of usable energy introduced by G. Gouy, largely used in the Anglo-Saxon world under the name availability, we could also speak of usable enthalpy or usable energy, depending on whether one considers an open or closed system.

Beyond the question of vocabulary, we note that this new function is characterized by the introduction of a variable outside the system, T_0. As such, it is no longer a state function in the strict sense. However, as T_0 is generally assumed to be constant, the exergy is sometimes presented as a linear combination of state functions.

Another remark can be made at this stage: exergy allows us, through the product of entropy by the environment temperature, to express variations of this state function in one dimension and by orders of magnitude similar to those usually encountered by the engineer in energy calculations, which facilitates the understanding of physical phenomena.

Finally, as discussed below, by this artifice, we manage to formally eliminate from the calculations the "free" heat source represented by the environment.

Exergy can be used to evaluate the quality of the changes made in actual processes, compared to ideal reversible ones.

You will sometimes encounter the notion of dead state used in exergy analyses. It can be defined as the state in which the system is in mechanical, thermal, and chemical equilibrium with its surroundings. Its exergy is then equal to zero.

Multithermal open steady-state system

Let us now consider an open system that exchanges work and heat with n external sources at constant temperatures T_k, and environment taken at temperature T_0 (Figure 6.10).

Applying the first law gives here

$$\Delta h + \Delta K = \tau + Q_0 + \sum_{k=1}^{n} Q_k$$

and entropy production is given by the second law:

$$\Delta s = \frac{Q_0}{T_0} + \sum_{k=1}^{n} \frac{Q_k}{T_k} + \Delta s_i$$

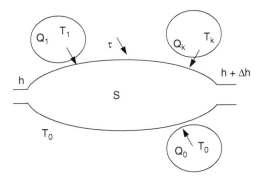

FIGURE 6.10

Multithermal open system.

with Δs_i, positive or zero, being the entropy generation. Neglecting kinetic energy changes, we get:

$$\Delta x_h = \Delta\left(h - T_0 s\right) = \tau + \sum_{k=1}^{n}\left(1 - \frac{T_0}{T_k}\right)Q_k - T_0 \Delta s_i \tag{6.4}$$

or, with $\Delta x_{hi} = T_0\,\Delta s_i \geq 0$:

$$\tau + \sum_{k=1}^{n}\left(1 - \frac{T_0}{T_k}\right)Q_k - \Delta x_h - \Delta x_{hi} = 0 \tag{6.5}$$

τ is the work received by the system, Δx_h the exergy variation of the fluid passing through it, and Δx_{hi} the exergy dissipation resulting from irreversibilities. We call the Carnot factor the term $\theta = 1 - T_0/T$. This is the factor by which we must multiply a quantity Q of heat available at temperature T to obtain the value of its exergy. If we call useful heat the quantity $x_q = \theta\,Q$, Equation (6.5) gets the form:

$$\tau + \sum_{k=1}^{n}\theta_k Q_k - \Delta x_h - \Delta x_{hi} = 0$$

$$-\tau = \sum_{k=1}^{n} x_{qk} - \Delta x_h - \Delta x_{hi} \tag{6.6}$$

The maximum engine work that can provide an open system is equal to the sum of heat-exergies of the sources with which it exchanges heat, less the change in exergy of the fluid passing through it and the exergy destroyed because of irreversibilities.

Application to a two-source reversible machine

Let us consider a cyclic two-source reversible machine operating between a source at temperature T_1 and environment at temperature T_0.

On a cycle, we have: $\Delta x_h = 0$, $\Delta x_{hi} = 0$ (reversible machine).

Applying formula (6.6) gives: $-\tau = x_{q1} = \left(1 - \frac{T_0}{T_1}\right)Q_1$

If $T_0 < T_1$, $x_{q1} \geq 0$: the system receives heat exergy from the hot source at T_1 and converts it into work τ.

Here, we find the Carnot cycle, whose effectiveness is in this case $\eta = 1 - \dfrac{T_0}{T_1}$

If $T_0 > T_1$, $x_{q1} \leq 0$: the system gives heat exergy to the cold source at T_1, taking it from the hot source at T_0. For this, we must provide work τ. It is then a refrigeration cycle or heat pump (Equations 6.1 and 6.2).

Special case: heat exchange without work production

In the special case where there is no work produced, Equation (6.6) can be rewritten as:

$$\Delta x_{hi} = \sum_{k=1}^{n} x_{qk} - \Delta x_h \tag{6.7}$$

The exergy destroyed in the process is equal to the sum of the heat-exergies of the various sources, minus the change in exergy of the fluid passing through the system.

Exergy efficiency

Exergy allows us to rigorously define the concept of system efficiency, and therefore to quantify its thermodynamic quality: it is the ratio of exergy uses to exergy resources. It is always between 0 and 1, and so is much greater than the irreversibilities are lower. Exergy resources is the sum of all exergies provided to the cycle from outside. Exergy uses represents the outgoing exergies that can be valued. Those that are lost, like those released into the atmosphere, are not included.

$$\eta = \frac{\text{exergy uses}}{\text{exergy resources}} \tag{6.8}$$

Energy and exergy balances

The performance analysis of various technologies discussed in this book leads in a conventional way to calculate their energy balances. We will also show the additional interest that building exergy balances may have, which poses no particular problem but needs to be done very carefully as otherwise errors can be committed.

In the sections dealing with the various applications, we will see how in practice the energy and exergy balances can be established, and how they differ. We limit ourselves here to present the principles, suggesting the reader wishing to deeper explore the question to refer to the literature, relatively abundant today, including the books of Borel and Moran cited in reference (Borel & Favrat, 1987; Moran, 1989).

Note that the introduction of exergy balance is a relatively difficult exercise, which is available only to experienced energy engineers and scientists by conventional methods. The one that we recommend presented below makes it possible to avoid a good part of these difficulties.

Energy balances

For a system in steady state (in which case we will restrict ourselves), establishing an enthalpy balance is simply to recognize rates of heat flux, useful work and enthalpy at boundaries, and possibly in special cases like a combustion reaction (although this seems to be in contradiction with the first law) to reintroduce a term of power generation in the volume. Thus, the general equation of balance for a control volume becomes:

$$\left\{ \begin{array}{c} \text{inward} \\ \text{transport by} \\ \text{the surface} \end{array} \right\} - \left\{ \begin{array}{c} \text{outward} \\ \text{transport by} \\ \text{the surface} \end{array} \right\} + \left\{ \begin{array}{c} \text{transfer} \\ \text{through} \\ \text{the surface} \end{array} \right\} + \left\{ \begin{array}{c} \text{generation} \\ \text{in the} \\ \text{volume} \end{array} \right\} = 0 \tag{6.9}$$

Building an energy balance does not generally pose a particular difficulty.

These balances are very useful and widely used, particularly in engineering and design departments. However, they have an important limitation: derived from the first law, they do not take into account the quality of energy, so that 1 kWh of electricity has the same value as 1 kWh of heat, and this is irrespective of temperature level. To reflect this quality, we must add the state function introduced by the second law, entropy.

The theory of exergy, succinctly presented above, provides a quite rigorous framework to quantify the thermodynamic quality of any system, open or closed, in steady state or not. It is increasingly accepted as the preferred tool to compare and optimize thermodynamic cycles, thanks to exergy balances.

Exergy balances

Exergy depends on both the system state and the chemical composition of its elements. It differs from energy in that the latter is preserved, while exergy is destroyed whenever irreversibilities exist. Since it is not a conservative quantity, Equation (6.9) cannot be applied and a specific approach should be used.

Let us recall some results established above. For an open multitherm system in steady state traversed by a constant flow of fluid, which exchanges work and heat with n external sources at constant temperatures T_k, and the environment at temperature T_0, the exergy Equation (6.6) can be rewritten as:

$$\Delta x_{hi} = \tau - \Delta x_h + \sum_{k=1}^{n} x_{qk} \tag{6.10}$$

Δx_{hi} represents the exergy dissipation resulting from irreversibilities, τ the work received by the system, Δx_h the exergy variation of the fluid passing through it. $\theta_k = 1 - T_0/T_k$ is called the Carnot factor. $x_{qk} = \theta_k Q_k$ is the heat exergy or useful heat received from source at temperature T_k.

Let us specify that, in what we call irreversibilities, there is at the same time the exergy destroyed and the exergy lost by the system.

Equation (6.7) shows that the maximum work that an open system can provide is equal to the sum of heat exergies of sources with which it exchanges heat, minus the change in exergy of the fluid passing through it and the exergy destroyed due to irreversibilities.

When the heat sources are no longer at constant temperature, when several shaft systems must be considered and the system is crossed by several fluids, Equation (6.10) must be replaced by a more complicated one.

To calculate irreversibilities of each component j, it becomes (6.11):

$$\Delta x_{hij} = \sum_{k=1}^{n} x_{qk} - m_j \Delta x_{hj} + \tau_j \tag{6.11}$$

If the heat exchange takes place at variable temperature:

$$x_{qk} = \int_{T_a}^{T_b} \left(1 - \frac{T_0}{T_k} \right) \delta Q_k \tag{6.12}$$

Assuming a linear variation of heat exchange function of T:

$$x_{qk} = \left(1 - \frac{T_0}{T_b - T_a} \right) \ln \left(\frac{T_b}{T_a} \right) Q_k$$

To calculate irreversibilities in a heat exchanger, simply add member by member equations (6.11) related to each fluid flowing through it. If it is adiabatic (no heat exchange with the surroundings), heat exergies are zero. As the useful work is zero, the irreversibility is equal to the sum of exergy variations of both fluids.

As you will be able to see in the exergy balances of systems involving combustions which will be presented in the different chapters of this book, combustion chambers and boilers are the seat of strong irreversibilities, so that the exergy efficiency of these components does not exceed approximately 75%, and often represents around 50% of the total irreversibilities of the systems studied.

It should be noted in this regard that the calculation of the exergy available at the combustion chamber outlet requires certain special precautions, which will be presented in Chapter 12, when studying the properties of moist gases. Note also that the fuel chemical exergy taken into account in combustion calculations is equal to its Higher Heating Value (HHV).

We introduced earlier the exergy efficiency of a system, always between 0 and 1, and higher than the irreversibilities are low given by Equation (6.8).

If the system consists of components satisfying Equation (6.11), this general definition is mathematically translated by Equation (6.13):

$$\eta_x = 1 - \frac{\Sigma \Delta x_{hij}}{\Sigma \Delta x_{qi}^+ + \Sigma \tau_j^+} \tag{6.13}$$

The denominator represents the exergy provided, i.e., the sum of positive heat exergies and useful work provided to the cycle. The numerator is the sum of the cycle irreversibilities.

Practical implementation in a spreadsheet

In practical terms, the exergy balance of a cycle can be established in the following manner when it is modeled in Thermoptim:

* Once the model is properly set and calculated (it is not actually enough to load a project and export its results: as the exergies are not saved in project files, they must be recalculated for each component), the results file can be exported (line "Export Results" menu "Results Files"). This is a text file that can easily be read in a spreadsheet, and in which different values are calculated for a number of point state functions (intensive variables) and the energy and exergy involved in processes;
* The first lines of this file can be copied into a pre-built spreadsheet. You should take care that the spreadsheet does not always recognize the Java thousands separator, and it may be necessary to make a global suppression of that separator (for this, copy the separator, select "Delete" from the menu "Edit" of the spreadsheet, paste the separator in the search field without putting anything in the replacement field, and then click "Replace All"). You may also have to modify the decimal separator;
* You must then carefully correct by hand the part of the worksheet that makes the exergy balance, as the lines for the different processes should be constructed differently (see Equations 6.14–6.16) depending on whether they relate to adiabatic expansion or compression, heat exchanges with the surroundings or throttling. For the second ones, you must also specify the value of the source temperature. Finally, you should take into account the power inputs received by the system from outside;
* If you use the demo version of Thermoptim, you must modify the spreadsheet by copying one by one the exergies brought into play from the values calculated in the point screens and multiply them by the appropriate process flow rates.

For simple processes, through which flows a fluid flow rate m, Equation (6.10) is rewritten as appropriate:

For adiabatic processes: ($Q = 0$, $\tau = m\,\Delta h$)

$$\Delta X_{hi} = m\Delta x_{hi} = m\Delta h - m\Delta x_h \tag{6.14}$$

For heat exchange processes without work and with heat exchange with a source at temperature T ($\tau = 0$, $Q = m\,\Delta h$)

$$\Delta X_{hi} = -m\Delta x_h + \left(1 - \frac{T_0}{T}\right)m\Delta h \tag{6.15}$$

For throttling processes without work nor heat exchange ($\tau = 0$, $Q = 0$)

$$\Delta X_{hi} = -m\Delta x_h \tag{6.16}$$

In the worksheet in Figure 6.11, the Thermoptim result file appears in the upper part, and in the bottom the lines that need to be corrected. The connection between the lower and the

BOX 6.6
Key issues

Exergy balance spreadsheet

For simple cycles, establishing an exergy balance poses no particular difficulty but needs to be done very carefully; otherwise, errors can be committed.

To facilitate this task, a spreadsheet, named ExerBalanceThopt.xls, has been prepared for you. Downloadable from the Thermoptim-UNIT portal, it gathers a number of worksheets related to the examples illustrating Thermoptim use. It is complemented by a detailed methodological note which explains how to use Thermoptim result files.

In addition, the Diapason e-learning module S06, which deals specifically with exergy balances, will guide you through your first steps, and modules S23, S28 and S32 will help you build the exergy balance of a gas turbine, a steam plant or a vapor compression refrigeration machine.

```
PROCESSES              6
name        inlet point outlet point type       m ?h        m ?Xh       type_ener  flow rate
condenser            4           1 exchange  -2050.7642  -68.268683 other              1
superheater 3b                   3 exchange   890.820093  529.71328 purchased          1
evaporator  3a          3b         exchange   898.274438  489.934436 purchased         1
economiser              2 3a       exchange  1569.38928  584.319094 purchased          1
pump                 1           2 compression  18.2939999  16.5282088 useful           1
turbine              3           4 expansion -1326.0137 -1552.2263 useful              1
```

```
         enthalpy balance                                exergy balance

component      dh         t          Q         Tk       dxq      dxh      dxhi    % overal losses
condenser    -2051               -2051     283.15        0      -68        68            5%
superheater    891                 891      1573        730      530       201           14%
evaporator     898                 898      1573        737      490       247           17%
economiser    1569                1569      1573       1287      584       703           49%
pump            18      18                                        17         2            0%
turbine      -1326   -1326                                     -1552       226           16%

cycle            0   -1308        1308                             0      1446          100%
                                        sigma(xq-)               2754
                                        sigma(tau+)              0.00

energy efficiency         38.94% exergy efficiency       47.5%
```

FIGURE 6.11
Exergy balance of a simple steam cycle.

upper is made by the cell formulas, those in red should not be modified, except in special cases discussed below in italics, while those in blue correspond to the mechanical power put into play, and those in black to heat exchange with outside sources. The important thing to check is first that each line applies to a suitable component (the order of processes in Thermoptim file is unpredictable because it depends on that of their creation), and second that the temperatures T_k are entered correctly. You should correct line by line the balance calculation by copying and pasting.

In addition to this formatting of the bottom of the spreadsheet, it may be necessary to perform a number of changes, depending on the case treated.

Additional explications are given in Section 10.2.3 of the first edition of this book.

Productive structures

Even using a spreadsheet, building the exergy balance of a somewhat complicated system can be difficult in practice, and this is for two main reasons:

*   First, exergy balances are not conservative, and there is no simple way to check the consistency of the result;
*   Second, as many specific settings must be made in the spreadsheet, even though each is relatively simple, and their number increases the risk of error.

It turns out that it is possible to automate the calculation of exergy balances, even for complex systems, by introducing a new type of diagram called by Valero a **productive structure** (Valero et al., 2000).

A productive structure is a graph of the exergy consumptions and productions of the system studied. It can be deduced from the physical diagram and the thermodynamic setting, to which it suffices to add some information on the mechanical power connections, the possibilities of valuing outgoing flows, and the temperatures of the external sources with which the system exchanges heat.

It is made up of productive or dissipative units PDUs, junctions and branches.

Each PDU is characterized by its own product flow, but can also have other outlet flows or subproducts. To produce them, it consumes external resources or products provided by other units. At each junction, a product flow is obtained by summing up inlets of the same nature but from different origins.

The productive structure of the steam plant is given in Figure 6.12.

It can be interpreted as follows: the steam power plant is a machine which receives an external exergy input in the boiler, which corresponds to the three productive units at the top-left of the screen, and by internal recycling, exergy is provided to the pump, at the bottom-left of the screen.

This exergy is partly converted into mechanical form in the turbine and partly dissipated in the condenser. The net power is the fraction of mechanical power not recycled.

Note that the values of exergy flows are displayed on the productive structure.

Once the productive structure is established, the exergy balance is automatically built.

The only information that has to be added to that of the usual Thermoptim project and the productive structure are the temperatures of the sources with which the power plant exchanges heat and the mechanical coupling between the turbine and the pump. We have assumed here that they are worth 10°C for the condenser and 1,300°C for the boiler.

Figure 6.13 shows the exergy balance of the steam plant. The presentation is different from that of Figure 6.11, but the results are the same.

As we have already indicated, one of the great advantages of exergy balances is that they make it possible to quantify the irreversibilities, whereas the comparisons that we have made so far with the Carnot cycle in the entropy chart provided only qualitative information.

In particular, the penultimate column of this balance provides the distribution of irreversibilities between the different components of the system.

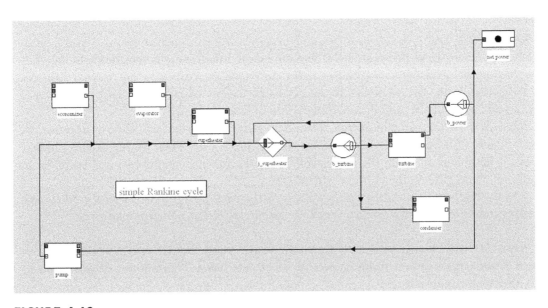

FIGURE 6.12

Productive structure of the steam cycle.

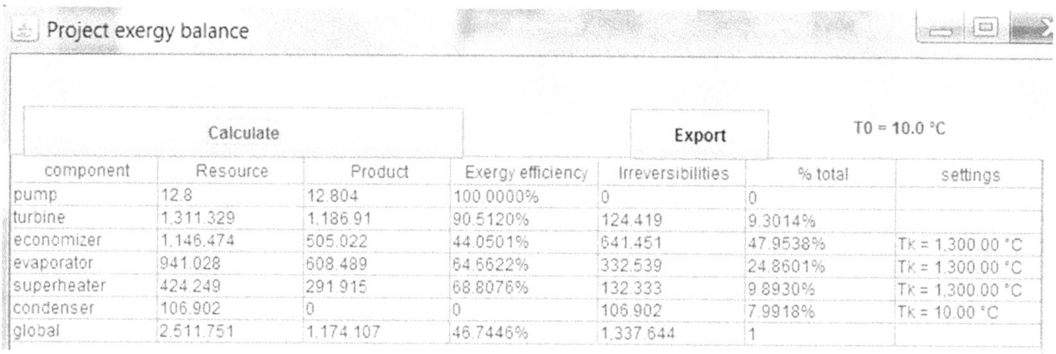

component	Resource	Product	Exergy efficiency	Irreversibilities	% total	settings
	Calculate			Export		T0 = 10.0 °C
pump	12.8	12.804	100.0000%	0	0	
turbine	1,311.329	1,186.91	90.5120%	124.419	9.3014%	
economizer	1,146.474	505.022	44.0501%	641.451	47.9538%	Tk = 1,300.00 °C
evaporator	941.028	608.489	64.6622%	332.539	24.8601%	Tk = 1,300.00 °C
superheater	424.249	291.915	68.8076%	132.333	9.8930%	Tk = 1,300.00 °C
condenser	106.902	0	0	106.902	7.9918%	Tk = 10.00 °C
global	2,511.751	1,174.107	46.7446%	1,337.644	1	

Project exergy balance

FIGURE 6.13

Steam cycle exergy balance.

In the present case of the simple steam cycle, it can be seen that 48.6% of the irreversibilities take place in the economizer.

We will see Chapter 8 how it is possible to reduce them.

Figure 6.14 shows the correspondence that exists between the physical diagram and the productive structure.

Guided explorations BESP1 and BESP2 explain how to operate on a practical level when you use productive structures to build up exergy balances.

Implementation in thermoptim

A specific editor with the various components used in production structures has been developed in versions 2.6 and 2.8 of Thermoptim (Figure 6.15).

It is available from the Special menu screen of the simulator (Ctrl B). You find on the palette icons for processes-points to represent the input and output flows, for PDUs, and for junctions and branches.

You construct a productive structure either manually by selecting the icons on the palette, by placing them on the work plan, and by connecting them with links, analogous to the way a diagram is built, or by automatically generating the productive structure from the simulator and the physical diagram.

Each component of the diagram corresponds to a PDU. These are connected to links that account for the fluxes of exergy variations in the system studied.

Additional explications are given in Section 10.3 of the first edition of this book as well as in the Thermoptim-Unit portal.[1]

Conclusion

The great advantage of this approach is that the exergy balances are automatically built, while the spreadsheet calculations are asking us each time to take great care if we are to avoid mistakes.

Indications in the last column of the exergy balance allow one to easily verify the setting of the productive structure, and thus reduce the risk of error.

We will show the value of exergy balances in Part 3, when we will seek to improve the performance of the simple cycles which we studied previously. Their exergy balances indeed make it possible to quantify the irreversibilities of the various components and therefore to know on which to focus the improvement efforts.

The method we present, based on productive structures, proves to be a useful guide which secures and simplifies the work to a very large extent. Indeed, productive structures:

- Enable a very eloquent graphical representation of exergy flows in energy systems;
- Can be largely automatically generated when one has a Thermoptim diagram and its thermodynamic settings;

---

[1] https://direns.mines-paristech.fr/Sites/Thopt/en/co/structures-productives.html.

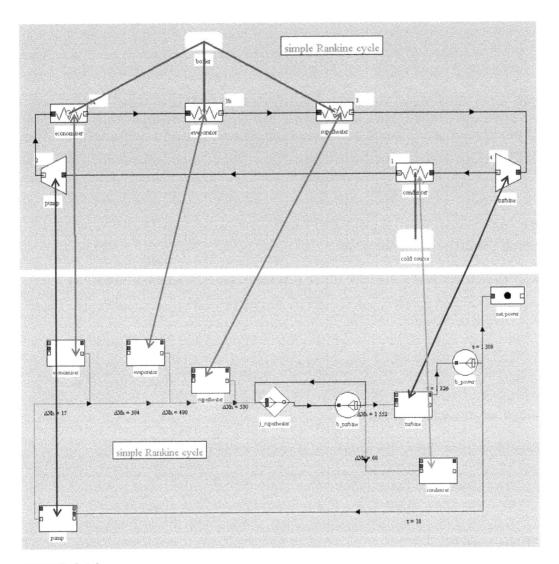

FIGURE 6.14

Correspondence between the physical diagram and the productive structure.

FIGURE 6.15

Productive structure editor.

- Facilitate and secure building up exergy balances;
- In addition lead to thermo-economic analyses for those who wish to perform them.

In the rest of this book, many examples of exergy balances will be provided, most of which have been established by calling on productive structures. As mentioned above, the BESP1 and BESP2 exploration guides will allow you to familiarize yourself with this very powerful method.

BOX 6.7
Guided educational exploration

Exergy balance and productive structure of a simple steam cycle (exploration BESP-1)

In this guided exploration, you will see how the productive structure of a steam plant cycle can be constructed and how it allows one to calculate the exergy balance of the modeled system.

BOX 6.8
Guided educational exploration

Exergy balances and productive structures of various cycles (exploration BESP-2)

In this guided exploration, you will analyze the productive structures associated with various cycles which have been the subject of guided explorations, with a view to establishing their exergy balances.

Bibliography

A. Bejan, G. Tsatsaronis, M. Moran, *Thermal design and optimization*, John Wiley & Sons, New York, ISBN: 978-0-471-58467-4 December 1995.

L. Borel, *Introduction aux nouveaux bilans énergétiques*, Entropie, Paris, n 153, 154, 1990.

L. Borel, D. Favrat, *Thermodynamique et énergétique*, Presses Polytechniques Romandes, Lausanne, Vol. 1 (De l'énergie à l'exergie), 2005, Vol. 2 (Exercices corrigés), 1987.

T.J. Kotas, *The exergy method of thermal plant analysis*, 2nd edition, Krieger Publishing Company, Malabar, FL, 1995.

A. Lallemand, Thermodynamique appliquée: Bilans entropiques et exergétiques, Techniques de l'Ingénieur, Traité Génie énergétique, BE 8 008.

M. Moran, *Availability analysis: A guide to efficient energy use*, Prentice Hall, New York, 1989.

M. Moran, *Exergy analysis, The CRC handbook of thermal engineering*, (Edited by F. Kreith), CRC Press, Boca Raton, 2000, ISBN 0-8493-9581-X.

A. Valero, L. Serra, J. Uche, Fundamentals of thermoeconomics, Lectures 1–3, Euro summer course on sustainable assessment of clean air technologies, 3–8 April 2000. Lisbon. Centro Superior Tecnico, available from CIRCE, Research Center for Energy Resources and Consumption: http://teide.cps.unizar.es:8080/pub/publicir.nsf/codigos/0153/$FILE/cp0153.

Optimization by thermal integration (pinch method)

Introduction

Thermal integration, or **pinch method,** is a relatively recent method (dating back to the eighties) for designing the most efficient heat exchanger and utility networks in an energy facility or a process plant.

It is based on thermodynamic laws and the study of heat exchanged between streams to be cooled (availabilities) and warmed (needs). The study of composite curves (CCs), in particular Carnot factor difference curve (CFDC), allows one to minimize the exchanger network internal irreversibilities, and thus improve its performance.

It is a visual and graphic method, which allows the engineer to keep a physical approach of the phenomena, while very often optimization methods are purely numerical. But most importantly, optimization is performed without any *a priori* assumptions about the heat exchanger network configuration, which is defined only at a later time.

We show how system integration is implemented in Thermoptim, and conclude by giving two concrete examples presented in two guided explorations (optimization of a heat exchange network and of a dual-pressure combined cycle).

Basic principles

Traditional approaches in thermodynamic systems optimization, valid for maximizing one by one the various components of a facility, remain insufficient to guide the designer in choosing the best configuration of the entire system.

As shown by the heat recovery steam generator (HRSG) example (Chapter 5), specific irreversibilities appear in addition to those of each element, since the various components are assembled together. These irreversibilities can be described as systemic because they depend mainly on relative positioning of components. Reducing these irreversibilities allows an increase in the effectiveness of the systems considered, including through better internal regeneration.

Consider a system in which matter is transformed by various processes, some exothermic and other endothermic. Each of these processes receives or provides heat. If you want to optimize the overall energy expenditure, to reduce external inputs, you must make the best of all heat internally available. Thermal or heat integration provides a rigorous method for this, to optimize the overall configuration of the facility, and ensure greater consistency between needs and energy availability.

For simplicity, we first present the basic principles of heat integration without using exergy analysis, and then we shall generalize our approach.

DOI: 10.1201/9781003175629-7

Pinch point

The basic principle of heat integration is to classify all heat needs depending on temperature levels to which they relate, in order to find the best match of hot and cold streams. For this, we use the graph already introduced in Chapter 4, where temperature is in ordinate and enthalpies are put into play in abscissa (enthalpy diagram of Figure 7.1).

Note that for purely thermal exchanges, enthalpies are equal to heat exchanged. In this diagram, an economizer–evaporator assembly without superheating appears. The fluid that evaporates corresponds to the lower curve (1–2–3), which presents an angular point 2 corresponding to the beginning of the boiling. The hot fluid is cooled from 4 to 5, by decreasing its sensible heat.

On segments (1–2) and (4–5), we have $\Delta H = \dot{m} C_p \Delta T$, and thus $\Delta T = \dfrac{\Delta H}{\dot{m} C_p}$

The slopes of segments (1–2) and (4–5) are equal to the inverse of the heat capacity rates of the corresponding fluids. On segment (2–3), we have obviously $\Delta T = 0$, and the vaporization takes place at constant temperature for a pure component or azeotrope.

We see at point 2 a minimum temperature difference P between both fluids, called pinch. This point plays a fundamental role in the design of heat exchanger networks, as it represents the smallest difference in temperature in the facility.

To reduce the cost of equipment, it is preferable that the pinch is not too low. However, the thermodynamic optimization of the complete system often requires it to be as small as possible. In general, the pinch point is the most constraint zone in the thermal system.

On this graph, it is clear that below the pinch, the slope of the segment representing the cold fluid is greater than that corresponding to the hot fluid:

$$\left(\dot{m} C_p\right)_h \geq \left(\dot{m} C_p\right)_c$$

Above the pinch, the reverse is true: $\left(\dot{m} C_p\right)_h \leq \left(\dot{m} C_p\right)_c$

The heat exchanger, sized according to the pinch, works outside this point with larger temperature differences between the two fluids, which indicates an increase in irreversibility, without any compensation, which is therefore not desirable. However, given the peculiarity of the exchanger considered here (it is an evaporator which therefore has a vaporization plateau), it is not possible to reduce these irreversibilities except by using a condensable vapor as hot fluid. Note that it is precisely what brings mechanical vapor compression (MVC) when possible (see Chapter 17).

As discussed below, the pinch point is a break in the thermal system and leads to an independent study of each zone. In most cases, facilities are designed so that such points are located between the various elements.

For a more complex system, heat integration leads one to rank needs and energy availability as a function of temperature, then plot them on a (T, h) diagram.

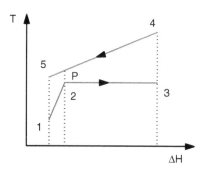

FIGURE 7.1

Pinch point.

Integration of complex heat system

We begin by illustrating thermal integration on a relatively simple case, from an article by Gourlia (1989), to which the reader may refer for details.

Consider a plant with the following energy availability, for example, related to exothermic processes:

Stream 1	220°C–40°C	1,800 kW
Stream 2	320°C–200°C	2,400 kW
Stream 3	140°C–40°C	800 kW
Stream 4	100°C	1,200 kW
Boiler	600°C	1,900 kW
And where the needs are:		
Stream 5	40°C–320°C	6,000 kW
Stream 6	220°C	600 kW

In this factory, enthalpies available in streams 1–4 correspond to releases, and so are somehow "free" calories whose value we seek to maximize, while the 1,900 kW provided by the boiler at 600°C is of another kind: it is purchased energy, provided by a "hot utility", that we generally seek to minimize. The value of 1,900 kW was determined in a conventional manner by an engineering and design department.

Thermal integration will allow us, as we shall see below, first to determine the minimum amount of heat required by utilities and their temperature levels, and second to define a suitable exchanger network for the plant to operate. Before explaining the methods that lead to these results, we first generalize the presentation that was made of the enthalpy diagrams.

Plot of the composite curve

The temperature level at which heat may be exchanged is a major constraint in the design of a heat exchanger network; we begin by sorting heat brought into play depending on this parameter.

For this, we usually assume a linear variation of enthalpy which can be transferred by sensible heat (assumption of constant heat capacities of streams that come into play). We perform a combination of availability and needs for the different temperature intervals of the system, whose bounds are determined by the original specifications, as summarized previously.

In order of increasing temperatures, this combination is as follows, needs and availability being assumed to be at first approximation linear functions of temperature between the different bounds:

Availability:

<40°C	0
<100°C	$1,800\dfrac{100-40}{220-40}+800\dfrac{100-40}{140-40}=1,080\,\text{kW}$ streams 1 and 3
≤100°C	$1,080+1,200=2,280\ \text{kW}$ streams 1, 3 and 4
<140°C	$2,280+1,800\dfrac{140-100}{220-40}+800\dfrac{140-100}{140-40}=3,000\,\text{kW}$ streams 1 and 3
<200°C	$3,000+1,800\dfrac{200-140}{220-40}=3,600\,\text{kW}$ stream 1
<220°C	$3,600+2,400\dfrac{220-200}{320-200}+1,800\dfrac{220-200}{220-40}=4,200\,\text{kW}$ streams 1 and 2
<320°C	$6,200\ \text{kW}$ stream 2
≤600°C	$8,100\ \text{kW}$ boiler

Needs:

<40°C	0

$<220°C \qquad 6,000\dfrac{220-40}{320-40}=3,857\,\text{kW} \quad \text{stream 5}$

$\leq220°C \qquad 3,857+600=4,457\ \text{kW} \quad \text{streams 5 and 6}$

$<320°C \qquad 6,600\ \text{kW} \quad \text{stream 5}$

We can plot these energies in a (T, h) diagram seeking to maximize the value of enthalpies according to their respective temperature levels. We overlap the needs and availability curves shifting them so that the first one is below the second.

This gives the diagram in Figure 7.2, the so-called composite curve (CC). The bold line represents the cumulative availability depending on the temperature, while the thin line, that of requirements. In the example shown, the temperature levels of availability are sufficient to ensure the provision of needs, and there remains a surplus of 1,500 kW between 40°C and 100°C, which must be removed by cold utilities, for example, by a cooling tower exchanging with ambient air.

The diagram shows the existence of a pinch at 100°C, 2,280 kW, which corresponds to about 24°C difference between the two curves. It also allows the temperature differences in each area to be known, and therefore the exact size of heat exchangers. It is unnecessary to size them too tight if they operate with large temperature differences. Note that nothing prevents that there are several pinches, the extreme case is that of a counter-flow heat exchanger between two fluids of the same heat capacity rate, in which case the temperature difference between them remains constant, which corresponds to an infinite number of pinches.

In the example shown, curves do not intersect on the chart. In other cases, they can do, which means we must provide more energy. Heat integration allows the best location for it to then be chosen.

Exergy representation of the composite curve

So far, we reasoned, for simplicity, in a (T, h) diagram, whose physical meaning is quite telling. A more general reasoning way, though less directly accessible, is to use as ordinate, not temperature, but the Carnot factor $\theta = 1 - T_0/T$.

Such a representation has a double interest:

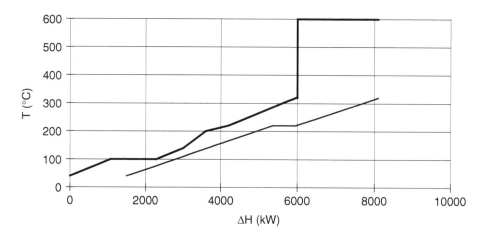

FIGURE 7.2

Composite curves.

- First, the area between the two curves corresponds, for purely thermal systems,[1] to the exergy destroyed, that is to say to irreversibilities;
- Second, it is well suited for comparison between thermal and noble energies (mechanical or chemical).

The diagram in Figure 7.3 relates to the thermal system studied. We see here that the irreversibilities are almost evenly distributed between 2,500 and 5,500 kW, but they are more important at low and high temperatures.

Although thermodynamically more instructive, the exergy approach is seldom used consistently in thermal integration studies. Two main reasons can explain this phenomenon:

- First, in engineering departments exergy culture is not generalized. Its implementation is a bit tricky, with some risks of error for the uninitiated;
- Second, we have already noted that thermal integration methods allow the user to reason physically, that is to say essentially qualitatively, as we shall see later. An energy approach of the type that we will now expose reveals itself usually sufficient.

Design of exchange networks

This brief overview has shown the usefulness of thermal integration to view irreversibilities in a complex system. It allowed us to highlight temperature differences within heat exchangers, but not to define the overall architecture of the network. That is what we shall see in this section. We content ourselves, for reasons that have just been explained, with a classical enthalpy approach.

On the CC (Figure 7.4), the pinch distinguishes two zones:

- The lower zone, where "cold" streams (needs) are unable to absorb all the available heat in "hot" streams (availability), a surplus having to be discharged through a complementary coolant. This area, excess heat, generally behaves as a heat source and is called exothermic zone or cold-end;

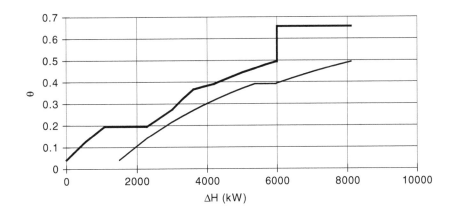

FIGURE 7.3

Exergy composite curves.

---

1 These are systems receiving no work from external forces, where streams exchanging sensible heat are assumed incompressible, whether gas or liquid, and where the flows are without pressure drop. In this case, $ds = c_p dT/T$, and $dx_{hi} = c_p T_0 (1/T - 1/(T + \Delta T))dT$, which corresponds to the area between the Carnot factors of both streams exchanging heat with a temperature difference ΔT.

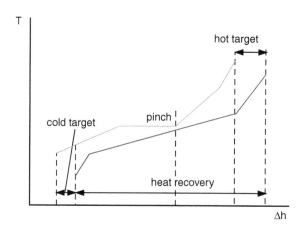

FIGURE 7.4

Endothermic and exothermic zones.

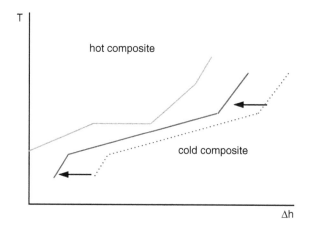

FIGURE 7.5

Shifting composite curves.

* The upper zone, where the situation is dual to the previous one: the need for heat of "cold" streams is above the enthalpy available in "hot" streams, and additional heating is necessary. This area, lacking heat, behaves as a heat sink area and is called endothermic or hot-end.

As shown in Figure 7.5, dragging the cold composite from right to left, we vary the gap with the hot composite, gap representative of thermal irreversibilities. When the pinch value is zero, no heat transfer exists between these two zones: the pinch separates the system in two thermally independent parts. This is obviously a borderline case. It corresponds to the case where the heat absorbed by the endothermic zone and the one rejected by the exothermic zone are minimal.

When the pinch value is not zero (Figure 7.6), heat absorbed at high temperature and that released at low temperature are both increased by the same amount α, which corresponds to heat passing through the pinch between both zones (it is called cross-pinch heat transfer). Similarly, heat provided in the exothermic zone (β), or a cooling in the endothermic zone (γ), increases the need for hot and cold utilities. So we see an additional advantage of thermal integration: highlighting the pinch point, it reduces α to minimum economic size and therefore allows design of the best heat exchanger networks in the upper and lower areas. Indeed, an exchanger design error in one of the areas, resulting in an increase of the enthalpy exchanged, leads inevitably to an oversizing of the other zone.

The method presented below allows for determining an exchange configuration of the network minimizing heat supply by utilities in the hot endothermic zone, and extraction of heat

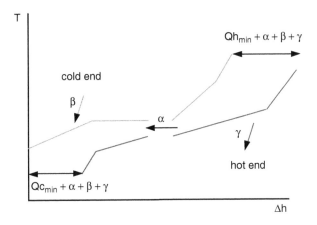

FIGURE 7.6
Plus–minus principle.

from cold utilities in the exothermic area. This configuration, optimal at least according to this criterion, may often reflect a greater number of heat exchangers than other arrangements. It may therefore be necessary to consider other alternatives, which is done by relaxing the criteria above.

Minimizing the pinch

As searching the pinch by dragging one from the other CCs is neither very easy nor very precise, Professor B. Linnhoff proposed a method whose main lines are given here. It is to work directly on the enthalpy balances by temperature levels, while when building the CC, we dissociated the curve of the availability from that of needs.

The pinch minimization algorithm is used to determine the minimum energy consumption corresponding to the minimum pinch, called the **hot target**. The original Linnhoff algorithm, of which it is derived, is known as the problem table algorithm (PTA).

The algorithm can be decomposed into four steps:

- Shifting all minimum and maximum temperatures by $\Delta T_{pinc}/2$, subtracting it for hot streams and adding it for cold streams (by doing this, it is certain that heat exchange between streams at the same shifted temperature can take place);
- Determining the problem temperature intervals by sorting the upper and lower temperature limits;
- Building enthalpy balances (by temperature interval) of hot and cold composites and their difference;
- Building the overall summary table and determining the minimum hot utility value (the hot target).

Implementation of the algorithm

Shifting temperatures

To be certain that the hot and cold streams may exchange heat when grouped in an appropriate temperature interval, we must make sure that a temperature difference ΔT_{pinc} always exists between them. To do so, we simply slightly modify their temperature levels:

- For hot streams, reducing them by $\Delta T_{pinc}/2$;
- For cold streams, increasing them by $\Delta T_{pinc}/2$.

The value of ΔT_{pinc} is arbitrary: it depends on the problem considered. We generally retain about 16°C–20°C for an exchange between gas and 8°C–10°C for liquids. For two-phase exchange, these values may be reduced.

It is also possible, when studying a complex system, to retain minimum pinch values differing according to the streams considered. It is in that the algorithm chosen in Thermoptim differs from Linnhoff's. However, in the example that follows, we will consider for simplicity a single value.

For example, let us come back to the problem studied above, but without showing the enthalpy supplied by hot utilities (boiler at 600°C) or extracted by cold utilities (cooling tower).

By setting an artificial $\Delta T = 1$°C for evaporation–condensation, to avoid having segments of zero slope, the specifications can be summarized in Table 7.1.

Taking into account ΔT_{pinc} (10°C here), we obtain Table 7.2 giving shifted temperature intervals. Obviously, the enthalpies that are put into play are not affected by this shift.

Search of temperature intervals

The second step determines the problem temperature intervals by sorting the bounds of the stream upper and lower temperatures.

We sort all bound temperatures T_{inf} and T_{sup}, in order of decreasing values, which in this case leads to the ten temperature intervals of Table 7.3.

Establishment of enthalpy balances

We then build up the enthalpy balances (by temperature interval) of hot and cold composites and their difference.

TABLE 7.1
DATA WITH ACTUAL TEMPERATURES

Stream	$\dot{m}\,C_p$ (kW/K)	T_s (°C)	T_t (°C)	ΔH (kW)
1	10	40	220	1,800
2	20	200	320	2,400
3	8	40	140	800
4	1,200	100	101	1,200
5	21	40	320	6,000
6	600	220	221	600

TABLE 7.2
DATA WITH SHIFTED TEMPERATURES

Stream	$\dot{m}\,C_p$ (kW/K)	Type	T_{inf} (°C)	T_{sup} (°C)	ΔH (kW)
1	10	Hot	35	215	1,800
2	20	Hot	195	315	2,400
3	8	Hot	35	135	800
4	1,200	Hot	95	96	1,200
5	21	Cold	45	325	6,000
6	600	Cold	225	226	600

TABLE 7.3

CLASSIFICATION BY TEMPERATURE INTERVAL

Interval	T_i (°C)	T_{i+1} (°C)	Streams	$T_i - T_{i+1}$ (°C)	$\Sigma(\dot{m}\, C_p)$ (kW/K)	ΔH_{na} (kW)
1	325	315	*5*	10	21.43	214
2	315	226	*5,2*	89	1.43	121
3	226	225	*5,2,6*	1	601.43	601
4	225	215	*5,2*	10	1.43	14
5	215	195	*5,2,1*	20	−8.57	−171
6	195	135	*5,1*	60	11.43	686
7	135	96	*5,1,3*	39	3.43	134
8	96	95	*5,1,3,4*	1	−1,196.60	−1,197
9	95	45	*5,1,3*	50	3.43	171
10	45	35	*1,3*	10	−18.00	−180

For each temperature interval, we identify the streams flowing through it (the fourth column of Table 7.3). The streams are so classified; we can group them by type and calculate enthalpy balances of each interval, using Equations (7.1–7.3), with index n referring to needs, a to availability, and na to the difference between them:

$$\Delta H_n = \left(T_i - T_{i+1}\right)\sum \dot{m}\, c_{pn} \tag{7.1}$$

$$\Delta H_a = \left(T_i - T_{i+1}\right)\sum \dot{m}\, c_{pa} \tag{7.2}$$

$$\Delta H_{na} = \Delta H_n - \Delta H_a = \left(T_i - T_{i+1}\right)\left(\sum \dot{m}\, c_{pn} - \sum \dot{m}\, c_{pa}\right) \tag{7.3}$$

with ΔH_n and ΔH_a representing the enthalpy changes of needs and availability in interval i. They thus correspond to hot and cold composite segments.

ΔH_{na} is positive if the heat requirements exceed the availability, and negative otherwise. It represents the net enthalpy balance of the interval, and is used, on the one hand, to determine the minimum pinch, and, on the other hand, to draw the grand composite curve (GCC) (Figure 7.7), which is given in Table 7.3.

Determination of the minimum target

Finally, we establish the overall summary in Table 7.3, which determines the minimum target.

Note that enthalpies available in an interval (negative by construction) are at a sufficient temperature to be used in all intervals of higher order: if, for example, in interval i, we have a surplus of energy, it can be used to heat a stream in interval i + 1.

It is thus possible to identify a cascade of temperatures and corresponding enthalpy balances, obtained by subtracting the values of ΔH_{na} calculated (Table 7.4).

Table 7.4 shows that the maximum enthalpy deficit is equal to 1,605 kW, in the absence of the boiler. It takes place between intervals 7 and 8, at 95°C (the value of 96°C is artificial, because it comes from the 1°C temperature difference introduced to avoid a zero slope vaporization segment).

The system pinch is located there. Given the value of ΔT_{pinc} chosen, this corresponds to 100°C for hot streams and 90°C for cold streams. The deficit must be offset by a $\Delta H_{min} = 1,605$ kW input. Its value is 295 kW lower than that retained by the engineering and design department (1,900 kW). If we provide the missing 1,605 kW at high temperature, we obtain Table 7.5.

TABLE 7.4
INTERVALS WITH HEAT BALANCES

Interval	T_i (°C)	T_{i+1} (°C)	ΔH_{na} (kW)	Sum (kW)
1	325	315	214	−214
2	315	226	127	−341
3	226	225	601	−943
4	225	215	14	−957
5	215	195	−171	−786
6	195	135	686	−1,471
7	135	96	134	−1,605
8	96	95	−1,197	−409
9	95	45	171	−580
10	45	35	−180	−400

TABLE 7.5
INTERVALS WITH HEAT INPUT

Interval	T_i (°C)	T_{i+1} (°C)	ΔH net (kW)	(kW)
				1,605
1	325	315	214	1,391
2	315	226	127	1,264
3	226	225	601	662
4	225	215	14	648
5	215	195	−171	819
6	195	135	686	134
7	135	96	134	0
8	96	95	−1,197	1,196
9	95	45	171	1,025
10	45	35	−180	1,205

Endothermic and exothermic zones are evident: the former comprises intervals 1–7, while the latter 8–12. At the pinch, the enthalpy flow that passes from one interval to another is zero by construction.

As can be seen, the great interest of the pinch minimization algorithm is to allow one to relatively simply determine the minimum value of the heat input to be provided to the system.

If we plot the curve on a graph whose abscissa is the fifth column of Table 7.5 and ordinate the third (with one additional point which has coordinates ((ΔH_{min}, T_1)), we obtain the shifted temperature GCC, which gives, for each temperature, the cumulative net enthalpy balance (Figure 7.7).

Establishment of actual composite curves

Once the target is known (here ΔH_{min} = 1,605 kW), we can return to the actual temperatures, and recalculate the actual needs and availability CCs, reusing part of the pinch minimization algorithm:

- Determining the problem actual temperature intervals by sorting the actual upper and lower temperature bounds. The procedure is the same as before, but we start from Table 7.1 instead of Table 7.2;

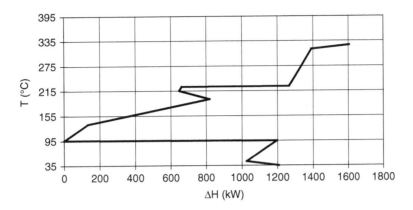

FIGURE 7.7
Shifted grand composite curve.

- Building hot and cold composite enthalpy balances (by temperature interval). The procedure is also similar, but there is no need to calculate ΔH_{na}, unless you want to draw the actual GCC. Then, you obtain the values of ΔH_n and ΔH_a for each actual temperature interval;
- *Composite curve plot*. The plot of the hot CC poses no problem: we know its origin, which corresponds to the lowest hot stream temperatures and a zero enthalpy. Interval by interval, by combining H_a values, it is constructed without difficulty.

Moreover, the maximum enthalpy of the cold CC is equal to the sum of the maximum enthalpy of the hot composite and the minimum target ΔH_{min}. Its maximum temperature is the highest temperature of cold streams. Interval by interval, by deducting values of ΔH_n from this maximum, the cold composite is built. Its minimum corresponds to the enthalpy value to evacuate by external cooling (the cold target). The actual CCs are shown in Figure 7.8.

Plot of the carnot factor difference curve (CFDC)

The construction of exergy (CCs) (Figure 7.9) is no problem, since it is a simple change of ordinate axis: $\theta = 1 - T_0/T$.

To plot the CFDC, things are somewhat more complex because you must reverse the two composites. Since they are composed of a series of segments, the procedure is as follows:

- Search all the values of enthalpies of the two composite interval bounds;
- Identify, by linear interpolation for each enthalpy bound of the cold composite, the corresponding value of the temperature of the hot composite;
- Identify, by linear interpolation for each enthalpy bound of the hot composite, the corresponding value of the cold composite temperature.

This results in a table showing the temperature values of the hot or cold composites, each time there is a slope change in one of them. The curve is inverted, and the CFDC is deduced by simple difference and change of ordinate (Figure 7.10).

Of course, if you do not know the temperature levels of the hot target and external cooling, you can only plot the CFDC on the part of the abscissa common to the two composites. Pinch is minimal when it is located on the locus of minimal pinches (LMP), here with $\Delta T_{pinc} = 10°C$.

The power supplied by hot utilities equals 1,605 kW, when it was 1,900 kW in the original system, which represents a saving of 15.6%. In addition, energy discharge is reduced from 1,500 to 1,205 kW, which represents a reduction of almost 20%. These savings are

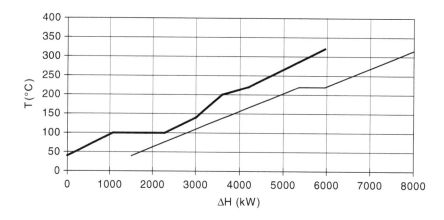

FIGURE 7.8

Actual composite curves.

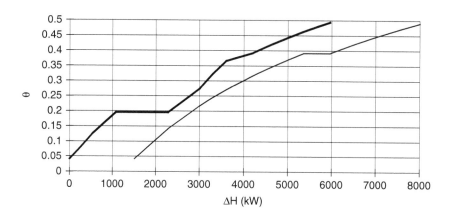

FIGURE 7.9

Exergy composite curves.

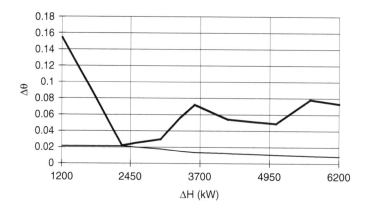

FIGURE 7.10

Carnot factor difference curve.

accompanied by a reduction in capital expenditure since the endothermic and exothermic zone exchange surfaces are also reduced accordingly. Thermal integration allows in cases like this, quite common in practice, achievement of both investment and operation savings, although frequently you can only obtain one at the expense of the other.

Matching exchange fluids

General procedure

Heat integration is a valuable guide to design the overall architecture of the exchanger network by choosing the best pairs of streams exchanging heat. For this, we should first represent the system in a grid diagram of the type presented in Figure 7.11. Streams are represented by oriented horizontal segments, and heat exchangers by vertical links connecting the pairs of streams matched, identified by a number. Utilities are represented by a circle marked H if they are hot, and C if they are cold.

In this example, a hot stream is cooled from 170°C to 120°C in a first heat exchanger, while stream 2 is heated from 25°C to 140°C. Stream 1 is then cooled at 100°C in heat exchanger 2, while stream 3 is heated from 30°C to 100°C. Finally, a cold utility cools stream at 1°C–30°C while streams 2 and 3 are heated by two hot utilities at the final temperature of 200°C.

Let us now study the previous energy system. It involves six streams, as shown in Figure 7.12. Pinch is located in the central division, and the actual stream temperatures (that is, not shifted by ΔT_{pinc}) are shown. Note that streams 2 and 6 are present only in the endothermic zone, and stream 4 in the exothermic area.

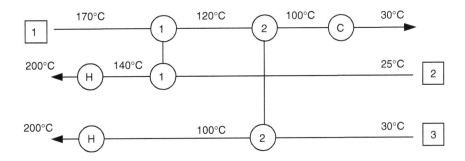

FIGURE 7.11

Grid diagram.

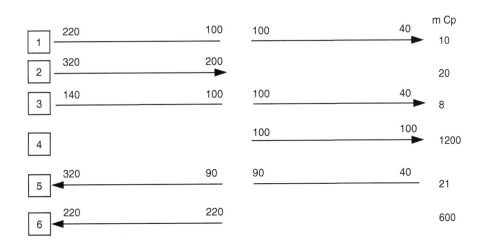

FIGURE 7.12

Initial grid diagram.

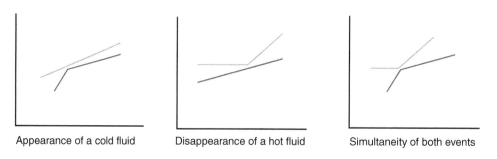

Appearance of a cold fluid   Disappearance of a hot fluid   Simultaneity of both events

FIGURE 7.13
Different types of pinches.

The problem now is to choose the best pairing of streams.

Note first that for a pinch to exist, if we accept the assumption of constant heat capacity rate, there must be, as shown in Figure 7.13:

- Either appearance of a new cold stream (need);
- Or disappearance of a hot stream (availability);
- Or simultaneity of both events: onset of a cold stream and disappearance of a hot stream.

This is because, at the pinch, heat flow constraints are automatically satisfied:

- Exothermic zone $\sum \dot{m}\, c_{ph} \geq \sum \dot{m}\, c_{pc}$;
- Endothermic zone $\sum \dot{m}\, c_{ph} \leq \sum \dot{m}\, c_{pc}$.

The procedure is as follows:

- Study separately endo- and exothermic zones;
- Begin the design at the pinch and depart gradually from it, to deal first with the most constrained problem;
- Import energy solely in the endothermic zone, and export only from the exothermic zone;
- Maximize the load of heat exchangers (to minimize their number).

In practical terms, the problem does not usually have a single solution, and a number of difficulties can arise. To find an acceptable solution, a number of constraints must first be met, as indicated below.

For the heat transfer to be done, it is obviously necessary that the heat capacity rate constraint is met in each heat exchanger, without infringing the second law of thermodynamics. If this rule is broken, it may be necessary to split some streams, as we shall see later.

Moreover, as we must not cool by utilities in the endothermic zone, this implies that at the interval level located just above the pinch, all hot streams must be cooled by cold streams, i.e., that the number of cold streams must be greater than or equal to that of hot streams. Again, if this rule is not respected, one or more cold streams must be split.

In a dual manner, the number of hot streams in the interval located just below the pinch must be greater than or equal to that of cold streams.

In our case, at the pinch, streams selected are given in Table 7.6 for the first interval of the endothermic zone. This corresponds to hot stream 4 disappearance.

Analysis of the endothermic zone

As shown in Table 7.6, the only cold stream in the first interval is 5, while there are two hot streams. We must then split this stream into two.

TABLE 7.6
STREAMS AT THE PINCH LEVEL

Number	ṁ C$_p$ (kW/K)	ΔH (kW)
1	−10	−390
3	−8	−312
5	21.4	835.7

TABLE 7.7
STREAMS ABOVE THE PINCH

Number	ṁ C$_p$ (kW/K)	ΔH (kW)
1	−10	−600
5	21.4	1,285.7

To find how, look at what happens in the interval above (No. 6). Streams selected are given in Table 7.7.

Stream 3 disappears. Therefore, the corresponding load must be supplied to stream 5, which gives: $\dot{m}\,c_{p52} = \dot{m}\,c_{p3} = 8$.

We deduce $\dot{m}\,c_{p51} = 13.4$, which, to exhaust stream 1 with a total of 1,200 kW, leads to an exit temperature of heat exchanger 1 equal to 179°C.

Streams 1 and 3 are exhausted, while only stream 2 still contains availability (2,400 kW), which must be shared between the divided stream 5 and stream 6.

Several possibilities exist: one is to transfer the entire load in 6 (600 kW) and part of the residual load on the first branch of 5 (1,800 kW on 1,888 required). However, this scenario leads to a failure in terms of temperature.

It is better to remix both stream 5 branches at the outlet of heat exchangers 1 and 2:

The temperature of the mixture is given by:

$$179\,\dot{m}\,c_{p51} + 130\,\dot{m}\,c_{p52} = T\,\dot{m}\,c_{p5} \qquad T = 160.9°C$$

A single exchanger then allows to exhaust stream 2. Its balance is given by:

$$2,400 = (T - 160.9)\,\dot{m}\,c_{p5}$$

We find a stream 5 outlet temperature equal to 273°C.

The grid diagram is given in Figure 7.14.

Note that the heat input to be provided by utilities is equal to 1,608 kW, i.e., slightly more than 1,605 kW demonstrated previously. This is because the 10°C pinch was located around 95°C, not 96°C, as the study of the energy cascade had indicated.

As shown by the analysis of the GCC, other possibilities exist if one wants utilities to be used at least in part at lower temperature. Indeed, in our mode of reasoning, we start from the pinch and we depart gradually, allowing utilities to match the load in the zone bounds.

If we wish to incorporate them minimizing their irreversibility, we must consider them as additional streams to be integrated into the heat cascade, which complicates things a bit and usually results in an increased number of heat exchangers, and therefore in a higher total installation cost.

The new GCC that is obtained in that case is shown in Figure 7.15.

It shows three new pinches, at 250°C, 225°C and 215°C, which are different from the first: they are utility pinches, introduced arbitrarily to optimize their thermal levels. As the latter two are very close, they can be initially aggregated.

The heat exchanger network design can be made using the procedure given above, starting from each successive pinch. To avoid problems, you should first deal with the most

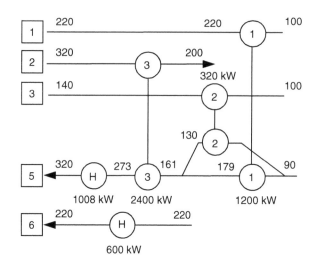

FIGURE 7.14

Endothermic zone grid diagram.

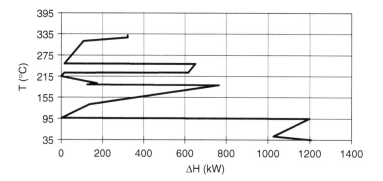

FIGURE 7.15

GCC with utility pinches.

constrained pinch, if any. In this case, the problem is relatively simple, and just treat them by decreasing temperature level.

Above the 250°C pinch, the constraint on the maximum heat is met with streams 2 and 5. We can further maximize the load on 2, which allows exchange of 1,300 kW in exchanger 1 and to heat 5 from 245°C to 305°C. Heat exchanger 2 can then provide from H_1 the 320 kW missing for 5 to reach 320°C.

Below this pinch, the constraint on the heat capacity rates is met between 5 and H_2. It is not possible to maximize the load on 5 because, slightly above the 215°C pinch, we have seen there exists a fourth 225°C pinch due to stream 6, where it is imperative to pair H_2 and 6. H_2 and 6 are both evaporation/condensation heat exchangers, and it may be advantageous to maximize the load on this fourth exchanger, which leaves only 35 kW available on No. 3, in which the inlet temperature of 5 is equal to 243°C.

Two streams are then to be matched, 2 and 5, for a 700-kW load. We can check the accuracy of calculations by the consistent level of inlet and outlet temperatures of heat exchanger No. 5.

Below the 215°C pinch, we can begin by matching 5 and 2, to deplete the 400 KW available in stream 2, which determines the inlet temperature of 5 in heat exchanger 6: 191°C. Given the levels of stream temperatures, it is then necessary for heat exchanger No. 7 to provide the coupling between 5 and H_3, which leads to an inlet temperature of 5 in this exchanger equal to 161°C.

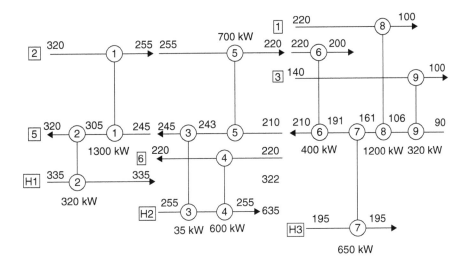

FIGURE 7.16
Endothermic zone grid diagram with utilities.

Among the streams remaining, only 1 can exchange heat with 5 at this temperature level. If the load on heat exchanger 8 (1,200 kW) is maximized, the inlet temperature of 5 in this exchanger becomes equal to 106°C, which is borderline, and the pinch is normally 10°C. To do so would require stream 3 to take over from 1 at a higher temperature, but it would add a heat exchanger, so we accept this additional constraint.

Finally, the remaining 320 kW can be exchanged between 3 and 5 in a heat exchanger No. 9. As before, a verification of the correctness of calculations can be performed at this level.

We obtain ultimately the result of Figure 7.16.

The total number of heat exchangers, utilities included, is now 9 instead of 5, an increase of over 50%, which is quite large, and, except in special cases, not generally justified. We note, however, that the proposed method can handle in a relatively simple manner even complex situations.

Analysis of the exothermic zone

Streams selected at the first interval are given in Table 7.8.

Throughout the exothermic zone, the only cold stream is 5 and, at the pinch, the constraint $\dot{m} \, c_{pc} \geq \dot{m} \, c_{pf}$ cannot be verified unless it is matched to stream 4. By maximizing the load of this exchanger (5), and knowing that the inlet temperature of 5 is 40°C, we obtain $h_5 = \dot{m} \, c_{p5}$ $(90-40) = 1\,071$ kW.

Stream 5 is completely warmed by that heat addition; it cannot provide cooling for the entire 4 nor for 1 and 3, which must be performed by cold utilities, for a total of 1,209 kW.

Coming back to the first analysis of the endothermic zone, we end up with the grid in Figure 7.17.

As demonstrated by studying the endothermic zone variant where utilities were placed so as to reduce irreversibilities, a number of difficulties can arise when determining the architecture of the exchanger network. In our case, we chose, on the one hand, to finally operate with a 6°C pinch to avoid being forced to introduce an additional heat exchanger, and, on the other hand, we took as main constraint the 225°C pinch, given the existence of the vaporization–condensation for streams 6 and H_2. Very frequently in practice, we are confronted with problems of this kind or other, leading to amend the procedure described above.

Various advanced methods are used to pursue further optimization of the network, depending on the configurations identified and prosecuted criteria (economic in most cases, including research of the minimum number of heat exchangers). You will find in the

TABLE 7.8

STREAMS SELECTED AT THE FIRST INTERVAL

Number	ṁ C_p (kW/K)	ΔH (kW)
1	−10	−10
3	−8	−8
4	−1,200	−1,200
5	21.4286	21.4286

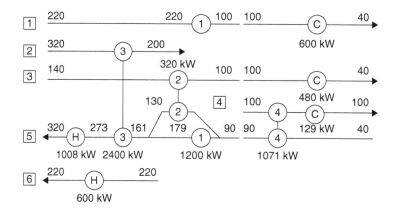

FIGURE 7.17

Grid diagram of the whole heat exchanger network, hot-end left and cold-end right.

literature, and especially in the book by Linnhoff (1982), developments relating to these methods (Trivedi, 2000).

Thermal machines and heat integration

Thermal integration allows to highlight the pinches, which determines an exchanger design constraint: a pinch occurs when the average slope of needs is higher on the left, and lower on the right than that of availability. Once these values are known, it is possible to make a technical and economic optimization calculation of the exchange process by a sensitivity analysis on cost variations around these points.

The pinch is created by changes in slope, we must, as far as possible:

- Pair streams of similar slopes;
- Carefully consider the position of latent heat exchangers;
- Not hesitate to export excess heat from the exothermic zone, if they can find a use elsewhere;
- Not hesitate to import heat in the endothermic zone if any are missing.

Thermal integration also provides as we will see useful insights on the placement of heat engines.

Placement of heat pumps or mechanical vapor compression (MVC)

You should refer to Chapter 11 for a detailed presentation of heat pumps and Chapter 17 for MVC. The opportunity to use a heat pump or MVC can be analyzed from the CCs or GCC:

this requires the existence of two zones with similar needs and availability, and availability temperature is slightly insufficient. The heat pump then raises the temperature level of availability, which is valued at the sole price of the energy consumed by the compressor. Thus, if conditions are met, this transfer can greatly reduce the need for both hot and cold utilities. This means that the heat pump evaporator is situated in the exothermic zone and the condenser in the endothermic area. If this rule is not satisfied, the device acts as a simple heat exchanger, its overall coefficient of performance (COP) is limited to 1 and, more often, its inclusion in the system has no interest at the economical level. A similar reasoning can be done on the placement of a MVC.

Placement of heat engines, cogeneration

Similarly, one can use these curves to assess interest to include in the system internal combustion engines to make combined heat and power (CHP) (see Chapter 10). It can be shown that only two configurations may in practice be retained: the engine must be placed either in the endothermic zone or in the exothermic area, but never in between. In the first case, we produce mechanical energy using exergy available from a hot source at high temperature (e.g., a boiler) and process requirements at a lower temperature, while in the second process, thermal effluents are given higher value, providing engine work between them and the environment. The latter case is economically feasible only if the discharge temperature is sufficiently higher than that of the environment.

Two guided explorations have been prepared so that you can familiarize yourself with the pinch method. The first deals with the optimization of a simplified heating network, while the second addresses the optimization of a dual-pressure combined cycle.

BOX 7.1
Guided educational exploration

Optimization of a heating network by the pinch method (OPT-1)

The objective of this exploration is to show you, in a simple example, how the pinch method can be applied to optimize a heat network.

BOX 7.2
Guided educational exploration

Optimization of a dual-pressure combined cycle by the pinch method (OPT-2)

The objective of this exploration is to show you how the pinch method can be applied to optimize a dual-pressure combined cycle.

There you will find a detailed presentation of the Thermoptim optimization window and explanations on how to set the fluids that must be taken into account in the optimization process.

Bibliography

C. Floudas, I. Grossmann, Synthesis of flexible heat exchanger networks for multiperiod operation, *Computers and Chemical Engineering*, Pergamon Press, Oxford, 10 (2): 153–168, 1986.

R. Gicquel, Méthode d'optimisation systémique basée sur l'intégration thermique par extension de la méthode du pincement : application à la cogénération avec production de vapeur, *Revue Générale de Thermique*, tome 34 405, octobre 1995.

R. Gicquel Modics, Généralisation de la méthode d'optimisation systémique aux systèmes thermiques avec échangeurs imposés, *Revue Générale de Thermique*, tome 35: 423–433, 1996.

J.P. Gourlia, La méthode du pincement ou exploitation des diagrammes température/enthalpie, *Notions de base, Revue Générale de Thermique,* n 327, Paris, 1989.

B. Linnhoff, *User Guide on Process Integration for the Efficient Use of Energy*, Pergamon Press, Oxford, 1982.

B. Linnhoff, E. Hindmarsh, The pinch design method for heat exchanger networks, *Chemical Engineering Science*, 38 (5), 745–763, 1983.

B. Linnhoff, Use pinch analysis to knock down capital costs and emissions, *Chemical Engineering Progress*: 3257, august 1994.

T. Trivedi, *Pinch Point analysis, The CRC handbook of thermal engineering*, (Edited by F. Kreith), CRC Press, Boca Raton, 2000, ISBN 0-8493-9581-X.

3 Main conventional cycles

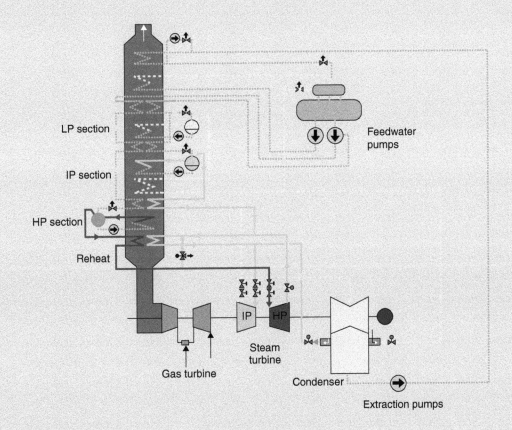

LP section

IP section

HP section

Reheat

Feedwater pumps

Gas turbine

Steam turbine

Condenser

Extraction pumps

Part 3 discusses how the foundations laid in the first two parts can be applied to main conventional cycles.

The guiding principle behind the analyses is the reduction of irreversibilities, special attention being paid to those that arise from temperature differences with external sources and during internal regenerations. In addition, the value of staged compression and expansion is highlighted whenever possible.

In this context, exergy analyses are of great interest, because they make it possible to quantify irreversibilities.

Classical energy conversion technologies are reviewed and analyzed as systems implementing the components whose operation has been studied previously. The link is made between scientific knowledge and technological achievements, which are presented in more detail than in the previous chapters.

DOI: 10.1201/9781003175629-3

Variants of steam power plants

Introduction

For various technological reasons that we mentioned in the first part, we do not know how to industrially manufacture components capable of both transferring heat and achieving effective compression or expansion.

This is why compressors and expansion machines are devices in which heat exchange with the surroundings is negligible, which is called adiabatic.

Furthermore, the work involved in a reversible compression or adiabatic expansion is, for an ideal gas, proportional to the absolute temperature T_s of the fluid at the suction of the machine.

It is therefore always advantageous to cool a gas before compressing it, and if the compression ratio is high and if it is technologically possible, one may be led to stage the compression and to cool the gas between two compression stages, thanks to a heat exchanger.

In the same way, the expansion work is all the more important as the temperature of the fluid is higher.

There is therefore always an interest in reheating a gas or a vapor before expansion.

This is why, if the expansion ratio is high and if it is technologically possible, we seek to stage the expansion and to heat the gas between two expansion stages.

We have so far studied two power cycles: that of the steam power plant and that of the gas turbine. We will see in this chapter how the first one can be improved: our goal is to minimize irreversibilities. Gas turbine cycles will be studied in Chapter 9.

In practice, we will see that the modifications of the cycles essentially relate:

- On the one hand, to the reduction of temperature differences both with the surroundings and internally;
- On the other hand, to staged compression and expansion.

In this chapter, we recall the main technological constraints to which the steam cycles are subject; then, we present the gains brought to the Rankine cycle by reheat and regenerative cycles with open and closed feedwater heaters.

The exergy balances and plots in the (h, ln(P)) and entropy thermodynamic charts of these cycles show the relevance of these tools.

We then study the main variants of conventional steam power plants: supercritical cycles, binary cycles, cycles of pressurized water reactor (PWR) plants, and we end with organic Rankine cycles (ORC), which are an alternative to conventional cycles increasingly used when the temperature of the hot source decreases or the installed capacity is less than 10 MW.

DOI: 10.1201/9781003175629-8

General technological constraints on steam cycles

In general, the objective pursued in the development of power cycles is to produce as much mechanical power as possible from the available heat source, and this takes into account existing technological constraints.

One of the major technological limitations of flame cycles is the strength of steel boiler tubes, subjected to both high pressure (HP) and high temperature. We have already presented in Chapter 5 the general implications of this constraint on the configuration of heat exchangers in boilers.

Water or steam flows through these tubes at pressures up to 150 or 300 bar. On the exterior, they are in direct contact with hot combustion gases at a pressure close to atmospheric or a few tens of bars at most in some cases. They are therefore subject to both a HP difference, and high convective and radiative flux. Unless we use very expensive special steels, the levels of pressure and temperature must be limited.

It was demonstrated a few years ago that the technological limit of conventional pulverized coal or oil plants is supercritical steam generator (SG) at 240 bar and a capacity of 1,300 MW. However, economically, the installed capacity in existing fleets is generally between 500 and 900 MW, and turbine inlet conditions are close to 165 bar and 560°C. Without desulfurization, the efficiency reaches 40% in these conditions. On average, however, the efficiency of national plant fleets is closer to 34%.

Other technological constraints of all steam plants are the following:

- First, steam quality at the end of expansion should not be too low; otherwise, too large liquid droplets form, which constitute a dangerous abrasive for mechanical blades. 0.7 is considered a lower limit that it is important not to reach, and generally one seeks not to exceed 0.85;
- Second, the steam specific volume at the low-pressure (LP) turbine outlet is extremely large, which induces very high volume flow rates, i.e., huge flow sections, high flow velocity and gigantic wheel diameters, especially in nuclear power plants;
- Finally, the condenser pressure is very low, so that air leakage is inevitable. We must therefore always extract air from the condenser, which results in significant energy consumption (up to 0.5% of the plant capacity), especially since, for reasons of simplicity and cost, low-efficiency steam ejectors are generally used.

BOX 8.1
Worked example

Extraction of noncondensable gases from a condenser

This example analyzes the use of ejectors to remove the noncondensable gases from the condenser of a steam propulsion engine of the Merchant Marine, to understand the mechanisms that come into play and to estimate its impact on the energy balance of the ship.

It is presented in the portal guidance pages (https://direns.mines-paristech.fr/Sites/Thopt/en/co/fiche-guide-td-fg21.html).

Reheat steam power plants

The reference cycle that we will use for steam power plants is a variant in terms of settings of that presented in the first part: the LP is set at 0.023 bar, the HP is set at 165 bar, and the superheating temperature is 560°C, as shown in the synoptic view of Figure 8.1. Its efficiency is 39%.

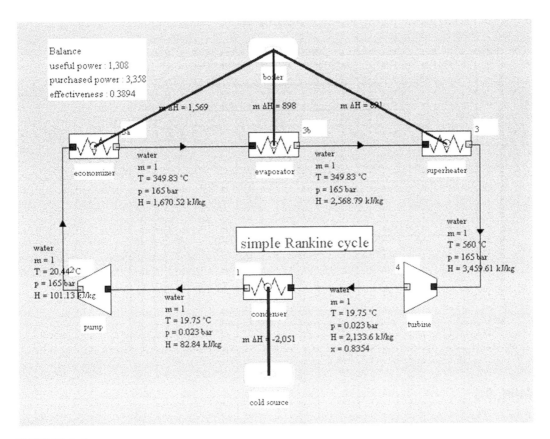

FIGURE 8.1

Reference steam power plant cycle.

Let us calculate for this cycle the value of Carnot's effectiveness, with $T_c = 15 + 273.15 = 288.15$ K and $T_h = 1,000 + 273.15 = 1,273.15$ K: $\eta = 78\%$. The actual efficiency is therefore close to 50% of the maximum theoretical efficiency.

As we have seen above, it is possible to improve such a cycle by using staged compressions and expansions.

A first idea to improve the Rankine cycle is to approach the Carnot cycle by conducting reheats. In this case, we begin by partially expanding the steam, and then it passes again into the boiler, where it is heated, at the new pressure, at the maximum cycle temperature (Figure 8.2). This operation can optionally be repeated several times.

The result is an increase in power and efficiency of a few percent.

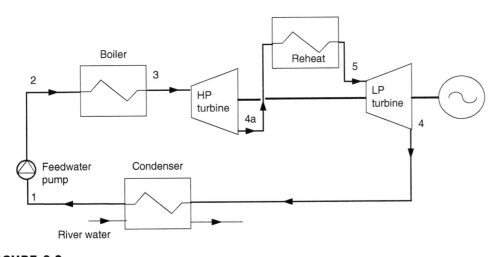

FIGURE 8.2

Reheat cycle.

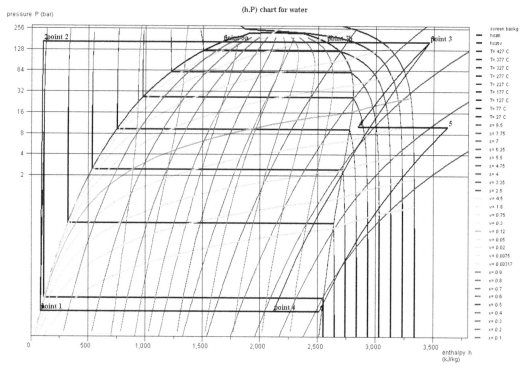

FIGURE 8.3

Reheat cycle in the (h, ln(P)) chart.

Figure 8.3 shows the cycle plot in the (h, ln(P)) chart. We have superimposed the new cycle in bold and the simple cycle in thin lines.

One can observe a clear increase in the quality at the end of expansion, which is always interesting to extend the life of the turbine blades, for which liquid droplets are abrasive as indicated above.

The price, however, is a higher complexity, but as the expansion must be staged anyway, this improvement has no major technological impact on the power plant.

Figure 8.4 shows this cycle in the entropy diagram. We can clearly see that the reheat increases the surface of the cycle, which brings it closer to that of Carnot.

To correctly model these reheats, we will have to use a new notion, complementary to that of isentropic efficiency.

Polytropic concept

Recall that, for a turbine, the isentropic efficiency is defined as the ratio of the actual work provided by the turbine to the work it would have provided if the adiabatic expansion was perfect.

Note that knowing the isentropic efficiency η_s provides no indication of the law followed by the fluid during the irreversible expansion. To know this law, you have to give yourself additional hypotheses.

One of the most common hypotheses leads to the widely used notion of polytropic, which can cover slightly different definitions depending on the authors.

The hypothesis here is to consider that irreversibilities are uniformly distributed throughout the expansion, which amounts to supposing that, during any infinitely small stage of the process, the isentropic efficiency keeps a constant value, by definition equal to the polytropic efficiency η_p which thus appears to be an infinitesimal isentropic efficiency (Figure 8.5 where τ_{ra} is the work value for the reversible adiabatic).

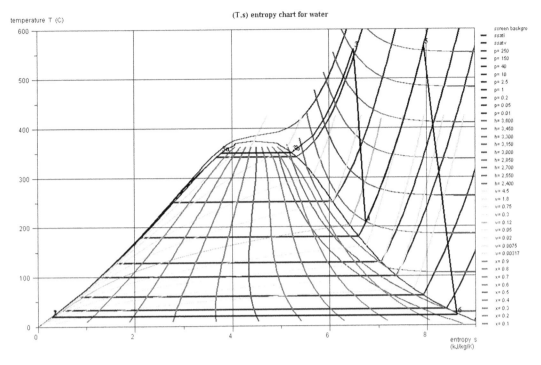

FIGURE 8.4

Reheat cycle in the (T, s) entropy chart.

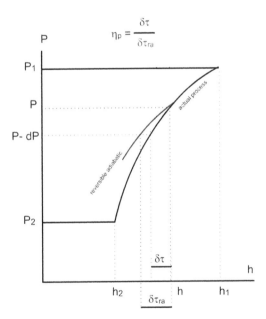

FIGURE 8.5

Polytropic efficiency.

We can then show that, for a perfect gas, the law of the process followed by the fluid is of the type $\mathbf{Pv}^k = C_{ste}$, but it is not as simple for a vapor.

For multistage machines with large numbers of similarly constructed stages, such as a steam turbine, this efficiency has a clear physical meaning: it is, in a way, the isentropic efficiency of a stage.

Modeling a multistage turbine according to a polytropic approach is therefore more realistic than assuming that its isentropic efficiency is constant whatever the expansion ratio.

Knowing the polytropic efficiency of the machine and the expansion ratio, it is possible to determine the isentropic efficiency of the machine.

Thermoptim allows you to set compression or expansion indifferently by choosing one approach or another, as you will see in guided exploration C-M1-V3.

Let us go back to the reference example. Figure 8.6 shows the turbine screen. Its setting for the expansion from 165 to 0.023 bar is as follows: isentropic reference and Set the efficiency and calculate the process, and the isentropic efficiency is equal to 0.85.

To calculate the polytropic efficiency leading to this value, choose polytropic reference and calculate the efficiency; the outlet point is known, then click on Calculate.

The value of the polytropic efficiency of the machine is determined as 0.80501 (Figure 8.7).

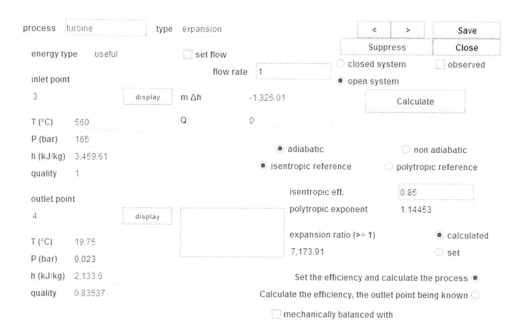

FIGURE 8.6

Turbine process screen in isentropic mode.

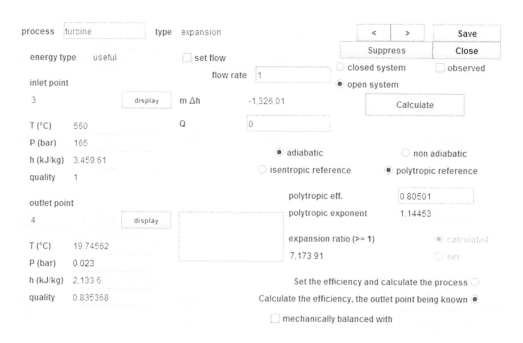

FIGURE 8.7

Turbine process screen in polytropic mode.

Now return to the "Set the efficiency and calculate the process" calculation mode.

In the modeling of the reheat cycle, we will use the value of 0.805 for both HP and LP turbines.

With this setting and an intermediate pressure of 10 bar, the result of the reheat cycle modeling is provided by Thermoptim: the efficiency goes from 38% for the reference cycle to 41.8%, and the power output from 1,174 kW to 1,720 (Figure 8.8).

This cycle is the subject of guided exploration (C-M1-V3).

Regenerative and reheat Rankine cycle

The second area of improvement in power cycles consists in reducing irreversibility by temperature heterogeneity.

In Chapter 6, we have analyzed the exchanges of the simple Rankine cycle with its external sources by comparing this cycle with the Carnot, and seen that the most significant thermal irreversibilities are located in the economizer.

Remember that, on the cold source side, isothermal condensation makes it possible to limit the temperature difference between the working fluid and the cooling fluid, so that the irreversibilities by temperature heterogeneity are low.

On the side of the hot source, on the other hand, a fuel is burned in the boiler capable of producing fumes at more than 1,500°C to heat the working fluid to much lower temperatures: in the economizer, the water comes out of the pumps at a few tens of degrees Celsius, while vaporization and superheating take place at several hundred degrees C. In all cases, the temperature differences with the hot source are very large, but it is clearly in the economizer that lays the greatest irreversibility due to temperature heterogeneity.

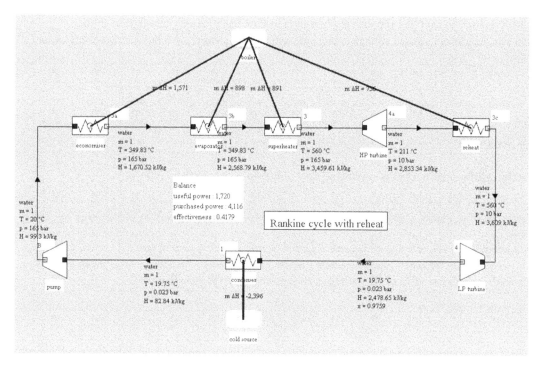

FIGURE 8.8

Reheat cycle synoptic view.

TABLE 8.1

EXERGY BALANCE OF THE RANKINE REFERENCE CYCLE

Component	Resource	Product	Exergy Efficiency (%)	Irreversibilities	% Total (%)	$T_0 = 283.15$ K
Pump	18.3	16.5	90.4	44,044.0	0.1	
Turbine	1,552.2	1,326	85.4	226.2	15.6	
Condenser	68.3			68.3	4.7	$T_k = 10°C$
Economizer	1,286.9	584.3	45.4	702.6	48.6	$T_k = 1,300°C$
Evaporator	736.6	489.9	66.5	246.7	17.1	$T_k = 1,300°C$
Superheater	730.5	529.7	72.5	200.8	13.9	$T_k = 1,300°C$
Global	2,754	1,307.7	47.48	1,446.3	100.0	

This is confirmed by the exergy balance of the reference Rankine cycle, as given in Table 8.1. If you analyze the distribution of irreversibilities, you see that almost half (48.6%) are located at the economizer. This balance allows us to quantify what had been qualitatively put in evidence in the (T, s) entropy chart.

Consider a cycle with reheat (Figure 8.2). If we extract a little quantity of steam, called bleed steam, at the outlet of the first expansion, at point 4a in the figure, its pressure remains high enough to be able to condense it at a temperature allowing to preheat the pressurized water leaving the pump in point 2 in a mixer called an open feedwater heater.

The enthalpy of the vapor is much greater than that of the liquid, due to the latent heat of vaporization; it is possible to preheat the liquid thanks to a small extraction of bleed steam during expansion.

This cycle is called regenerative Rankine cycle also known as extraction and reheat steam cycle (Figure 8.9).

Of course, for the operation to be possible, the bleed steam must be at a temperature higher than that of the liquid, which means that in practice, we must limit ourselves to a partial water preheating.

Table 8.2 gives the exergy balance of the new cycle. Its overall efficiency has increased as compared with the reference one (52.7%), and above all, the irreversibilities in the economizer have been greatly reduced (31.1% instead of 48.6%). Note also that the share of irreversibilities

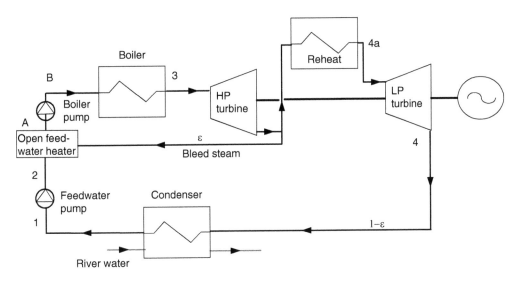

FIGURE 8.9

Regenerative reheat steam cycle.

TABLE 8.2

EXERGY BALANCE OF THE RANKINE REGENERATIVE REHEAT CYCLE

Component	Resource	Product	Exergy Efficiency (%)	Irreversibilities	% Total (%)	$T_0 = 283.15$ K
Feed pump	16.4	16.4	100.0			
Extraction pump	0.8	0.8	100.0			
HP turbine	677.7	606.3	89.5	71.4	5.1	
LP turbine	1,092.4	960.8	88.0	131.6	9.5	
Open feedwater heater	142.9	71.2	49.8	71.7	5.2	
Condenser	67.8			67.8	4.9	$T_k = 10.00°C$
Economizer	947	513.9	54.3	433.1	31.1	$T_k = 1,300.00°C$
Superheater	730.5	529.7	72.5	200.8	14.4	$T_k = 1,300.00°C$
Reheat	526.7	358.8	68.1	167.9	12.1	$T_k = 1,300.00°C$
Vaporizer	736.6	489.9	66.5	246.7	17.7	$T_k = 1,300.00°C$
Global	2,940.8	1,549.9	52.70	1,390.9	100.0	

of the HP and LP turbines is 14.6% instead of 15.6% in the reference cycle. This reduction is due to the staged expansion with reheat.

It should be noted that due to extraction, the mass flow rate of the fluid that evolves is not the same in the different parts of the machine. If ε is the bleed steam rate, the fluid flow between points 4a, 3c, 4, 1 and A is $(1-\varepsilon)$ kg/s, and the flow rate that evolves between points A, 3a, 3b, 3 and 4a is equal to 1.

The flow rate in the LP turbine is lower than in the reheat cycle; the capacity of the installation decreases slightly.

There are two kinds of feedwater heaters:

* Open feedwater heaters in which the hot fluid (the extracted steam) and the cold fluid (the condensed water) are mixed;
* Closed feedwater heaters in which the heat transfer between the two fluids takes place in a conventional exchanger. The condensed steam leaving the exchanger is then expanded to the condensing pressure and mixed with the steam leaving the LP turbine before entering the condenser.

In the first case, it is necessary to provide two pumps in the water circuit, while in the second, one is enough but you must install heat exchangers.

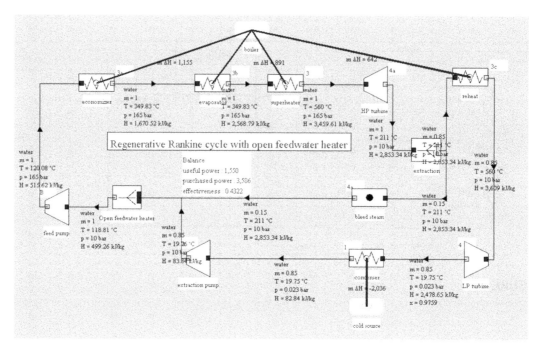

FIGURE 8.10

Synoptic view of the regenerative Rankine cycle with open feedwater heater.

In practice, in high-capacity power plants used in electricity generation, several feedwater heaters of both kinds are used (from 6 to 9), operating at temperatures ranging from 30°C to 50°C.

Overall, extractions may contribute to an improvement of almost five points of the internal efficiency of the Rankine cycle. Combined with reheat, the gain is about 7 points, a 20% higher efficiency than the basic cycle.

The synoptic view of a reheat and extraction cycle using an open feedwater heater modeled in Thermoptim is given in Figure 8.10. We note the slight decrease in power: 1,550 kW instead of 1,727 kW of the reheat cycle, due to the reduction of the flow in the LP turbine.

To correctly model the extraction in Thermoptim, you need to use a divider and a mixer, which are nodes. These nodes make it possible to model the elements of the diagram where the mixtures and the divisions of fluids take place.

This cycle is the subject of guided exploration (C-M1-V5).

BOX 8.4

Guided educational exploration

Steam power plant regenerative and reheat rankine cycle with open feedwater heater (C-M1-V5)

The steam plant cycle with reheat, by staging the expansion, slightly improves the performance of the simple cycle.

The second area of improvement in power cycles consists in reducing irreversibility by temperature heterogeneity, here thanks to regeneration.

You will learn how to set a mixer and a divider.

The synoptic view of a reheat and extraction cycle using a closed feedwater heater modeled in Thermoptim is given in Figure 8.11. Note that the efficiency is slightly lower than with an open feedwater heater, due to additional irreversibilities in the exchanger and the throttling of the condensed bleed steam.

Figures 8.12 and 8.13 give the plot of this cycle in the (h, ln(P)) and entropy charts, superimposed on the simple cycle, in dotted lines.

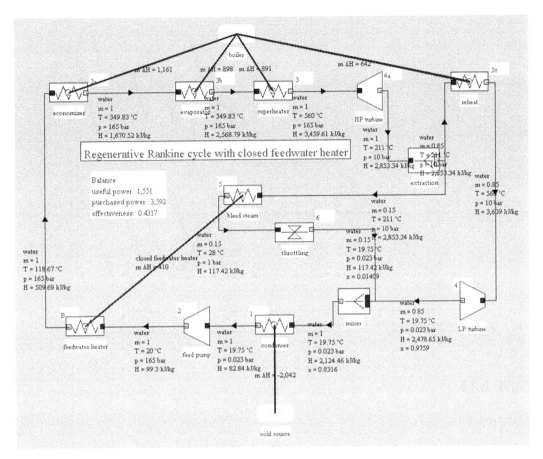

FIGURE 8.11

Synoptic view of the regenerative Rankine cycle with closed feedwater heater.

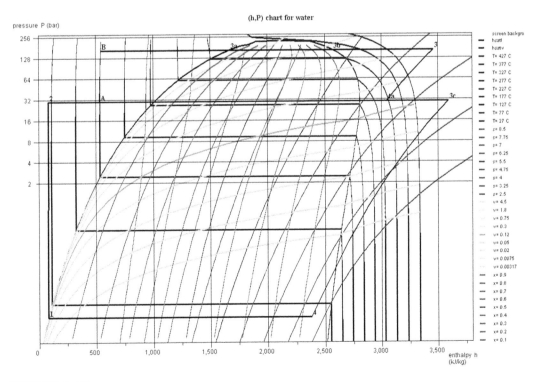

FIGURE 8.12

Regenerative and reheat Rankine cycle in the (h, ln(P)) chart.

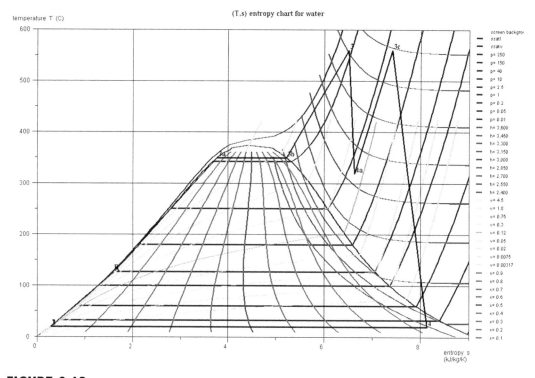

FIGURE 8.13

Regenerative and reheat Rankine cycle in the (T, s) entropy chart.

Supercritical cycles

The optimum technical and economic capacity of flame Rankine cycles until recent years matched with boiler conditions of about 560°C and 165 bar, leading, with a reheat and without extraction, to a thermodynamic efficiency close to 40%. To significantly increase the efficiency, it is possible to use the so-called supercritical cycles in which water pressure exceeds the critical pressure of 221.2 bar (Figures 8.14 and 8.15).

The result is obviously much higher stress for boiler tubes. Progress on the creep tube resistance can provide technological solutions unthinkable a short time ago.

Another constraint faced by supercritical boilers is the following: due to the absence of evaporator, you cannot cool the furnace by pipe screens traversed by boiling water with very high heat transfer coefficients. This is why a different technology is used, without drum separator, in the so-called mono-tubular boilers (improperly because in reality the layers of tubes are arranged in parallel), or in English "once-through" to indicate the absence of recirculation. These tubes are fluted with internal and external fins, mounted in spiral bundles.

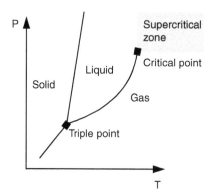

FIGURE 8.14

Supercritical zone.

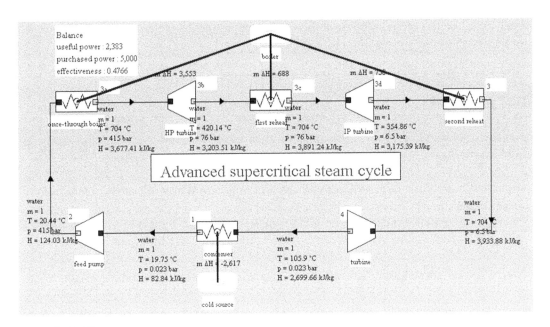

FIGURE 8.15

Supercritical cycle synoptic view.

Supercritical cycles are not new (40% of the former Soviet Union plants are supercritical, as well as more than 150 plants in the United States). The recent trend is to seek out increasingly high boiler conditions and two reheats.

To fix ideas, we modeled in Thermoptim such an advanced supercritical steam cycle (Figure 8.15). The efficiency reaches 47.7% (against 40% for a conventional subcritical cycle). Of course, such a cycle can also include feedwater heaters. We didn't model them for simplicity.

Figure 8.16 gives the plot of this cycle in the (h, ln(P)) chart.

Thus, this technology provides efficiency gains between 6% and 10% depending on the steam pressure and temperature conditions, for a cost increased by 3%–5%. As the addition

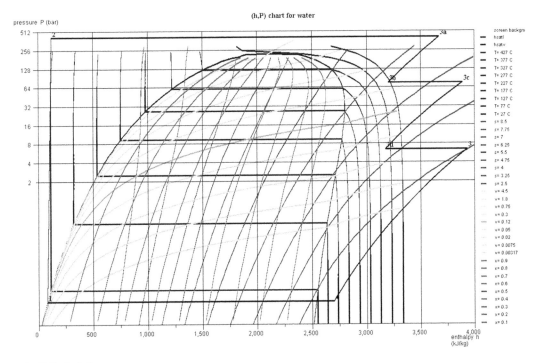

FIGURE 8.16

Supercritical cycle in the (h, ln(P)) chart.

of supercritical boilers offers more flexibility at the operational level than that of conventional drum boilers, their use is spreading more and more.

Binary cycles

One of the drawbacks of steam is the great value of its specific volume at LP and temperature, which results in:

* First, LP stages have large exhaust sections and huge fins that work at the strength limit of the material;
* Second, losses by residual velocity of the exiting steam may exist when the condensing temperature is very low.

To overcome these drawbacks, we consider the use of binary water/ammonia cycles in which steam is replaced, in the terminal portion of the LP expansion (when its specific volume is very high), by ammonia which, under the same conditions of temperature, would be about 120 times more dense. Thermoptim model of such a cycle is shown in Figure 8.17.

In this cycle, ammonia plays a dual role:

* It transfers heat between the LP steam turbine outlet and the cold source, i.e., ambient air;
* It serves as a working fluid expanding in a second turbine that replaces the LP steam turbine stages.

Steam is condensed in a condenser/boiler consisting of bundles of tubes on which occur steam condensation and ammonia vaporization; the temperature difference between the two fluids is close to 7°C. Ammonia vapor is then expanded to produce work in an ammonia turbine driving an alternator.

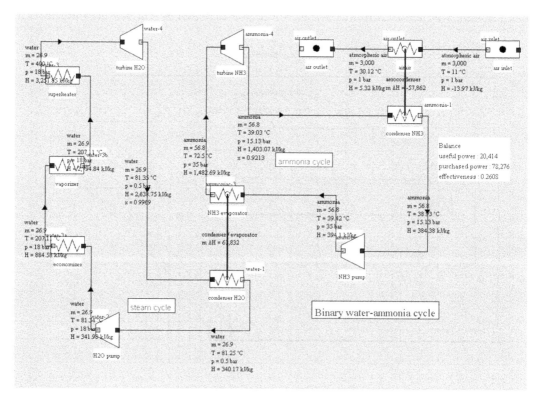

FIGURE 8.17

Synoptic view of a binary water/ammonia cycle.

After expansion, the vapor condenses in contact with the cold source. Heat transfer between the condensing ammonia vapor and air takes place in an air condenser made up of finned tubes. After condensation, ammonia is pressurized, and the bottom cycle is thus closed.

The cycle which is presented here is a simplified cycle; the efficiency of which is low (26.1%). To put it simply, a simple steam cycle with an air-cooled condenser of the same effectiveness would have an efficiency of 26.7%. The loss of efficiency due to the insertion of the ammonia cycle is therefore in this case of the order of 2.2%, i.e., relatively low compared to the potential advantages.

If the use of cooling towers is called into question because of the growing environmental constraints (scarcity of water resources and global warming), the use of binary cycles of this type could become widespread. Recent summer heat waves have shown that the solutions currently being implemented will not continue to be used in the future if temperature continues to rise. Other fluids than ammonia may also be considered.

Nuclear power plant cycles

In nuclear power plants using steam cycles, heat can be provided either directly in the reactor (boiling water reactor (BWR)) or by an intermediate heat transfer fluid which transfers heat from the nuclear reactor core (PWR).

In this section, we look more specifically at the PWR, the most industrially developed. In Chapter 15, we will expand our analysis by presenting the main types of cycles that are considered to be used in future nuclear reactors.

In a PWR, for safety reasons, the maximum steam cycle temperature and pressure are limited at levels well below those used in flame plants. In current PWR plants, pressure in the generator is about 60 bar and the steam temperature rarely exceeds 275°C.

The block diagram of a PWR is shown in Figure 8.18. On the left side of the diagram appears the containment structure comprising three main organs:

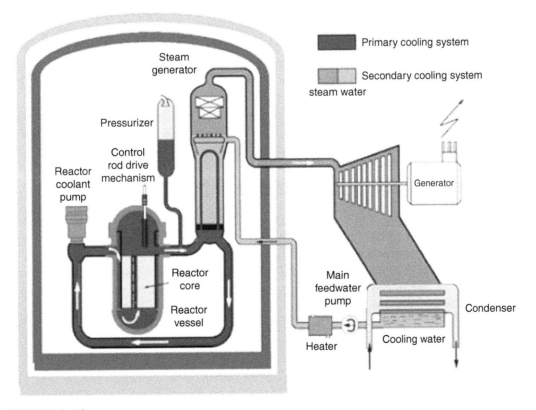

FIGURE 8.18

Sketch of a pressurized water reactor, Documentation Framatome.

- The reactor with its control system (control rod drive mechanism);
- The SG;
- The pressurizer.

These three devices are connected by the primary circuit, including connecting pipes and reactor coolant pumps, which circulate the coolant in the direction of the arrows.

The SG is connected to the secondary circuit located outside the containment structure, which corresponds to the thermodynamic cycle followed by steam, symbolized in the diagram by a turbine, a condenser, a feedwater pump and a heater.

Primary circuit

In a PWR, extraction of heat from the nuclear core involves two circuits for safety reasons, the cooling fluid (water under pressure) in contact with the reactor core being radioactive, due to fission products migrating through the cladding and products dissolved in water (mainly due to corrosion), which capture neutrons. To prevent contaminated water from being in contact with the outside and passing through electricity production cycle components, the working fluid is separate from the coolant.

The choice of this fluid is based on the qualities of water as a coolant (high heat capacity), environmentally and for use (stability, safety, availability). However, it imposes a strong constraint: the need to ensure that water remains in liquid state in the reactor vessel, to avoid local superheating of the nuclear fuel due to the presence of steam inducing low heat exchange coefficients.

For this, water must be maintained at a pressure higher than the saturation pressure at its maximum temperature in the reactor core. The whole primary circuit needs to withstand this maximum pressure, which results in severe mechanical stresses. The whole of this circuit is sized accordingly, and the system pressure is regulated with great precision to avoid either an overpressure that could cause circuit leakage or rupture, or a pressure decrease, given the risk of boiling in the core and fusion of the fuel. The pressurizer role is precisely to ensure this function.

A compromise must be found between safety constraints, thermodynamic cycle efficiency and installation costs. In existing plants, the maximum temperature of the thermodynamic cycle is set at about 280°C, and the primary circuit at about 330°C. Specifically, the temperature of the primary circuit changes from about 290°C (zero power) to 325°C (maximum power).

To ensure nonboiling primary water, the coolant system pressure is set at 155 bar, corresponding to a saturation temperature of 345°C, which provides a small safety margin. Such pressure is already high and imposes severe technological constraints at all levels.

To accurately control the primary circuit pressure, the pressurizer is a reservoir containing water in two-phase state: pressure and temperature are linked by the saturation pressure law (Chapter 3).

To control the pressure, one simply regulates the temperature, which is done by heating or cooling (by spray) water in the pressurizer. It is in communication with the entire primary system and sets its pressure (Figures 8.18 and 8.19).

The rest of the primary circuit (Figure 8.19) consists of three pump unit/SG sets. Pumps have just the role of ensuring that the primary water circulates throughout the reactor, and simply offset the pressure drops (about 8 bar).

Steam generator

The SG is able to transfer the total power of the reactor to the secondary circuit, with a very low temperature difference, as the performance of the thermodynamic cycle increases as its temperature does.

PWR Primary circuit

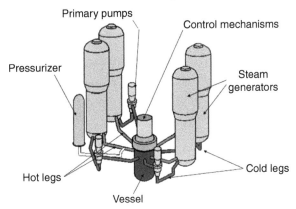

FIGURE 8.19

Nuclear boiler, Documentation Framatome.

We have seen that the primary water temperature varies between 290°C and 325°C. The presence of feedwater heaters in the secondary circuit makes the secondary water enter the generator at a temperature of about 220°C.

The current performance of the SGs used in PWRs (Figure 8.20) leads to a maximum output temperature of about 275°C.

Given the low temperature differences between primary and secondary circuits, the need to transfer significant power forbids in practice achieving any superheating in the SG, because the exchange coefficients between the primary liquid and superheated steam would be too small.

Therefore, the type of SGs used by virtually all manufacturers of nuclear boilers is shown in Figure 8.20.

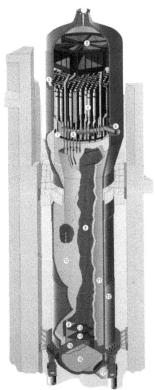

framatome

FIGURE 8.20

Nuclear steam generator, Documentation Framatome.

Primary water enters the SG in the lower part of the unit via a pipe hidden in the figure and symmetric of port 17, from which it exits cooled. This water passes, from the bottom up in the back of the figure, and down in its front, inside perforated plates designed to distribute the flow between the inverted U-shaped tubes (bundle 8).

Water from the secondary circuit enters in pipe 6 in the liquid state at a temperature of about 220°C and a pressure of about 70 bar. It is distributed in the periphery of the SG by torus 5, provided with tubes that allow it to flow into jacket 11 located between the outer wall 12 and the bundle 10, which acts as economizer. It then follows an upward flow around the bundle, inside envelope 10, and partially vaporizes with a steam quality between 0.2 and 0.4, in a regime of nucleate boiling leading to very high heat exchange coefficients.

The two-phase emulsion then passes through cyclone separators 4 and then dryers 2. The liquid fraction falls to the bottom of the SG and is recirculated with the feedwater (the recirculation rate is between 2 and 4.5). The vapor fraction reaches a quality above 0.997, and exits through top port 1, to be directed towards the turbine HP section.

Secondary circuit

A peculiarity of PWR nuclear power plant SGs is, as we have seen, that there is no initial superheat. Complete expansion of steam from this state would lead to too low a steam quality, which would be both disadvantageous in terms of performance, and fatal to the mechanical strength of turbine blades. The solution adopted consists in using a special organ called the moisture separator reheater (MSR), to stage the expansion by providing a reheat at a pressure of about 11 bar, which can increase efficiency and meet the quality constraint at the end of expansion.

The MSR (Figure 8.21) receives steam partially expanded with a quality close to 0.87, the liquid phase of which is separated and directed to feedwater heaters, while the steam passes

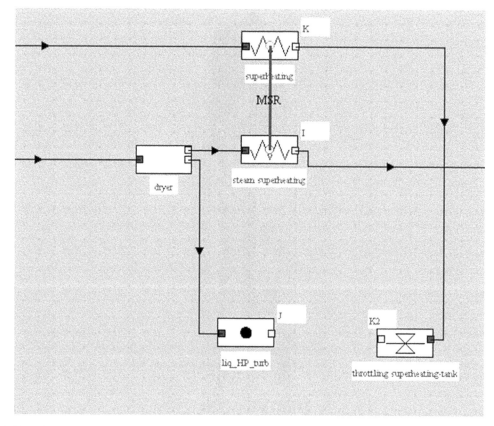

FIGURE 8.21

Moisture separator reheater in Thermoptim.

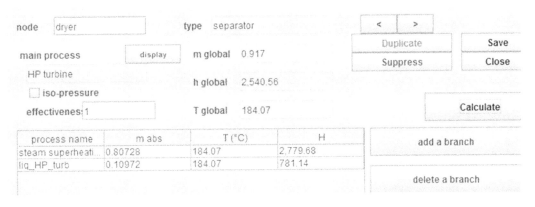

FIGURE 8.22

Separator screen.

through a heat exchanger traversed internally by a small flow of saturated steam at HP (and therefore higher temperature), which condenses. For a 1,500 MW unit, two MSRs of 370 tons each are needed: their length is 24.8 m, their height is 6 m and their width is 5.3 m.

To model this device in Thermoptim, we need a phase separator, which is a special node, and a heat exchanger.

The MSR (Figure 8.21) has three components:

* The separator, which is a phase separator;
* The "superheating" process, which represents the fraction of the flow of superheated steam leaving the SG at HP, which condenses;
* The "steam superheating" process, which represents the medium-pressure (MP) steam which is superheated.

The link between these two processes corresponds to the heat exchanger called MSR.

The separator screen is shown in Figure 8.22. It receives most of the flow, namely, 91.7% of the total, leaving the HP turbine at MP and at the corresponding saturation temperature, with a quality equal to 0.88.

The result of the calculation is that 11% of the total flow rate of the SG is found in the liquid state and is directed towards the feedwater tank, while 80.7% corresponds to saturated steam at MP and is directed towards the superheater.

The screen of the superheater heat exchanger is shown in Figure 8.23.

In the left part appear the characteristics of the hot fluid, which is the vapor coming from the condensing SG, and in the right part are displayed those of the cold fluid, leaving the separator, whose temperature goes from 184°C to 253.4°C, which corresponds to a superheat of approximately 69.6°C.

To set the superheater, we set all the known elements, namely, the flow rates of the two fluids, the state of the water entering and leaving the hot process (the HP steam that condenses), as well as the state of the water entering the cold process (the MP steam that we want to superheat).

We have thus set five constraints, so that the exchanger can be calculated.

The superheated MP steam is directed to the LP turbine, where it produces mechanical power, while the two other flows leaving the MSR are mixed in the feedwater tank with the expanded and condensed HP steam, before to be compressed by the pump (Figure 8.24).

PWR cycle

Figure 8.25 represents the diagram of a PWR cycle, modeled in Thermoptim, excluding feedwater heaters.

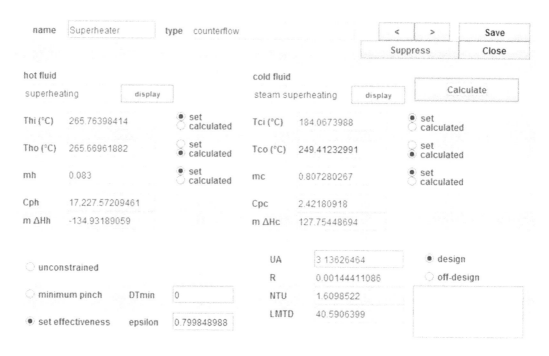

FIGURE 8.23

Superheater screen.

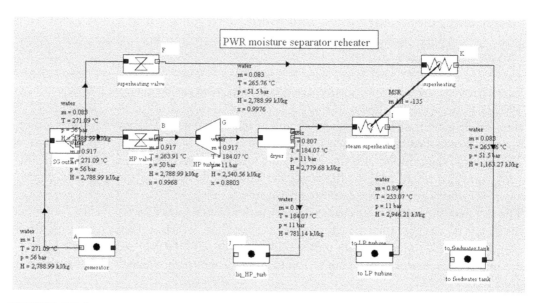

FIGURE 8.24

Moisture separator reheater synoptic view.

BOX 8.5

Guided educational exploration

Pressured water reactor (PWR) nuclear power plant (C-M1-V8)

This example shows how a PWR cycle with MSR can realistically be modeled with Thermoptim.
This model is rather simple insofar as no extraction or reheat is taken into account, except for the MSR.

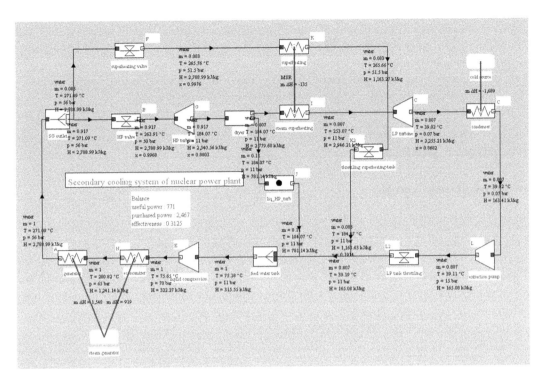

FIGURE 8.25
Synoptic view of the PWR cycle.

We find the model of the MSR in the upper-right part of the diagram.

With the feedwater heaters that are actually used, the cycle efficiency can reach about 33.5%.

This cycle is the subject of a guided exploration (C-M1-V8). Refer to Chapter 4 for additional information on exchangers in Thermoptim.

Cycle plot in the water (h, ln(P)) chart is shown in Figure 8.26.

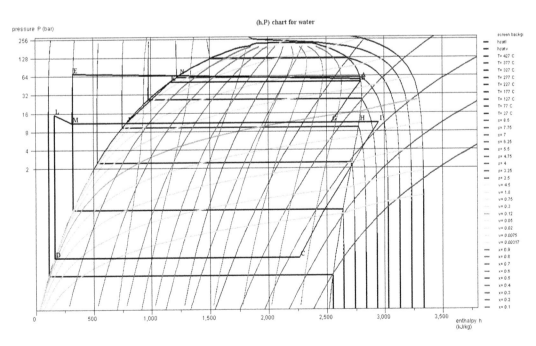

FIGURE 8.26
PWR cycle in the (h, ln(P)) chart.

The two expansions are clearly shown in the diagram. The first (B-G) takes place entirely in the liquid–vapor equilibrium (LVE) zone, while the second (I-C) begins in the vapor zone and ends in the LVE zone.

Interpreting the diagram is a bit tricky, since the flow rate varies depending on the process.

The inlet point into the separator is G, and the two exit points H and J.

ORC power plants

Limits of steam cycles

Steam power plants use water, which turns out to be an excellent thermodynamic fluid, with its critical temperature of 374°C and its high latent heat of vaporization at ambient pressure and temperature. Its low viscosity makes it possible to limit the consumption of auxiliaries, and its reduced price and its nontoxicity finish placing it in a very good position compared to all its competitors. However, for certain applications, other fluids may sometimes prove more suitable.

The main field of application for steam cycles is that of producing electricity from either fossil fuels, for capacities of around 300–800 MW, or nuclear reactors, for even higher capacities.

Advantages of ORC cycles

ORC are variants of steam cycles, which are used when the hot source from which one wishes to produce mechanical power is at low or medium temperature, or when the installed capacity is low and steam installations are no longer economical.

When the temperature of the hot source drops, or that the installed capacity decreases, typically below 10 MW, the performance of steam cycles deteriorates, and it becomes preferable to resort to other thermodynamic fluids.

As many of these are organic in nature, it is customary to label these cycles organic, but other types of fluids, such as ammonia or carbon dioxide, can be used.

In large installations, the expansion machine is generally a turbine. In middle range, volumetric devices are preferably used, such as screw machines, and, for very small capacities, scrolls or piston machines.

ORC have aroused growing interest worldwide since the end of the 1980s, particularly in Anglo-Saxon countries, for four main classes of applications, which account for most of the achievements, in terms of installed capacity and number of installations:

* Geothermal power plants;
* Biomass combustion plants;
* Heat recovery from effluents;
* Thermodynamic solar power plants.

Unlike water, the dew curve of some organic fluids has a slope such that the expansion takes place entirely in the vapor phase, and even deviates from the saturation curve.

Under these conditions, it may be advantageous to use a regenerative cycle.

In this cycle (Figure 8.27), the steam leaves the turbine in 4, in the superheated state. The energy corresponding to the desuperheating (4–4bis) is recovered to preheat the condensed liquid before entering the SG (2–2bis).

Example of an ORC cycle: the closed OTEC cycle

We will present here an example of an ORC cycle intended to produce electricity from the thermal gradient of the oceans.

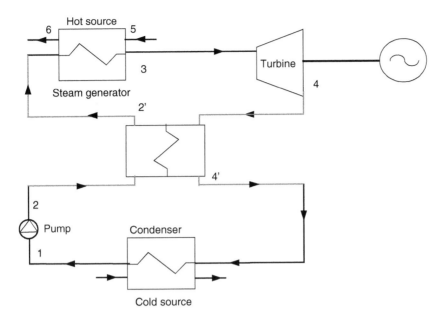

FIGURE 8.27

Regeneration ORC cycle.

OTEC means ocean thermal energy conversion. OTEC cycles are designed to generate electricity in warm tropical waters using the temperature difference between water at the surface (26°C–28°C) and in depth (4°C–6°C), from 1,000 m (Figure 8.28 in which water temperature is given on the abscissa, and the depth on the ordinate).

In all cases, the need to convey very high water flow rates and pump cold water at great depth induces significant auxiliary consumption. Optimization of an OTEC cycle is imperative to take into account those values.

Two main types of cycles are used: closed cycles and open cycles, invented by two French, respectively, Jacques d'Arsonval in 1881 and Georges Claude in 1940, who proceeded to a first experiment.

Although technically valid, OTEC cycles are not yet economically viable. Prototypes of various capacities have been realized or are being considered, including in Hawaii and Tahiti.

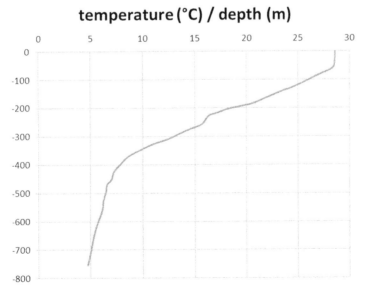

FIGURE 8.28

Ocean thermal gradient.

Closed cycles use hot water at about 27°C to evaporate a liquid that boils at a very low temperature, such as ammonia or an organic fluid. The vapor produced drives a turbine, then is condensed by heat exchange with cold water at about 4°C from deeper layers of the ocean.

The thermodynamic cycle is similar to a steam cycle, except for the working fluid (Figure 8.29). The sizing of heat exchangers is of course even more critical given the very small temperature difference between hot and cold sources. Pinch values should be as low as possible while remaining realistic.

A possible Thermoptim model of this cycle is shown in Figure 8.30. Its efficiency is 2.84%.

The hot water flow rate is here equal to 27,000 kg/s, its temperature 26°C and the cold water temperature 4°C.

Let us calculate for this cycle the value of the Carnot effectiveness, with $T_c = 4 + 273.15 = 277.15$ K and $T_h = 26 + 273.15 = 299.15$ K: $\eta = 7.35\%$. The actual efficiency is therefore equal to around 39% of the theoretical one, which is not that bad.

This cycle is the subject of guided exploration (C-M1-V9). An example of OTEC open cycle will be presented in Chapter 16.

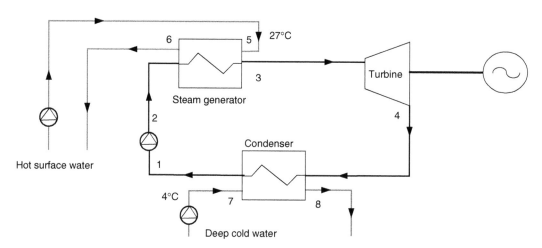

FIGURE 8.29

OTEC closed cycle.

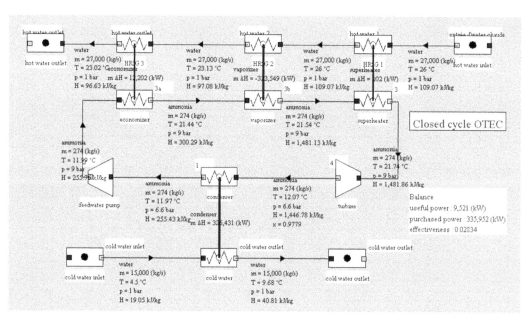

FIGURE 8.30

OTEC closed cycle Thermoptim synoptic view.

BOX 8.6
Guided educational exploration

Closed ammonia ORC cycle (C-MI-V9)

This example shows how can be modeled a closed ORC cycle intended to generate electricity from the thermal gradient of the oceans.

Bibliography

S. Jayet-Gendrot, F. Arnoldi, P. Billard, C. Dorier, Y. Dutheillet, L. Lelait, D. Renaud, Des matériaux innovants pour un sujet brûlant, Revue Epures, DRD EDF 1999, Paris.

B. Liu, P. Rivière, C. Coquelet, R. Gicquel, F. David, Investigation of a two stage Rankine cycle for electric power plants. *Applied Energy*, 100, 285–294, 2012, Elsevier. doi: 10.1016/j.apenergy.2012.05.044. ⟨hal-00748172⟩

H. Struchtrup, *Thermodynamics and Energy Conversion*, ISBN 978-3-662-43714-8, Springer, Berlin, Heidelberg, 2014.

Conventional internal combustion engines

Introduction

In this chapter, we continue to apply the reflections of the previous chapter to conventional internal combustion engines: gas turbines as well as reciprocating gasoline, gas and diesel engines.

The variants of gas turbine cycle are essentially regeneration and staged compression with intercooling or staged expansion with reheat. We also study turbojets which are in fact variations of the gas turbine open cycle, with a diffuser and a nozzle.

We then present the operating principles of reciprocating internal combustion engines: four stroke and two stroke spark ignition and diesel motors. After discussing briefly the analysis of their cycles and then their overall performance, we study in more detail the consequences of combustion chemical kinetics on the operation and design of these two categories of engines. A brief discussion on turbochargers follows, and then we examine how emissions of pollutants can be controlled, thanks to modern injection systems including lambda sensor and three-way catalytic purification converters for gasoline engines.

Gas turbine cycles and variants

Combustion gas turbine

Reference cycle

We will consider as a reference the cycle called the Brayton cycle presented in Chapter 4 to illustrate the combustion calculations, with the proviso that the air inlet temperature is 15°C instead of 25 (Figure 9.1).

Its setting is as follows: the turbine inlet temperature is 1,150°C, the high pressure 16 bar, and the compressor and the turbine are assumed to have polytropic efficiencies equal to 0.85.

The efficiency of this cycle is equal to 35.8%.

In Chapter 6, we have examined the exchanges of this gas turbine cycle with its external sources in order to determine where the most significant thermal irreversibilities are located.

On the side of the hot source, a fuel is burned in the combustion chamber capable of producing fumes at more than 2,000°C, but, for technological reasons, we must limit the turbine inlet temperature below 1,200°C–1,400°C in terrestrial gas turbines, higher values being reached in certain turbojets. It is not possible to change the cycle at this level.

Concerning the cold source, the release into the atmosphere of the gases leaving the turbine corresponds to a very great irreversibility. It is sometimes possible to reduce it.

DOI: 10.1201/9781003175629-9

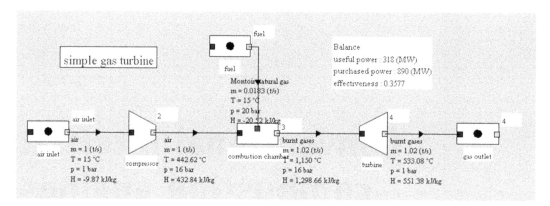

FIGURE 9.1

Reference gas turbine cycle.

Analysis of the exergy balance

Let us now have a look at its exergy balance (Table 9.1), which provides quantitative elements that are quite interesting when we seek to improve this cycle. Its exergy efficiency is 33.5%. If you analyze the distribution of irreversibilities, you see that more than half are located in the combustion chamber, which is not surprising, due to the conversion of the chemical exergy of the fuel into heat, and that 38.4% correspond to the gas release in the atmosphere.

The rest is roughly distributed between the turbine and the compressor.

As it is impossible to reduce irreversibilities taking place in the combustor, improving the Brayton cycle first implies using the enthalpy available in the gases released to the atmosphere.

Cycle with regeneration

Principle

The exhaust gases exit at 551°C, while the temperature of the air leaving the compressor is 476°C. It is therefore possible to partially heat this air before entering the combustion chamber, which further reduces fuel consumption. All you have to do is insert a heat exchanger between the exhaust gases and the compressed air.

This variant of the simple cycle is called regeneration cycle (Figure 9.2).

We call regenerator effectiveness the ratio $\epsilon = \dfrac{h_{5'} - h_{2'}}{h_5 - h_{2'}}$, which characterizes the fraction of the available enthalpy actually recoverable.

The heat to be provided to the cycle is reduced in this case to $(h_3 - h_5)$.

TABLE 9.1

EXERGY BALANCE OF THE REFERENCE CYCLE

Component	Resource	Product	Exergy Efficiency (%)	Irreversibilities	% Total (%)	$T_0 = 288.15$ K
Fuel	990.8					
Air inlet	0.2					
Compressor	442.7	402.2	90.9	40.5	6.1	
Turbine	796.4	760.9	95.5	35.5	5.4	
Combustion chamber	0	1,062.8	76.3	330	50.1	
Gas outlet				252.8	38.4	Loss
Global	991.1	332.3	33.53	658.8	100.0	

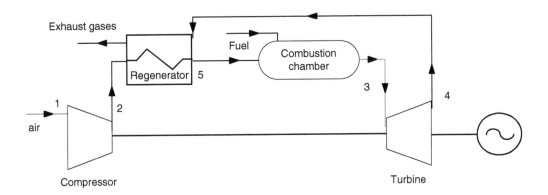

FIGURE 9.2

Sketch of a regenerative gas turbine.

Introducing the ratio $\theta = \dfrac{T_3}{T_1}$ of the temperatures at the inlet of the turbine and of the compressor, it is possible to plot the theoretical efficiency η of this cycle as a function of the compression ratio $r = \dfrac{P_2}{P_1}$, assuming that the working fluid is unique (air) and perfect.

Figure 9.3 shows the shape of the efficiency of the regenerative gas turbine for two values of effectiveness: 0.85 and 0.5. The four curves in each family correspond to different values of θ: 3, 3.5, 4 and 4.5.

We note that the efficiency improvement brought by the regenerator is even better than the compression ratio is low. Indeed, beyond a certain limit, itself a function of θ, the fluid heating in the compressor is such that temperature $T_{2'}$ at the output of the compressor becomes higher than temperature $T_{4'}$ at the end of expansion. The regenerator can no longer function and becomes useless.

As shown in the graph of Figure 9.3, all curves corresponding to the same θ and different values of ε intersect at the point where $T_{2'} = T_{4'}$.

The results above are of course no more valid when we no longer assume that the working fluid is unique and perfect, but the existence of a limit remains true.

In the case of the gas turbine that interests us, the introduction of a regenerator with an effectiveness of 0.85 leads to the results shown in Figure 9.4.

The efficiency has increased and is now worth 39.5% instead of 35.8%.

This cycle is subject to a guided educational exploration (C-M2-V2).

BOX 9.1

Guided educational exploration

Exploration of a regenerative gas turbine (exploration C-M2-V2)

In this guided exploration, you will start by studying the setting of combustion in a simple gas turbine cycle, and then you will be interested in the regeneration cycle.

Regenerative brayton cycle exergy balance

The exergy balance of this cycle is given in Table 9.2. Its exergy efficiency (36.1%) has slightly increased as compared to the reference one (33.5%).

The loss corresponding to the gases released in the atmosphere is smaller than before but still the major source of irreversibilities, except for the combustion chamber.

Given that the current trend consists in increasing the compression ratio in gas turbines, regeneration is poorly suited for modern machinery using advanced technologies,

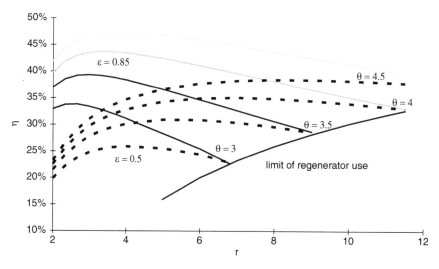

FIGURE 9.3

Limits of the regenerative Brayton cycle.

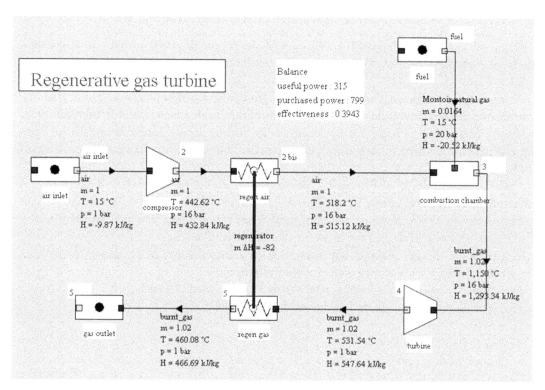

FIGURE 9.4

Synoptic view of a regenerative gas turbine.

TABLE 9.2

EXERGY BALANCE OF THE REGENERATIVE CYCLE

Component	Resource	Product	Exergy Efficiency (%)	Irreversibilities	% Total (%)	
Fuel	910.1					
Compressor	442.7	402.2	90.9	40.5	7.0	
Turbine	793.9	758.5	95.5	35.4	6.1	
Combustion chamber	0	1,056.3	77.5	306.7	52.7	
Gas outlet				198.4	34.1	Loss
Regen air	51.6	51	98.8	0.6	0.1	Regenerator
Global	910.3	328.8	36.11	581.58	100.00	

except sequential combustion turbines. Instead, it is almost always used for small turbines (25–500 kW), called microturbines. We shall see in Chapter 10 that combined cycles can valorize the residual enthalpy available in the exhaust gases when regeneration cannot be used.

Cycle with intercooling and regeneration

Another idea to improve the cycle is to use staged compression or expansion. This operation can be repeated several times if necessary.

Let us examine the case of a so-called intermediate cooling or intercooling cycle, in which the compression is fractionated (Figure 9.5). At the outlet of the first compression stage, the air at 185°C is cooled by exchange with the outside air at 15°C, which makes it possible to lower its temperature to 20°C. It is then recompressed at the final pressure, the compression work being lower.

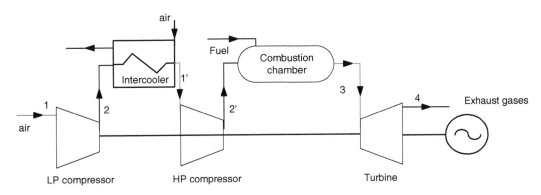

FIGURE 9.5

Two-stage compression gas turbine with intermediate cooling.

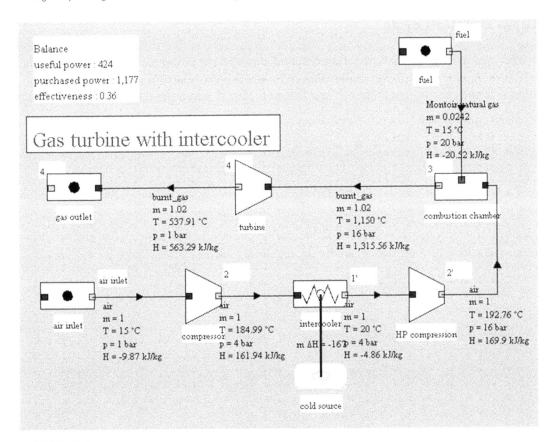

FIGURE 9.6

Synoptic view of a two-stage compression gas turbine with intermediate cooling.

As the outlet temperature of the second compression stage has also dropped, more heat has to be supplied to the combustion chamber. Globally however, the overall balance remains beneficial.

This mainly results in an increase in power and a very small gain in efficiency (Figure 9.6). Note that the end of compression temperature being much lower than in the simple cycle (193°C instead of 442°C), additional regeneration would be beneficial.

The difficulty is that the morphology of aeroderivative machines that lead to the best performance is generally not well suited to intermediate cooling, so this solution is rarely used.

This cycle is subject to a guided educational exploration (C-M2-V3).

BOX 9.2
Guided educational exploration

Exploration of a staged compression gas turbine (exploration C-M2-V3)

This guided exploration presents a staged compression gas turbine cycle, which improves the basic cycle.

However, it is often easier to insert a sequential combustion than an intercooling, as the combustion chamber is much smaller than a cooling exchanger (Figure 9.7).

Another cycle improvement involves combining the above changes by performing a two-stage compression with intermediate cooling and expansion with multi-stage sequential combustion, and a regenerator which can then be used without temperature crossover.

The efficiency thus increases accordingly, but at the cost of increased complexity and higher cost. It thus departs significantly from the initial simplicity of the gas turbine.

Steam injection cycle

Another way to improve the cycle efficiency, and above all the power available on the turbine shaft, is to inject steam into the combustion chamber. It is thus possible to recover part of the enthalpy of the exhaust gases and increase the mass flow through the turbine. Moreover, this

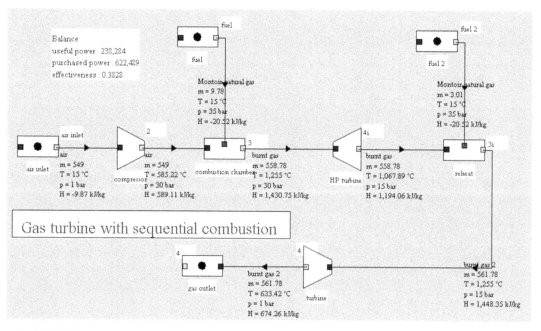

FIGURE 9.7
Synoptic view of a staged expansion gas turbine with sequential combustion.

mode of operation reduces emissions of nitrogen oxides NO_x. This type of cycle can achieve efficiencies near 45% in practice but has the disadvantage of requiring water.

The thermodynamic analytical study of these cycles is more complex than that of the regenerative cycle previously studied and will not be treated here. Their modeling in Thermoptim poses no particular problem, at least insofar as there is no need to detail the operation of the heat recovery steam generator (HRSG).

Cooled gas turbine

The performance of gas turbines being very sensitive to the intake air temperature (see Chapter 2), it may be advantageous to cool the air in an artificial way, using a compression refrigeration cycle. This is particularly the case in hot countries where power needs are time-varying, and where the peak call is often due to air conditioning needs during the hottest hours. In such circumstances, it may be economically attractive to produce ice during off-peak hours and use this ice as a cold source to cool the intake air during peak hours. The machine efficiency is slightly improved, but its capacity is substantially increased.

Example of Figure 9.8 corresponds to a gas turbine where the ambient air at 30°C is cooled at 1°C. The gain in efficiency is 1 point, but power output is increased by 12.5%.

Mechanical configurations

The simplest gas turbine is, as we noted above, a machine where the compressor and turbine are coupled on the same shaft, which directly drives the recipient machine.

An analysis of the transient operation shows that a **single shaft turbine** is poorly suited for operation at partial load, especially if the speed is set (as in electricity generation, for example). However, in case of sudden discharge of the receiving machine, the runaway speed of the gas turbine will be low, the compressor absorbing, as we have seen, almost two thirds of the power supplied by the turbine.

The limits of adaptation of the single shaft turbine lead to the idea of separating it into two parts according to their respective functions: first the **auxiliary turbine**, usually located upstream, whose role is solely to drive the compressor, and second the **useful power turbine**, driving the receiver machine (Figure 9.9). We can distinguish **the gas generator**, upstream, and the mechanical energy generator, downstream.

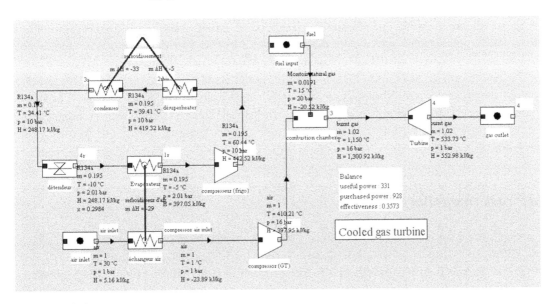

FIGURE 9.8

Synoptic view of a cooled gas turbine.

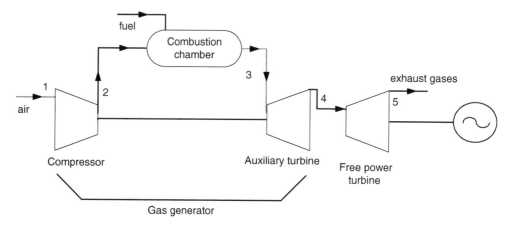

FIGURE 9.9
Two shaft gas turbine.

The gas generator being mechanically independent of the receiver machine, its speed can vary without constraint.

Part load operation is obtained by simultaneously playing on the fuel injection and the rotation speed, which reduces the gas generator flow rate. The power output can be adjusted by maintaining acceptable efficiency in a much wider range than with a single shaft turbine (80% of nominal output at 60% load, 60% at 30% load). However, the nominal efficiency is slightly lower than that of the single shaft gas turbine, due to increased friction losses, and the installation cost is slightly higher.

Twin shaft facilities are mainly used when seeking a good performance over a wide range of power variation of the receiving machine. This is the case in transport, in particular aeronautical propulsion, which we will study further.

An additional advantage of two shaft gas turbines is that it is possible to run the gas generator at high speed (between 20,000 and 30,000 rpm), which allows for very compact structures. In particular, manufacturers of aircraft engines have developed very powerful techniques, using hollow shafts to reduce congestion caused by the presence of two shafts.

Finally, startup problems are greatly simplified since we can start the gas generator at its rated speed before coupling the useful turbine to the receiving machine. This provides a reserve power to overcome the starting torque of the latter, which is not the case in single shaft systems.

Some manufacturers even offer three shaft machines, such as Rolls Royce with its Trent industrial turbine, of 50–60 MW capacity, which reaches an efficiency of 42.5% with a compression ratio equal to 35. The mass of the Trent being equal to about 29 t, this machine has one of the largest specific capacities of the market (1.8 kW/kg). Its design has benefited from the experience gained on the RB211, built in nearly 300 units over the past fifty years.

The new Siemens' SGT-A45 TR gas turbine model is an advanced aeroderivative development based on proven technology from Siemens and Rolls Royce Aero-Engine. The core engine is derived from components of Siemens' Industrial Trent 60 that have been adapted to the proven free power turbine. It utilizes two-shaft architecture and provides 44 MW at 42% efficiency (Figure 9.10).

Aircraft propellers

For years, airplanes have been propelled by propellers driven by gasoline engines. Even today, this is the best solution for small airplanes. When, for larger aircraft, the propeller is kept in view of its excellent performance, it is often driven by a turboprop engine using an open cycle gas turbine. However, turbojet engines have replaced the propeller for the propulsion of many aircraft, including most long-haul airliners. It is also the only engine that is suitable for supersonic flight, and it thus equips mostly military aircraft. At very high speeds, and for

FIGURE 9.10
Siemens' SGT-A45 TR gas turbine. (Courtesy of Siemens Energy.)

the propulsion of long-range missiles, the jet engine reaches its limits, and the ramjet is used because it provides very good efficiency. However, it cannot operate autonomously at take-off, which must then be provided by a turbojet or rocket engine.

Let us start with a little bit of thermodynamics.

In the previous sections, we assumed the kinetic energy variations of fluids undergoing processes were negligible. In the diffusers and nozzles that equip turbojets, this assumption is no longer valid, the practical effect being obtained by converting into pressure the kinetic energy of the fluid.

We restrict ourselves to the case of steady flows where the absolute pressure and the three velocity components are assumed to be constant over time. The flow occurs in fixed pipes, which implies the absence of moving walls. These conditions are obtained in the convergent and the diffuser of a turbomachine.

We assume the fluid is perfect from the hydrodynamic point of view, that is to say without viscosity (note that an ideal gas is not necessarily a perfect fluid from this point of view). We obtain the law of a compressible fluid flow applying the first law of thermodynamics to a duct with changing cross section (Figure 9.11).

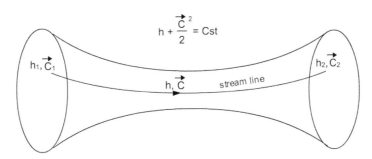

FIGURE 9.11
Duct with changing cross section.

The fundamental law of adiabatic flow in a fixed reference follows directly from the first law. It reflects the conservation of total enthalpy $h + K$, and is written along a duct with changing cross section, C being the fluid velocity:

$$h + \frac{C^2}{2} = \text{Const} \qquad (9.1)$$

Assuming further that the flow is reversible adiabatic ($ds = 0$) then:

$$dh = vdp, \quad \text{and}$$

$$vdP + Cdc = 0$$

It is possible to change the fluid pressure acting on its velocity and vice versa.

This equation simply results from the generalization of the first law taking into account kinetic energies.

A jet engine of the type that is commonly used in aviation is a simple modification of a gas turbine open cycle studied previously: the turbine is sizted to only drive the compressor.

At the turbine outlet, the excess energy available in the gas at medium pressure and temperature is converted into kinetic energy in a nozzle. The thrust results from the difference of momentum between intake and exhaust gases (Figure 9.12).

Recall that the momentum is the product of mass flow by speed.

By definition, the thrust is equal to

$$F = \dot{m}\left(C_1 - C_5\right) \qquad (9.2)$$

C_1 being the speed of the aircraft and C_5 that of the gas at the nozzle exit.

The turbojet also includes an inlet diffuser, which is used to create a static precompression at the compressor inlet (Figure 9.13)

A turbojet is therefore the combination of a diffuser, a gas generator and a nozzle.

We speak of a gas generator because the function of the "compressor, combustion chamber and turbine" assembly is solely to generate hot gases at a pressure higher than ambient pressure, so that these gases can then be converted into kinetic energy in the nozzle.

It is quite possible to model with good precision various cycles of turbojets with Thermoptim.

However, Thermoptim core components are not sufficient to build such models: to represent the inlet diffuser and the outlet nozzle, it is necessary to use two external classes, that is to say two extensions of the software package.

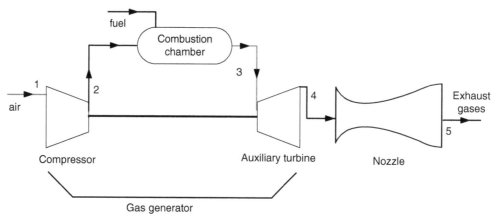

FIGURE 9.12

Sketch of a turbojet.

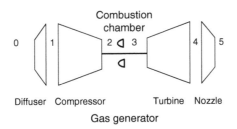

FIGURE 9.13

Single-flow turbojet.

The nozzle screen[1] is given in Figure 9.14. The upstream point corresponds to the outlet of the gas generator. Its pressure here is around 2.4 bar, and its temperature is 891°C.

The downstream point is at an ambient pressure of 0.265 bar because the aircraft is aloft.

The right part of the screen includes the nozzle parameters. For the nozzle, two calculation methods are possible: to determine the outlet pressure knowing the outlet velocity, or to determine the outlet velocity knowing the outlet pressure, which is generally the case.

Model parameters are as follows:

● The gas inlet velocity (m/s), generally low;
● The isentropic efficiency of the process (0.95);
● Either the gas outlet velocity (m/s), or the gas pressure at the exit of the nozzle, depending on the calculation option chosen.

The Thermoptim model of a simple turbojet engine is given in Figure 9.15.

It also uses a diffuser[2] component similar to the nozzle.

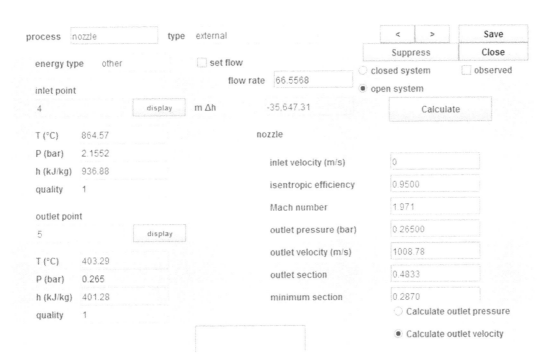

FIGURE 9.14

Nozzle screen.

---

[1] https://direns.mines-paristech.fr/Sites/Thopt/en/co/modele-tuyere.html.
[2] https://direns.mines-paristech.fr/Sites/Thopt/en/co/modele-diffuseur.html.

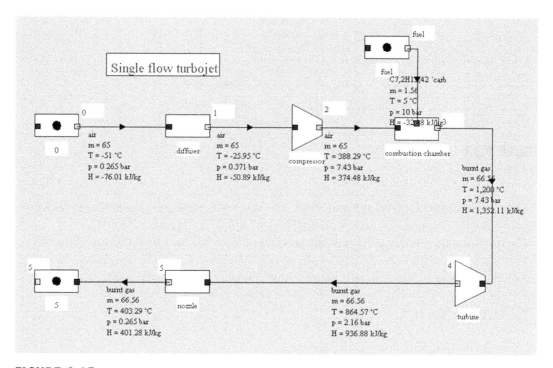

FIGURE 9.15

Synoptic view of a single flow turbojet.

We may at first be satisfied with a simple model where we assume that we know the compression ratio and the flow passing through the machine and polytropic or isentropic compressor and turbine efficiencies. Calculating the model then poses no particular problem.

Once the model is built, it becomes possible to calculate the engine performance. The specific thrust and the ratios of power and consumption to thrust are among magnitudes most commonly used for this.

The expression of the thrust is here, taking into account the variation of the flow rate through the engine due to fuel injection:

$$F = \dot{m}_0 C_0 - \dot{m}_5 C_5$$

The power output W is the product of thrust by the aircraft speed:

$$W = FC_0$$

This cycle is subject to guided educational exploration (C-M2-V4).

BOX 9.3
Guided educational exploration

Exploration of a turbojet (exploration C-M2-V4)

This guided exploration presents the cycle of a single-flow turbojet engine. As the components of the Thermoptim core are not sufficient to make such models, to represent the inlet diffuser and the outlet nozzle, it is necessary to use two external classes, i.e., two extensions of the software package.

The overall efficiency of a turbojet remains low because there is a tradeoff between a good thermal efficiency, which involves providing gases with a high kinetic energy in the nozzle, and a good propulsive efficiency, which requires a low relative gas velocity at the nozzle exit.

The turbofan engine (Figure 9.16) reconciles these two apparently contradictory requirements by accelerating to a speed barely higher than that of the plane, an air flow (called secondary flow) complementary to that which flows through the gas generator. This operation is made possible, thanks to an additional low-pressure compressor, called a fan, also driven by the turbine.

The dilution ratio is the ratio of secondary flow to primary flow. It can reach 10 for some jet engines. Propulsion is then ensured by a large air flow at low speed and a small flow of exhaust gases from the primary classical cycle.

The need to withdraw from the turbine the power consumed by the fan alters however the thermal efficiency by reducing the available enthalpy at the turbine outlet.

Figure 9.16 also shows that the assembly is made of two concentric axes, that of the low-pressure section rotating inside that which carries the stages of compression and expansion at high pressure.

It is possible to model a turbofan, with high or low bypass ratio.

We must split the airflow entering the engine at the outlet of the diffuser. One part is directed towards the combustion chamber and then two turbines in cascade, one being balanced with the engine compressor, and the other with that of the fan, as shown in the diagram of Figure 9.17. Two nozzles are then used to calculate the velocity of ejected gases and therefore the engine thrust.

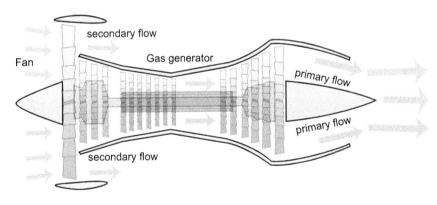

FIGURE 9.16

Cross section of a turbofan.

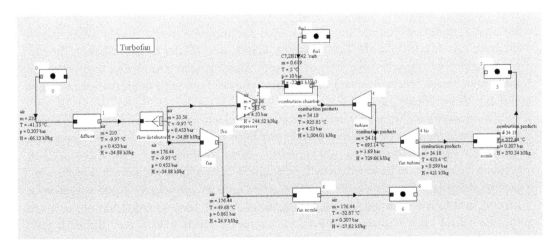

FIGURE 9.17

Synoptic view of a turbofan.

Conclusion

The first part of this chapter has enabled us to show the importance of gas turbines as a means of producing mechanical power on land or for aeronautical propulsion.

As we will see in Chapters 10 and 14, gas turbines can also be combined with other cycles, especially steam, to form high efficiency combined cycles and cogeneration plants.

Reciprocating internal combustion engines

Now let us look at reciprocating internal combustion engines, which we have not yet studied given the modeling difficulties they present. We will now be able to explain why.

Reciprocating engine capacity ranges for usual application from less than 1 kW to about 1 MW. The crowning achievement use of these machines is the propulsion of road vehicles.

We will develop more than in the other chapters certain technological aspects linked to combustion in fast reciprocating gasoline and diesel engines, because it is at this level that the most important constraints are today, in terms of efficiency and pollutant emissions.

There are two main types of reciprocating internal combustion engines:

* Spark ignition engines, whose principle was established by the French Beau de Rochas in 1860, and the first project carried out by the German Otto in 1876. Although some of these engines burn gaseous fuels, we will qualify them in the following by gasoline engines;

* Compression ignition engines, called Diesel, by the name of their German inventor who patented them in 1892.

Piston machines occupy a prominent place among internal combustion engines. This situation stems from two causes:

1. By their periodical operation, these machines are particularly suitable for processes where the temperature reaches high values. Being in contact with the fluid at various stages of process, the walls of the machine are subject to an average temperature well below the maximum temperature, whereas in a continuous flow machine such as gas turbine, some parts are constantly subjected to this temperature. In addition, the velocities being much lower in piston machines, the fluid/wall exchange coefficients are smaller. Finally, the average heat flux received by the walls is low enough that, with adequate cooling, we can keep them in good working condition while producing inside the cylinder combustion at more than 2,500 K, which would be quite impossible in a continuous flow machine.

2. Internal combustion engines are widely used for propulsion of vehicles of small and medium capacity. Indeed, the piston engine adapts much better to this use than the turbine engine, which is feasible only with very high characteristic speed and therefore must rotate at considerable speed when power is moderate, with interposition of fragile and expensive reducing gear.

   It is customary to distinguish the modes of operation of spark ignition and diesel engines by characterizing each by a different cycle; the Beau de Rochas cycle (or Otto depending on the authors) is characterized by a constant volume combustion and the diesel cycle by combustion at constant pressure. In fact, especially in fast engines, we will see that the combustion delays are such that more complex cycles must be considered if one wishes to be precise. In these circumstances, what distinguishes the two types of engines is not so much the theoretical cycle than combustion characteristics which follow very different laws depending on whether the fuel is volatile or not. So we understand that the complexity of the physicochemical phenomena taking place during combustion in a piston engine is such that the basic ideal cycles just allow approaching reality in a relatively simplified way.

General operation mode

All reciprocating internal combustion engines operate on the same general process described schematically in Figure 9.18. A variable volume is defined by a cylinder, one of the bases of which is fixed, called head, and the other is a movable piston in the cylinder bore, driven by a connecting rod system. In a four-stroke engine, the organs that control the inlet or exhaust valves are actuated by push buttons coupled to the drive shaft by a camshaft.

In various ways, depending on whether the engine is two or four stroke, fresh gas is introduced into the cylinder at atmospheric pressure during the intake phase (fuel mixture formed in advance in conventional gasoline engines, clean air in diesel engines).

The piston being at a distance from the bottom of the cylinder, the intake port is closed, the volume V between the piston and the bottom being occupied by a certain charge of fresh gas.

Approaching the bottom of the cylinder, the piston compresses the charge in the volume v of the combustion chamber, that is to say the remaining space when the piston reaches the end of its stroke, called top dead center or TDC. This compression is substantially adiabatic and occurs without appreciable internal friction. The key factor is the operating volumetric compression ratio $\rho = V/v$, a geometric characteristic of the cylinder.

The combustion reaction is then triggered, either by local ignition of the mixture in gasoline engines or by injecting fuel into the compressed air in diesel engines. The combustion occurs during a relatively short time, while the piston continues its stroke. In practice, it occurs in a mode intermediate between the constant volume combustion and combustion at constant pressure. The piston continuing to move away from the bottom of the cylinder, the burned gases expand until the end of the stroke (bottom dead center or BDC) and then are evacuated and replaced by a new charge of fresh gas.

An important difference between a gasoline engine and a diesel engine is not in the mode of introducing fuel, which in some gasoline engines is also injected but when the fuel is introduced, which determines the nature of gas when the reaction starts.

The piston-rod-crankshaft system transforms the reciprocating movement of the piston into the rotary movement of the motor shaft (Figure 9.19).

In **four-stroke** cycles, the most common, the head of the cylinder is pierced by two holes, controlled by valves, which put it in communication with the intake and exhaust manifolds (Figure 9.19). Note the difference with reciprocating compressors, where valves are not controlled, but opened by differences in pressure between the cylinder and these manifolds.

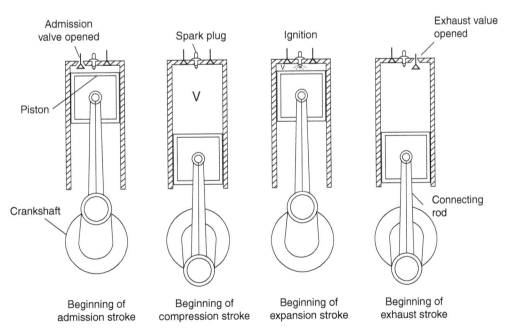

FIGURE 9.18

Four-stroke operation mode.

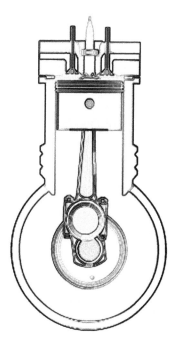

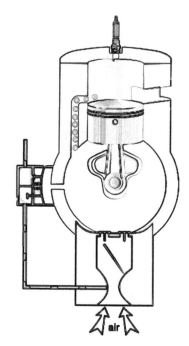

FIGURE 9.19
Four-stroke engine.

FIGURE 9.20
Two-stroke engine.

In **two-stroke** engines (Figure 9.20), exhaust occurs at the end of the expansion stroke through holes made in the side wall at such a level that they are unmasked by the piston at BDC. At the same time or shortly after, intake ports are opened which connect the cylinder with a manifold filled with fresh gas at a pressure slightly higher than that prevailing in the exhaust manifold.

The cylinder sidewalls and bottom of gasoline engines are always strongly cooled, usually by circulating water in holes in the walls (Figure 9.21), and sometimes in small-capacity motors with highly developed outer fins subject to a violent air current

In the gasoline engine, fuel is introduced well in advance so that the cylinder is full, when ignition occurs, of a substantially homogeneous mixture. In the diesel engine, fuel is injected at the last moment and burned as and when it is introduced.

Watt diagram

The evolution of gas pressure and specific volume in the cylinder is often represented in the Watt diagram (Figure 9.22). At the end of expansion in 3, the exhaust valve opens, pressure drops to atmospheric pressure and the piston performs a full stroke to the head, thus driving gases out.

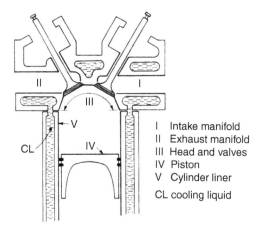

I   Intake manifold
II Exhaust manifold
III Head and valves
IV Piston
V Cylinder liner

CL cooling liquid

FIGURE 9.21
Cylinder water cooling. (Courtesy Techniques de l'Ingénieur.)

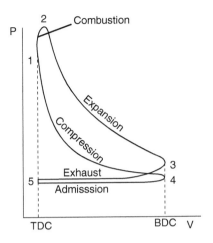

FIGURE 9.22

Watt diagram.

When it reaches the TDC in 5, the exhaust valve closes, and the inlet opens. Moving away, the piston draws a fresh charge of gas. In 4, the BDC, the intake valve closes, and compression 4–1 begins followed by combustion 1–2 and expansion. It is therefore a simple cycle in four strokes, hence the name four-stroke engine.

As we can see, the operation of these engines is much more complex than that of a gas turbine, so their modeling is not as simple.

Volumetric efficiency

A number of factors are reducing the mass of fresh gas entering the cylinder during the intake phase: the residual presence of smoke, the high temperature of the walls, which has the effect of dilating the fresh gas, admission losses, which cause the pressure in the cylinder to be slightly less than atmospheric pressure, and finally the exhaust counter-pressure. Moreover, acoustic phenomena related to the charge pulsating flow may, depending on the case, either decrease it (known as filling holes), or instead increase it (this is called natural supercharging). This pulsating flow naturally stems from the reciprocation of the piston and the opening and closing of valves.

To characterize these losses, we define a volumetric efficiency C_r, such that the mass of the fresh gas charge admitted by cycle is equal to

$$m_a = C_r \rho_a V_s = C_r V_s \frac{P_a}{r T_a}$$

We can show that, in the absence of acoustic phenomena that we have just mentioned, if f is the residual gas rate, ρ the volumetric compression ratio and if admission conditions are identified by index 1, the volumetric efficiency is given by

$$C_r = (1-f)\frac{\rho}{\rho-1}\frac{r_a}{r_1}\frac{T_a}{T_1}\frac{P_1}{P_a}$$

In practice, C_r is about 0.80–0.85 for a gasoline engine, and around 0.9 for a diesel engine.

Theoretical and actual cycles

In its classic form, the elementary study of cycles is based on rather crude approximations: the working fluid is equated with air, and itself equated to a perfect gas, and processes are considered perfect. To be more precise, we must consider the following:

- The mass and the actual chemical composition of the working fluid, which varies according to the phases of the cycle, due to the introduction of fuel and combustion;

- Variations of the specific heat with temperature;
- Molecular dissociation at high temperature;
- Losses related to the renewal of the charge;
- Combustion kinetics;
- Heat exchanges between the working fluid and the engine walls;
- Mechanical friction and auxiliary consumption.

We begin by studying ideal cycles that lend themselves to easy thermodynamic analysis, and then we give guidance on how to take into account main nonidealities.

Although highly simplified, these cycles allow us indeed to draw some important conclusions that are not challenged by further analysis, such as those relating to the influence on engine performance of volumetric compression ratio, combustion conditions (constant volume, constant pressure), the initial pressure in the cylinder, etc.

As we have just seen, the operation of these engines alternates phases in a closed system (closed valves) and in an open system (intake and exhaust), which has the effect of complicating their analysis. Their representation in a thermodynamic diagram is far from being as simple as that of the other cycles that we have studied.

Otto or beau de rochas cycle

Analyses of ideal cycles exclude intake and exhaust phases of two or four stroke real cycles. As a first approximation, the operation of gasoline engines can be represented by a cycle called Beau de Rochas or Otto, which is reduced, as indicated below, to four simple processes shown in Figure 9.23 in a Watt or entropy diagram.

The first phase is a reversible adiabatic compression 4–1.

This compression phase in a closed system begins after the inlet valves are closed and ends before ignition.

The second phase is a constant volume combustion 1–2;

Triggered at the end of compression, when the piston speed is zero, the combustion is supposed to be fast enough to be considered instantaneous, and therefore at constant volume, which is especially warranted in slow engines.

The third phase is a reversible adiabatic expansion 2–3;

This expansion in a closed system begins at the end of combustion and ends before the opening of the exhaust valves.

Finally, the fourth phase is 3–4 constant volume cooling.

By the end of expansion, opening the exhaust valve makes the pressure in the cylinder suddenly drop. It is assumed here that the discharge is instantaneous.

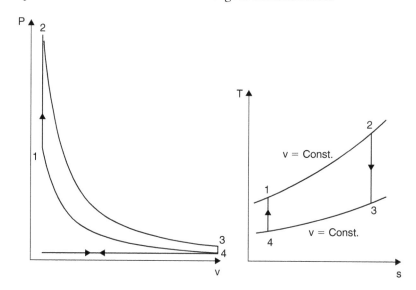

FIGURE 9.23

Beau de Rochas cycle.

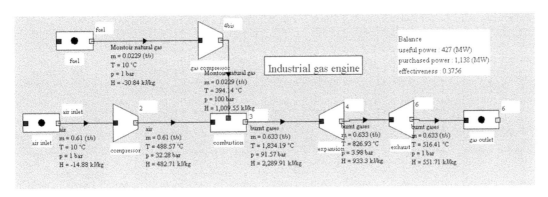

FIGURE 9.24

Synoptic view of an industrial gas engine.

Figure 9.24 shows an example of such a cycle model for an industrial gas engine: model E 2842 gas engine, manufactured by MAN Dezentrale Energiesysteme.

This cycle is the subject of guided exploration (C-M2-V5b). Here are some indications on its settings.

The compression setting is as follows: closed system, with a compression ratio set to 12 and not calculated, and an isentropic efficiency equal to 0.95 (Figure 9.25). This compression ratio is equal to the operating volumetric compression ratio $\rho = V/v$ introduced at the beginning of this section.

The setting of the combustion chamber is different from those we have considered for the gas turbine: combustion takes place in a closed system, the air factor is known, equal to 1.6, and the downstream mass volume is set by the upstream point, combustion being assumed to be instantaneous (Figure 9.26).

Taking into account the strong cooling of the engine, necessary for technological reasons, the efficiency of the combustion chamber is set to 0.79, which means that 21% of the thermal power is lost.

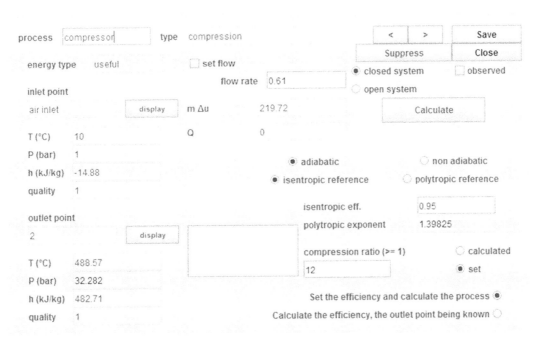

FIGURE 9.25

Compression settings.

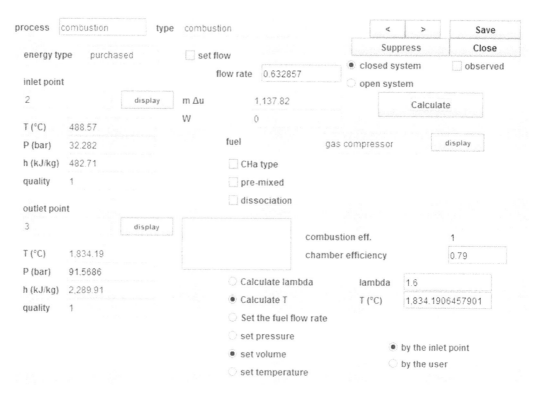

FIGURE 9.26

Combustion settings.

BOX 9.4

Guided educational exploration

Exploration of an industrial gas engine (exploration C-M2-V5b)

This guided exploration presents an industrial gas engine modeled with a so-called Beau de Rochas cycle.

The end of combustion pressure and temperature are calculated: 91.6 bar and 1,834°C.

The expansion takes place in two phases: first in closed system, until the valves open, and then open system. The expansion is therefore modeled by two expansion processes.

The setting of the expansion phase in a closed system is also specific: as for compression, the expansion ratio equal to 12 is set, and an isentropic efficiency of 0.95 is taken into account (Figure 9.27).

The pressure at the opening of the valves is calculated: 3.98 bar.

At the end of expansion in closed system, the exhaust valves are open, and expansion continues according to the same adiabatic law, up to atmospheric pressure, in open system this time, and without production of work, which justifies that the type of energy chosen is "other" (Figure 9.28). The isentropic efficiency is taken equal to 1.

Diesel cycle

The fundamental difference between the diesel cycle and the Beau de Rochas cycle is the replacement of the constant volume combustion by combustion at constant pressure, as shown by Watt diagram of Figure 9.29.

It is assumed here, hypothesis mostly valid for slow engines, that the expansion of gases due to combustion exactly compensates, in terms of pressure, the volume expansion due to the stroke.

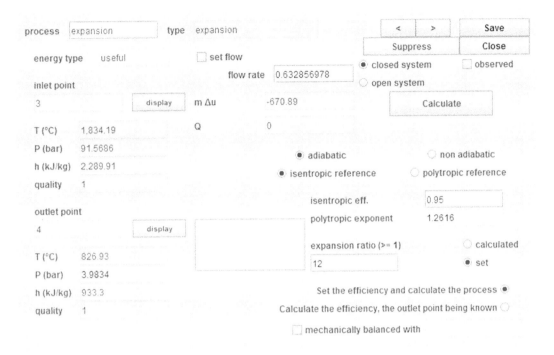

FIGURE 9.27

Closed system expansion settings.

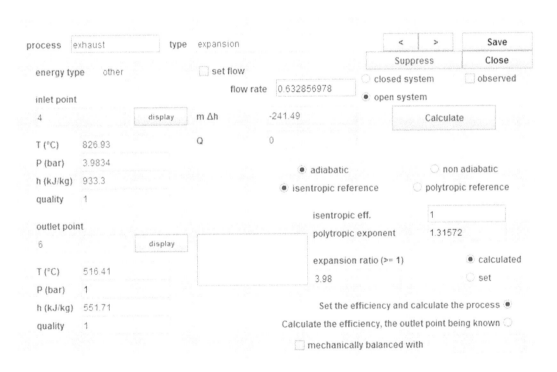

FIGURE 9.28

Open system expansion settings.

The model in Figure 9.30 represents such a cycle, with a volumetric compression ratio of 20 instead of 12. The efficiency is better than that of the gasoline engine following the Beau de Rochas cycle, but the gain comes mainly from the increased compression ratio.

This example highlights the important fact that the functions of components can differ from their geometric configuration. In the case of reciprocating internal combustion engines (gasoline or diesel), the same set of parts (the cylinder and the piston) successively plays the role of compressor, combustion chamber and then expansion device. The functional

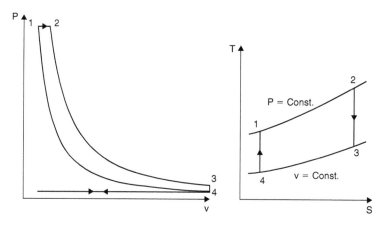

FIGURE 9.29

Diesel cycle.

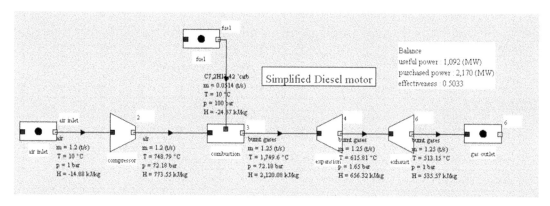

FIGURE 9.30

Simplified diesel engine model.

representation of such a system leads to connecting together three elements representing these different functions.

Actually, whether it is a gasoline or diesel engine, combustion takes place neither at constant volume nor at constant pressure but appears as shown in the diagram in Figure 9.22. This difference is mainly due to the very high speed of rotation of most of these motors.

The Beau de Rochas and constant pressure diesel models that we have presented therefore deviate greatly from reality.

A better approximation can be obtained by considering that combustion takes place in three stages: it begins at constant volume, continues at constant pressure and ends at constant temperature. However, this model is also open to criticism (Figure 9.31).

As a result, unlike the functional processes that we have studied so far, combustion in reciprocating internal combustion engines calls for at least three reference processes, which makes their thermodynamic analysis much more complex.

Figure 9.31 shows the combustion in three stages, the cooling of the engine and the recirculation of part of the burnt gases due to the presence of a dead space in the cylinders.

This model calls for some comments:

* The intake charge includes ambient air and a fraction of recirculated gas coming from the dead space that still exists in the cylinder (3.3% by mass). Its composition is determined in the "inlet" mixer;
* Combustion takes place in three phases, as proposed by Thelliez (1989), the fuel being injected three times;
* Expansion occurs in two steps: in closed system between points 5 and 6, then in open system from 6 to 7 in valves and exhaust pipes.

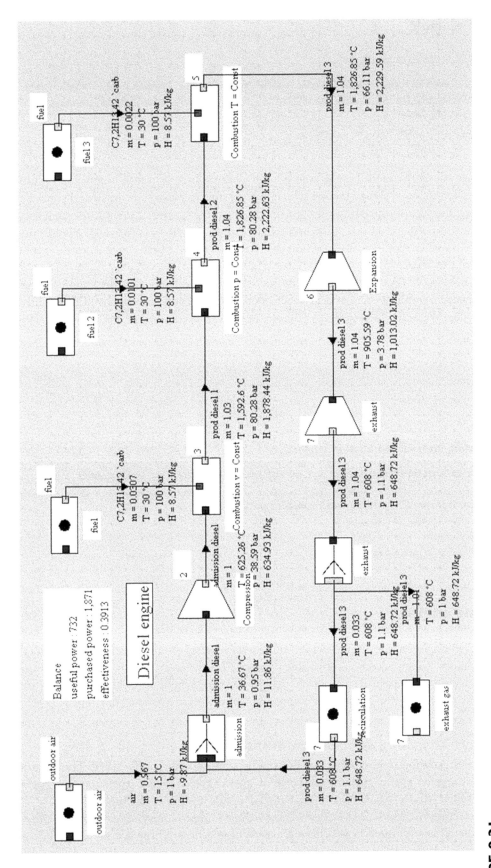

FIGURE 9.31

Synoptic view of the model with a combustion in three steps.

BOX 9.5
Worked example

Modeling of a diesel engine cycle

This example shows how a Diesel engine cycle can realistically be modeled with Thermoptim. It is presented in the portal guidance pages (http:// https://direns.mines-paristech.fr/Sites/Thopt/en/co/fiche-guide-fg20.html) as well as in the Diapason session 38 (https://direns.mines-paristech.fr/Sites/ Thopt/en/co/session-s38en-exercise.html).

Such a model is a little difficult to set, given its complexity.

The analysis of theoretical cycles and overall performance of piston engines have allowed us to identify the main parameters on which it is possible to play to optimize and control their operation. Once the compression ratio and fuel (and therefore its LHV) are selected, the variables that can be used to control the engine are essentially the heat provided (fuel flow) and the intake pressure. We have already said that a key difference between gasoline and diesel is precisely at this level.

To go further, it is necessary to separate the study of gasoline and diesel engines, to reflect the influence of combustion chemical kinetics on the operation and design of these different engine types. We refer the reader to Chapter 4 for a presentation of the basic mechanisms governing combustion phenomena.

BOX 9.6
Key issues

Reciprocating internal combustion engines

The following self-assessment activities will allow you to check your understanding of reciprocating internal combustion engines:

* General mode of operation of a reciprocating internal combustion engine, gfe
* Differences between gasoline and diesel engine, gfe
* Locating a gasoline engine cycle in a Watt diagram, ddi
* Locating a Diesel engine cycle in a Watt diagram, ddi

Gasoline engine

In a conventional gasoline engine, when the reaction starts, the combustion chamber is filled with an almost homogeneous combustible mixture. The basis for the study of combustion in these engines is the chemical kinetics of fuel gas mixtures, the main problem encountered in gasoline engines being the control of combustion conditions to prevent knocking.

Limits of knocking and octane number

The simplest solution is to set an upper limit to the compression ratio $\rho = V/v$. The experience shows that a given engine running at specified speed and burning a particular fuel begins to detonate when ρ exceeds a certain value, which depends mainly on the fuel.

It is therefore necessary to classify fuels according to their ability to detonate. For this, we proceed as follows in order to get rid of the type of engine used for testing: we test at given speed in an experimental engine with adjustable head, a continuous series of fuels obtained by mixing in variable proportions a highly explosive fuel, normal heptane C_7H_{16} and a low explosive fuel, iso-octane C_8H_{18}.

We thus get the law of variation V/v = f(x), of the limit compression ratio as a function of the iso-octane content x of the mixture. To classify any fuel F, we test the same engine at the same speed and determine the limit volumetric ratio ρ_c for which knocking occurs (measured by an electromechanical system recording pressure changes in the cylinder). Expressed as a percentage, value x_c such as $\rho_c = f(x_c)$ is called **octane number** i of the fuel. This characteristic presents the advantage of being independent of the engine used. ρ values range from 4 (for i = 0) to 12 (for i = 100) for standard test motor (CFR engine of the Cooperative Fuel Research Committee). Note that the value of i may exceed 100 for some fuels such as ethanol or benzene, less explosive than the iso-octane.

We are in practice led to define two octane numbers: the Research Octane Number or RON, characteristic of the behavior in mild operating conditions (low load, urban), and the Motor Octane Number or MON, characteristic of the behavior in severe operating conditions (high loads, high speed traffic on highway etc.).

For conventional fuels, the MON is 10–12 points less than the RON, the values of the latter being between 97 and 99 for premium fuels and between 89 and 92 for regular gasoline.

In order of increasing octane number, fuels usually rank as follows: acetylene, normal saturated hydrocarbons (alkanes), branched chain hydrocarbons, alcohols, ethers and carbon monoxide. In a series of hydrocarbons, octane number decreases when the number of carbon atoms increases.

We can considerably increase the octane number of gasoline by incorporating the following:

* Either significant proportions (5%–20%) of low explosive fuels, such as alcohols (methanol CH_3OH or ethanol C_2H_5OH) or methyl ether (MTBE methyltertiobutylether C_4H_9-O-CH_3 and TAME tertioamylmethylether C_5H_{11}-O-CH_3);
* Or very low quantities (parts per thousand) of an organometallic compound, tetraethyl lead, Pb $(C_2H_5)_4$, which presumably acts as a catalyst inhibiting the production of volatile compounds. However, the dosage of these additions must be limited to prevent corrosion of exhaust valves by lead. In addition, these additives are incompatible with catalytic converters that are spreading increasingly to limit emissions of nitrogen oxides NO_x.

For a given octane number, the limit value of V/v depends on the motor and its regime. It increases when the speed increases and is strongly influenced by the design of the combustion chamber. It can be raised substantially by increasing the turbulence of the mixture when it burns.

In slow industrial reduced turbulence motors, the V/v limit can reach 10 when the fuel is produced by a gas furnace or dry coke gasifier containing only carbon monoxide. In modern, high speed and enhanced turbulence traction motors, it reaches 6.5 for a fuel of octane number 80, 8 with alcohol premium grade fuels, and can exceed 10 when the fuel is pure alcohol.

In any event, whatever the progress made, the compression ratio of a gasoline engine is limited, because of the nonknocking condition, well below the values achievable in a diesel engine. As the internal efficiency η_i is a function of that ratio, the efficiency of gasoline engines is significantly lower than that of diesel engines. This condition has also limited the possibilities for use in gasoline engines of various provisions allowing the performance of internal combustion engines (supercharging, insulation of cylinder and piston walls, etc.) to be improved.

Strengthening of turbulence

It can be shown that, in the absence of significant turbulence, the deflagration speed V_d is low, a few cm/s to few m/s. V_d is the speed of propagation of the flame in the combustion mixture. Given these values, one of the hardest problems in the development of fast gasoline engines, after ruling out the knocking, is to accelerate deflagration so that the combustion process takes place in a timely manner. It seems very difficult to solve if we compare the time required to travel the distance between the spark plug and the farthest point of the combustion chamber (at speed V_d), with the combustion duration time.

To fix ideas, consider an engine of 100 mm bore, running at 3,000 rpm. The flame has to travel a 100 mm to sweep the combustion chamber. With $V_d = 5$ m/s, very high value if the fluid is at rest, this corresponds to a lag of one-fiftieth of a second. However, at 3,000 rpm, it is precisely the duration of a complete revolution, and that corresponding to a 30° crank rotation (to which it would be desirable to limit the combustion duration) is twelve times lower.

To limit the ignition delay, a first arrangement, still used, is to practice some ignition advance, characterized by the angle A between the axis of the crank and that of the cylinder before TDC, when the spark occurs.

But the advance A should not exceed a certain value, beyond which the combustion becomes explosive (due to the expansion of gases already burned, the early combustion significantly increases the temperature of the unburned gas at the end of the compression phase, so as to exceed the autoignition limit). In addition, the spark advance has the serious defect to create a variable delay in inverse proportion of the engine speed, so that a given value of A is too large at low speed (engine knocks) and inadequate at high speed.

Actually, it is by communicating high turbulence to the mix during combustion that we get to accelerate the deflagration speed in high-speed motors. This is achieved by an appropriate arrangement of outlets of the intake valves in the chamber, so that gases enter the engine at high speed which is preserved during compression. The design of the combustion chamber is generally made so that the piston movement may generate the sought turbulence.

Note, however, the following:

* That the turbulence has the defect of substantially increasing heat exchange coefficients and harmful wall actions, first heating during admission, which reduces the charge and the mean pressure, and also increasing heat losses during combustion and expansion;
* That it does not raise the lower richness limit (beyond which the combustion becomes impossible), quite the contrary;
* However, the variation of turbulence with speed plays in the desired direction and therefore has a self-regulating effect because it increases with speed, which automatically decreases the combustion duration.

A second serious defect specific to gasoline engines is that it is impossible to control power, as would be most rational, by decreasing the amount of fuel, because of the lower richness limit. To reduce it, we use more or less defective means, e.g., in traction motors, creating a throttling at admission (by closing the throttle) which lowers the efficiency a lot.

Overall, the efficiency of gasoline engines drops quickly when the load decreases and the idling consumption is high.

Formation of fuel mix, fuel injection electronic systems

In a spark-ignition engine, the fuel mixture formation can take various forms depending on the fuel and the technique used.

When the fuel is gaseous mixture, formation presents no difficulty. The mixer must simply be located at a sufficient distance from the intake valves, and adjusting richness can be easily carried out by a valve controlling the gas flow.

It is much more difficult to obtain a homogeneous mixture in gasoline engines burning a volatile liquid fuel. The conventional method for forming the mixture is to use a carburetor consisting essentially of a float chamber feeding a jet, which injects the fuel sprayed into the admitted air stream in a throttling of the vein. Since they are no longer used for environmental reasons, we will not discuss them here. Refer to Section 13.4.3.1 of the first edition for an analysis of their operation.

In modern gasoline engines, electronic fuel injection is increasingly used, mainly to better regulate the combustion by controlling very precisely the amount of fuel injected. This carburization mode is particularly imperative when using catalytic converters, which require that richness be held in a very narrow range around stoichiometry, as we will see below.

Electronic fuel injection

The development of electronics and smart sensors capable of providing to a microprocessor real-time information on the instantaneous state of the engine made possible the realization of extremely sophisticated carburetion and ignition engine management systems, capable, for an acceptable cost, of controlling the engine operation and in particular to sharply reduce emissions of pollutants.

These systems have gradually replaced carburetors by fuel injection electronics, enabling more specific dosing than the former. Figure 9.32 shows a cutaway of a high-pressure-injection arrangement, with the gasoline injector at the top of the cylinder head. We present these systems in more detail below, after dealing with emissions of different pollutants.

Among the various advantages of injection near or into the cylinder, let us say by the way that it solves the problem of fuel vaporization, if properly adjusted to evenly distribute the liquid in the intake air. It can thus improve the homogeneity of the mixture and make it unnecessary to preheat the air, which increases the maximum charge and average pressure.

Real cycles of gasoline engines
Differences between theoretical and real cycles

Real cycles deviate significantly from theoretical cycles presented above, for several reasons:

* The working fluid is not a perfect gas. Actually, there are two fluids of different chemical compositions (Figure 9.33): the air/fuel mix during the compression phase and the mixture of combustion products during the expansion phase. To increase the accuracy of calculations, we should at least adopt the assumption of ideal gas behavior and calculate the thermodynamic properties of working fluid based on Dalton's law taking account of specific heat changes with temperature;
* At the very high temperatures reached in the Beau de Rochas cycle, molecular dissociation plays a role. Under these conditions, state 2 can be rigorously determined only by combined iterations on the temperature and composition of combustion products.

Figure 9.34 shows that the effect of molecular dissociation is to significantly lower peak pressure reached in the theoretical cycle and in turn increase the pressure at point 3 because of molecular recombinations that occur during expansion.

FIGURE 9.32
High-pressure-injection arrangement. (Courtesy Bosch.)

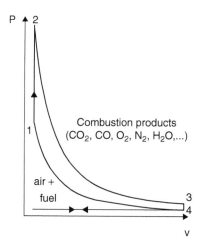

FIGURE 9.33

Beau de Rochas cycle.

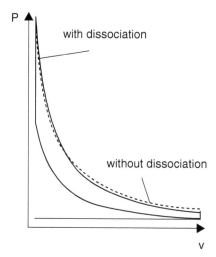

FIGURE 9.34

Influence of dissociation.

Molecular dissociation has also the effect of slightly lowering the internal efficiency of the cycle, with a particularly noticeable effect when the combustion temperature is very high, i.e., for a richness close to 1 or an excess air close to 0.

Among other causes of cycle deformation, we can mention the following:

- Pumping power, or fraction of the power consumed internally to ensure the renewal of the charge (pressure drop at fresh gas admission and burned gas exhaust);
- The P(v) law during combustion, which is not comparable to an isochoric process, as assumed in the Otto cycle;
- Wall actions that make compression and expansion not strictly follow the isentropic law.

These three causes make that the indicator diagram of an engine on a bench take the form shown in Figure 9.35.

In gasoline engines, since combustion is not instantaneous, it is necessary to trigger the ignition before the piston reaches TDC. This is called ignition advance, which we seek to optimize so that the decrease of the diagram area be as small as possible (areas A and B in gray in Figure 9.35).

The pressure drop across the inlet and outlet valves induces a pumping power corresponding to a negative area in the diagram (shaded area C). To minimize the drawbacks of these

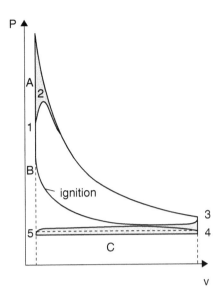

FIGURE 9.35
Theoretical and real cycle.

losses, valves slightly overlap, that is to say exhaust valves are opened before the piston reaches the BDC, and admission valves closed after it has crossed the BDC.

Finally, the cylinder walls play a thermal shunt role between the upper and lower parts of the diagram and distort isentropes (2–3) and (4–1), reducing the effective area of the diagram.

Moreover, the overall efficiency drops due to the partial filling of the cylinder during the intake phase, and this is because of three main reasons: the residual presence of smoke, the warming of the charge in contact with hot surfaces and admission pressure drop.

In practice, unless one uses very sophisticated two- or three-dimensional models, which require in particular to provide combustion laws taking into account the flame propagation and wall losses, one can hardly make calculations beyond the consideration of changes in the chemical composition of the working fluid and dissociation as suggested by Thelliez (1989).

Figure 9.36 shows the results provided by a gasoline engine Thermoptim model with a theoretical associated cycle of the type introduced above. This model, presented in detail in the Diapason session S39,[3] is far from precise, but it takes realistically into account the combustion of the fuel mixture, as well as changes in the specific heat capacity depending on the composition of the working fluid and temperature. It calls for a few comments:

* The fuel mixture is prepared here once and for all in the "admission" mixer, whose upstream branches are ambient air, fuel and a fraction of recirculated gas coming from the dead space that still exists in the cylinder (3.3% by mass);
* The fuel mixture combustion is calculated by checking the option "premixed" in combustion Thermoptim screens (Figure 9.37);
* It occurs in three phases, as proposed by Thelliez (1989), gas composition in the cylinder varying accordingly;
* Expansion occurs in two steps: in closed system between points 5 and 6, then open system from 6 to 7 in valves and exhaust pipes.

Abnormal combustion

Combustion in engines may have various abnormalities, such as knocking, preignition and reignition.

---

[3] https://direns.mines-paristech.fr/Sites/Thopt/en/co/session-s39en-exercise.html

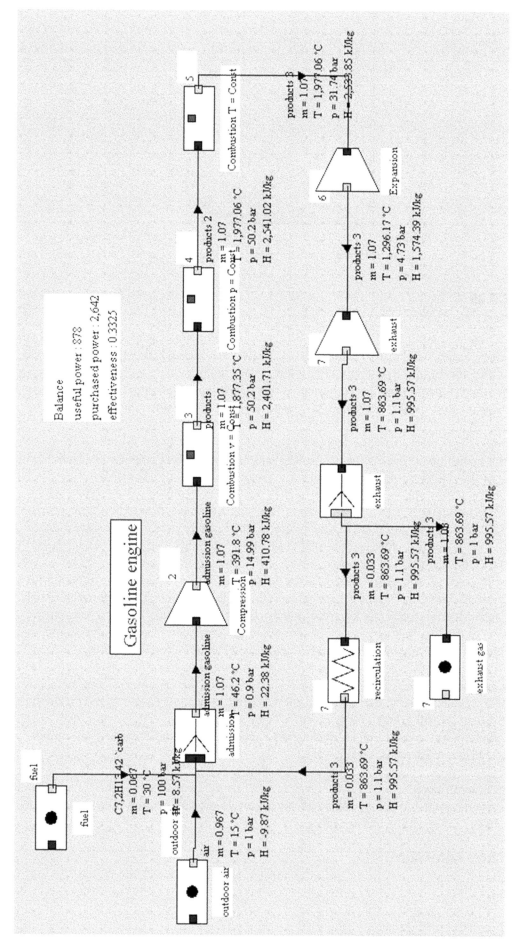

FIGURE 9.36

Synoptic view of a gasoline engine.

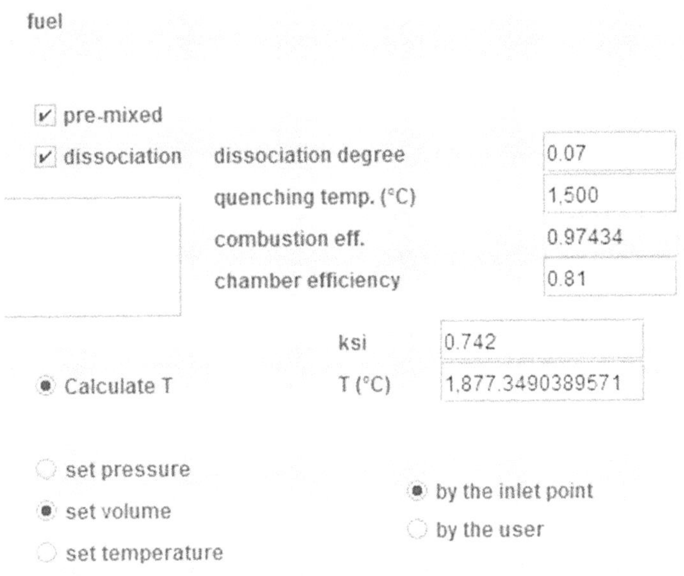

fuel

☑ pre-mixed

☑ dissociation   dissociation degree   0.07

    quenching temp. (°C)   1,500

    combustion eff.   0.97434

    chamber efficiency   0.81

    ksi   0.742

● Calculate T   T (°C)   1,877.3490389571

○ set pressure

● set volume   ● by the inlet point

○ set temperature   ○ by the user

FIGURE 9.37

Combustion settings.

Knocking is caused by an explosion that causes a sudden pressure increase in the cylinders. At low speed, it appears that knocking is related to the RON, and at high-speed to the MON supplemented by a term characteristic of the olefin content and cracking mode used for refining.

Preignition is self-ignition caused by a hot spot in the cylinder, which can be formed by the electrodes of the spark plug if they are too insulated. Hot spots, however, generally result from deposits formed on the cylinder head or piston during combustion. When these deposits stem from the ashes of the engine oil (low fuse barium and calcium salts), they can cause more than 1,000°C hot spots and degenerate into a knock of catastrophic consequences for the engine (piston melting).

Reignition is characterized by an ability of the engine to run when the ignition is off. It is a low-temperature self-ignition mainly due to the RON.

Performance of gasoline engines

The effective thermal efficiency of gasoline engines is lowered relative to the theoretical efficiency by two main causes:

- First, waste heat, as a result of incomplete combustion as well as wall losses, reduces the power output;
- Secondly, mechanical losses absorb a fraction of the work which is variable but never less than 12% and much higher in small capacity engines.

Efficiency losses vary widely from one engine to another. Loss by incomplete combustion is much more important in liquid fuel engines than in those that use a gaseous fuel, due to incomplete vaporization problems mentioned above. Losses ranging between 5% and 20% may exist, according to operating conditions.

Wall losses are even higher than the dimensions, and the piston reduced linear velocity is small. As a guide, the cooling water discharges generally between 25% and 33% of the heating value of fuel.

Under these conditions, the actual efficiency of gasoline engines never exceeds 32%, falling to about 25% in average capacity traction gasoline motors.

Diesel engines

In a diesel engine, liquid fuel is injected at the end of compression in a combustion chamber filled with clean air and should immediately burn afterwards. This characteristic has two important consequences:

* The absolute necessity of self-ignition: indeed, a spark would likely be ineffective in bursting in air, and, moreover, the mixture in the chamber being not homogeneous, the conditions of combustion propagation would be very bad;
* The need for high excess air, due to the heterogeneity of the mixture: introduced at the last moment, and therefore poorly distributed, the fuel must have enough oxidizer.

Compression ignition conditions

In a diesel engine, air temperature at the time of injection, i.e., at the end of compression, depends essentially on the compression ratio $\rho = V/v$ and incidentally on air temperature T_0 and wall actions, these consisting of heating during the admission and start of compression and cooling at the end of compression. These effects roughly offset each other in a cooled wall chamber, and air temperature at the time of injection differs little from what would give an isentropic compression.

As this temperature is an increasing function of V/v, the need for self-ignition has the effect of setting a lower compression ratio limit, the opposite of what happens in gasoline engines, where the nonknocking condition sets an upper limit to this ratio.

As to the practical limit of V/v for a particular fuel for proper combustion in all cases, care must be taken that an adequate margin of safety exists, as self-ignition must occur under the worst conditions, i.e., as low as the outside temperature at start might be (the walls being cold), and secondly be frank enough to allow combustion even at maximum speed. The lower limits of V/v are thus dependent on engine speed and are significantly higher in high-speed motors than in slow engines. In the latter, the limit V/v can be lowered to 13 for fuels easy to ignite (light distillates), reaches 14 for medium fuel oil, and up to 15 and even 16 for very heavy and slightly hydrogenated oils.

Note that these values significantly exceed those of most gasoline engines, even burning low explosive fuels. Therefore, pressure at the end of compression is much higher in diesel engines: it can exceed 50 bar. Under these conditions, to avoid overloading the joints, one tries to minimize the pressure at the end of combustion by strictly controlling it. The question presents itself very differently in slow and fast engines.

Ignition and combustion delays

The overall ignition and combustion delay is the order of several milliseconds. In slow diesel engines (marine engines running between 80 and 250 rpm), it is very small compared to the cycle duration, and in this case, there is no particular difficulty in controlling the combustion: injection can be adjusted so that the pressure does not exceed 10%–20% of P_2, and we are very close to a constant pressure combustion.

The situation is quite different in high-speed diesel engines, where speeds exceed 3,000 rpm. The duration of one revolution is then less than 20 ms, and that corresponding to a rotation of 30°, in which it would be desirable to confine the combustion is less than 1.7 ms. Without special precautions, it is much less than the overall ignition and combustion delay. It becomes very difficult to produce combustion on time while controlling the pressure. If, for compensating ignition delay, injection is triggered too early before TDC, there is a serious risk: the fuel vaporizing partially in the air before catching fire, forms a combustible gas mixture that can detonate when combustion starts. Combustion is extremely brutal, engine joints deteriorate rapidly, and furthermore, the decomposition of explosive molecules of fuel produces black smoke that fouls the engine.

It is therefore essential in high-speed diesel engines to reduce to an absolute minimum the ignition delay. A first step in this direction is to burn in these engines only fuel with self-ignition temperature and ignition delay as small as possible.

From this point of view, fuels are classified by testing them in an experimental engine operating with a given V/v, at a given speed. The ignition delay is measured on a pressure-time diagram. The point where injection starts is known from the timing of the fuel pump, and the point where combustion starts is characterized by a very definite change of slope on the curve. We can thus establish an intrinsic characterization of fuels following the cetane number (cetane number of diesel fuel is the cetane (a highly flammable hydrocarbon) fraction of a mixture of cetane and mesitylene (a low flammable hydrocarbon), which has the same ignition delay as the fuel processed).

In practice, high speed diesel engines burn almost exclusively light distillates, which is, in the range of products of petroleum distillation, the easiest fraction to ignite and burn in diesel. However, the specification of the cetane number for diesel is much less common than that of octane for gasoline, because, ultimately, the ignition delay depends mainly on structural arrangements adopted to accelerate ignition and combustion.

These provisions are very varied.

First, one generally takes for high speed diesel a V/v well above the limit: the latter is close to 13 in slow engines and, in high-speed diesel, it commonly reaches 16 and sometimes 18 or even 20.

A second method, widely used, is to produce injection, or derive a fraction of it, in a part of the combustion chamber called the auxiliary chamber, whose walls are systematically poorly cooled. This is called a divided chamber. The auxiliary chamber is connected to the rest of the combustion chamber through a narrow opening. The air that is expelled during compression is heated by the action of the walls and throttled through the orifice and is thus heated to a higher temperature than the rest of the chamber. In addition a high turbulence reigns, and, conversely, the expulsion of burnt gases from the auxiliary chamber during combustion and the beginning of expansion creates a significant turbulence in the rest of the combustion chamber. These progressive combustion conditions have another interest: that of reducing the combustion noise, which can be quite high in a direct injection chamber, as ignition occurs abruptly when the self-ignition conditions (temperature and delay) are met.

The method has two variants:

- In **prechamber engines** (Figure 9.38), the injector opens directly into the prechamber. The fuel burns very quickly and undergoes incomplete combustion, and then under the effect of pressure, incompletely burned gases are expelled at high speed into the main chamber, where combustion ends. An obstacle is placed in the path of the fuel jet to help divide and mix it with air. This type of chamber is quieter, but its efficiency is lower than others. Until recently, it was the type of configuration chosen for passenger cars for which comfort is an important criterion;
- In **swirl chamber engines** (Figure 9.39), geometry is optimized to minimize pressure drops (larger discharge section, tangential outlet in the main chamber), which improves efficiency, but at the price of a slightly higher combustion noise.

All these provisions are more or less in default at startup, the action of cold walls lowering the temperature. This is why some engines are equipped with electric heaters that heat up before starting (glow plugs) either a specific area of the chamber, or the air stream admitted.

It is obviously impossible to accurately analyze the effect of provisions as complex as those mentioned, and the development of high-speed engines is the result of empirical trial and error. Nevertheless, despite the reduction in ignition delay, combustion is less well controlled in these engines than in slow engines. The pressure is much higher, so that maximum pressures of 70 and even 80 bar are reached.

In high-speed direct injection diesel engines, the piston and the chamber are simply designed so that a high turbulence is maintained during injection. Fuel must be mixed as homogeneously as possible without recourse to a prechamber, which induces particularly severe production constraints (Figure 9.40). The piston geometry is optimized to increase turbulence, and the injector is multi-hole. The efficiency of this type of engine is 20% higher than prechamber engines, but its noise is much higher because the combustion pressure is more difficult to control.

Prechamber process

Swirl chamber process

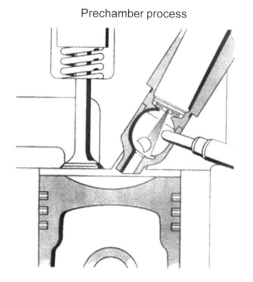
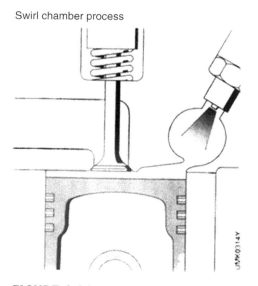

FIGURE 9.38
Prechamber. (Courtesy Bosch.)

FIGURE 9.39
Auxiliary chamber. (Courtesy Bosch.)

Direct injection process

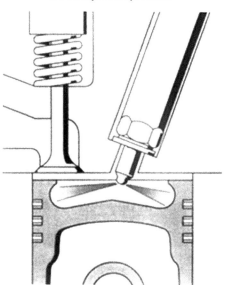

FIGURE 9.40
Direct injection. (Courtesy Bosch.)

Due to the development of "Common Rail" injection systems allowing achievement of very fine control of the injection, and therefore better management of these pressures, direct injection is spreading today in many diesel engines, even small displacement ones.

Air utilization factor

The injection of fuel at the last moment does not, as we have seen, allow it to spread evenly in the chamber, so as to fully utilize the air, and when the mass m of fuel exceeds a certain value m_m always substantially less than the mass m_{max} that could burn the air admitted, the combustion is incomplete, and the engine produces black smoke and fouls.

A key objective of diesel, whether slow or high speed, is to raise as much as possible the ratio m_m/m_{max}, or maximum air utilization factor (note that this rate is equivalent to richness, but the term richness is for homogeneous mixtures). This is achieved by giving the jets the greatest possible penetration and increasing turbulence during combustion.

Currently, the air utilization factor drops to 40% in slow imperfect scavenging two-stroke engines, with opposite blowdown holes, rarely exceeds 60% in slow four-stroke engines without significant turbulence and can reach 75% or even 80% in divided chamber engines.

If the air utilization factor cannot exceed a certain value, it can be as small as desired. In contrast to what happens in a gasoline engine, the mass of fuel injected can be reduced near zero in a diesel, and the operation remains perfectly regular. The adjustment of diesel is thus always done by varying m and is much more economical than that of gasoline engines.

The result is that diesel is better suited than gasoline engine for operation at reduced load, and its average efficiency is closer to maximum efficiency.

Thermal and mechanical fatigue

If the theoretical thermal efficiency of diesel is much better than that of a gasoline engine, this advantage has a counterpart: the diesel organs are subjected to exceptionally high thermal and mechanical fatigue, which complicates the construction and increases the costs.

With regard to mechanical fatigue, the normal maximum pressures are much higher in the case of diesel, but also the joints and the cylinder must be designed to withstand much higher occasional accidental pressures. Three phenomena may indeed generate significant overpressure:

* First, the excess air would burn a much more significant quantity of fuel;
* Secondly, the peak pressure is limited, even in high-speed motors, at a fraction of that produced by combustion at constant volume;
* Finally, there may be unintended presence of an excess of fuel, resulting in a much more intense combustion which can lead, at constant volume, to a pressure three times the rated pressure.

It is almost impossible to prevent abnormal combustion from occurring accidentally, in the case of leakage of fuel during compression or massive movement of oil. The excess fuel introduced into the air vaporizes prematurely, and combustion is at constant volume, thus explosive. It is imperative that such events, so exceptional be they, do not lead to the engine destruction. Ultimately, the components of a diesel engine should be calculated with a wide safety margin, much higher than that required in a gasoline engine.

Thermal fatigue is also higher in diesel, in contrast to what the theory may indicate. Indeed, the theoretical calculation assumes the fuel is evenly distributed in excess air, and it leads to moderate combustion temperatures, substantially lower than those achieved in the gasoline engine. Actually, the fuel is concentrated in a fraction of the air where combustion takes place without appreciable excess air, and, in the flame jet coming out of the injector, the temperature is extremely high. As pressure and speed are also very high, the exchange coefficients are significant, thus subjecting the parts in contact with the jet to heat flux well above that given to the walls of a gasoline engine.

Finally, the proper functioning of a diesel engine requires very good maintenance, as an efficient self-ignition is compromised by a minimal sealing defect type roundness of the cylinder or ring rupture, which does not significantly disturb the operation of a gasoline engine.

For all these reasons, for a given speed and capacity, a diesel engine is much heavier and expensive than a gasoline engine, and its maintenance costs are also higher. The increase in depreciation expense and maintenance offsets to some extent the reduction of fuel expenses.

Cooling of walls

There is an important difference between the cooling conditions of gasoline and diesel engines: whereas in the former case, all internal walls, as well as the cylinder of the combustion chamber, must be maintained at a moderate temperature to avoid self-ignition of the mixture by a hot spot, in diesel, only the side walls exposed to ring friction must be cooled.

The walls of the combustion chamber, the bottom of the piston and the cylinder head may be covered with a heat-resistant alloy and insulated from cooled parts. This method reduces wall losses and improves combustion. However, it increases the heating of the intake air.

Fuels burnt in diesel engines

Slow motors can burn the less volatile liquid fuels, heavy fuel oil, oil shale and tar, with the sole condition that these liquids are refined, that is to say are the result of distillation. The experience shows that a diesel powered by crude liquid fuel fouls rather quickly, because of tarry deposits erasing rings and requiring frequent complete revisions.

Attempts to burn in diesel pulverized solid fuel (lignite, coal) have been unsuccessful so far due to rather fast fouling of the engine, especially from abrasion caused by mineral ashes which mix with the lubricating oil.

You can burn in a diesel gaseous fuel compressed in advance at a pressure sufficient for injection, provided there is an auxiliary injection of liquid fuel to start the combustion (dual-fuel engines).

Finally, we have seen that high-speed diesels burn almost exclusively light distillates.

Actual cycles of diesel engines

The origin of the differences between actual cycle and theoretical cycle is almost the same for diesel as for gasoline engines. However, the effect of dissociation is less sensitive because of the presence of large excess air. Under these conditions, it can legitimately be neglected in almost all cases.

The causes of efficiency reduction in diesel engines are similar to those outlined for gasoline engines: heat waste and mechanical losses.

However, the actual thermal efficiencies are well above those of gasoline engines. For non-turbocharged engines, they can reach 40% in good conditions, and they usually exceed 32% in the most adverse conditions (low-capacity engines, two-stroke engines with imperfect scavenging).

These results can be improved by supercharging, studied in the next section.

Supercharging

General

Supercharging is to feed an engine at a pressure greater than atmospheric pressure, using an auxiliary compressor. Its main purpose is to increase the mass rate and thus reduce the overall size and weight of the engine.

Indeed, the engine capacity is roughly proportional to the mass flow of working fluid passing through it. In a piston engine, for a given speed of rotation, the volume flow is defined by geometry. To increase capacity, it is sufficient to reduce the specific volume of the working fluid, which is precisely what supercharging does.

In other words, to refer to the previous analysis, supercharging can increase the volumetric efficiency C_r to values greater than unity, when it is naturally limited to values between 0.8 and 0.9.

Originally, superchargers have been developed for aviation, to compensate the power loss in altitude due to the low air pressure. Since then supercharging has become very widespread, particularly in diesel engines, where it is often provided by a self-powered turbocharger driven by exhaust gases, such as those presented in Figures 9.41 and 9.42.

This supercharging mode has indeed many advantages: a turbine/compressor assembly is mounted between the intake and exhaust engine manifolds. The turbine (2 in Figure 9.42) and compressor (1) are coupled and form a completely independent system, mechanically unconnected with either the main engine, or with any receiver, so that on the one hand the turbine shaft work automatically balances the compressor work, and secondly the speed of the unit can be adjusted at will, independently of that of the engine.

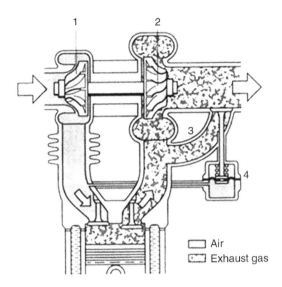

FIGURE 9.41

Turbocharger. (Courtesy Bosch.)

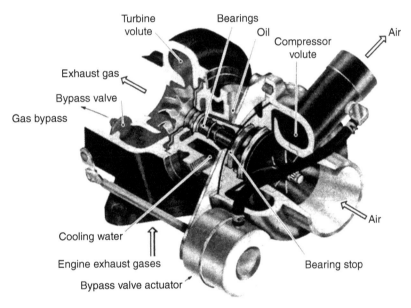

FIGURE 9.42

Cutaway of a turbocharger. (Courtesy Techniques de l'Ingénieur.)

Basic principles

In the final expansion phase, the exhaust valves are opened earlier than in a normally aspirated engine and used to drive a turbine directly coupled to a compressor, usually centrifugal, placed between the air filter and the intake manifold. An intercooler helps prevent potential gains due to the pressure rise to be lost due to the increase of temperature.

The cycle is modified as shown in Figure 9.43. The work output, equal to the net work, since the group is mechanically independent, is thus increased by work (P_s-P_c) (v_s-v_1) produced during the intake and expansion strokes.

We call volumetric efficiency coefficient s the ratio of the specific volume of air at standard conditions (P_a, T_a), to that of the air in the intake manifold v_s.

$$s = \frac{v_a}{v_s} = \frac{P_s}{P_a}\frac{T_a}{T_s}$$

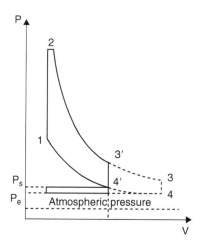

FIGURE 9.43
Supercharged cycle.

If we assign an index 0 to the magnitudes relative to aspirated (not supercharged) engine, we obtain for the supercharged engine:

$$C_r = sC_0$$

The volumetric efficiency coefficient s is naturally less than the compression ratio achieved by the supercharger, especially if it does not include an intercooler.

For gasoline and diesel car engines, volumetric efficiency coefficients take values between 1.2 and 1.8 approximately. They exceed 2 for diesel truck engines and can reach up to 4 for heavy duty diesel engines greatly supercharged.

Supercharging significantly reduces specific losses of the main engine in relative value, so that the effective efficiency can be increased by almost 10%. Indeed, heat loss decreases, because the exchange coefficients grow more slowly than the density, and mechanical losses are also lowered due to the reduction of the swept volume.

However, the main interest of supercharging often resides in reducing the cost associated with the increase of the mean pressure.

It is important to note that the benefits are lower for supercharged gasoline engines than for diesel engines, except particular application such as aviation or competition. Indeed, we saw that one of the most limiting constraints for these engines is the nonknocking condition, which sets an upper limit to the volumetric compression ratio. This condition is necessary in the supercharged gasoline engines, whose compression ratio must usually be reduced, which has the effect of lowering the internal efficiency of the engine. Supercharging can be justified only if power gain is imperative, for a given weight and size.

Engine and pollutant emission control

Emissions of pollutants: mechanisms involved

Pollution from automobile engines stems from incomplete burning. Indeed, if combustion was perfect, the exhaust would only include water vapor, carbon dioxide and nitrogen gas, completely innocuous vis-à-vis air pollution and health, except as regards the greenhouse effect. In this section, we are mainly interested in gasoline engines, and we then give some indication on diesel engines.

Main gasoline engine pollutants are

* carbon monoxide CO (1%);

- unburned hydrocarbons UHC (0.1%–0.3%);
- nitrogen oxides NO$_x$ (0.1%–0.3%).

These three pollutants are harmful to health, and the last two are involved in the phenomena of acid rain. The accurate determination of emissions of pollutants can be approached only by the application of the law of mass action, taking into account the kinetics of combustion, but accurate calculations are limited because these assumptions are only very roughly verified in an engine cylinder. Studies show, however, that richness and its reverse air factor λ have a significant impact on emissions of various pollutants, as shown in Figure 9.44.

Emissions of carbon monoxide increase with richness, especially when its value exceeds 0.93, which also corresponds to the maximum emission of nitrogen oxides. The content of unburned hydrocarbons in turn passes through a minimum for a rather lean mixture (R = 0.8) and is growing quite substantially with richness when it exceeds 1.

Depending on the air factor, these evolutions obviously follow inverse laws.

To reduce emissions, you can do the following:

- **Limit the formation of pollution** during combustion. This solution had been first retained by European manufacturers, who had adopted it since the early 1970s, reducing about 60% the volume of specific emissions, achieving energy conservation and optimizing combustion. This solution has its limits, because it is impossible, due to dissociation, to achieve a perfect combustion;
- **Destroy the pollution** caused by combustion before discharging the exhaust gases into the atmosphere. This is the route chosen first in the U.S. and Japan, and subsequently in Europe. This is done using a catalytic converter, i.e., a treatment system catalyzed by precious metals (platinum, palladium and rhodium) which is placed on the exhaust line, in order to destroy almost all pollutants. Today this is the most efficient way, but it is expensive and tends to increase by a few percent (2–5) the engine consumption.

In addition, to be effective, catalytic converters forbid the use of leaded fuels. Catalysts are indeed "poisoned" by lead (as well as phosphorus). As lead is also in itself a pollutant with harmful effects on health, refiners have sought to develop lead-free fuels. Since this element had been introduced to increase the gasoline octane number, it was necessary to compensate the decrease of octane due to abandoned lead by additives and other treatments. To do this, refiners have had to step up new refinery processes, as addition of oxygenated organics (e.g. ethanol) is limited because of the need to vaporize the fuel. The solution therefore requires a further refining of gasoline, which is an additional manufacturing cost varying from a few cents per liter for unleaded gasoline (MON 85/RON 95) to about 10 cents per liter for unleaded high octane number premium (MON 88/RON 98).

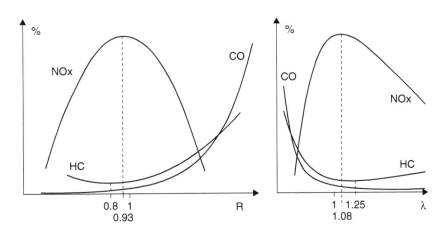

FIGURE 9.44
Gasoline engine pollutant emissions.

Combustion optimization

In order to reduce pollutant emissions and maximize fuel use, it is necessary to control combustion in the best possible way. This is achieved firstly by optimizing the engine design, partly by improving carburetion systems, and finally mastering ignition conditions at best.

Optimization of engine design

The engine design has a significant influence on its performance and the emission of pollutants: efficiency is an increasing function of the volumetric compression ratio, but it is the same for NO_x emissions, since they depend directly on temperature in the combustion chamber. The geometry of the latter strongly influences the formation of unburned hydrocarbons, so manufacturers are moving to small surface compact chambers. The position and number of spark plugs or valves play a significant role on the quality of the ignition as well as pressure drop and turbulence in the chamber. Intake pipes, well sized, may allow a natural boost favorable for performance.

Changes in fuel systems

Tests on gasoline engines have shown they can develop a maximum power for a slight lack of air ($R \approx 1.15$ or $\lambda \approx 0.85$), and their maximum efficiency is obtained with a slight excess air ($R \approx 0.8$ or $\lambda \approx 1.2$). Moreover, the best idle is obtained for stoichiometric conditions ($R = \lambda = 1$), and a lack of air of 15%–25% ($R \approx 1.25$ or $\lambda \approx 0.8$) is favorable for good acceleration.

It follows that there is no optimal value of richness or air factor that allows all user requirements to be met. In practice, values between 0.9 and 1.1 are generally accepted. To keep richness in a narrow range, it is necessary to determine with reasonable accuracy the amount of air sucked in and mix the corresponding amount of fuel. Manufacturers have therefore initially focused their efforts on improving fuel systems depending on the engine load.

Today most motors use fuel injection systems in the intake manifolds, a development that reflects the following reasons:

* Injection allows very precise metering of fuel depending on load condition and engine speed and therefore allows better control emissions of pollutants;
* Injection can be performed in the immediate vicinity of valves, thus limiting the risk of fuel condensation on the intake manifolds. Moreover, if we use one injector per cylinder, we are certain to get a good distribution of the mixture;
* Removing the carburetor, we can optimize air flow in the intake manifolds, which allows a better volumetric efficiency, and therefore improves the engine performance;
* Finally, we solve with the required accuracy the various difficulties associated with carburetors: providing fuel for acceleration, cold start, idling.

We will see later that the most efficient pollution control systems (multifunctional "three ways" catalytic converters) require that the air ratio be maintained within a very narrow range around 1, which only injection systems are capable of achieving.

Adaptation of the ignition

However, the best fuel system is not sufficient to ensure perfect combustion: it is also necessary to best fit the ignition point depending on the engine load and speed.

Given combustion propagation delays in the chamber, it is necessary both to create turbulence and secondly to cause the ignition before the piston reaches the top dead center TDC.

It is indeed fundamental that ignition is performed at a particular time (point Z_a, curve 1 in Figure 9.45), which varies with load and engine speed. If a spark is created too early (point Z_b, curve 2), there is a risk of knocking, the temperature at the end of compression being too high. If it is delayed (point Z_c, curve 3), the combustion begins only when the piston descends,

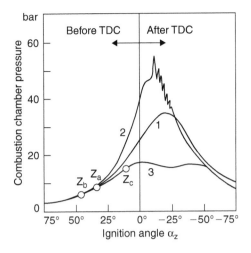

FIGURE 9.45

Pressure inside cylinder function of advance ignition. (Courtesy Bosch.)

and the cycle is highly truncated. In both cases, efficiency and engine power drop, and many disorders may appear, among which, in case of premature ignition, knocking, rapidly fatal to the engine. We characterize the ignition advance by the angle of the crankshaft from its position at TDC.

In a first approximation, we can consider that the combustion delay varies little with load and engine speed for a given mixture richness. In fact, if the chamber design effectively promotes turbulence, these delays decrease as speed increases, but this variation is small compared to the crankshaft rotation speed.

It is therefore necessary to constantly adjust the ignition advance based on load and engine speed.

Engine control

Until recent years, the mechanical and vacuum advancement of the timing was the only way to modulate advance according to engine conditions. These mechanical systems used weights progressively deviating from the igniter rod under the action of centrifugal force, depending on engine speed. A further correction was made by a suction capsule connected to the intake manifold and controlled by pressure extraction on both sides of the throttle.

Although already very efficient, these mechanical devices could only generate advance correction curves of simple shape, like the one shown on the right side of the illustration in Figure 9.47. They were therefore unable to take into account finer ignition adjustments, in particular to reduce emissions of pollutants.

Indeed, the composition of exhaust gases, in particular the concentration of NO_x and unburned hydrocarbons, can be significantly influenced by the ignition advance angle, as shown by the curves of Figure 9.46.

This shows that the optimum advance angle increases with air factor λ, due to the increased ignition delay due to fuel dilution. The NO_x concentration increases when $[O_2]$ increases and the temperature is high and then drops for high λ when the temperature decreases.

These curves also show the need to find a compromise between environmental constraints, which lead to choose low ignition advances, and specific fuel consumption, for which values a little higher are preferable.

To optimally control combustion, it is necessary to regulate the ignition advance by taking into account not only the load (the more it increases, the more the temperature of the exhaust gas rises, which allows afterburning in expansion and expulsion phases, with reduction of CO and unburned hydrocarbons but increased NO_x) and the engine speed (when the engine speed increases, friction increases, as well as pollutants) but also the temperatures of intake air, coolant and exhaust gas, as well as throttle position.

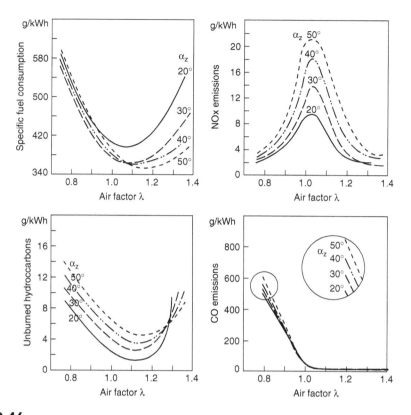

FIGURE 9.46

Influence of air factor and ignition angle α_z on emissions and specific consumption. (Courtesy Bosch.)

It is now possible to do this by using electronic ignition control, which uses a network of characteristic curves called ignition map, determined on a test bench and implemented on a microprocessor.

Figure 9.47 compares the fine mapping obtained with an electronic control to that provided by a mechanical drive.

Catalytic purification converters

A catalytic converter comprises the following:

- The catalytic converter itself;
- Auxiliary devices which should be added to a conventional engine to ensure proper operation of the process (e.g., electronic fuel injection for three-way converters or injection of additional air for one-way catalysis).

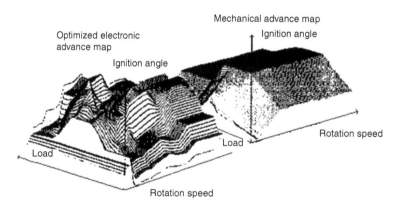

FIGURE 9.47

Mechanical and electronic combustion maps. (Courtesy Bosch.)

We distinguish now two techniques:

- Simple oxidation catalysis (one way catalysis) that can act on CO and HC;
- Multifunctional catalysis, more commonly known as three-way, that allows to act on CO, HC and NO_x by simultaneous chemical reactions of oxidation and reduction (Figures 9.48 and 9.49).

The catalytic converter, of size similar to that of a conventional exhaust line, is placed on the pipe near the engine in order to process hot exhaust gases.

It contains a catalyst which is in the form of a deposit of precious metals (1–3 g per converter) on a porous monolithic ceramic honeycomb, to develop a large surface area.

Catalytic converters are a very effective technique of remediation: over 90% of conversion in new condition when the engine and catalyst are warm.

The one-way catalysts require oxygen in the exhaust gases, which can be achieved by an air pump fitted to the vehicle. However, this technique does not require sophisticated carburetion or injection device. Being without action on the nitrogen oxides, the oxidation catalyst must be coupled to a device for reducing such pollution. This is achieved by recycling a fraction of exhaust gases to the admission.

With three-way catalyst converters, a perfectly balanced and uniform air fuel mixture must be ensured, which can now be performed by controlled electronic fuel injection. In fact, for achieving simultaneous removal of the three pollutants, the composition of exhaust gases should remain within a very narrow range.

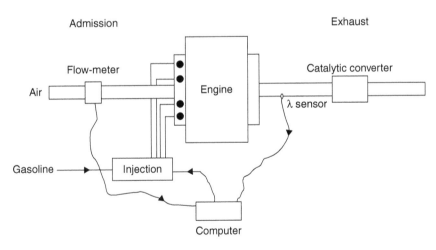

FIGURE 9.48

Three-way catalytic conversion.

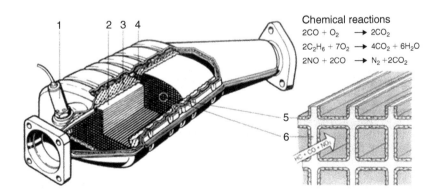

Chemical reactions

$2CO + O_2 \rightarrow 2CO_2$

$2C_2H_6 + 7O_2 \rightarrow 4CO_2 + 6H_2O$

$2NO + 2CO \rightarrow N_2 + 2CO_2$

FIGURE 9.49

Cutaway of a three-way catalytic converter. (Courtesy Bosch.)

Figure 9.49 shows the cutaway of a catalytic converter, with the oxygen sensor (1), the ceramic monolith (2), a flexible metallic grid (3), a jacketed heat insulation (4), a platinum coating (5) and a ceramic or metal support (6).

This is only possible by controlling fuel injection by a sensor (the said Lambda sensor by reference to air factor) that continuously doses the oxygen content of exhaust gases.

This sensor consists of a body in solid electrolytic ZrO_2 material (porous ceramic, allowing diffusion of oxygen). As shown in Figure 9.51, the two-point lambda sensor has a very steep voltage characteristic, to detect the value 1 of the air factor λ.

Figure 9.50 shows the configuration of tube-type lambda oxygen sensor in the exhaust pipe, with the special ceramic (1), the electrodes (2), the contact (3), the exhaust pipe (5), the porous protective layer ceramic (6), the exhaust gases (7) and air (8).

Several types of sensors are available: two-point sensors, possibly heated to better control their operating temperature and better reduce emissions, planar sensors, obtained by screen printing, which, using a special processing circuit, deliver a continuous measure of the air factor between 0.7 and 4, which allows for even finer control (Figure 9.51).

The effectiveness of catalysis is highlighted by the curves of Figure 9.52, which show, as a function of the air factor, the reductions in emissions of the various pollutants: it shows that if we want both to reduce NO_x emissions and HC and CO emissions, it is imperative that the value of λ stays within a very narrow band (between 0.99 and 1 for a conversion of 90% of pollutants).

This means a slight loss of engine efficiency relative to the optimum (obtained, as we have seen, for $\lambda \approx 1.2$).

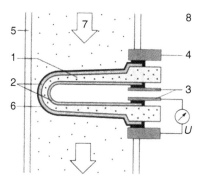

FIGURE 9.50

Lambda sensor in exhaust line. (Courtesy Bosch.)

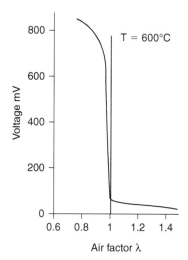

FIGURE 9.51

Lambda sensor characteristic curve.

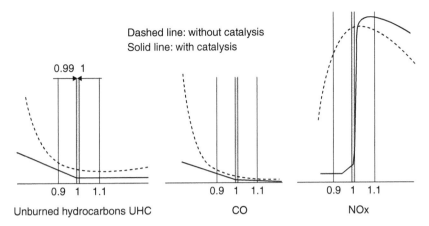

FIGURE 9.52
Catalytic conversion effectiveness.

This comes from the fact that we are pursuing two seemingly contradictory goals: on the one hand to reduce nitrogen oxides, which imposes work in the absence of oxygen, and also to further convert in CO_2 and H_2O the unburned hydrocarbons and carbon monoxide, which requires oxygen.

With the catalysts, it is possible to combine these different operations, but with the imperative of fully controlling the oxygen dosage, which can be done thanks to the Lambda sensor because of its very steep voltage characteristic for $\lambda = 1$.

If the value of λ drops below 0.99 or exceeds 1, the efficiency of catalytic converters falls rapidly: 65% for $\lambda = 0.98$ or 1.01, 40% for $\lambda = 0.97$ or 1.02.

It is for this reason that the use of a "three ways" catalytic converter can only be justified with a very precise control of fuel and combustion, that is to say with injection and electronic ignition control devices of the type we presented above.

Lambda control completes the control by ignition mapping, associating it with a closed loop on the oxygen content of exhaust gases. It is thus possible to further refine the settings for the engine by maintaining λ in the desired [0.99–1] band. However, the Lambda sensor gives a reliable signal only above a temperature of about 350°C. At startup, therefore, the lambda control is inoperative, and the controller must operate in open loop.

Case of diesel engines

In a diesel engine, pollution problems occur slightly differently. Indeed, because of the large excess air, complete combustion is more easily achieved, provided that the conditions of penetration of the fuel jet do not favor the formation of CO and soot, as a result of local oxygen deficit (the presence of inert gases – or excess air – reduces the partial pressures of the constituents and therefore increases the degree of reaction).

In principle, CO and soot are normally oxidized during the main phase of combustion and even during expansion. However, sometimes adverse conditions prevent this combustion from taking place, leading to pollution of exhaust gases by the excess soot, which then forms an opaque cloud characteristic of diesel engines incorrectly set: one speaks of particulate emissions.

Emissions of gaseous pollutants

For the rest, the gas pollution problems occur similarly in diesel and gasoline engines. However, due to excess air, emissions of carbon monoxide and unburned hydrocarbons are reduced, and the lower combustion temperature limits those of nitrogen oxides. Without after-treatment, these emissions are somewhat lower than in a gasoline engine.

The essential problem is that the three-way catalytic converters cannot be used because of the value of lambda. There is also the specific problem of sulfur dioxide, which must be removed at the refinery.

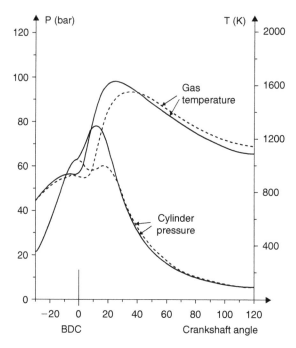

FIGURE 9.53

Pressure in diesel cylinder. (Courtesy Techniques de l'Ingénieur.)

A compromise must again be made between performance and emissions of pollutants. This is particularly the case when one seeks to limit nitrogen oxides, whose formation is favored (compared to gasoline engines) by the presence of excess air. Maximum power and efficiency are obtained by using injection advance, that is to say, by injecting the fuel before the piston reaches the TDC, to account for the ignition delay (of the order of 2 ms) and ensure that the combustion actually starts at about the end of compression.

By renouncing this injection advance, we greatly limit the levels of pressure and temperature in the cylinder during combustion, thereby significantly reducing the formation of nitrogen oxides but at the cost of a deterioration of efficiency and power.

Figure 9.53 shows the shape of changes in temperatures and pressures for a heavy-duty engine set:

* In the first case (solid lines) with a slight injection advance, which leads to a specific consumption of 220 g/kWh and a NO_x concentration of 1,000 ppm;
* In the second case (dashed curves) without injection advance (specific consumption of 235 g/kWh, NO_x concentration of 600 ppm).

NO_x emission reduction is remarkable (40%), but it comes at the cost of almost 7% over-consumption.

Particulate emissions

The particles emitted by diesel engines are mainly soot (carbon) and to a lesser extent, hydrocarbons and aerosols from fuel, lubricant or sulfates.

Chemically, they are often unpleasant odorous aldehydes that could harm human health due to the presence of aromatic compounds. Their small size (diameter of 1 μm) allows them to remain suspended in the atmosphere, generating white, blue or black smoke characteristic of the exhaust gas of a diesel engine incorrectly set.

To convert these particles, the most effective solution today is to use a particulate filter that ensures separation of exhaust gases and their subsequent removal. It is possible, due to excess air in the gas, to further oxidize CO, HC and particulates, but NO_x treatment involves the use of a specific device. One can for example use the SCR process (Selective Catalytic

Reduction), where an aqueous solution of urea is sprayed on the gas, which is transformed into ammonia by hydrolysis, and subsequently reduces nitrogen oxides.

Particulate filters are generally made of extruded ceramics shaped as honeycomb, filled alternately with plugs, to force gas through the material. Self-cleaning filter is provided by the combustion of soot, which takes place if temperature exceeds 550°C. To avoid clogging, additives allow this limit to be lowered to 200°C–250°C.

Bibliography

H. Abdallah, Analyse énergétique, exergétique et économique des cycles de turbine à combustion, Thèse de Doctorat, Université de Nantes, 27 novembre 1998.

J. Andrzejewsky, M. Thelliez, Coefficient de remplissage et taux de gaz résiduels. *Entropie*, 134, 95–100, 1987.

R. Bidard, J. Bonnin, Energétique et turbomachines, Collection de la DER d'EDF, Eyrolles, Paris, 1979.

G. Bidini, S. Stecco, Motori a combustione interna, Pitagora Editrice, Bologne, 1993.

Bosch, Commandes pour moteurs à essence, Cahiers Techniques, Stuttgart, 1999.

Bosch, Commandes pour moteurs diesel, Cahiers Techniques, Stuttgart, 2000.

R. Carreras, M. Quera, A. Comas, A. Calvo, Maquinas térmicas, LMTA, ETS d'Enginyeria Industrial, Universitat Politécnica de Catalunya, Terrassa, 2001.

C. Clos, Technologie des moteurs alternatifs à combustion interne, Techniques de l'Ingénieur, Traité Mécanique et chaleur, B 2800.

T. de Neef, B. Mouille, P. A. Destailleur, La production d'énergie au moyen de systèmes diesel modernes, Revue de la Société des Electriciens et des Electroniciens, Paris, décembre 1999.

A. F. El-Sayed, *Aircraft propulsion and gas turbine engines*, CRC Press, Boca Raton, FL, 2008, ISBN 978-0-8493-9196-5.

B. Geoffroy, Distribution à soupapes, Techniques de l'Ingénieur, Traité Mécanique et chaleur, B 2805.

M. Giraud, J. Silet, Turbines à gaz aéronautiques et terrestres, Techniques de l'Ingénieur, Traité Mécanique et chaleur, B 4 410.

M. Gratadour, Application de la suralimentation aux moteurs, Techniques de l'Ingénieur, Traité Mécanique et chaleur, B 2 630.

J. C. Guibet, Carburants et moteurs, Publications de l'IFP, Editions Technip, Paris, 1987.

A. Haupais, Combustion dans les moteurs diesel, Techniques de l'Ingénieur, Traité Mécanique et chaleur, B 2700.

IHPTET, Air dominance through propulsion superiority, site Web: http://www.pr.afrl.af.mil/divisions/prt/ihptet/brochure/Intro.htm.

L. S. Langston, Introduction to gas turbines for non engineers. *Global Gas Turbine News*, 37(2): 9, 1997.

J. P. Moranne, Refroidissement des moteurs à combustion interne, Techniques de l'Ingénieur, Traité Mécanique et chaleur, B 2830.

A. Parois, Suralimentation par turbocompresseur, Techniques de l'Ingénieur, Traité Génie Mécanique, BM 2 631.

J. Rauch, Insonorisation et remplissage des moteurs à piston, Techniques de l'Ingénieur, Traité Mécanique et chaleur, B 369,9.

B. Raynal, Moteurs thermiques et pollution atmosphérique, origine et réduction des polluants, Techniques de l'Ingénieur, Traité Mécanique et chaleur, B 378,1.

F. Ribes, J. L. Meyer, Microturbines pour la production décentralisée, Revue de la Société des Electriciens et des Electroniciens, Paris, décembre 1999.

Rolls Royce, *The jet engine*, 5th edition, Derby, 1996, ISBN 0-902-121-2-35.

F. Roux, Graissage des moteurs thermiques alternatifs, Techniques de l'Ingénieur, Traité Mécanique et chaleur, B 2750.

P. Stouffs, S. Harvey, Energétique avancée des cycles à turbomachines, Cours de DEA Thermique, Energétique et Génie des Procédés, Nantes, septembre 1996.

K. Tasadduq, A. Lasalmonie, Les matériaux structuraux chauds pour turbomachines, Annales des mines, Paris, février 1995.

M. Thelliez, Analyse des produits de combustion d'un moteur à combustion interne, VIIe Congreso de Ingeniera Mecanica. Valencia, décembre 1988.

M. Thelliez, Analyse énergétique des cycles des moteurs à combustion interne au moyen d'un cycle théorique associé. *Entropie*, 148, 41–49, 1989.

J. Trapy, Moteur à allumage commandé, Techniques de l'Ingénieur, Traité Génie Mécanique, BM 2540.

Vision 21 Program Plan, Clean Energy Plants for the 21st Century, Federal Energy Technology Center, Office of Fossil Energy, US Department of Energy.

L. Vivier, Turbines à vapeur et à gaz, Ed. Albin Michel, Paris, 1965.

Combined cycle, cogeneration or CHP

Introduction

In this chapter, we will continue to apply the reflections of the previous sections to the high-performance hybrid energy systems made up of combined cycles and cogeneration or CHP plants.

We will start by studying the combined cycles coupled to a gas turbine, and we will show that the optimization of the heat recovery steam generator (HRSG) may require several pressure levels.

We will then talk about other combined cycles, in particular coupled to diesel or gas engines.

Most of the chapter will deal with cogeneration. After introducing the performance indicators used to characterize cogeneration facilities, the main types of configuration will be studied: boilers and steam turbines, reciprocating internal combustion engines and then gas turbines.

The last part deals with trigeneration, that is to say the combined production of heat, mechanical power and cold.

Combined cycles

The excellent efficiencies reached today by combined cycle power plants (above 60% LHV) are the result of integration into a single production unit of two complementary technologies in terms of temperature levels: gas turbines, which operate at high temperature (in an aero-derivative turbine, gases typically enter at 1,300°C in the expansion turbine and come out at around 500°C), and steam plants, which operate at lower temperatures (between 450°C and 30°C in this case).

We saw that regeneration can significantly increase the performance of the Brayton cycle, but the percentage of energy recovered is even lower than the temperature, and pressure levels of this cycle are higher. In modern gas turbines, regeneration is rarely possible or economically worthwhile. Another way to enhance the residual enthalpy of the exhaust gases is to use them as a heat source for a second cycle of production of mechanical energy (Figure 10.1). **Combined cycles** correspond to this new generation of thermal power plants.

The combined cycle thus obtained is a particularly successful marriage in the search for improved thermal performance: with currently available machines, efficiencies exceed 55% and are higher than those we can hope, even in the medium term, of the most advanced future steam plants.

DOI: 10.1201/9781003175629-10

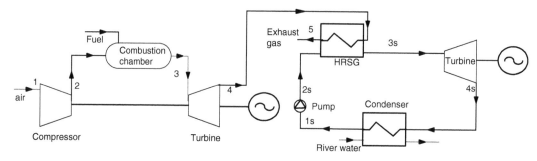

FIGURE 10.1

Sketch of a combined cycle.

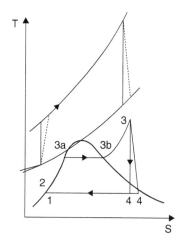

FIGURE 10.2

Combined cycle in the entropy chart.

In a simple combined cycle of this type, the gas turbine provides two-thirds of the total capacity. The steam turbine, fueled by superheated steam conditions of 85–100 bar and 510°C–540°C, provides the remaining third.

The heat exchanger which ensures thermal coupling between the two cycles is called the HRSG.

The simplest combined cycle (that is to say, without afterburner) is shown in Figure 10.1: as the temperature of the gas turbine exhaust gas can exceed 550°C, the maximum temperature level reached in a steam cycle, it is quite possible to recover the enthalpy available at the outlet of a gas turbine to heat a steam cycle.

With some simplifying assumptions, it is possible to construct an entropy chart allowing, for a set of suitable scales, to superimpose the two thermodynamic cycles (Figure 10.2). In this diagram, where the work done is proportional to the area of the cycle, the gas turbine provides more power than the steam engine (two-thirds of the total in practice).

We can sometimes improve the cycle efficiency by using the various changes discussed during the presentation of the steam cycle: reheat and regeneration.

However, as discussed below, the problem of steam cycle optimization differs substantially from that of large steam power plants, due to the pinch that appears in the HRSG. This is a point that we have already covered in Chapter 5, and which we will develop below.

Overall performance

The enthalpy exchange in a combined cycle can be summarized by the diagram in Figure 10.3.

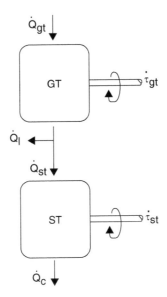

FIGURE 10.3
Block diagram of a combined cycle.

* The gas turbine receives heat Q_{gt} from the hot source. It provides on the one hand a useful work τ_{gt} and secondly a heat $(Q_{st} + Q_l)$. The first term is the heat supplied to the steam cycle, the second losses;
* The steam cycle produces useful work τ_{st}, and the condenser rejects heat Q_c.

Let us call η_{gt} the gas turbine efficiency, η_{st} that of the steam cycle, η_{cc} that of the combined cycle and ε the HRSG effectiveness, that is to say the ratio of Q_{st} to $Q_l + Q_{st}$:

$$\varepsilon = \frac{Q_{st}}{Q_l + Q_{st}} = \frac{Q_{gt}}{Q_l + Q_{st}} \frac{Q_{st}}{Q_{gt}} = \frac{1}{1 - \eta_{gt}} \frac{Q_{st}}{Q_{gt}}$$

$$\eta_{cc} = \frac{\tau_{gt} + \tau_{st}}{Q_{gt}} = \eta_{gt} + \eta_{st} \frac{Q_{st}}{Q_{gt}} = \eta_{gt} + \varepsilon\left(1 - \eta_{gt}\right)\eta_{st}$$

$$\eta_{cc} = \eta_{gt} + \varepsilon\left(1 - \eta_{gt}\right)\eta_{st} \tag{10.1}$$

The combined cycle efficiency is equal to the sum of that of the gas turbine and the product of its complement to 1 by the HRSG effectiveness and the steam cycle efficiency.

For example, with $\eta_{gt} = 0.29$, $\eta_{st} = 0.32$, $\varepsilon = 0.83$, we obtain $\eta_{cc} = 0.48$.

This expression shows that it is as important to optimize the steam cycle as the recovery steam generator and thus its effectiveness ε. Difficulties arise because the problem is highly constrained, and there may be conflicts between these two objectives.

The optimization of such a combined cycle is based on the reduction of its internal irreversibilities, which can be grouped into three broad categories: mechanical irreversibilities, that take place in the compressor and turbines, combustion irreversibilities, and purely thermal irreversibilities, related to temperature differences in the heat exchangers.

Much has already been done to limit the mechanical irreversibilities and reducing combustion irreversibility is directly related to the maximum temperature of the fumes, which itself depends on the strength of the combustion chamber materials and above all of the initial expansion stages in the gas turbine (stator and rotor).

So we focus on what follows only on the reduction of thermal irreversibilities, i.e., on the optimization of plants whose turbine outlet temperature is set. These irreversibilities result from differences in temperature between the hot and cold parts of the cycle.

In cogeneration plants (CHP) studied below, problems arise in a similar manner, especially if steam needs at medium and high pressure are important.

Analysis of a single pressure combined cycle

In a combined cycle plant, the vein of hot gases exiting the gas turbine must be cooled by water of the steam recovery cycle. In a single pressure cycle, water enters the heat exchanger in the liquid state at about 30°C after being compressed by the feedwater pumps downstream of the condenser.

It is heated at the boiling temperature corresponding to its pressure (economizer), then vaporized at constant temperature and superheated before being expanded in the steam turbine. Figure 10.4 shows the heat transfer within the heat exchanger between hot gases and water. The associated enthalpy diagram shows that if we set for technical reasons a pinch minimum value (temperature difference between both fluids, see Chapters 4 and 7) between points 6 and 9 on the one hand, and between points 4 and 11 on the other hand, heat exchanges take actually place with much larger differences in nearly all of the heat exchanger. This stems from the need to vaporize water, which induces a very important "plateau" at a constant temperature.

The example in Figure 10.5 corresponds to such a combined cycle. It is subject to a guided exploration (C-M3-V1).

BOX 10.1
Guided educational exploration

Single pressure combined cycle (exploration C-M3-V1)

This guided exploration presents a single pressure combined cycle. Emphasis is placed on the setting of the internal exchanger which allows the residual enthalpy of gases leaving the turbine to be transferred to the steam cycle, and which is called a HRSG.

You will learn how to set a triple heat exchanger and study the concept of pinch.

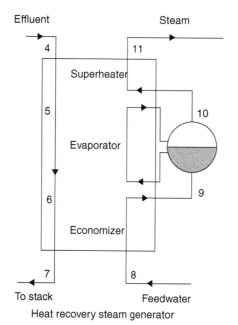

Heat recovery steam generator

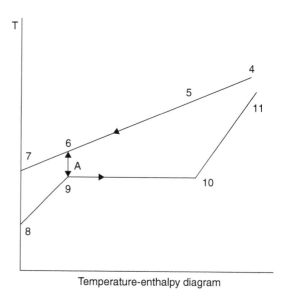

Temperature-enthalpy diagram

FIGURE 10.4
Steam generator exchange configuration.

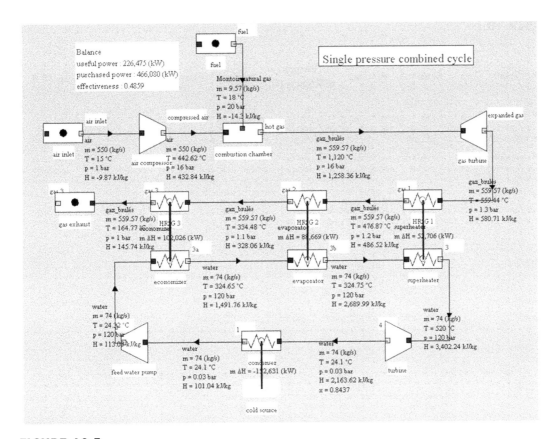

FIGURE 10.5
Synoptic view of a combined cycle.

Hot gases exit the gas turbine at 580°C, and the maximum pressure of the steam cycle is equal to 120 bar. In these circumstances, it is impossible to cool gases below 169°C, which represents a small loss as compared to the simple gas turbine.

Combined cycle exergy balance

The exergy balance of this cycle is given in Table 10.1. Its exergy efficiency (45.4%) has significantly increased as compared to that of the reference gas turbine (33.5%) given in Chapter 9.

Losses in the HRSG and those in the exhaust are linked, as we have seen in Chapter 5. They represent 11.4% and 7% of irreversibilities. Their reduction is therefore an important issue. The ideal heat exchange corresponds to the case where the curve of gas cooling and that of water heating would be parallel. The heat exchanger would then operate in counter-flow, and irreversibility would be minimal. This is not feasible with water, and the single pressure cycle has strong internal irreversibilities.

To improve the cycle performance, we use multiple steam cycles at different pressure levels (two, three or even four). Figure 10.6 shows the value of using multilevel pressure: with some simplifying assumptions and a choice of scales, we can superimpose on an entropy chart gas turbine and steam power plant cycles. In all three cases, the grey surface represents work provided for the same heat input in the gas turbine. The rectangle in the dashed line is the Carnot cycle.

The optimization of such cycles is a complex problem, because to get the better cooling of the hot gas stream, there are many degrees of freedom on the pressure levels, on the corresponding flow rates and on placement of heat exchangers (in series or in parallel). Figure 10.7 gives an example of an industrial three-pressure HRSG, with the whole combined cycle.

This optimization problem is quite new. It did not arise in old power plants, in which very large irreversibilities occurred for technical and economic reasons related to the thermal

TABLE 10.1

EXERGY BALANCE OF THE COMBINED CYCLE

Component	Resource	Product	Exergy Efficiency (%)	Irreversibilities	% Total (%)	$T_0 = 288.15$ K
Fuel	544,889					
Air inlet	130					
Air compressor	243,492	221,231	90.9	22,262	7.5	
Gas turbine	406,069	388,407	95.7	17,662	5.9	
Combustion chamber	0	584,511	76.3	181,411	61.0	
Feed water pump	930	930	100.0	0	0.0	
Turbine	112,414	95,993	85.4	16,421	5.5	
Gas exhaust				20,748	7.0	Loss
Condenser	4,895			4,895	1.6	$T_k = 15.00°C$
Economizer	52,770	36,667	69.5	16,103	5.4	Economizer
Evaporator	58,199	48,100	82.6	10,098	3.4	Vaporizer
Superheater	39,301	31,612	80.4	7,689	2.6	Superheater
Global	545,019	247,730	45.45	297,289	100.0	

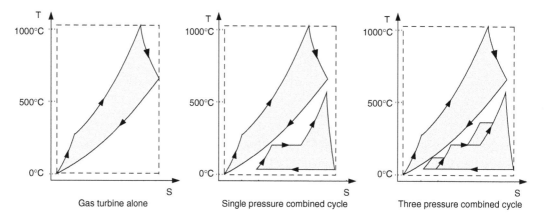

FIGURE 10.6

Comparison of work provided by a gas turbine and combined cycles.

resistance of steel boilers and sulfur content of fumes. Guided exploration OPT-2 will show you how this problem can be addressed using the pinch method implemented in Thermoptim.

BOX 10.2

Guided educational exploration

Optimization of a dual pressure combined cycle by the pinch method (OPT-2)

The objective of this exploration is to show you how the pinch method can be applied to optimize a dual pressure combined cycle.

There you will find a detailed presentation of the Thermoptim optimization window and explanations on how to set the fluids that must be taken into account in the optimization process.

Gas turbines are not the only engines that can operate in a combined cycle.

In recent years, the increase in diesel engine temperatures has resulted in the following gains, particularly suitable for use in combined cycle:

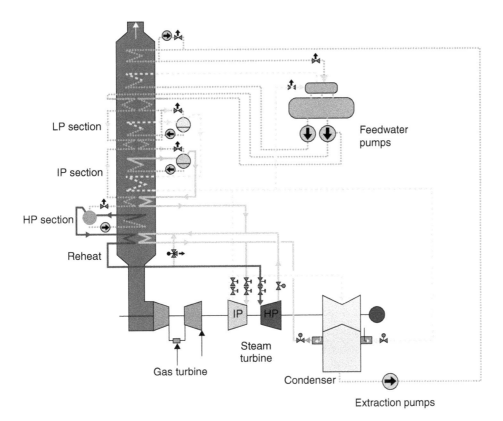

LP section

IP section

HP section

Reheat

Feedwater pumps

Gas turbine

IP   HP

Steam turbine

Condenser

Extraction pumps

FIGURE 10.7

Three-pressure HRSG.

- Mechanical efficiency increased from 45% to 47%;
- Enthalpy of exhaust gas increased from 27% to 32%;
- Cooling losses reduced from 24% to 16%.

Today, a less than 100 MW diesel combined cycle based on medium-speed diesel engine reaches an overall efficiency of 55%, which makes it competitive with gas turbines in this capacity range.

Conclusions and outlook

The very high efficiencies provided by combined cycles explain the enthusiasm for these plants and the rapid development of their market: Mitsubishi Hitachi Power Systems, Ltd. (MHPS) has commissioned in April 2020 T-Point 2, its new combined cycle power plant validation facility at Takasago Works in Hyogo Prefecture, Japan. T-Point 2 is expected to achieve power output of over 566 MW (60 Hz), with nearly 64% efficiency, 99.5% reliability and a turbine inlet temperature of 1,650°C.

Among the many benefits of combined cycles, we can mention the following:

- As for gas turbines, combined cycles are normally designed standardized and modular, so that different components are factory built and assembled quickly on site, while a steam power plant must be calculated and built case by case;
- Thus, combined cycles have almost no effect of scale. It is not necessary to construct a single unit of large capacity: we can start with small units and add others as demand grows;
- Due to the excellent efficiencies that are achieved and also the use of fuels with low sulfur and nitrogen, the environmental impact of these technologies is much lower than that of their competitors: CO_2 emissions are only equal to 40% of those of coal steam plants, and they require three times less water cooling;

- In the same way, the combined cycle footprint is close to $80\,m^2/MW$, against about $200\,m^2/$ MW for a steam power plant. It is therefore easier to locate them close to consumption areas.

One major limitation is that gas turbines require the use of clean fuel (expensive), such as natural gas or light distillates, which excludes the use of heavy fuel oil or coal, traditional basic fuels for power plants. However, development of coal gasification would allow this energy source to fuel these efficient combined cycles.

In the coming years, efficiencies should rise from 50% to 60%, and prices fall further, thereby increasing the economic competitiveness of these machines.

Cogeneration or CHP

General

We call **cogeneration** or **CHP for Combined Heat and Power** the combined production of thermal energy and mechanical energy or electricity.

The basic idea of cogeneration is that combustion takes place at very high temperatures (above 1,000°C), while the need for heat in industry or for heating occurs at lower temperatures, generally between 80°C and 300°C (Figure 10.8).

In these circumstances, it is quite possible, when using combustion to meet heating needs, to take advantage of this temperature difference to generate electricity through a power cycle. The heat source of the power cycle is the boiler or the combustion chamber, and the cold source corresponds to the heat needs.

The main advantage of cogeneration cycles is that they are among the most efficient in terms of energy use.

Generally, the objectives pursued by CHP are twofold: firstly to achieve economies of operation and secondly to ensure security of electricity supply at least for part of the units. Given their purposes, cogeneration plants can be grouped into three classes:

- "Heat and power" installations where heat is the commodity, electricity being a byproduct allowing to give a better value to the fuel. This is the case of large plants using heat or district heating or garbage incineration facilities. Priority is given to the provision of heat, electricity, easily transportable, being valued by selling power surpluses to utilities. In the event of mains utility failure, the plant operates in island mode;
- "Total energy" systems seeking to ensure electrical autonomy, heat being the byproduct. They are generally off-grid plants and ships;
- Not autonomous facilities, undersized for economic reasons, for which a supplement is provided by utilities for electricity and by a conventional boiler for heat. The installation works only when electricity prices are high and heat needs are high. This type of installation is quite common and is often the one that leads to the best financial results for the company.

Technically, it is customary to classify CHP into two families, depending on the power cycle used:

FIGURE 10.8

Cogeneration principle.

- Boiler and steam turbine systems, which are widespread, as the benefits of this configuration has been known for over a century. They can use a wide variety of fuels, including coal or waste;
- Internal combustion engine systems, which use either gas turbines or reciprocating engines (especially diesel and gas engines). Heat is recovered from exhaust gases as well as coolants and lubricants. Only liquid and gaseous fuels can be used in these engines.

A cogeneration plant produces both heat and electricity. To describe its performance in both regulatory and technical terms, we introduce a number of indicators, defined below.

Performance indicators

Let us call Q_c the heat supplied to the cogeneration machine, that is to say, released by the combustion reaction, Q_u the useful heat, τ the mechanical energy or electricity produced. In what follows, these different energies are expressed in the same units, usually kWh or MJ.

Let us call:

- Mechanical efficiency the ratio $\eta_m = \dfrac{|\tau|}{Q_c}$;

  It characterizes the performance of the facility as a generator of electricity. The best mechanical efficiencies are obtained in conventional power plants where $Q_u = 0$.

- Overall efficiency the ratio $\eta_g = \dfrac{|\tau + Q_u|}{Q_c}$;

  It characterizes, in terms of energy, the overall efficiency of the facility.

- Exergy efficiency the ratio $\eta_x = \dfrac{\left|\tau + \left(1 - \dfrac{T_0}{T}\right)Q_u\right|}{Q_c}$;

  T_0 is the temperature of the environment, and T is the temperature at which heat is provided. It allows, through the introduction of the Carnot factor, to characterize the temperature level at which heat is provided. Q_c being the heat released in the combustion chamber, we assimilate it here to the exergy of the fuel.

- Heat-power ratio the ratio $C_F = \dfrac{Q_u}{|\tau|}$;

  It is representative of the distribution of energy between heat and electricity.

- Specific equivalent consumption the ratio $C_E = \dfrac{Q_c - \dfrac{|Q_u|}{\eta_c}}{|\tau|}$.

  η_c being an average conventional boiler effectiveness, usually taken equal to 0.9.

It represents the primary energy consumption leading to the production of 1 kWh of electricity. In fact, this is not quite the case, because heat provided Q_c is final energy and not primary energy, which induces a slight bias. The indicative value of CE for a conventional power plant is greater than 2.5, when it is only 1.7 for a combined cycle plant of 60% efficiency.

These indicators, as defined above for a given operating point, are likely to vary depending on operating conditions, including the environment temperature. To estimate their average values over a long period, such as the heating season or a year, they are calculated from the cumulative values of variables considered.

The regulator has used some of them to determine if an energy facility may or may not be considered as a cogeneration unit, which determines its ability to sell electricity to utilities.

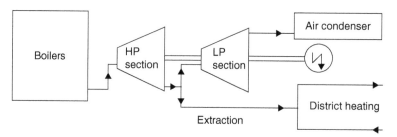

FIGURE 10.9
Boiler and steam turbine cogeneration unit.

Boilers and steam turbines

In a boiler and steam turbine cogeneration plant, there is, depending on the case, a single turbine called back-pressure or two turbines called extracting and condensing (see Figures 10.9 and 5.11).

In back-pressure turbines, well adapted as heat needs change little, steam produced in the boiler at an initial pressure generally between 30 and 50 bar, is expanded at a pressure (called back-pressure) of about 2–6 bar and temperatures of 130°C–160°C. This steam is then used directly in processes or district heating.

In extracting and condensing turbines, all the steam passes through the high-pressure section, which behaves like a back-pressure turbine. A fraction is then extracted to feed processes or district heating, while the remainder is expanded in a low-pressure section and finally condensed either by air (Figure 10.9) or by water cooling.

This type of turbine can largely decouple the production of electricity from that of heat and is therefore very well suited for cogeneration facilities used for space heating. In summer, extraction is minimal and electricity production maximal, and in winter, it is the opposite.

The overall plant efficiency of back-pressure facilities, however, is slightly higher than that of extracting and condensing units, because part of the heat is discharged to the condenser in the latter. These configurations are widely used for applications where heat needs are important, as in garbage incinerators (municipal solid waste incineration or MSWI), district heating networks and heavy industries.

Reciprocating engines

Now let us look at cogeneration systems with reciprocating internal combustion engines.

The simplest and most common solution is the production of either hot water at a temperature of 100°C, or superheated steam at 110°C–120°C, as auxiliary to a classical boiler (Figure 10.10). Depending on its purity, water can be directly heated in the engine or must pass through a low temperature heat exchanger. It then recovers the exhaust gas heat in a heat exchanger placed in series. Hundreds of such units of all capacities (a few kW to several MW) are installed worldwide.

A second solution is to cool the engine by an air flow which, in series, provides a convective cooling, and then passes through a recovery heat exchanger on oil, the supercharger intercooler if it exists, the classic radiator cooling system and finally an air/exhaust flue gas exchanger. Hot air is then used for drying, its enthalpy being, if necessary, raised by a supplementary burner.

The engine can also be used for air conditioning, driving a compressor directly connected on the shaft, to obtain industrial cooling or chilled water, heat recovered being used for purposes of either heating or cooling in an absorption machine. An alternator can at times be coupled to the engine instead of the compressor, which allows, according to the tariffs of electricity and refrigeration needs, to modulate the production.

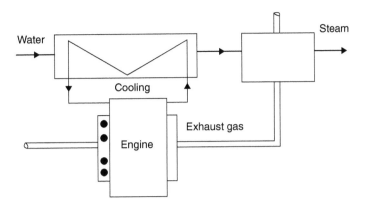

FIGURE 10.10

Internal combustion engine cogeneration unit.

The reciprocating engine may finally directly drive a heat pump compressor. The overall efficiency of the system can be very high, given the coefficients of performance of heat pumps.

The heat-power ratio CF is quite low, between 0.5 and 1.5. Overall efficiency is generally very good, above 70%. Mechanical efficiency is usually very high, between 30% and 35% for small gas engines, and up to 45% for large diesel and gas engines. Specific equivalent consumption CE is of the order of 1.6–2.

Figure 10.11 gives an example of a cogeneration installation using a model E 2842 gas engine, manufactured by MAN Dezentrale Energiesysteme. The Thermoptim model is a variant of the one we presented when studying alternative internal combustion engines, by supplementing two cogeneration exchangers. This cycle is the subject of guided exploration (C-M3-V2).

BOX 10.3
Guided educational exploration

Industrial gas engine used in cogeneration (exploration C-M3-V2)

This guided exploration presents a cogeneration installation using the industrial gas engine that we modeled with a Beau de Rochas cycle in another guided exploration (C-M2-V5b).

Emphasis is placed on the calculation of the performance indicators of the cogeneration system.

You will learn how to set a thermocoupler.

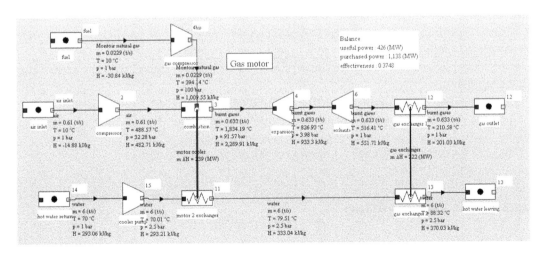

FIGURE 10.11

Gas engine cogeneration.

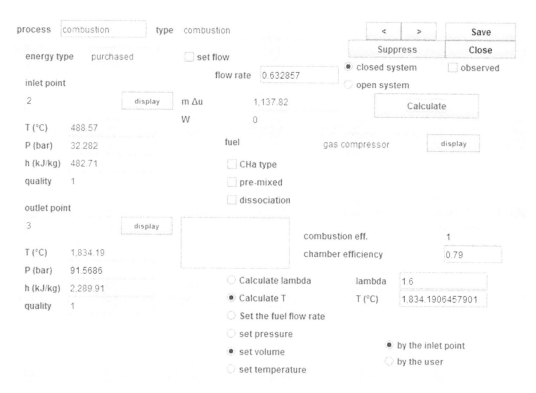

FIGURE 10.12

Screen of the combustion process.

The coupling to the combustion chamber is carried out using a Thermoptim functionality that we have not yet used, that of a thermocoupler.

As explained in Chapter 4, a thermocoupler allows components other than "exchange" processes to connect to one or more "exchange" processes to represent thermal couplings.

For a combustion chamber, the calculation is carried out in order to balance the thermal power transferred to the cooling fluid, itself defined by the combustion efficiency chosen.

In this example, its value is 0.79, which means that 21% of the power released by combustion is transferred via the thermocoupler (Figure 10.12).

The screen of the thermocoupler (Figure 10.13) is similar to that of an exchanger, the exchange process appearing on the left of the screen under the name of "thermal fluid" and the combustion chamber on the right, under the name "process".

In this case, the user has only the choice between two calculation modes: either calculate the outlet temperature of the exchange process at a given flow rate, or calculate the flow rate, this temperature being known.

In this example, we have chosen to determine the outlet temperature, the flow rate of the coolant being known.

The thermal power communicated to the water is 239 kW.

The second exchanger on exhaust heat is entirely conventional. By choosing an effectiveness equal to 0.7, an additional power of 222 kW is recovered.

Table 10.2 gives the performance indicators of the cogeneration facility.

Gas turbine cogeneration

Now let us look at gas turbine cogeneration facilities.

In gas turbines, the total residual heat is found in the exhaust. The performance of the cogeneration system is directly related to the recovery of these gases.

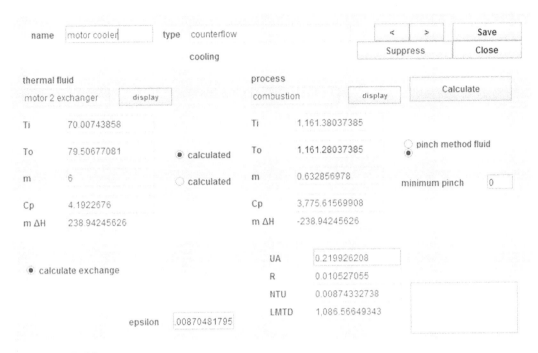

FIGURE 10.13

Thermocoupler screen.

TABLE 10.2

PERFORMANCE INDICATORS OF THE COGENERATION FACILITY

Purchased power Q_c	1,138
Mechanical power τ	426
Thermal power Q_u	461
Mechanical efficiency	37.4%
Thermal efficiency	40.5%
Overall efficiency	77.9%
Heat-power ratio	1.08
Specific equivalent consumption	1.47

One solution is to cool exhaust gases in an air-fumes heat exchanger which can heat the air which is then used for many applications. If the turbine is stopped, an auxiliary boiler ensures the supply of heat to meet the plant needs. One generally uses several cascading recovery exchangers on the exhaust, so that it can be cooled as much as possible, and air is available at different temperatures for various uses.

Another very efficient scheme is to directly use the exhaust gases as hot fluid in a dryer (see Chapter 17). As in the case of the reciprocating engine, the turbine itself is placed in the airflow, so that all losses can be recovered, leading to an overall efficiency close to 1. In addition, the outlet pressure of the exhaust gas is sufficient to avoid any fan. The hot gas temperatures (400°C–500°C) being compatible with many industrial requirements, applications of this method are numerous.

Another solution, widely used today, especially to replace an existing boiler, is to install a HRSG at the outlet of the gas turbine. The problem of optimizing the recovery exchangers is very similar to the one we have discussed before: the best configuration must indeed both cool at best the GT exhaust and provide heat at the highest temperature level possible as needed.

When the gas turbine used is a micro-turbine rated less than 100 kW, it is enough to heat water, either upstream of an existing boiler, or for hot water uses. The recovery exchanger is then simpler and less expensive than a HRSG. A gas micro-turbine generally operates with a low compression ratio and with a regenerator to improve performance.

Criteria for selection

In general, CHP leads to a better use of primary energy than is allowed for separate production of heat and mechanical power. However, the decision-maker rarely views the problem in terms of primary energy savings: he must justify his choices based on the micro-economic context in which they operate.

Before deciding to use a cogeneration facility, it is necessary to make an extensive study of power and thermal needs and their evolution over time. Indeed, the corresponding investments are generally high, and, to amortize them, the facility must operate at its optimum economic point as long as possible when the energy price is justified, and if possible, at about 80%–90% of the maximum engine capacity, range leading to the best technical performance.

The profitability calculation depends fundamentally on operating conditions and their evolution over time, each facility representing a special case that requires detailed study. It is particularly important to properly ensure that the statutory criteria authorizing the resale of electricity to utilities will be respected.

If security considerations require the use of an independent mechanical power production unit, it is almost certain that its use in cogeneration will be profitable, the overhead being limited to expenses for heat recovery, often well below those corresponding to the engine and alternator.

Moreover, the technical solution choice depends on many factors. For purposes of comparison, Table 10.3 provides approximate values of the different performance indicators for key possible technologies. The first two rows correspond to what is now called micro-cogeneration, based on Stirling engines or small capacity micro turbines. The configuration GT + ST is representative of hybrid solutions between cogeneration and combined cycle, an example of which is given below.

Examples of industrial plants

We give below two examples of industrial facilities involving gas turbines. The first is relatively simple: a micro-turbine is used to generate electricity and to preheat water in liquid form in an economizer, while the second, more complex, uses an aero-derivative gas turbine of average capacity, with or without afterburner, a steam cycle and a boiler. The system configuration changes depending on the heating requirements (which vary with outside temperature).

The two cases presented correspond to actual installations in operation. Their name is omitted at the request of manufacturers who exploit them.

Micro-gas turbine cogeneration

In the plant we are interested in, a micro-turbine of 100 kW is used for warm 1.82 kg/s of water from 70°C to 90°C. The turbine sucks 0.78 kg/s of air that is compressed at 5 bar and then passes through a regenerator before being heated to 950°C in the combustion chamber

TABLE 10.3

COMPARISON OF DIFFERENT TECHNOLOGIES

	Capacity	η_g (%)	η_m (%)	CF	CE
Stirling	0.5–100 kW	>70	15–30	1.2–7	1.8–2.6
Micro-GT	25–75 kW	>80	25–32	1.5–2.2	1.5–1.7
Engines	0.05–50 MW	>70	25–45	0.5–1.8	1.6–2
GT	5–200 MW	>80	35–40	0.8–1.3	1.4–1.6
ST	0.5–200 MW	>80	6–22	3–12	1.6–3
GT + ST	20–200 MW	>80	>40	0.8–1.2	1–1.4

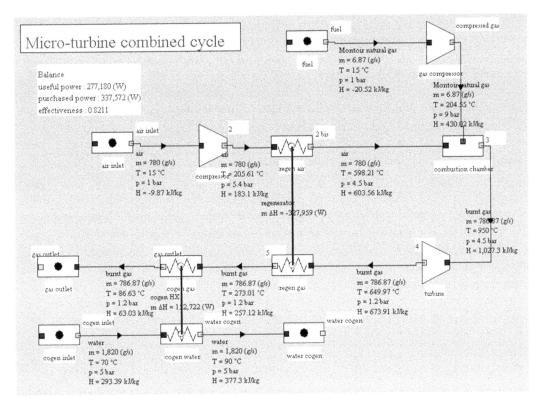

FIGURE 10.14

Synoptic view of a micro-gas turbine cogeneration plant.

burning natural gas. Gases are expanded at a temperature of 650°C and then pass successively through the regenerator and the cogeneration exchanger (Figure 10.14). A gas compressor is required to raise the pressure of natural gas from the distribution network.

Modeling such a facility in Thermoptim poses no particular problem and leads to the synoptic view of Figure 10.14. Here useful energy includes not only compression and expansion work, but also thermal energy supplied to the water circuit, which leads, for the use conditions adopted, to an overall efficiency of nearly 82%: about 125 kW of electrical power and 153 kW of heat for a fuel consumption of 340 kW. Mechanical efficiency is 36.7%, and heat-power ratio is 1.23.

The exergy balance of the cycle is given in Table 10.4.

TABLE 10.4

MICRO-GT COGENERATION EXERGY BALANCE

Component	Resource	Product	Exergy Efficiency (%)	Irreversibilities	% Total (%)	$T_0 = 288.15$ K
Fuel	369,896					
Air inlet	184.87					
Compressor	150,516	143,940	95.6	6,576	3.1	
Turbine	289,565	278,068	96.0	11,497	5.3	
Combustion chamber	0	561,945	81.6	127,015	59.1	
Cogen inlet	36,210					
Water cogen	64,284	64,284				
Gas compressor	3,095	2,708	87.5	387	0.2	
Cogen gas	54,287	28,074		26,213	12.2	Cogen HX
Gas outlet				19,052	8.9	Loss
Regen air	196,838	172,625	0.0	24,213	11.3	Regenerator
Global	406,291	191,339	47.09	214,952	100.0	

The exergy efficiency is only 47%, that is to say a little more than half the overall cogeneration efficiency. This is partly explained by strong irreversibilities in heat exchangers, working with important log-mean temperature difference (LMTD, cf. Chapter 4), 60°C for the regenerator and 70°C for the CHP exchanger.

Exergy balance properly highlights the distribution of irreversibilities: about 60% in the combustion chamber, and between 11% and 12% in the two heat exchangers, which, as we have seen, are not optimized. Exhaust gas losses are lower (8.8%).

Industrial gas turbine cogeneration

The second cogeneration plant that we present is much more complex and of a higher capacity.

The heating system is a circuit of pressurized water whose temperature is between about 90°C and 130°C. The circuit exits the facility at its maximum temperature and then passes through the city exchange substations ensuring the heating and/or domestic water heating of homes, buildings, hospitals, school groups, etc. Finally it returns to the plant at its minimum temperature to be heated again.

City heat requirements vary throughout the year depending on climatic conditions. The plant must adapt to these needs and adjust the thermal power it provides to the network.

Facility description

The facility has three independent circuits which exchange only heat (Figure 10.15):

• A **gas turbine** (GT), connected to a generator, provides about 80% of site electricity production. Gases exiting the GT at about 450°C are burned again with a small amount of fuel. Flue gases, whose temperature is about 600°C, warm steam circuit water in a recovery

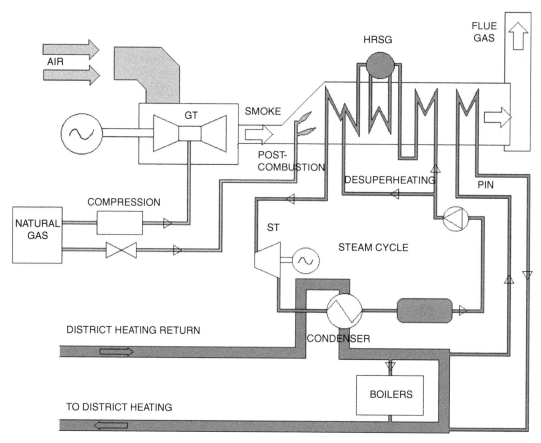

FIGURE 10.15

Schematic overview of the cogeneration plant.

boiler and then the water of the urban network. Finally, they are released into the atmosphere at about 150°C, through the chimney;

- A **steam circuit** provides the remaining 20% of electricity production. Water is heated into steam and then superheated in the recovery boiler by exhaust fumes from the GT. A desuperheating is performed in the middle of the boiler to regulate the steam turbine inlet temperature. It is then expanded in a steam turbine (ST) connected to a generator. In the condenser, in which pass both the steam circuit and the urban network, steam is condensed. Finally, liquid water is directed to the feedwater tank and pumped into the recovery boiler;

- The **urban network** is traversed by hot liquid water under pressure. The latter returns at minimum temperature. It is then reheated in the condenser, where it acts as a heat sink. Part of this water is heated in the recovery boiler in a heat exchanger called "pin". The flow in the pin is limited. If the thermal contributions from the condenser and the pin do not meet the network needs, the remainder is provided by boilers in derivation.

In this example, we modeled all three circuits and their interactions by heat exchangers and then determined the energy balance of the plant and the various indicators of the cogeneration plant: mechanical (or electric), heat, global and heat-power ratio in different operating cases.

Operating cases

The plant must adapt to the needs of the district heating system providing the hot water circuit with a thermal power which depends on climatic conditions and, in average, on outdoor temperature (from 38 MW, 1,130 t/h of water between 90°C and 120°C at 5°C to 64 MW, 1,370 t/h of water between 90°C and 130°C at −5°C).

The plant can supply heat to the hot water circuit in three ways: through the steam circuit condenser, the pin and the by-pass boilers. The maximum flow in the steam circuit is 70 t/h. For low heat requirements, boilers are not used, and a lower steam flow rate is circulated. It therefore requires less heat input at the recovery boiler (the steam turbine inlet temperature must always be the same: 485°C). For this, the temperature of the smoke is controlled by adjusting the fuel flow in the afterburner (the GT still operating at substantially the same regime, we may alter neither the exhaust gas flow nor its temperature). Direct heating of water from district circuit in the pin then decreases. The more important the heat needs, the greater the steam circuit flow and the fuel flow in the afterburner to achieve the maximum steam throughput of 70 t/h in the circuit. When the thermal needs of the urban system can no longer be provided by the condenser and the pin, the boilers are put in operation.

Two operating cases and a limiting case may be distinguished:

- **Very important heat demand $T_{ext} < T_{lim}$:** the flow in the steam circuit is maximum, i.e., 70 t/h. The afterburner is at maximum. The thermal energy provided in the condenser and the pin is at maximum. The boilers are in operation and provide the necessary input;

- **Smaller heat demand $T_{ext} > T_{lim}$:** boilers are off. The thermal energy provided by the condenser and the pin is not at maximum. The flow in the steam circuit is less than 70 t/h, and the afterburner is not at maximum;

- **The limiting case** corresponds to maximum afterburning and steam output of 70 t/h. Thermal power is then 43.9 MW in the condenser and 7.4 MW in the pin. The total power recovered by the hot water circuit is 51.3 MW. This roughly corresponds to $T_{ext} = T_{lim} = 0°C$.

Modeling in thermoptim

The model can be decomposed into four interconnected modules: the gas turbine and steam circuit are simple variants (due to the afterburner and desuperheating vein) of examples presented in the previous chapters; the recovery boiler which connects them (Figure 10.16), and the district heating network (Figure 10.17) are the other two submodules.

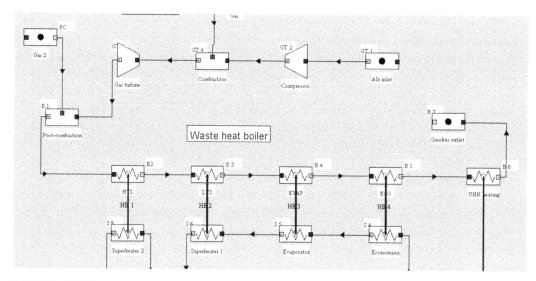

FIGURE 10.16

Recovery steam generator.

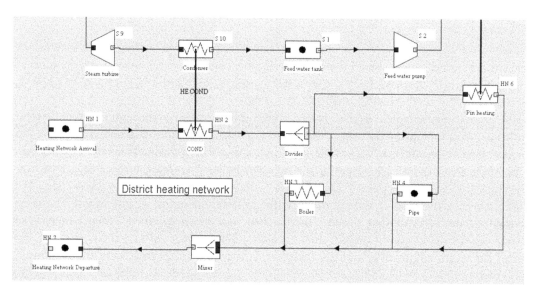

FIGURE 10.17

District heating network.

The whole model leads to results shown in Figure 10.18 for an outside temperature of −1°C. Table 10.5 shows the influence of this parameter on the main global indicators used in cogeneration.

BOX 10.4

Worked example

Cogeneration plant producing electricity and providing heat to a district heating

This example which corresponds to a real-world case is presented in the Diapason session S47En (https://direns.mines-paristech.fr/Sites/Thopt/en/co/session-s47en-district.html).

This cogeneration plant produces electricity and provides heat to the district heating network of a town of 30,000 inhabitants.

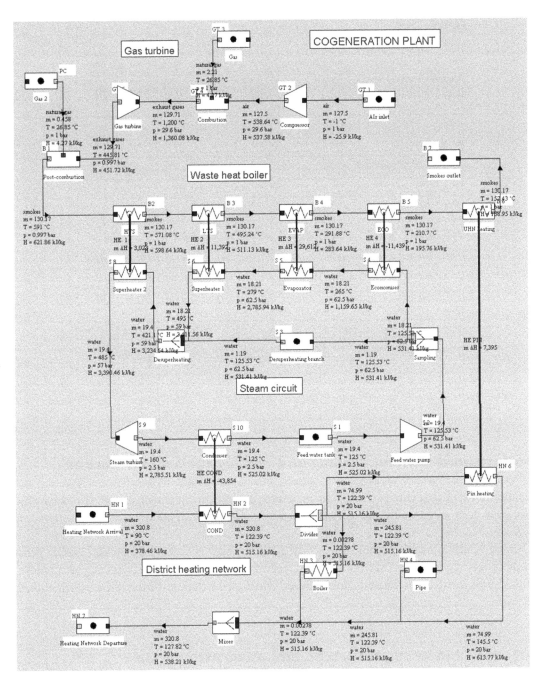

FIGURE 10.18

Industrial cogeneration plant.

Trigeneration

We talk of trigeneration to describe methods for simultaneous production of mechanical power (usually electricity), heat and cooling. Trigeneration is thus somehow a generalization of the CHP.

So that we can use trigeneration, it is imperative that on the one hand we may need in one place heat and cooling, which is quite rare, and that on the other hand we have an electricity production system.

We speak here of two types of trigeneration plants:

- Plants for producing cooling and heating for hypermarkets;
- Absorption installations powered by a micro-gas turbine and also producing hot water.

TABLE 10.5
COGENERATION ENERGY BALANCE

Outside temperature	−5°C Power (MW)	−1°C Power (MW)	2°C Power (MW)
GT combustion	107.9	107.9	107.9
Afterburner	22.4	22.4	14.3
Boilers	11.5		
TOTAL purchased power	**141.7**	**130.2**	**122.2**
GT turbine	117.8	117.8	117.8
GT compressor	71.8	71.8	71.8
Net GT power	46.0	46.0	46.0
ST power	11.7	11.7	9.8
TOTAL mechanical power	**57.7**	**57.7**	**55.8**
Condenser	43.9	43.9	36.7
Pin heating	7.4	7.4	7.5
Boilers	11.5		
TOTAL useful thermal power	**62.7**	**51.2**	**44.2**
TOTAL useful power	**120.5**	**109.0**	**100.0**
Mechanical efficiency	40.7%	44.3%	45.7%
Overall efficiency	85.0%	83.7%	81.8%
Heat-power ratio	1.09	0.89	0.79

Production of central heating and cooling for a supermarket

In supermarkets, it is common, at least in winter, to have simultaneously needs to heat the store and cool rooms and refrigerated display cases, at temperatures either positive or negative.

It may be economically feasible, especially if the hypermarket has its own emergency generator, to use trigeneration. Most of the time, related facilities are diesel or gas engines, whose cooling heat is used for heating, and whose shafts drive vapor compression refrigeration machines.

Trigeneration by microturbine and absorption cycle

We are interested here in a trigeneration plant where gases coming out of a 125 kWe micro-turbine are used both to provide the necessary heat to the desorber of a LiBr-H_2O absorption machine (Figure 10.19) and secondly to produce 0.5 kg/s of hot water at about 80°C. The turbine is a variant of the CHP example presented above. Expanded gases pass successively through the regenerator, the desorber and the cogeneration heat exchanger.

The modeling of this facility uses thirty components, representing several hundred coupled equations. All these components are available in Thermoptim core, with the exception of sub-system (absorber-solution exchanger-desorber) of the absorption refrigeration cycle (the three lower elements of the sketch in Figure 10.19), which replaces the compressor of a vapor compression cycle. We thus understand the interest of the external class mechanism in a case like this: by adding a specialized component to represent the missing module, the work involved is much smaller than if we had to write a program to model the whole cycle.

So we just have to create an external component to represent the module which does not exist, which firstly makes use of the LiBr-H_2O pair, whose properties can be modeled either directly in the external component or as a particular external substance (see Chapter 13 how in practice to create such a component). Furthermore, this module requires both a heat sup-

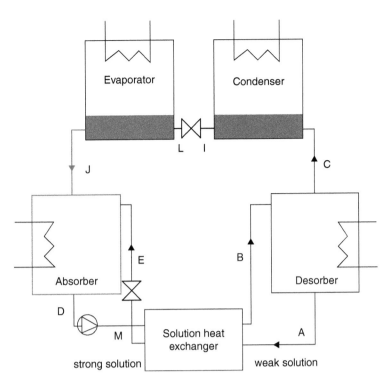

FIGURE 10.19

Sketch of the absorption cycle.

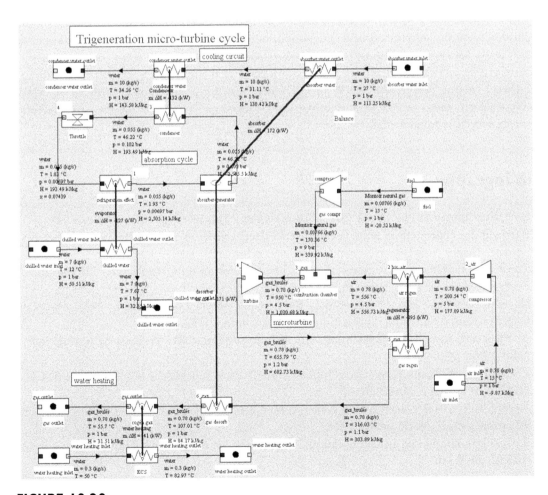

FIGURE 10.20

Synoptic view of the trigeneration plant.

ply at high temperature in the desorber, and heat extraction at medium temperature in the absorber. Representation of thermal coupling is possible using two thermocouplers, called "absorber" and "desorber" on the synoptic view of Figure 10.20: the external component "generator-absorber" calculates the thermal energy to be exchanged, and each thermocoupler recalculates the "exchange" process to which it is connected.

BOX 10.5
Worked example

Modeling of a trigeneration plant

The modeling of a micro-turbine LiBr-H_2O trigeneration plant is presented in a guidance page of the Thermoptim-UNIT portal (https://direns. mines-paristech.fr/Sites/Thopt/en/co/fiche-guide-td-fg12.html).

The resolution of the thermodynamic model presented is explained, with its implementation in the external class LiBrAbsorption.

Once the missing component is created, modeling the trigeneration plant poses no particular problem and leads to the diagram in Figure 10.20, in which we recognize in the top left the refrigeration cycle with its cooling circuit, in the middle-right the micro-turbine and in the lower left the hot water production exchanger.

The setting of the micro-turbine is that of an existing industrial machine, but the trigeneration plant itself is fictitious: we have simply sought to recover the available enthalpy in the gas turbine exhaust to produce cooling and heat hot water.

In the synoptic view of Figure 10.20, useful energy includes not only the compression and expansion work but also the cooling energy and thermal energy supplied to the water system, which leads, for these conditions, to an overall efficiency of about 78% which corresponds to approximately 125 kW of electrical power, 176 kW of heat input to the desorber, providing 127 kW of cooling power, and 40 kW of hot water heat, for a fuel consumption of 375 kW. Mechanical efficiency is 32.7%, and heat-power ratio 1.76. The COP of the single-effect absorption cycle is equal to 0.72.

Bibliography

Ashrae, Fundamentals Handbook (SI), Thermodynamics and Refrigeration Cycles, 2001.

P. Barroyer, La cogénération pour la production décentralisée d'énergie, Revue de la Société des Electriciens et des Electroniciens, Paris, décembre 1999.

S. Candelier, Modélisation d'une installation de cogénération industrielle avec Thermoptim, École des Mines de Paris, Paris, juillet 2001.

P. Daverat, Moteurs à gaz et cogénération, Revue Générale de Thermique, n° 383, n° spécial Cogénération, Etat de l'art, applications, Paris, novembre 1983.

A. F. El-Sayed, Aircraft propulsion and gas turbine engines, CRC Press, Boca Raton, 2008, ISBN 978-0-8493-9196-5.

R. Gicquel, M. Williams, K. Aubert, Optimisation du cycle eau-vapeur d'une centrale IGCC, HPC'01, Paris, septembre 2001.

C. Kempf, Les centrales électriques à cycle combiné, Revue de la Société des Electriciens et des Electroniciens, Paris, décembre 1999.

C. Levy, Cogénération en génie climatique, Techniques de l'Ingénieur, Traité Mécanique et chaleur, BE 9 340.

Compression refrigeration cycles

Introduction

In a power cycle, heat is provided to produce mechanical energy. A refrigeration cycle operates in reverse: it receives mechanical energy which is used to raise the temperature level of heat.

In this chapter, we will transpose analyses on improving power cycles to refrigeration cycles, the optimization of which also calls for the reduction of irreversibilities.

Three types of refrigeration cycles are commonly used:

- Refrigeration cycles;
- Heat pump cycles;
- Mechanical vapor compression cycles (Chapter 17).

The first two, which will be presented in this chapter, differ only by the levels of operating temperature and the desired effect. In refrigeration cycles, we try to cool a cold chamber, while a heat pump is used for heating. They put into play refrigeration fluids of various compositions, which allow one to transfer heat at low temperature to a medium at high temperature.

We analyze the main thermodynamic cycles which are generally used, including ejector cycles. A specific section is devoted to cryogenic cycles, in particular Linde and Claude.

General

Now that we have studied the main power cycles, let us come to the refrigeration ones.

Let us recall that we have presented the principle of operation of these cycles in Chapter 2 and that we have studied in Chapter 3 and in Guided Exploration S-M3-V9 some possible settings of the simple refrigeration cycle. We also compared in Chapter 6 the plots in the entropy chart of this cycle and that of Carnot and discussed the differences between them.

Let us recall also that the useful energy of a refrigeration cycle is the heat extracted from the evaporator, and the purchased energy is the work provided to the compressor. As the ratio of the two is generally greater than 1, the term efficiency is no longer suitable, and this is why we speak of the cycle coefficient of performance (COP).

Although the problem of optimizing refrigeration cycles is very different from that of power cycles, in both cases, we seek to minimize irreversibilities, so that the approaches meet.

To fix ideas, consider a refrigeration cycle intended to cool to a temperature of −10°C a cold enclosure placed in a room at 20°C.

$$COP = \frac{T_2}{T_1 - T_2}$$

DOI: 10.1201/9781003175629-11

$$T_2 = -10°C = 263.15 \text{ K}$$
$$T_1 - T_2 = 30°C$$
$$COP = 263.15/30 = 8.8$$
Reverse Carnot COP is 8.8.

In practice, the thermodynamic fluid evolves between temperatures $T_1 = 75°C$ and $T_2 = -20°C$ approximately, which corresponds to a COP equal to 253.15/95 = 3.7, much lower than the theoretical value of 8.8.

In the cycles that we have studied so far, the refrigerant evaporation pressure was equal to 1.78 bar, which allowed a cold chamber to be cooled slightly below 0°C. Such a value is acceptable for a simple refrigerator, but not for a fridge-freezer, the temperature of the cold enclosure of which must be at least –15°C.

We will therefore use a much lower evaporation pressure, equal to 1 bar, for our reference cycle. For the rest, we will keep the settings of Guided Exploration S-M3-V9: superheat 5°C, subcooling –10°C, with the proviso that the compressor isentropic efficiency will be 0.8.

The reference cycle that we consider is then given in the synoptic view of Figure 11.1, and its plot in the (h, ln(P)) chart in Figure 11.2.

The evaporation and condensation pressures are 1 and 12 bar. Its COP is worth 2.04.

Reference cycle exergy balance

Table 11.1 gives the exergy balance of this cycle.

This exergy balance is very interesting: while, with the machine's COP being 2.04, one might think that the machine is very efficient, its exergy efficiency is only 27.7%.

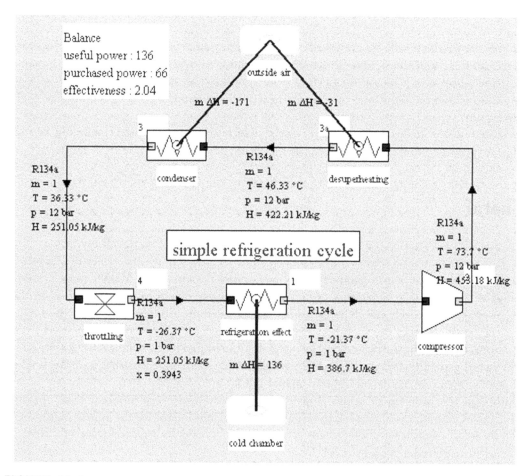

FIGURE 11.1

High pressure ratio simple cycle.

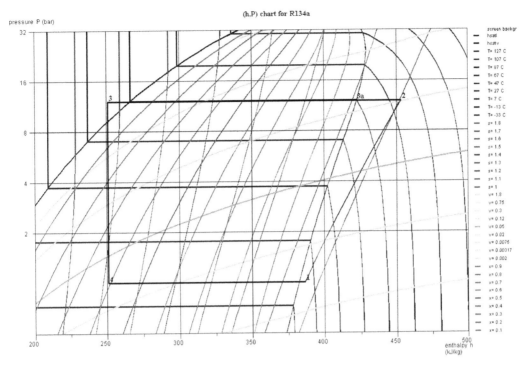

FIGURE 11.2

Plot of the cycle in the (h, ln(P)) chart.

TABLE 11.1

EXERGY BALANCE OF THE REFERENCE CYCLE OPERATING BETWEEN 1 AND 12 BAR

Component	Resource	Product	Exergy Efficiency (%)	Irreversibilities	% Total (%)	$T_0 = 293.15$ K
Throttling	12.26			12.26	25.5	
Compressor	66.48	54.98	82.7	11.50	23.9	Compr
Refrigeration effect	25.46	18.39	72.2	7.06	14.7	$T_k = -15.00°C$
Desuperheating	3.46			3.46	7.2	$T_k = 20.00°C$
Condenser	13.80			13.80	28.7	$T_k = 20.00°C$
Global	66.48	18.39	27.7	48.09	100.0	

An exergy balance is much better to qualify the performance of a refrigeration cycle than a simple enthalpy balance!

Irreversibilities are fairly evenly distributed: the throttling and the compressor represent around 25% each, the evaporator 15% and the condenser 29%.

Improvement of the simple refrigeration cycle

To improve the refrigeration cycles, we are also led on the one hand to minimize the irreversibilities coming from temperature heterogeneities both outside the system and internally and on the other hand to use staged compression.

The expressions of the reverse Carnot cycle show that the value of the COP deteriorates when the temperature difference $(T_1 - T_2)$ increases.

It is clear that the value of COP is greater, the smaller the difference $(T_1 - T_2)$.

When this difference increases, the compression ratio increases accordingly, which has the effect of:

- Lowering the isentropic efficiency;
- Increasing the compressor outlet temperature to very high values, with the risk of oil decomposition.

We know that staged compression with intercooling can reduce compression work.

This complication of the cycle is justified when the temperature difference $(T_1 - T_2)$ increases.

In practice, as soon as the compression ratio exceeds 6, the single-stage cycle reaches its limits and must be replaced by multistage cycles. In most cases, the refrigeration systems are two-stage.

Two-stage cycles

We know that when it is necessary to fractionate a compression, it can be advantageous to cool the fluid between two stages. Cooling can be ensured by the condenser external heat source or another cooler if exists.

Consider our reference cycle, i.e., with subcooling of 10°C and 5°C superheat, working between 1 and 12 bar (Figures 11.1 and 11.2). Its COP is equal to 2, the refrigeration effect being 135.6 kJ/kg and the compression work 66.5 kJ/kg.

Consider what can be done by using a staged compression. The compression ratio being equal to 12, the intermediate pressure can be chosen equal to 3.5 bar at first guess.

Maintaining an isentropic efficiency equal to 0.8, the compression end temperature is equal to 23.6°C, that is to say just above that of condenser cooling air (assumed to be at 20°C).

With intermediate cooling from 24°C to 10°C, the COP is slightly improved and passes to 2.1 (Figure 11.3). The gain is low, and this cycle will not work without an additional heat sink.

To ensure both the internal cooling of the vapors exiting the low-pressure compressor and also increase in the vaporization plateau, it is interesting to stage the expansion. The simplest and most effective cycle is called a total injection cycle or refrigeration cycle with flash chamber (Figure 11.4).

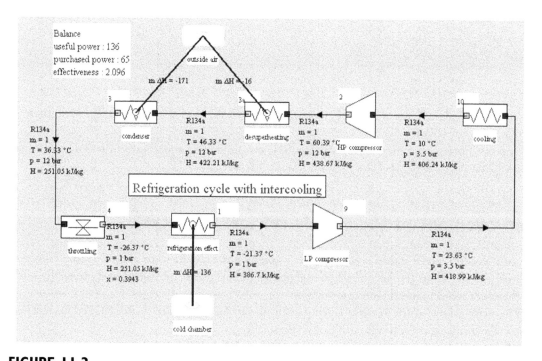

FIGURE 11.3

Cycle between 1 and 12 bar.

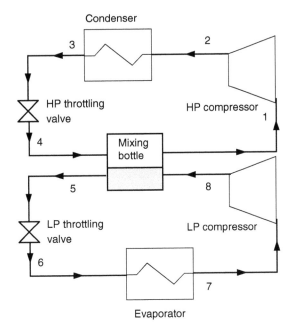

FIGURE 11.4

Sketch of a total injection two-stage compression cycle.

In this cycle, vapor exiting the LP compressor and two-phase fluid leaving the HP expansion valve are mixed in a bottle which acts like a capacitor and a separator, the vapor being sucked into the HP compressor, while the liquid phase passes through the LP expansion valve (Figures 11.4–11.6).

Table 11.2 gives the exergy balance of this new cycle. Its exergy efficiency has significantly increased, reaching 31.8% instead of 27.7%.

The components whose part in irreversibilities have changed the most are the HP and LP throttlings. The mixing bottle introduces few irreversibilities.

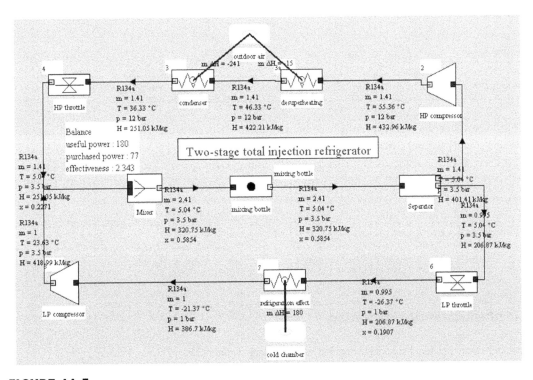

FIGURE 11.5

Synoptic view of a total injection two-stage compression cycle.

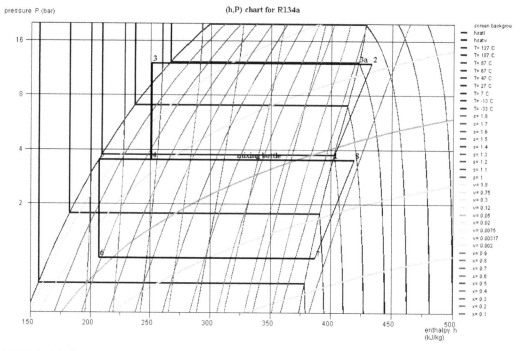

FIGURE 11.6

Total injection two-stage compression cycle in the (h, ln(P)) chart.

TABLE 11.2

EXERGY BALANCE OF THE TOTAL INJECTION CYCLE

Component	Resource	Product	Exergy Efficiency (%)	Irreversibilities	% Total (%)	$T_0 = 293.15$ K
HP throttle	4.60			4.60	8.8	
LP throttle	3.07			3.07	5.9	
Mixer	80.99	80.41	99.3	0.58	1.1	
HP compressor	44.47	36.81	82.8	7.67	14.6	Compr
LP compressor	32.29	26.00	80.5	6.29	12.0	Compr
Refrigeration effect	33.76	24.38	72.2	9.38	17.9	$T_k = -15,00°C$
Desuperheating	1.32			1.32	2.5	$T_k = 20,00°C$
Condenser	19.45			19.45	37.2	$T_k = 20,00°C$
Global	76.76	24.38	31.8	52.36	100.0	

This results in a significant improvement of the refrigeration cycle, whose COP reaches 2.34, which is 17% better than that of the single-stage cycle.

The refrigeration effect is 180 kJ/kg and the compression work 76.8 kJ/kg. The compression work is more important than in the basic cycle, because of the circulation, in the HP cycle, of a mass flow rate (1.4 kg/s) greater than in the LP cycle (1 kg/s).

Figures 11.6 and 11.7 show the plot of this cycle in the (h, ln(P)) and entropy charts. You will study it in a guided exploration (C-M3-V3).

BOX 11.1

Guided educational exploration

Total injection refrigeration installation (exploration C-M3-V3)

In this guided exploration, you will see how a total injection two-stage compression cycle can be modeled.
You will learn how to set a phase separator and a mixer.

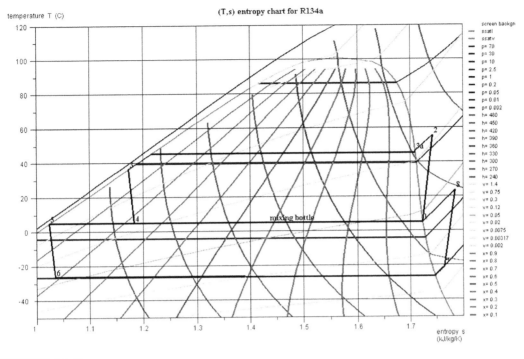

FIGURE 11.7

Total injection two-stage compression cycle in the entropy chart.

Other variants of this cycle can be used: partial injection cycle, medium pressure evaporator cycle, etc.

In the partial injection cycle, only the vapor phase after the high-pressure expansion valve is mixed with the vapor exiting the low-pressure compressor (Figure 11.8).

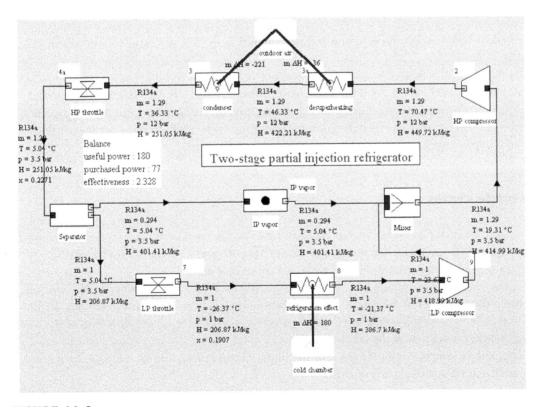

FIGURE 11.8

Synoptic view of a partial injection two-stage compression cycle.

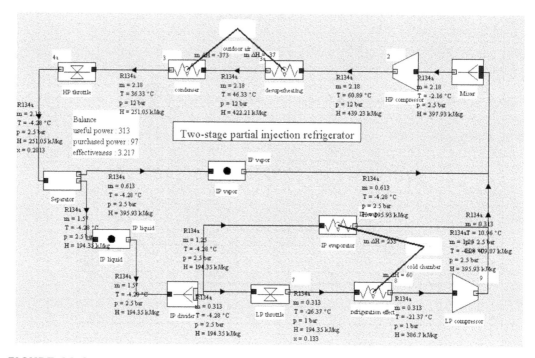

FIGURE 11.9

Synoptic view of a medium pressure evaporator cycle.

The performance of this cycle is slightly worse than that of the previous (COP equal to 2.33) because the compression work is slightly higher, the IP vapor being less cooled before recompression.

In the medium pressure evaporator cycle, some of the liquid leaving the IP separator is directed to a second evaporator allowing cooling at a temperature corresponding to the intermediate pressure IP. It is thus possible to cool two media at two different temperatures.

The average COP is increased since a fraction of the cooling effect takes place at the intermediate temperature. It is 2.49 when 1/5 of total liquid flow expansion valve output is directed to HP through the evaporator, but it would be 3.22 if the ratio was reversed, as in Figure 11.9.

By changing the mass flow-rate value passing through the IP evaporator relative to the total mass flow in the IP divider, the balance of the load between the two heat exchangers varies, allowing the machine to adapt to the demand. Note that, when the ratio of these two flow rates varies from 0 to 1, we switch from the two-stage cycle configuration between 1 and 12 bar to a single-stage cycle configuration between 3.5 and 12 bar.

Cascade cycles

In cascade refrigeration cycles, two or more refrigerants are used in separate cycles but coupled in pairs by exchangers which serve as an evaporator for one and a condenser for the other.

Figure 11.10 shows a configuration with two cycles in series.

In the example of Figure 11.11, we have superimposed two cycles using R134a, working between the pressures of 3.2 and 12 bar for the topping cycle and of 1 and 4 bar for the bottoming cycle. The evapo-condenser is configured so as to determine the flow rate of the topping cycle knowing that of the other one.

The performances are close to those of the total injection cycle, and the plots in the thermodynamic diagrams would be very close.

In this example, the objective of which was to illustrate how a simple cycle could be improved, we used the same refrigerant in both cycles, but we generally use two different refrigerants, each of which is best suited to the pressure and temperature levels of its cycle.

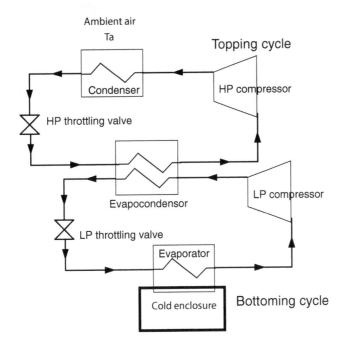

FIGURE 11.10

Sketch of a cascade cycle.

When the temperature difference $(T_1 - T_2)$ exceeds 80°C–100°C, it is not possible to cool with a single fluid. Indeed, there is no fluid that is well adapted to such different temperature levels, and it is preferable to use two-cycle cascade, the evaporator of one of them serving as the condenser of another. The two cooling circuits are then independent from the hydraulic but thermally coupled with the evaporative condenser.

The optimization of the global cycle is a complex problem, the temperature choice in the evaporative condenser being an important parameter. We give here an example corresponding to a milk sterilization cascade, a system used to simultaneously sterilize milk at 72°C and cool it at 5°C (Figure 11.12).

The system involves a cascade of two cycles, one at high temperature using R245ca, and one at low temperature using ammonia. The R245ca has been modeled with the external mixture library CTP Lib.

Milk comes from a tank at 5°C. It begins by being preheated at 66°C in a heat exchanger and is then pasteurized at 72°C in the condenser (90°C) of the topping cycle. At the outlet of the sterilizer, it is precooled at 11°C by exchange with the incoming milk and then cooled at 5°C in the evaporator of the bottoming cycle.

Figure 11.13 shows the synoptic view of such a facility, the topping cycle being two-stage total injection.

Special cycles

Cycles using blends

Blends have a behavior that deviates quite substantially from that of pure substances in the liquid-vapor equilibrium zone, because of the appearance of a fractional distillation (see Chapter 13 for further explanations).

Previously, we considered a simple R134a compression refrigeration cycle, and we presented the look of this cycle in different pure substance charts. Let us now consider a R407C cycle, with an evaporation pressure of 5 bar and a condensation pressure of 18 bar, used in air conditioning. The isentropic efficiency of the compressor is equal to 0.8.

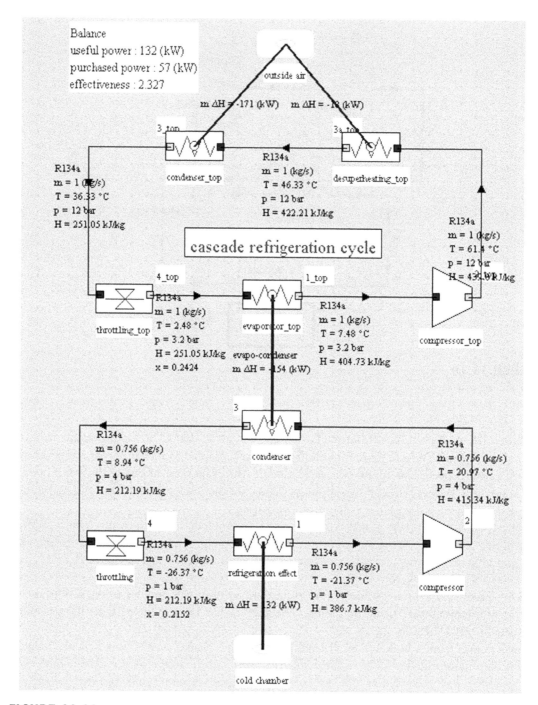

FIGURE 11.11

Synoptic view of a cascade cycle.

At point 1, before entering the compressor, the gas is superheated by 5°C above the dew point temperature equal to 2.47°C.

It therefore enters the compressor at 7.47°C and leaves it at 75.15°C. It is then desuperheated at 46.05°C, dew point at 18 bar, and then condensed at 41.29°C, bubble temperature at this pressure and subcooled by 5°C in liquid state before entering the expansion valve. Its quality at the throttling outlet is equal to 0.276.

Cycle modeling with Thermoptim leads to the results presented in Figure 11.14.

On the entropy diagram (Figure 11.15), to increase readability, we have not shown iso-quality curves. Point 1 slightly superheated compared to the saturated vapor is placed on isobar 5 bar. Irreversible compression results in an increase of entropy. Cooling with outside air has three stages: de-superheating (2–3a) in the vapor zone, condensation on line segment

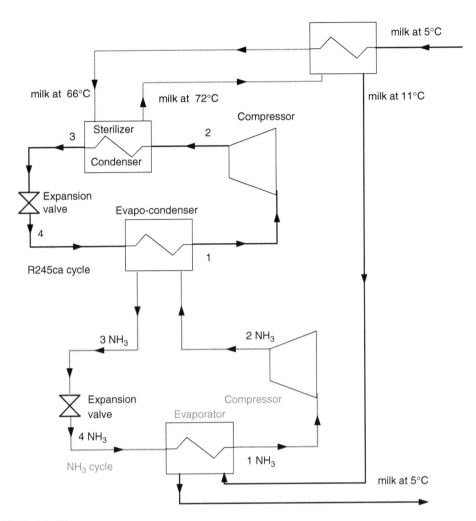

FIGURE 11.12

Sketch of the milk sterilization cascade.

(3a–3b) inclined because of temperature glide and a slight subcooling (3b–3) which almost coincides with the liquid saturation curve.

Isenthalpic throttling (3–4) leading to an increase of entropy, point 4 is located within the zone of vapor-liquid equilibrium (equal to 0.27).

On the (h, ln(P)) chart (Figure 11.16), to increase readability, we have not shown iso-quality curves. Irreversible compression results in an increase of entropy.

If we neglect the pressure drop in the condenser, the whole cooling is still here in the horizontal (2–3), the three parts representing desuperheating, condensation itself and subcooling appearing distinctly. The only difference with the pure substance chart is that in the area of liquid-vapor equilibrium, isotherms are inclined rather than horizontal segments.

Cycles using ejectors

An ejector[1] or injector (Figure 11.17) receives as input two fluids normally gaseous but which may also be liquid or two-phase (Chunnanond & Aphornratana, 2004):

* The high-pressure fluid called primary or motive;
* The low-pressure fluid, called secondary fluid or aspirated.

The primary fluid is accelerated in a converging-diverging nozzle, creating a pressure drop in the mixing chamber, which has the effect of drawing the secondary fluid. The two fluids are

---

[1] https://direns.mines-paristech.fr/Sites/Thopt/en/co/modele-ejecteur.html

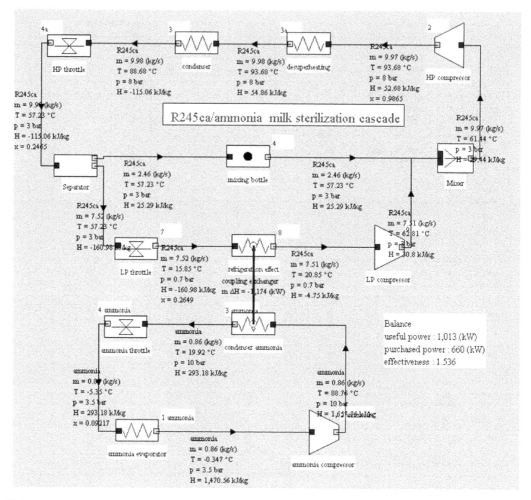

FIGURE 11.13

Synoptic view of the milk sterilization cascade.

then mixed, and a shock wave may take place in the following zone (throat in Figure 11.17). This results in an increase in pressure of the mixture and reduction of its velocity which becomes subsonic. The diffuser then converts the residual velocity in increased pressure.

The ejector thus achieves a compression of the secondary fluid at the expense of a decrease in enthalpy of the primary fluid.

An ejector model has been implemented in a Thermoptim external class for simulating cycles, including refrigeration, involving this component. It is an external mixer.

The value of introducing an ejector in a refrigeration cycle is mainly to reduce, or even eliminate, the compression work, relatively large as the fluid is compressed in the gaseous state.

BOX 11.2
Key issues

Ejectors

The following self-assessment activity will allow you to check your understanding of ejectors:

What is an ejector, gfe

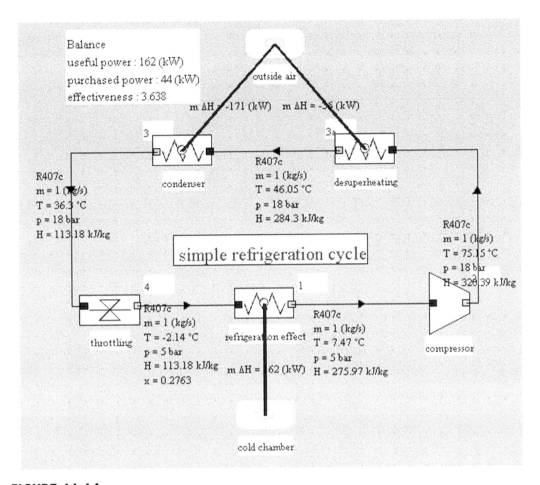

FIGURE 11.14

Synoptic view of a R407c refrigeration cycle.

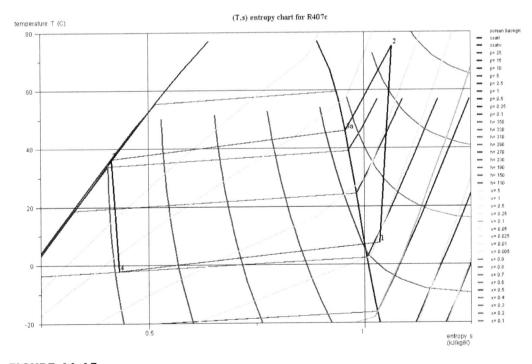

FIGURE 11.15

R407c refrigeration cycle in (T, s) chart.

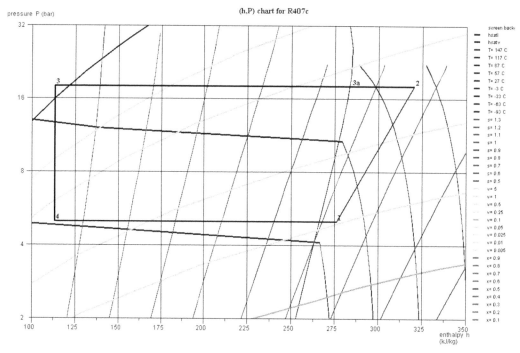

FIGURE 11.16

R407c refrigeration cycle in (h, log P) chart.

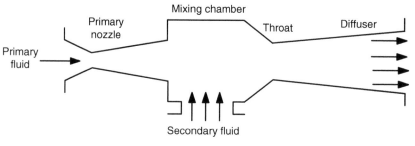

FIGURE 11.17

Cross section of an ejector.

Cycles without compressor

An ejector refrigeration cycle without compressor (Figure 11.18) is as follows:

- At the condenser outlet, part of the flow is directed to a pump that compresses the liquid, at the price of a very low work;
- The liquid under pressure is vaporized in a generator at a relatively high temperature (about 100°C) and possibly superheated, the temperature depending on fluid thermodynamic properties. Heat supplied to the generator is a purchased energy;
- This superheated vapor is then used in the ejector as motive fluid;
- The part of the liquid that was not taken up by the pump is expanded in the evaporator and then headed to the ejector as secondary fluid;
- The mixture leaving the ejector is condensed in the condenser, and the cycle is complete.

The advantage of this cycle is to replace compressor work by a much smaller work consumed by the pump and by heat supplied by a generator at medium or high temperature, which can be done using thermal effluents or solar collectors.

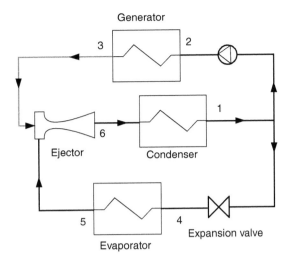

FIGURE 11.18

Ejector refrigeration cycle.

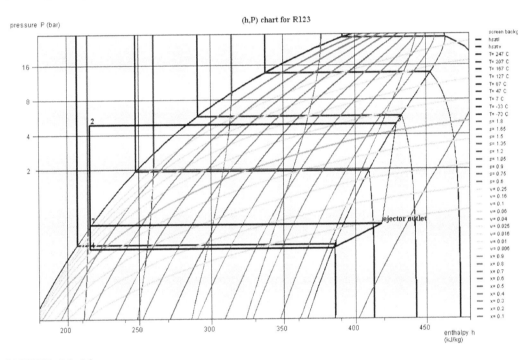

FIGURE 11.19

Ejector refrigeration cycle in the (h, ln(P)) chart.

In Figure 11.19, we have plotted in a (h, ln(P)) chart an example of R123 cycle suggested by Sun and Eames (1996). The plant diagram and the ejector screen are given in Figures 11.20 and 11.21. The ejector has four parameters:

First, the factor P_c/P_b of pressure drops at the entry of the secondary fluid into the ejector, which determines the minimum pressure in the ejector.

The second parameter is the friction factor to possibly take into account a pressure drop in the mixing zone.

The third parameter is the isentropic efficiency of the two nozzles (working fluid and entrained fluid).

The fourth parameter is the isentropic efficiency of the outlet diffuser.

These parameters only play second order on the calculations, which mainly depend on the enthalpies of the two fluids.

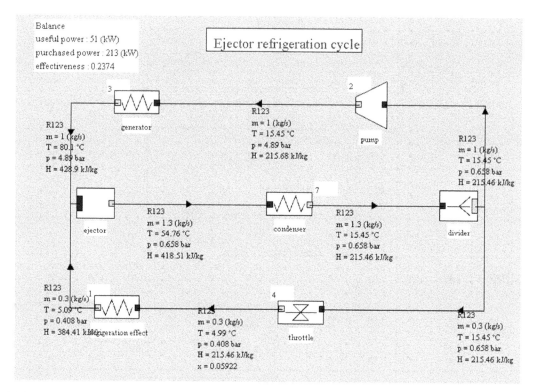

FIGURE 11.20

Synoptic view of an ejector cycle.

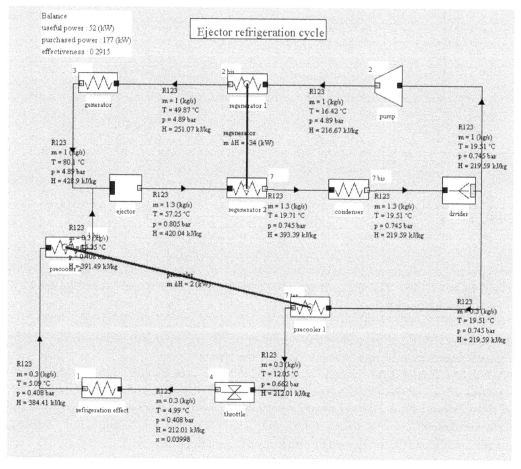

FIGURE 11.21

Synoptic view of the improved ejector cycle.

Thermoptim model results seem quite close to those (fairly synthetic) provided by the authors.

The efficiency remains very low as compared to that of an absorption cycle, even single-effect (see Chapter 13), but the system is very simple technologically.

However, one point to consider is that an ejector only works in good conditions if the ratio of internal sections is adapted to boundary conditions it faces. If the ratio of primary to secondary pressures departs from the rated values, there is an important risk that it is not well adapted, the performance of the cycle dropping then.

This cycle can be slightly improved by the incorporation of a precooler, which especially allows an increase in the cooling effect, and a regenerator, which reduces heat to provide to the generator (Figure 11.21).

Cycles with compressor

The ejector may also simply be used to reduce the throttling irreversibility of a conventional refrigeration cycle, creating a slight pressurization before compression (Figure 11.22), as proposed by Kornhauser (1990). In this case, the motive fluid is a liquid that expands and becomes diphasic, carrying over and compressing the aspirated fluid. The compression ratio achieved by the ejector is then much lower than in the previous case.

Consider as an example, the single stage R134a cycle studied above, working between 1 and 12 bar, whose COP was 2. The insertion of an ejector allows, as shown in Figure 11.23, to obtain an increase in COP of about 20%, similar to the benefit of a two-stage cycle.

This cycle is the subject of guided exploration (C-M3-V4).

BOX 11.3
Guided educational exploration

Ejector refrigeration installation (exploration C-M3-V4)

This exploration presents an ejector refrigeration cycle with compressor.
You will learn how to set an ejector, a phase separator and a mixer.

Note that the reduced flow of refrigerant in the evaporator is offset by the decrease in quality at the expansion valve output, leading to a cooling capacity almost constant while the compression work drops.

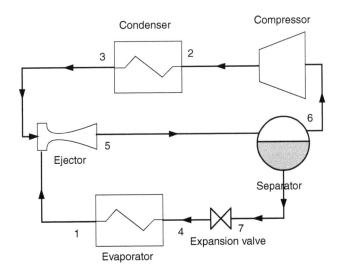

FIGURE 11.22

Cycle with compressor.

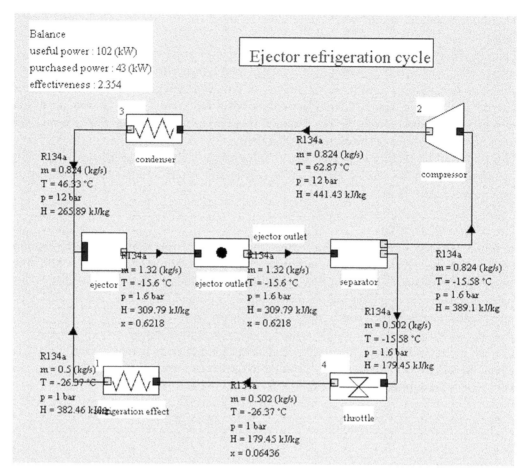

FIGURE 11.23

Synoptic view of a cycle with ejector and compressor.

A variant of this cycle is obtained using as primary fluid, not the liquid under pressure, but a supercritical fluid, especially CO_2 (Li & Groll, 2005), Figure 11.24. The principle remains the same, but the pressure obtained is higher than in the case of liquid. In this case also, adapting the operation of the ejector to a variation of its use conditions may be problematic.

This cycle with compressor and ejector however presents a major constraint: the flow distribution between the branches at high and low pressure must always be consistent with the

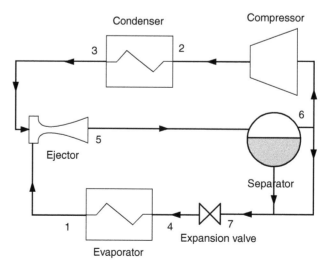

FIGURE 11.24

Cycle with ejector and compressor.

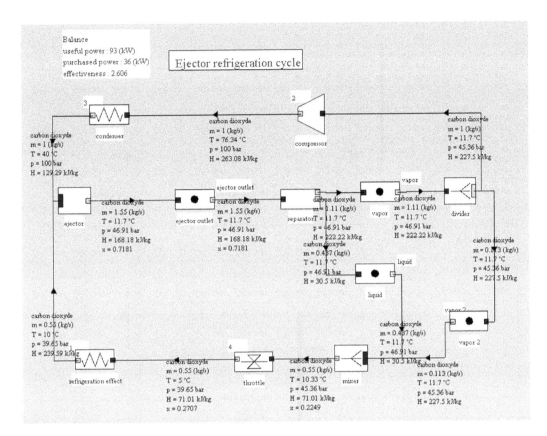

FIGURE 11.25

Synoptic view of the cycle with supercritical CO_2 ejector.

quality of the refrigerant at the outlet of the ejector, which requires the implementation of special controls. For this reason, the diagram shows a divider and a mixer at the separator output in Figure 11.25.

We consider that at point 3 flows 1 kg/s of supercritical CO_2 at 40°C and 100 bar and that at point 1, at the evaporator outlet, CO_2 vapor is at 40 bar and 10°C, which represents super-heating of about 5°C.

Supercritical CO_2 is expanded in the ejector, which allows it to entrain the vapor, and rec-ompress it slightly.

The two-phase mixture at the ejector outlet is separated, the complement to 1 kg/s vapor flow being remixed with the liquid stream before entering the expansion valve and then the evaporator. The main flow of vapor is compressed at 100 bar and then condensed at 40°C.

The Thermoptim synoptic view of the facility is given in Figure 11.25, and, for compari-son, the diagram of a cycle without ejector is given in Figure 11.26. The gain on the COP is clear.

Reverse brayton cycles

As its name implies, a reverse Brayton cycle achieves a cooling effect by reversing the gas turbine Brayton cycle studied in Chapter 2: a gas is compressed, cooled and then expanded (Figure 11.27).

As we have already pointed out, in all the refrigeration cycles, we find the three functions of this cycle, succeeding each other in this order: we compress, we cool and we expand.

The end of expansion temperature being low, this gas can be used to cool an enclosure, either by direct contact, especially if it is air, or through a heat exchanger.

Consider for example an open reverse Brayton cycle used to cool the passenger compart-ment of an automobile to reduce CO_2 emissions.

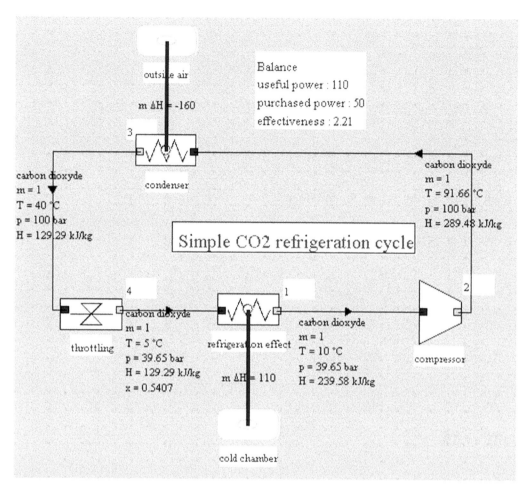

Balance
useful power : 110
purchased power : 50
effectiveness : 2.21

Simple CO2 refrigeration cycle

FIGURE 11.26

Classical cycle without ejector.

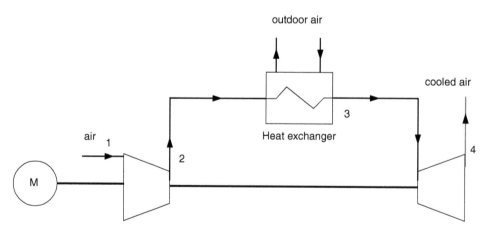

FIGURE 11.27

Sketch of an open reverse Brayton cycle.

The working fluid is air which undergoes the following processes:

- 1–2 adiabatic compression at 2.5 bar in a compressor of polytropic efficiency 0.875;
- 2–3 isobaric cooling of the compressed air with outside air in a heat exchanger;
- 3–4 adiabatic expansion in a turbine of polytropic efficiency 0.875.

The Thermoptim synoptic view of such a cycle is given in Figure 11.28.

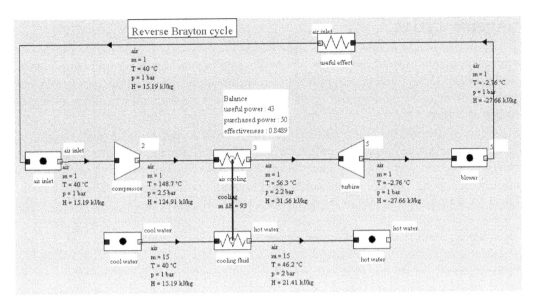

FIGURE 11.28

Synoptic view of an open reverse Brayton cycle.

The interior of the vehicle is represented by process "useful effect". Air at 40°C is sucked by the compressor and exits at about 150°C. External cooling via a water heat exchanger allows the air flow to be cooled at 56°C. It is then expanded in the turbine and exits at –3°C and at a pressure of 2.2 bar, the pressure drops being significant. This cold air is then blown into the cabin.

As can be seen, the cycle efficiency is not very high (0.85), but the system is relatively simple, and has moreover the advantage of not releasing any greenhouse gases in the event of accidental breakage of pipes.

Figure 11.29 shows the plot of the reverse Brayton cycle in the (h, ln(P)) chart.

Air reverse Brayton cycle was until recently widely used in aircraft for in flight cabin air conditioning.

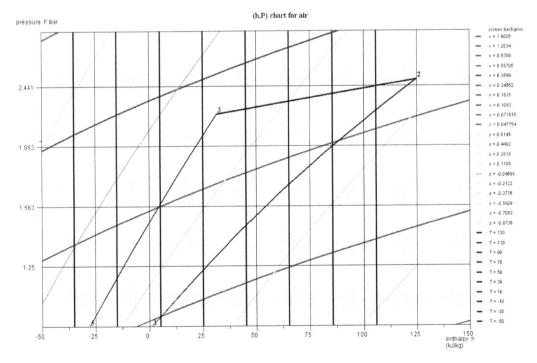

FIGURE 11.29

Open reverse Brayton cycle in the (h, ln(P)) chart.

Cryogenic cycles

The term cryogenics is used to describe methods of refrigeration at very low temperatures (typically below 125 K) and distinguish them from ordinary refrigeration cycles. Many of these methods relate to the liquefaction of gases known as permanent, like air, natural gas, hydrogen or helium.

Cryogenics is the field of engineering that focuses on systems operating at very low temperature, which poses special problems, particularly in terms of fluids and materials.

Refrigeration and cryogenic liquefaction cycles involve combinations of paraisothermal compressions, cooling, thermal regeneration and isenthalpic or adiabatic expansion of fluids.

There are four major families of cryogenic thermodynamic processes:

* Isenthalpic expansion Joule-Thomson processes;
* Isentropic expansion reverse Brayton cycles;
* Mixed processes involving isenthalpic and isentropic expansion (Claude cycle);
* Conventional or integrated cascades.

It is possible to model with Thermoptim some cryogenic cycles, but the exercise is often difficult because the fluid properties are rarely defined for the entire temperature range considered. You will find in this section some examples of liquefaction cycles for methane and nitrogen, and a reverse Brayton helium refrigeration cycle at very low temperatures. These examples are drawn from the documents cited in the bibliography, particularly from the 3,600 J leaflet from Techniques de l'Ingénieur, by P. PETIT.

Joule-thomson isenthalpic expansion process

We illustrate this process by some examples of cycles for liquefying natural gas and form Liquefied Natural Gas (LNG), considered here as pure methane.

Basic cycle

This first example is adapted from the exercise proposed by (Sandler, 1999, p. 129).

To liquefy natural gas, methane taken at 1 bar and 280 K is compressed to 100 bar and then cooled to 210 K (it is assumed in this example that a refrigeration cycle is available for that).

Isentropic compression is assumed, but the very high compression ratio requires the use of several compressors (3 in this example) with intermediate cooling to 280 K. Intermediate pressures are equal to 5 and 25 bar.

The gas cooled at 210 K is isenthalpically expanded from 100 to 1 bar, and gas and liquid phases separated. As shown in the diagram in Figure 11.30, the methane enters in the upper left, and liquid and gaseous fractions exit in the bottom right.

With the settings chosen, the synoptic view is given in Figure 11.31.

The compression work required per kilogram of methane sucked is 798.5 kJ, and 0.179 kg of liquid methane is produced, which corresponds to a work of 4.46 MJ/kg of liquefied methane.

Linde cycle

The Linde cycle (Figure 11.32) improves the previous on two points:

* Gaseous methane is recycled after isenthalpic expansion;
* We introduce a heat exchanger between the gaseous methane and methane exiting the cooler in order to cool the compressed gas not at 210 K but at 191 K.

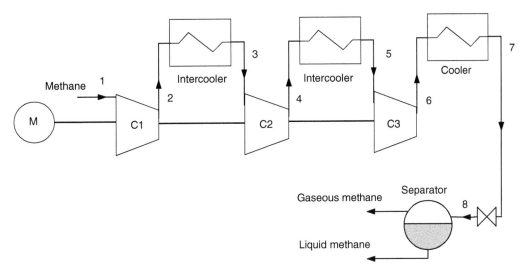

FIGURE 11.30

Simple liquefaction cycle.

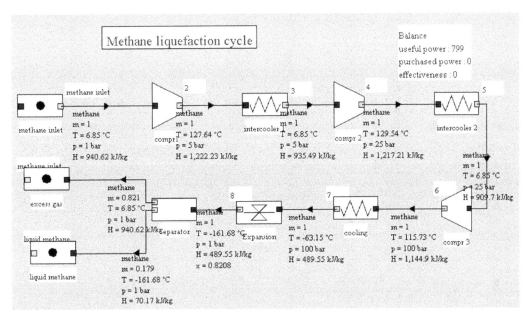

FIGURE 11.31

Synoptic view of the facility.

For these new conditions, the compression work per kilogram of liquefied methane becomes equal to 1.91 MJ, i.e., just 43% of the previous one (Figure 11.33).

The calculation of this cycle requires some caution, given the sensitivity of the balance of the heat exchanger to flow variations set by the separator.

The solution we found was to set the flow rate in the first intercooler and then to recalculate the entire project by repeatedly recalculating the regenerator. A stable solution could then be found, but it led to an inaccurate flow recirculation at the beginning. The correct rate value setting was obtained by successive iterations.

The performance gain comes mainly from the decrease of the expansion valve inlet temperature, which reduces the exit quality and thus increases the flow of the liquid phase. Moreover, the decrease in temperature at the inlet of the first compressor reduces the compressor work, but this effect is less important than the first.

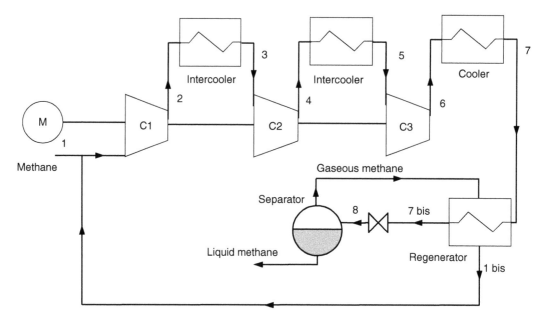

FIGURE 11.32

Linde cycle.

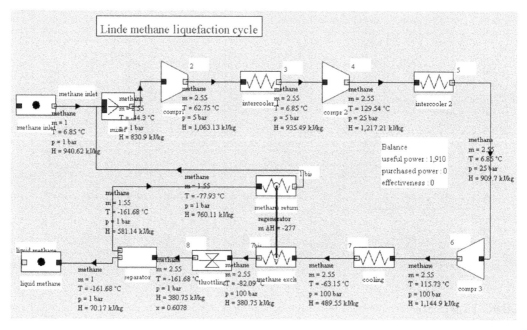

FIGURE 11.33

Synoptic view of the Linde cycle.

The plot of both cycles in (h, ln(P)) and entropy diagrams (Figures 11.34 and 11.35) illustrates the benefits of the second: the leftward shift of point 7 in 7bis in the Linde cycle more than doubles liquid quality[2] at the valve outlet (point 8).

Linde cycles for nitrogen liquefaction

As air is a mixture of gases which cannot be modeled in Thermoptim as condensable, we will consider here the liquefaction of nitrogen.

---

[2] The liquid quality is the complement to 1 of the vapor quality x that we have used so far.

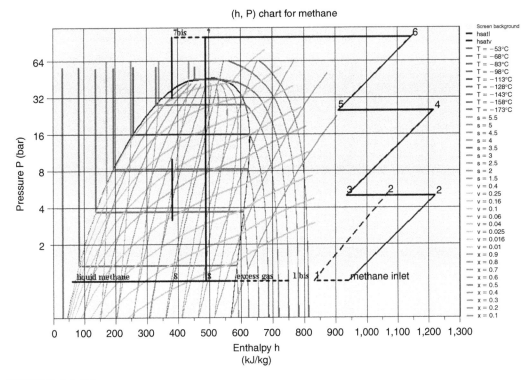

FIGURE 11.34

Linde cycle in dashed line and basic cycle in continuous line in the (h, ln(P)) chart.

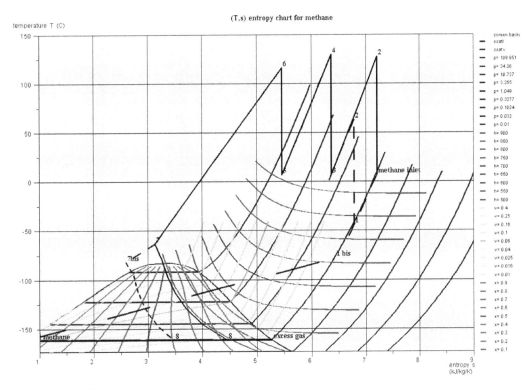

FIGURE 11.35

Linde cycle in dashed line and basic cycle in continuous line in the (T, s) chart.

Gaseous nitrogen at 1 bar and 280 K is compressed at 200 bar (cooled compression at 50°C) and then cooled at 280 K.

Nitrogen is then cooled in a heat exchanger with the non-liquefied part, allowing it to reach a temperature of about −110°C. It is then isenthalpically expanded at 1 bar, thereby about 8% is liquefied.

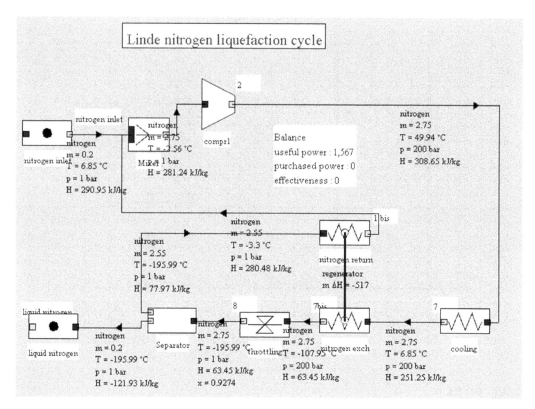

FIGURE 11.36

Synoptic view of nitrogen Linde cycle.

The liquid fraction is extracted, and the remainder in the form of saturated vapor is recycled in the regenerator and mixed with atmospheric nitrogen at 280 K, which closes the cycle.

The cycle can be modeled without difficulty in Thermoptim (Figure 11.36). Its performance is small, about 5.7 MJ/kg.

It can be improved with a double expansion cycle, but the setting is much more difficult to obtain because instabilities can occur in flow rates, because of the role played by the phase separators (Figure 11.37).

Reverse brayton cycle

The reverse Brayton cycle has been presented above, which we recommend you to refer to for details.

This cycle can be used for cryogenic applications. For example, Figure 11.38 shows the block diagram obtained for the model of a plant for producing cold below 20 K to refrigerate a liquid hydrogen bubble chamber.

In this cycle, helium is compressed at 20 bar and then cooled at 30°C before being divided into two streams which are expanded in parallel, the main stream following a conventional reverse Brayton cycle, while the secondary contributes to cooling the total flow.

Note that in this example, point 12bis temperature must be forced by hand, equal to that of point 12, because of the need to change the substance, the helium vapor being defined in Thermoptim only for T < 200 K.

Mixed processes: claude cycle

The Linde cycle uses isenthalpic expansion which has two drawbacks: firstly the expansion work is lost, and secondly cooling cannot be achieved if the fluid thermodynamic state is such that the Joule Thomson expansion leads to a temperature reduction.

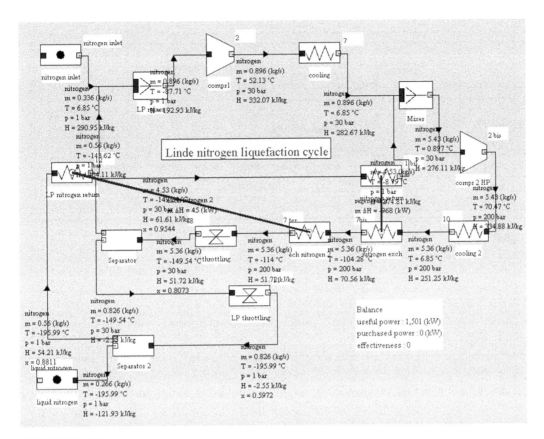

FIGURE 11.37

Synoptic view of two-stage nitrogen Linde cycle.

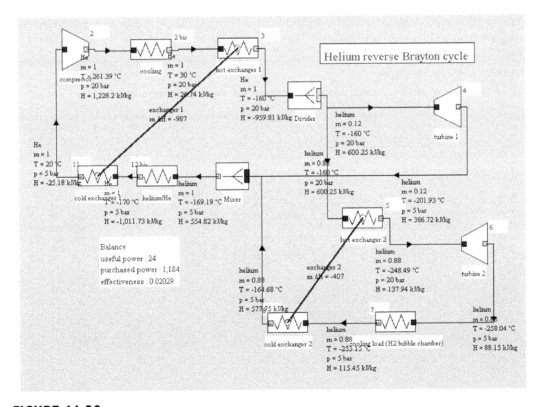

FIGURE 11.38

Synoptic view of the helium reverse Brayton cycle.

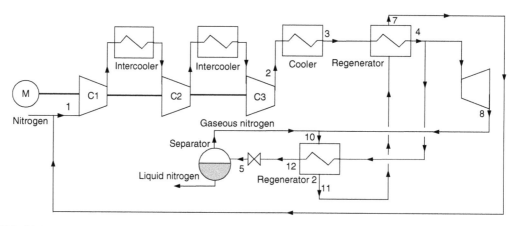

FIGURE 11.39

Sketch of the Claude cycle.

Claude has proposed a cycle that involves a turbine and an expansion valve and has the peculiarity that the plant operates with a single fluid compressed at a single pressure level, as shown in Figure 11.39.

The advantage of this cycle is that the compression ratio can be significantly lower than in the case of the Linde cycle. One difficulty is that the expansion machine cannot operate with good efficiency if the fluid remains in the vapor zone or keeps a high quality. The originality of the Claude cycle is to combine isentropic expansion in the turbine and isenthalpic expansion only in expansion leading to the gas liquefaction.

The beginning of the cycle is the same as that of Linde: compression of gas to liquefy and then cooling to about room temperature (1–3). The gas then passes through a regenerator which allows it to cool at about –105°C (3–4). The flow is then divided, about 15% being expanded in a turbine (4–8). The main flow passes through a second regenerator of which it is released at very low temperature (4–12). It undergoes isenthalpic expansion (12–5), and the liquid phase is extracted. The vapor is mixed with the flow exiting the turbine and serves as a coolant in the second regenerator (10–11) and then in the first (11–7) before being recycled by mixing with the gas entering the cycle.

This cycle can be modeled with Thermoptim. In the example of Figure 11.40, the compression power is about 3.8 MJ/kg.

The Claude cycle has been used in many facilities for the liquefaction of air.

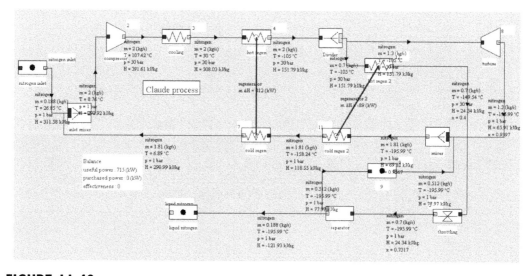

FIGURE 11.40

Synoptic view of the Claude cycle.

Heat pumps

In refrigeration cycles that we studied so far, the practical effect is the extraction of heat by the evaporator. We can also design a machine whose useful effect is heating by waste heat available in the condenser. We call such a machine a heat pump, whose cycle is very similar to that used in refrigeration. It differs only by temperature levels and therefore the working fluid.

For a heat pump, we define a heating COefficient of Performance (COP_h):

$$COP_h = \frac{T_1}{T_1 - T_2} = \frac{T_1 - T_2}{T_1 - T_2} + \frac{T_2}{T_1 - T_2} = 1 + COP_{refrig}$$

Indeed, the first law states that all the energies involved are discharged at the condenser, that is to say, the heat energy taken from the cold source plus the mechanical energy of compression.

Heat pumps can raise the level of temperature of a cold source with excellent efficiency, as long as the temperature difference is not too important.

This heating method is very attractive if one has a source of free heat at a sufficient temperature. For space heating, it has one drawback: the COP_h decreases gradually as the heating needs are increasing, as the temperature difference between the cold source and heating system increases simultaneously.

Heat pump applications in industry are as follows, T_e and T_c being, respectively, the temperatures of evaporation and condensation:

* $T_e < 20°C$, $T_c < 80°C$: standard conventional heat pumps, marketed and available in a catalog;
* $20°C < T_e < 80°C$, $T_c < 130°C$: specific heat pumps, though derived from refrigeration equipment but adapted to higher operating temperatures;
* for higher temperatures, such material is no longer appropriate. Mechanical vapor compression studied in Chapter 17 may be a solution, if the cold source is in the form of vapor.

Compressors used in practice are of three types: reciprocating ($P < 200\,kW$), screw ($100\,kW < P < 1\,MW$) or centrifugal ($P > 800\,kW$).

Currently, the main markets for heat pumps in industry concern drying of temperature-sensitive products or slow migration of moisture and energy recovery in plants where there are simultaneous and comparable requirements of heat and cooling.

Since the cycle of a heat pump is similar to that of an air conditioner, it is possible to design reversible machines that can provide these two functions. Thus, in the residential and tertiary sector, the main market for heat pumps is the case where they can be used for winter heating and summer air conditioning. In Japan, for example, where climatic conditions are particularly suited, millions of heat pumps are used. They are equipped with a reversing valve which allows one to shift from summer to winter use and vice-versa.

The whole problem of using a heat pump is to have a cold source for free at a suitable temperature and with sufficient heat available. For space heating, various solutions are possible:

* A river, lake or sea;
* The ground around a house, in which case the evaporator is buried;
* A pair of wells to power a geothermal heating system, such as for heating the Maison de la Radio in Paris;
* Gas or liquid waste in residential buildings or factories;
* Outdoor air, but evaporator frosting problems exist when air temperature is between 0°C and 5°C.

Defrosting can take various forms:

- Hot air blowing;
- Heating resistors;
- Circulation of hot gas by-passing gas leaving the compressor;
- Reversing the cycle by inverting the roles of the condenser and evaporator.

Basic cycle

The Thermoptim synoptic view of a heat pump is given in Figure 11.41, and its cycle is shown in Figure 11.42 on the entropy chart.

Exergy balance

To build the exergy balance of a heat pump, we must begin by giving a reference temperature and pressure. We take here $T_0 = 273.15$ K (0°C) and $P_0 = 1$ bar, corresponding to winter conditions in France. Table 11.3 can then be built, the losses being given as percentage of total irreversibilities. They are almost evenly divided between the compressor (25%), the condenser (37.5%), the expansion valve (20%) and the evaporator (13.5%).

Again, the big difference with the energy balance, which showed a COP_h of 4.3 is that the cycle efficiency is actually very low: two-thirds of the exergy provided by the compressor to the fluid are dissipated in losses.

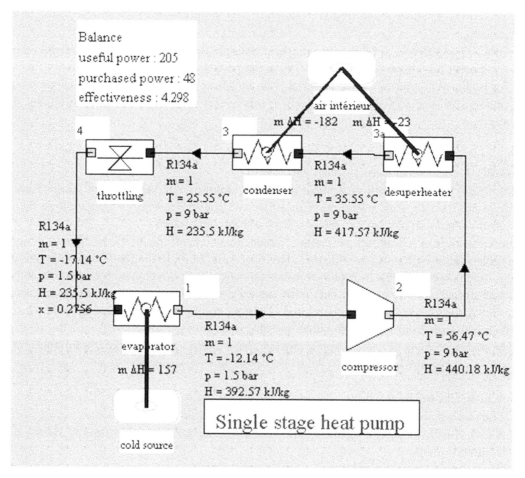

FIGURE 11.41

Synoptic view of a heat pump.

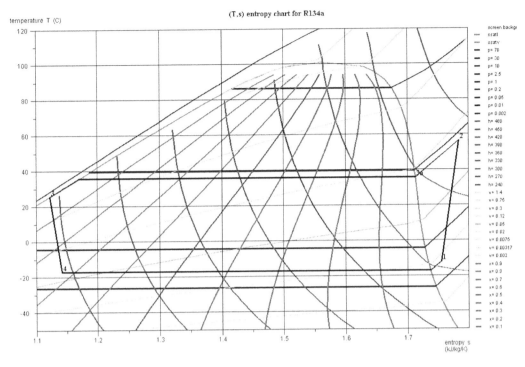

FIGURE 11.42

Heat pump cycle in the entropy chart.

TABLE 11.3

HEAT PUMP EXERGY BALANCE

Component	Resource	Product	Exergy Efficiency (%)	Irreversibilities	% Total (%)	$T_0 = 273.15$ K
Throttling	5.38			5.38	17.3	
Compressor	47.62	39.65	83.3	7.97	25.7	Compr
Evaporator	10.48	5.97	56.9	4.52	14.6	$T_k = -10.00°C$
Desuperheater	3.14	1.18	37.4	1.97	6.3	$T_k = 15.00°C$
Condenser	20.65	9.48	45.9	11.17	36.0	$T_k = 15.00°C$
Global	47.62	16.62	34.9	31.00	100.0	

Now consider what would be the exergy balance of a conventional heating boiler providing the same power of 205 kW. Consider for this a condensing boiler which would cool the smoke from 1,120°C to 77°C, to heat water for the heating circuit from 30°C to 50°C.

This is a kind of boiler whose exergy balance is shown in Table 11.4. Exergy efficiency is equal to 16.4%, about half that of the heat pump.

This balance illustrates very well the low exergy performance of the boilers: the irreversibilities are more or less evenly distributed (49%) between the combustion itself and the exchanger with the heating circuit.

If the heating was done by Joule effect in an electric boiler, for example, the exergy efficiency of the cycle would be even lower, equal to the ratio of the exergy gained by the heating circuit (26 kW) to the exergy consumption (205 kW), i.e., 12.7%.

These figures illustrate the value of the heat pump, much more efficient than electric heating or combustion, since, for space cooling, it is unnecessary to provide heat at high temperature.

TABLE 11.4

CLASSICAL BOILER EXERGY BALANCE

Component	Resource	Product	Exergy Efficiency (%)	Irreversibilities	% Total (%)	$T_0 = 273.15$ K
Fuel	242.97					
Combustion chamber		137.16	56.1	107.21	49.4	
Heating network	42.57	42.57				
Network return	16.55					
Heater	131.42			106.62	49.1	Heater
Gas exhaust				3.11	1.4	Loss
Global	259.52	42.58	16.4	216.94	100.0	

Bibliography

Ashrae, Fundamentals Handbook (SI), Thermophysical properties of refrigerants, 2001.

K. Chunnanond, S. Aphornratana, Ejectors: Applications in refrigeration technology. *Renewable and Sustainable Energy Reviews*, 8, 129–155, 2004.

D. Clodic, Y. S. Chang, Nouveaux fluides frigorigènes, caractéristiques et performances, MAD l'outil froid, n° 16, novembre 1999.

J. G. Conan, Réfrigération industrielle, Eyrolles, Paris, 1988.

T. Destoop, Compresseurs volumétriques, Techniques de l'Ingénieur, Traité Mécanique et chaleur, B 4 220.

M. Duminil, Machines thermofrigorifiques, Techniques de l'Ingénieur, Génie énergétique, BE 9. 730-9 736.

A. Gac, Conduite automatique des installations frigorifi ques, Techniques de l'Ingénieur, Traité Mécanique et chaleur, B 9 765.

A. A. Kornhauser, The use of an ejector as a refrigerant expander. *Proceedings of the 1990 USNC/IIR—Purdue Refrigeration Conference*, Purdue University, West Lafayette, IN, 1990, pp. 10–19.

D. Li, A. Groll, Transcritical CO$_2$ refrigeration cycle with ejector-expansion device. *International Journal of Refrigeration*, 28, 766–773, 2005.

S. Sandler, *Chemical and engineering thermodynamics*, 3rd Edition, John Wiley & Sons, Inc., New York, 1999.

H. Struchtrup, *Thermodynamics and energy conversion*, Springer, Berlin, Heidelberg, 2014, ISBN 978-3-662-43714-8.

D. W. Sun, I. W. Eames, Performance characteristics of HCFC-123 ejector refrigeration cycle. *International Journal of Energy Research*, 20, 871–885, 1996.

G. Vrinat, Production du froid, Technologie des machines industrielles, Techniques de l'Ingénieur, Traité Mécanique et chaleur, B 2 365.

S. K. Wang, Z. Lavan, P. Norton, *Air conditioning and refrigeration engineering*, CRC Press, Boca Raton, FL, 2000, ISBN 0-8493-0057-6.

Thermodynamics of moist mixtures and air conditioning

Introduction

Rising living standards in industrialized countries has been accompanied by increasingly large demands in terms of thermal comfort conditions in residential and commercial buildings. Air conditioning has progressively developed in recent decades, leading to continued progress in the various disciplines that make up what is known as HVAC (heating, ventilation and air conditioning).

Air conditioning is the best known of the thermodynamic applications of moist mixtures, but it is not the only one. When the water vapor contained in a gas is likely to vary, whether due to the condensation of the water it contains or the vaporization of water with which it is in contact, its moist properties vary. We have already mentioned the case of water contained in the combustion gases, which has a direct impact on the heating value of the combustion reaction. This is also what happens in air cooling coils where takes place condensation or even icing. Certain innovative cycles also use the variation of the humidity of the gases, as we will see in Part 4.

In this chapter, we begin by introducing thermodynamics of moist mixture and in particular the psychrometric diagram in which the air conditioning cycles can be represented, and then we limit ourselves to air conditioning processes (heating, cooling, adiabatic and steam or water humidification, desiccation, determination of supply conditions, condition line) and applications. We also explain how to calculate the exergy of moist gases, including the gases produced during combustion.

Two examples of building air conditioning will be dealt with, one for summer conditions and the other for winter.

Moist mixture properties

A number of commonly used gases are mixtures whose composition may vary due to condensation or vaporization of one of their components. In practice, the component that changes state is generally water, which justifies that a special chapter in this book is dedicated to mixtures of gases and water vapor, hereinafter referred to as moist mixtures.

Dry air in particular is hardly ever found: it almost always includes water, in the form of gas, liquid or ice crystals. Controlling the air humidity determines the comfort conditions and the preservation of food and many products. It is obtained by air conditioning techniques that will be presented at the end of this chapter.

DOI: 10.1201/9781003175629-12

In this section, we limit ourselves to the calculation principles of thermodynamic properties of moist mixtures. While it is certain that the most numerous practical applications regard air, our presentation will cover all gases for which our assumptions are valid, that is to say the vast majority of cases. For example, the humidity properties of combustion products, which almost always contain water, may be determined by the methods presented here, all implemented in Thermoptim.

Definitions and conventions

When looking at a moist mix, we are dealing with a mixture of a gas that does not condense, which we call the dry gas, and water that could condense. Note that we will not discuss mist here; we can assume that the volume occupied by any condensed water is still very small compared to that occupied by the gas phase. We will therefore neglect mist. Under these conditions, for a given volume, the total mass of the gas phase can vary, while the mass of noncondensable constituents remains constant. Therefore, it is customary to refer to the invariant mass which is the dry gas, all the thermodynamic properties of the mixture comprising the gas phase.

Everything happens as if somehow the moist mix was a mixture of two substances: the dry gas, whose composition is fixed, and water that may be present in one or more phases. As discussed later, it is essential, in order to carry out calculations, to clearly specify which substance the thermodynamic quantities used relate to, and what are the reference units and origin. The notations will be chosen accordingly.

We call **specific humidity** w the ratio of the mass of water contained in a given volume of moist mixture to the mass of dry gas contained in this volume.

Water in a moist mixture can be characterized in various ways: through its specific humidity, which we just introduced, but more typically by its mole or mass fraction, or by its partial pressure which is of particular interest here. As long as the partial pressure of water remains below its saturation pressure at the gas temperature, the water is in the form of vapor. Otherwise, it exists at least partly in the condensed liquid or solid state. There is therefore an upper limit to the amount of water vapor that may be contained in a moist mixture. It depends on the temperature and pressure. When, at a given pressure, the temperature drops below the saturated vapor temperature of water, water vapor begins to condense as a mist or on cold walls that define the system if they exist.

Given its practical importance, the saturated state is the reference, and we call **relative humidity** ε the ratio of the partial pressure of water vapor to its saturation vapor pressure at the temperature of the mixture. This ratio is equal to 1 (or 100%) when the vapor begins to condense. Otherwise, it is less than 1.

We call, respectively, **specific enthalpy** q' and **specific volume** v_{spec} the enthalpy and the volume of moist mixture referred to 1 kg of dry gas.

Principles of calculation

When its pressure is much lower than its critical pressure, a condensable fluid remains comparable to an ideal gas at the immediate vicinity of the liquid state. When this fluid is a mixture component, this rule applies taking into account the partial pressure.

As long as the liquid state does not appear, the mixture behaves like an ideal gas. When the mixture temperature drops below the condensation temperature corresponding to the partial pressure of the condensable component in the dry gas, liquid begins to appear. In most cases, although this is not strictly true, it is reasonable to consider that this liquid phase is pure, made up exclusively of the condensable component.

At that time, provided that equilibrium is established, experience shows that the situation is as follows:

- The vapor-liquid equilibrium relationship is met by the condensable component, as if the other constituents did not exist. Its saturated vapor (index sv) partial pressure, given by the law of saturation pressure, is therefore $P_{sv} = P_{sat}(T)$;
- Dalton's law applies to the calculation of thermodynamic functions of the gas phase;
- The law of phase mixture applies to the calculation of thermodynamic functions of the liquid and gas phases (assuming that the interfacial tensions are negligible).

To calculate the thermodynamic properties of the mixture in the presence of a liquid phase, the procedure is thus as follows:

- Knowing the temperature of the medium, one can determine the partial pressure of the condensable component. We deduce the sum of partial pressures $(P-P_{sv})$ of the other constituents. As we know their relative molar fractions, we can completely determine the composition of the gas phase, and the number of moles of the condensable component in the gaseous state. By difference, we know the number of moles liquefied;
- The Dalton and phase mixture laws are then used to calculate all the properties of the mixture. It is obvious that in this case, the composition of the mixture changes as a function of the quality of the condensable component, and this must be taken into account in the calculation of thermodynamic properties.

In Thermoptim, the calculation of moist mixtures is performed using this method. It is thus possible to study the behavior of a vapor and gas mixture of any composition.

Table 12.1 shows the equivalence between notations used in this book and Northern America.

Key relationships

We will use indices dg to describe the dry gas and mm for the moist mixture.

Case where the moist mixture composition is known

By definition of the partial pressure of water:

$$P_{vap} = x_{H_2O} \cdot P \tag{12.1}$$

The molar mass of the dry gas is

$$M_{dg} = \frac{1}{1-x_{H_2O}} \Sigma M_{nc} x_{nc} \tag{12.2}$$

nc representing the various noncondensable constituents.

TABLE 12.1
MOIST AIR NOMENCLATURE

	This Book	Northern America
Humidity ratio or specific humidity	w	W
Relative humidity	ε	Φ or RH
Dry-bulb temperature	t	t
Dew-point temperature	t_r	t_d
Wet-bulb temperature	t'	t*
Specific enthalpy	q'	h
Specific volume	v'	v_{spec}

The specific humidity is then by definition

$$w = \frac{y_{H_2O}}{y_{dg}} = \frac{M_{H_2O} x_{H_2O}}{M_{dg} x_{dg}} \tag{12.3}$$

If P_{sv} is the saturation vapor pressure of water:

$$\text{If } P_{vap} \leq P_{sv} \quad \epsilon = \frac{P_{vap}}{P_{sv}}$$

$$w = \frac{M_{H_2O}}{M_{dg}} \frac{P_{vap}}{P - P_{vap}} \tag{12.4}$$

If $P_{vap} > P_{sv} \quad \epsilon = 1$

$$w = w_{sat} = \frac{M_{H_2O}}{M_{dg}} \frac{P_{sv}}{P - P_{sv}} \tag{12.5}$$

$$w - w_{sat} = \text{condensate} = \frac{M_{H_2O}}{M_{dg}} \left[\frac{P_{vap}}{P - P_{vap}} - \frac{P_{sv}}{P - P_{sv}} \right] \tag{12.6}$$

The values of specific volume v_{spec} and specific enthalpy q' can then be determined, the latter being calculated with an enthalpy zero at 0°C.

Generally (and this is the case in Thermoptim) the reference state for calculating the gas enthalpies is taken at the standard value of 298 K. Now, HVAC engineers by custom choose 0°C as the reference state for dry air, and 0°C for saturated liquid water. The result is a mismatch between the enthalpies of moist gas as they are **usually calculated and those represented on the usual psychrometric charts**. This difference varies depending on the specific humidity.

In practice, as the enthalpies can be referred to three different references, you need to know how to distinguish them in order to avoid making mistakes. In what follows, we will use as appropriate:

- The moist mixture enthalpy h_{mm} ($h_{mm} = 0$ at $T = 298$ K)
- The dry gas enthalpy h_{dg} ($h_{dg} = 0$ at $T = 298$ K)
- The moist mixture specific enthalpy q' ($q' = 0$ at $T = 0$°C, liquid water).

In general, unless otherwise indicated, the first is used by default, whereas in the calculations specific to moist mixtures, it is usually the third form that is used.

In what follows, the index water indicates that the calculation is performed with the equations of water as a real fluid, and the index H_2O that it is conducted with the equations of water treated as an ideal gas. h_{vwater} is the enthalpy of water in the vapor state, and L_{0water} represents the enthalpy of vaporization of water at 0°C.

$$q'(t,w) = h_{dg}(t) - h_{dg}(0°C) + w h_{vwater}(t, P_{vap}) \tag{12.7}$$

or by making the approximation that water vapor behaves like an ideal gas:

$$q'(t,w) = h_{dg}(t) - h_{dg}(0°C) + w \left[h_{H_2O}(t) - h_{H_2O}(0°C) \right] + w L_{0water} \tag{12.8}$$

By introducing h_{mm}, enthalpy of the (ideal) moist mixture:

$$h_{mm}(t) = \frac{h_{dg}(t) + w h_{H_2O}(t)}{1 + w} \quad \text{we get:}$$

$$q'(t,w) = \left(h_{mm}(t) - h_{mm}(0°C)\right)(1+w) + wL_{0water} \tag{12.9}$$

Similarly, $v_{spec} = v_{mm}(1+w)$

In the supersaturated zone, the gas is saturated with water in the vapor state, the rest being water in the liquid state. The specific enthalpy is then given by:

$$q'(t,w) = h_{dg}(t) - h_{dg}(0°C) + wh_{vwater}(t, P_{sv})$$

Case where the dry gas composition is known

When we know the dry gas composition, the calculations are done slightly differently, because w is then given.

We deduce, M_{dg} being the dry gas molar mass:

$$P_{vap} = P \frac{w \dfrac{M_{dg}}{M_{H_2O}}}{1 + w \dfrac{M_{dg}}{M_{H_2O}}} \tag{12.10}$$

$$\epsilon = \frac{P_{vap}}{P_{sv}} \tag{12.11}$$

Other calculations are carried out as before.

Temperatures used for moist mixtures

The study of moist mixtures leads us to introduce several temperatures, whose definition is important to know, because for a given state of the gas, their values may be substantially different.

We call **dry-bulb temperature t** (°C) or T (K), the temperature indicated by a thermometer whose sensing portion is completely dry and is placed into the moist mixture. This is the temperature in the usual sense.

We call **dew point tr** (°C) or T_r (K) the temperature at which the first drops begin to appear when the moist mixture is cooled at constant humidity and pressure. If water condenses as ice, it is also known as ice temperature. This is the saturation temperature of the water at partial pressure P_{vap}.

With the previous notations, $T_r = T_{sat}\left(P_{vap}\right)$ if $P_{vap} \leq P_{sv}$, and $T_r = T_{sat}\left(P_{sv}\right) = T$, otherwise.

We call **wet-bulb temperature t'** (°C) or T' (K) the temperature indicated by a thermometer whose sensing element is covered with a thin film of water being evaporated due to strong mixing of the gas. In practice, the bulb of the thermometer is covered with a wick soaked with water. If measured in good conditions, this temperature is substantially equal to the adiabatic saturation temperature that would be obtained by moistening the gas to saturation in an adiabatic device, by spraying water at the equilibrium temperature.

One can show that its value is given by the following equation:

$$\text{enthalpy of liquid water at } t' \left(\text{saturated moisture at } t' - \text{initial moisture}\right) =$$

$$= \text{enthalpy of saturated mixture at } t' - \text{enthalpy of the initial mixture at } t.$$

Relative to the dry gas, the equation is

$$h_{lwater}(t')(w_{sat} - w) = h_{dg}(t') - h_{dg}(t) + w_{sat}h_{vwater}(t') - w_{hvwater}(t) \tag{12.12}$$

Relative to the specific units, it becomes

$$q'\left(t',w_{sat}\right)-q'\left(t,w\right)=\left(w_{sat}-w\right)h_{lwater}\left(t'\right)$$

This is an implicit equation in t', involving the mixture formed in part by the dry gas, and secondly by water. It can be reversed in a temperature range between a value close to the dew point as a minimum, and the dry bulb temperature as the maximum.

As stated above, we must take care that the reference values of enthalpies are consistent. Moreover, when t or t' falls below 0°C, the equations of liquid water must be replaced by those of solid water (ice).

Exergy of moist mixtures

The application of Dalton's law to a moist mixture states that the entropy of the mix is equal to the sum of the mass fractions of its two components multiplied by their specific entropies, calculated with their partial pressures.

The calculation of their exergies therefore theoretically does not pose a problem, with the exception that the reference state must be defined with precision and that the calculation of the partial pressures must be done rigorously.

In practice, therefore, these calculations turn out to be quite delicate, and approximate formulas have been published. We have retained that proposed by Wepfer et al. (1979).

The Wepfer exergy equation is however based on the assumption that the specific heat C_p of both components is constant, which is a bit restrictive. We have implemented a variant in Thermoptim, which uses much more accurate developments for C_p.

The specific exergy xh' of the mixture is thus calculated with precision by Equation (12.13).

$$xh' = xh_{dg}\left(t,p\right)+wxh_{lwater}\left(t,p\right)$$
$$+r_{dg}T_0\left[\left(1+w\_\right)\ln\left(\left(1+w\_0\right)/\left(1+w\_\right)\right)+w\_\ln\left(w\_/w\_0\right)\right] \qquad (12.13)$$

where $w\_ = w\, M_{dg}/M_{water}$

w_0 being the value of $w\_$ for the environment reference state or dead state.

The dead state which is generally adopted is T = 25° C, P = 1 bar, ε = 0.5.

As the gases produced by combustion most often contain water, this is how the exergy available at the combustion chamber or boiler outlet is calculated. The fuel chemical exergy taken into account in the calculations is equal to its HHV.

The use of exergy balances to analyze the performance of moist processes is less obvious than for the processes of the systems that we have so far mainly studied. For these systems, exergy was used to determine the maximum work that could be produced, while for moist processes, this is no longer the objective pursued.

One of the difficulties is to define which are the incoming exergies and which are the products sought in the considered system.

Various analyses have recently been published on the subject, a fairly complete review being available in (Ratlamwala & Dincer, 2013).

We will discuss this point on the occasion of various examples.

Moist mixture charts (psychrometric charts)

The main thermodynamic relations that we have given show that moist mixture variables and state functions are connected by relatively complex equations, which justify seeking an easy-to-use graphical presentation.

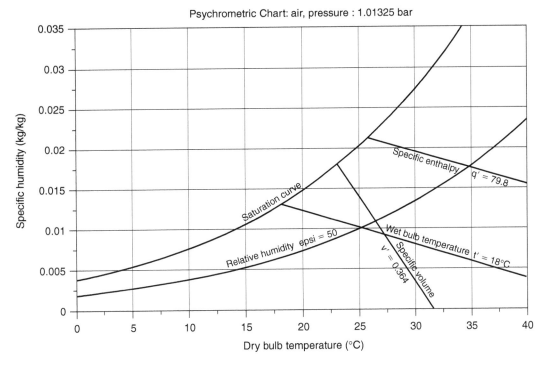

FIGURE 12.1

Carrier chart.

Two broad families of charts exist: the charts derived from that proposed by Carrier, with the dry-bulb temperature as the abscissa and the specific humidity as the ordinate (Figures 12.1 and 12.3), and Mollier charts (Figures 12.2 and 12.4), with specific humidity as abscissa and the specific enthalpy as ordinate.

The gas water saturation curve is the smooth curve concave upwards in the Carrier chart (Figure 12.1) and downwards in the Mollier chart (Figure 12.2).

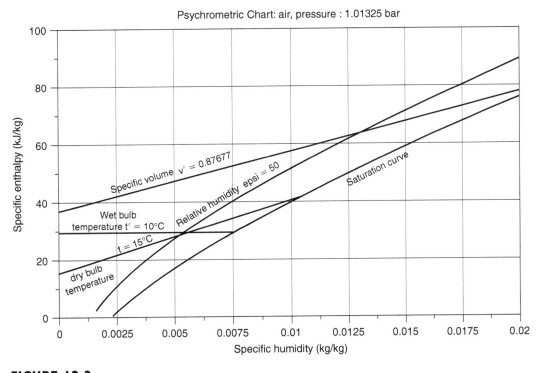

FIGURE 12.2

Mollier chart.

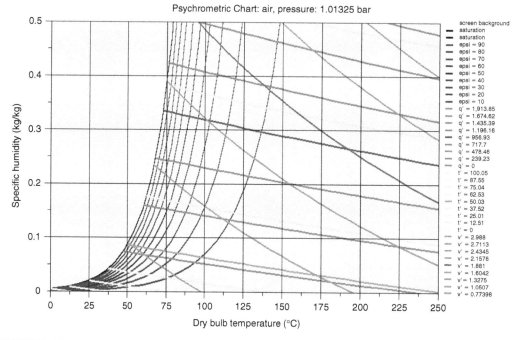

FIGURE 12.3

Carrier air psychrometric chart.

In the Carrier chart, the area above the saturation curve corresponds to cases where water is in excess and exists as a condensed liquid or solid: the gas is supersaturated. In the Mollier chart, the reverse is true: this zone is located below the saturation curve.

Carrier charts in Figures 12.1 and 12.3 relate to air at atmospheric pressure. They are equipped with different iso-value curves:

* Curves of equal relative humidity ε concave upward;
* Isenthalpic curves are very close to straight lines and wet bulb isotherm* curves are almost parallel to them* (usually not shown);
* Isovolume curves are very close to straight negative lines of slope* stronger than the previous.

The Mollier chart in Figures 12.2 and 12.4 also relates to air at atmospheric pressure. It is equipped with different isovalues:

* Curves of equal relative humidity ε are curves with concavity facing downwards;
* Dry bulb temperature curves are very close to straight lines* and wet bulb temperature* curves are almost parallel to the x-axis;
* Isovolume curves are very close to straight lines* of positive slope.
* At atmospheric pressure and above 100°C, it is impossible to condense water. On the Carrier chart, this translates into an almost vertical tangent on the saturation curve for 100°C, and on the Mollier chart by an asymptote to the saturation curve, corresponding to dry isotherm 100°C.

Water vapor/gas mixture processes

The main practical application of the previous developments is air conditioning to control temperature and humidity in indoor environments, but the problem is more general.

---

* By equating the dry gas to an ideal gas and neglecting in some cases the contribution of water vapor, these expressions are simplified and correspond to those of straight lines

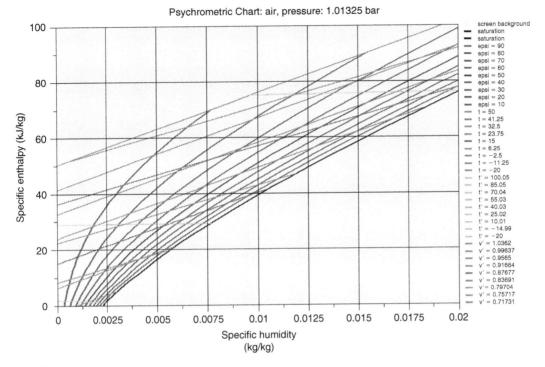

FIGURE 12.4
Moist air Mollier chart.

In this section, we present the basic treatments that undergo moist mixtures, and we illustrate each with examples pertaining to air. Air processing cycles involve different processes discussed below (Wang et al., 2000).

To represent these processes, professional practice has established the use of Carrier or psychrometric charts, or Mollier charts for moist mixtures (see Figures 12.2–12.4) of the kind that are built into Thermoptim.

The main processes that can undergo moist mixtures are mixture, heating, cooling, with or without water condensation, humidification, by water or steam, dehumidification, dessication, and finally, for air conditioning applications, it is often useful to determine the supply conditions for obtaining desired comfort.

In what follows, we present these basic operations and associated technologies, recall the equations that come into play and explain how they can be determined with Thermoptim and represented in moist mixture charts.

Moist process screens

To study changes that may undergo a moist mix, a "moist" process has been introduced. In fact, it corresponds to six different processes, which are distinguished by their category. The screen looks like the one given in Figure 12.5, but it varies slightly depending on the category. You should refer to the end of Chapter 3 for an overview of process screens.

This is a "supply" moist process, used to calculate the supply conditions for maintaining desired comfort, given water and thermal loads.

Before detailing how to use this screen, which will be done below, let us point out some points valid for all moist processes.

First, Thermoptim calculates moist mixture processes to the extent that they are represented by their dry gas and specific or relative humidity. The reason is simple: this mode of representation eliminates the need to introduce a new substance for each value of relative humidity.

It should be noted that this way of working is quite exceptional in Thermoptim: for all other processes, the calculations are made from the exact composition of the substance considered. If then the coupling must be made between moist processes and other processes, care should be taken to link them with moist mixtures of appropriate composition.

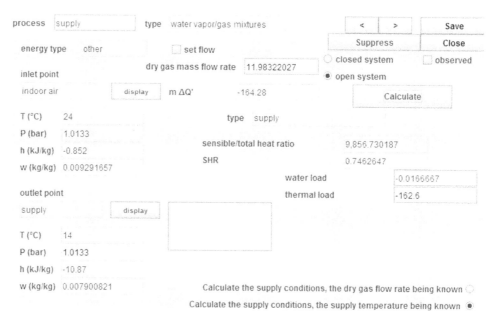

FIGURE 12.5

Supply process in Thermoptim.

In addition, the quantities being preferentially expressed in specific units, flows that appear on the moist processes are dry gas flows. Note that it is possible to configure Thermoptim so that the specific flow rates are displayed in the diagram editor (Menu Help/Global settings).

The concept of effectiveness is often used to describe the actual processes versus reference processes. For moist mixtures, these generally correspond to changes whose final state is saturated (full humidification, cooling to saturation).

The upper right corner of the screen recalls the general characteristics of the process, while the left shows the inlet and outlet points, indicating their temperature, pressure, enthalpy and specific humidity.

The settings of the various categories of moist processes and calculation methods are located in the central and lower right parts.

Moist mixers

Principle and Equations

A moist mixer is used to determine the specific properties of the mixture of several moist gases. This operation takes place, for example, when two ducts containing mixtures of different humidities come together to form a single vein. In practice, it is most often air mixtures in air conditioning systems, for example a mixture of outdoor air and recycled inside air (see examples at the end of the chapter).

The operation being adiabatic, the total enthalpy and the total mass flow rate are always conserved. For two branches, one obtains:

$$\dot{m}q'_{mix} = \dot{m}_1 q'_1 + \dot{m}_2 q'_2$$

$$\dot{m} = \dot{m}_1 + \dot{m}_2$$

There is usually conservation of the total specific humidity:

$$\dot{m}w_{tot} = \dot{m}_1 w_1 + \dot{m}_2 w_2 \tag{12.14}$$

except where there is supersaturation, in which case a portion of the water is condensed. Then, the liquid water must be subtracted from the specific humidity.

The three previous equations show that in a (w, q') coordinate system, and when there is no supersaturation, the point representative of the mixture is the barycenter of the two points representing the moist mixtures, the coefficients being equal to their mass flow rates. It follows that the three points are aligned, and that the determination of the mixture can be done simply graphically in a moist mixture chart using this system of axes. In the Carrier (w, t) chart, this is only true in first approximation.

The representation of the mixture of two moist air flows 1 and 2 in the chart is given in Figure 12.6. The mixture m is located at the junction of the two light and dark aligned segments.

The enthalpy being conserved, we begin by looking at the saturation conditions q'_{mix}. If the saturated specific humidity w_{sat} exceeds w_{tot}, $w_{mix} = w_{tot}$. Otherwise, $w_{mix} = w_{sat}$.

The dry temperature is in the first case that corresponding to q'_{mix} and w_{mix}, and in the second that of saturation.

Calculation of a mixture in thermoptim

The moist mixer node can determine the properties of a moist mixture of several dry or moist gases.

A moist mixer (Figure 12.7) differs from a simple mixer as follows:

- Firstly, it accepts as branches moist as well as nonmoist processes. Since the flow of moist processes is referred to the dry gas, the moist mixer makes the necessary corrections;
- Second, its main vein must be a moist process. If this is not the case, a message warns the user, and the calculations are done as if it were a simple mixer.

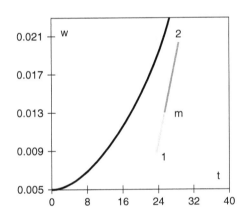
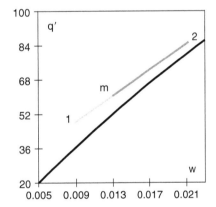

FIGURE 12.6

Moist mixer in Carrier and Mollier charts.

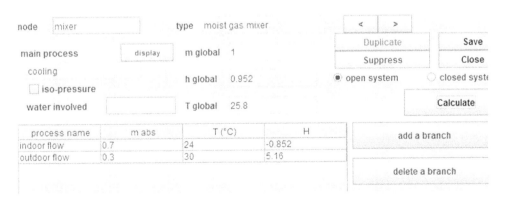

FIGURE 12.7

Moist gas mixer.

In general, the outlet point is located on the mixing line. In case of supersaturation, a message informs the user and Thermoptim searches the mixing point on the saturation curve, ensuring conservation of enthalpy. Excess water is then displayed in the upper right.

The values that are displayed are not expressed relative to dry gas.

Heating a moist mixture

Principle and Equations

The heating of a moist mixture is a constant moisture operation. In the Carrier chart, it is represented by a horizontal segment oriented to the left and in the Mollier chart by a vertical segment oriented upwards (Figure 12.8).

The heating is generally carried out in heating coils that can be of several types:

- Air coils traversed by hot water supplied by boilers;
- Air coils forming the condenser of a refrigerating machine;
- Electric battery heated by Joule effect.

The equation is

$$\Delta Q' = m_{air}\left(q'_2 - q'_1\right) \qquad (12.15)$$

After processing, it reduces to

$$\Delta Q' = m_{air}\left(1+w\right)\left(h_{mm2} - h_{mm1}\right) \qquad (12.16)$$

h_{mm} being the enthalpy of the moist mixture.

If $\Delta Q'$ is known, we reverse the equation, which gives t_2.

Calculation of heating in thermoptim

This process (Figure 12.9) can perform various calculations of changes undergone by a moist mixture. So that this process can also be used in special cases, the water content of the inlet and outlet points can be the same or different.

Two actions are possible here:

- Calculate the process, assuming the inlet and outlet points are known, in which case $\Delta Q'$ and the water involved are determined;
- Determine the state of the outlet point, the enthalpy change and the water involved being known.

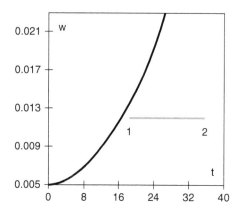
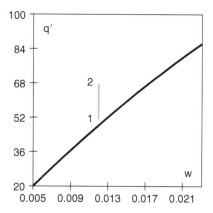

FIGURE 12.8

Heating process in Carrier and Mollier charts.

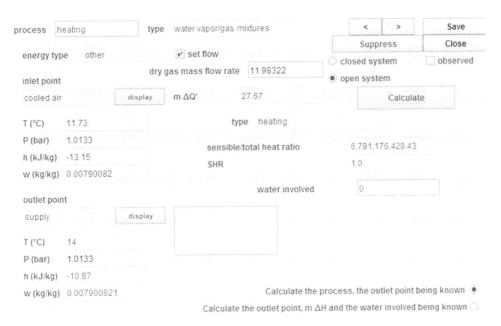

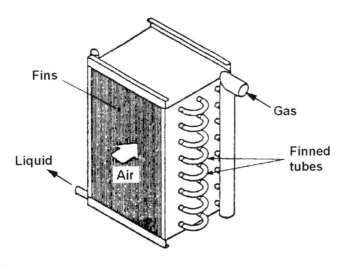

FIGURE 12.9

Moist heating process.

Cooling of moist mixture

Principle and Equations

To cool a moist mixture, it is passed through a special heat exchanger called a cooling coil, which can be cooled by ice water or by direct evaporation of a refrigerant (Figure 12.10). The mixture being in contact with the cold surfaces sees its temperature decrease.

Depending on circumstances, there may be condensation or not. If there is no condensation, specific humidity remains constant, and cooling can be represented by a horizontal segment oriented to the left in the Carrier chart and vertically oriented downwards in the Mollier chart. If condensation occurs, which is very often the case, the segment is oriented to the bottom left in both charts (Figure 12.11).

Theoretical perfect cooling in a cooling coil of infinite size would cool the moist mixture at the coil saturation temperature. It is customary to characterize a real process in taking this cooling as theoretical reference, introducing effectiveness α of the cooling coil and its average surface temperature t_s.

FIGURE 12.10

Cooling coil. (Courtesy Techniques de l'Ingénieur.)

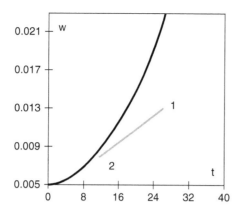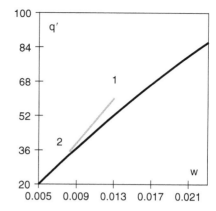

FIGURE 12.11

Moist cooling in Carrier and Mollier.

In the U.S., another approach is generally retained: the bypass factor method is based on the assumption that a part of the total flow which goes through the coil is never in contact with the cold surface: it bypasses it. Thus, the final state of the air exiting the cooling coil corresponds to the mix between both streams. Calling b the bypass factor, we could easily show that $\alpha = 1 - b$.

The advantage of both methods is that values of α and b of a given coil are characteristics largely independent of conditions of use.

To calculate point 2, we first search the saturation conditions $w_{sat}(t_s)$:

$$w_{sat}(t_s) = \frac{M_{H_2O}}{M_{dg}} \frac{P_{sv}}{P - P_{sv}}$$

and $q'(t_s, w_{sat})$

We calculate the outlet moist properties in view of effectiveness:

$\alpha = \dfrac{w_2 - w_1}{w_{sat} - w_1}$, which gives

$$w_2 = \alpha w_{sat} + (1 - \alpha)w_1 \qquad (12.17)$$

$$q'_2 = \alpha q'(t_s, w_{sat}) + (1 - \alpha)q'_1$$

We then search t_2 such that $q'_2 = q'(t_2, w_2)$.

A problem may arise when, in the psychrometric chart, the line from point 1 intersects the saturation curve at two points. Indeed, in this case, the point calculated from effectiveness may be in the saturated zone. This is called early condensation. Two possibilities exist: if one sets the outlet point specific humidity w_2, the end point lies on the saturation curve for $w_{sat} = w_2$; if we set the effectiveness, it is on the saturation curve for $q'_{sat} = q'_2$. We must therefore ensure that the point is not found in the saturated zone.

Calculation of cooling in thermoptim

This process (Figure 12.12) allows one to study the cooling of a moist mixture on a cooling coil, with or without condensation.

The cooling conditions are specified on the right of the screen: surface temperature and effectiveness of the cooling coil, if both are known. If the effectiveness is not known, the outlet point humidity must be given to calculate the process.

Two actions are possible here:

* The effectiveness being set, calculate the outlet point moisture. If early condensation occurs (when the line connecting the inlet point at the point where the saturation curve

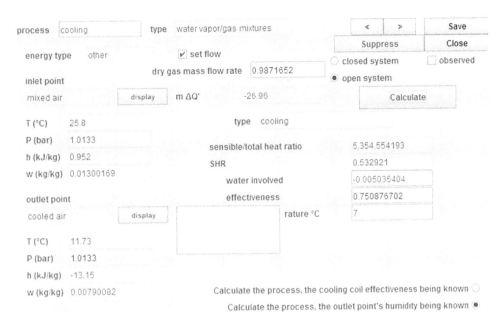

FIGURE 12.12

Moist cooling process.

at the surface temperature intersects the saturation curve at two points), the outlet point is sought on the saturation curve, the enthalpy being calculated from the effectiveness of the coil

* The humidity of the outlet point being set, calculate its temperature and the coil effectiveness. In case of early condensation (see above), the outlet point is searched on the saturation curve for the specific humidity desired.

Humidification of a gas

Principle and equations

To humidify a gas, we can proceed in two ways:

* Injecting steam into the gas, usually slightly superheated, or spraying water in very fine droplets with a diameter of 5–20 μm by ultrasound or compressed air, making sure to avoid condensation on the walls (Figure 12.13);

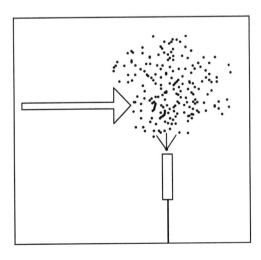

FIGURE 12.13

Steam humidification.

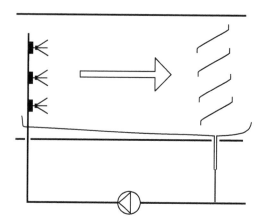

FIGURE 12.14
Humidification by water spray.

- Spraying water to form rain that wets the gas (Figure 12.14). The contact surface between gas and water being very important and the gas being far from the saturated state, the water vaporizes, and the gas humidity increases. In this case, the vaporization of water requires heat input, which can be supplied by both water and gas, or gas alone, in which case it is called adiabatic humidification. With this type of device, there is however a risk of bacterial growth, and it is therefore less used.

Humidification is often used in air conditioning, for example, to humidify in winter a very dry outside air before blowing it in a warm interior or in summer to provide passive cooling.

As has been presented for the cooling of a moist mixture, the reference being the theoretical wetting leading the gas to saturation, the real humidification is characterized by the previously defined effectiveness α.

Cases of nonadiabatic humidification (vapor or liquid water)

If we know the humidification effectiveness α, we first find the boundary saturation conditions w_{sat}:

$$\text{such as } w_{sat} = w_1 + \frac{m_{water}}{m_{air}}$$

and

$$q'(t, w_{sat}) = q'_1 + (w_{sat} - w_1) h_{H_2O} \tag{12.18}$$

$$\text{with } w_{sat} = \frac{M_{H_2O}}{M_{dg}} \frac{P_{sv}(t)}{P - P_{sv}(t)}$$

We then calculate the final moist conditions in view of effectiveness:

$$w_2 = \alpha w_{sat} + (1 - \alpha) w_1$$

$$q'_2 = \alpha q'(t, w_{sat}) + (1 - \alpha) q'_1 \tag{12.19}$$

We then seek t_2 such that $q'_2 = q'(t_2, w_2)$.

If we know the water absorbed by the fluid m_{water}, w_2 is directly calculated:

$$w_2 = w_1 + \frac{m_{water}}{m_{air}}$$

We then seek t_2 such as $q'(t_2, w_2) = q'_1 + (w_2 - w_1) h_{water}$

If we seek the boundary saturation conditions, we deduce the effectiveness α.

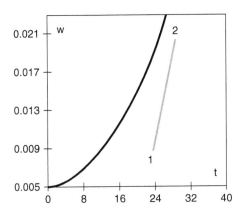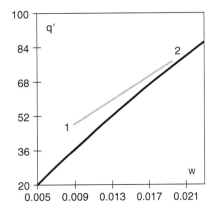

FIGURE 12.15

Spray humidification in Carrier and Mollier charts.

Humidification by water or steam spraying is reflected in the Carrier chart by a line segment tilted to the right or left according to the enthalpy of the fluid injected and in the Mollier chart by a segment inclined towards the right. Figure 12.15 shows a humidification with steam superheated at 115°C.

Case of adiabatic humidification

We begin by looking at the saturation conditions at the wet-bulb temperature $w_{sat}(t')$:

$$w_{sat}(t') = \frac{M_{H_2O}}{M_{dg}} \frac{P_{sv}(t')}{P(t') - P_{sv}(t')}$$

and $q'(t', w_{sat})$

We calculate the final moist conditions in view of effectiveness:

$$w_2 = \alpha w_{sat} + (1-\alpha)w_1 \qquad (12.20)$$

$$q'_2 = \alpha q'(t_s, w_{sat}) + (1-\alpha)q'_1$$

We then seek t_2 such that $q'_2 = q'(t_2, w_2)$.

Adiabatic humidification is reflected in the Carrier chart by a line segment tilted to the left corresponding to $t' = $ Const, and in the Mollier chart by a nearly horizontal segment, q' being approximately constant (Figure 12.16).

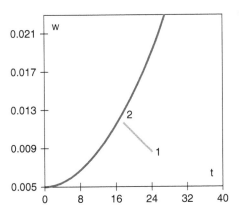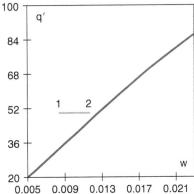

FIGURE 12.16

Adiabatic humidification in Carrier and Mollier charts.

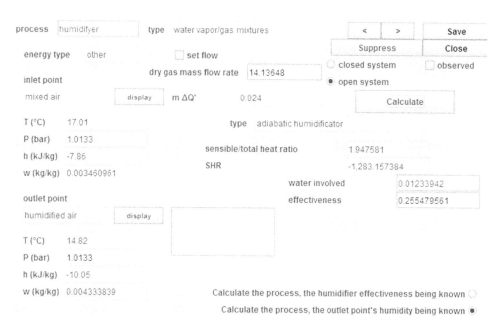

FIGURE 12.17

Adiabatic humidificator screen.

Calculation of humidification in thermoptim

Two distinct but similar types of processes are used to study either water or steam humidification or adiabatic humidification (Figure 12.17).

The humidification conditions are specified on the right of the screen: water or steam temperature and pressure, and humidifier effectiveness, if known. If effectiveness is not known, the outlet point humidity must be given to calculate the process.

Two actions are possible here:

* The effectiveness being set, calculate the outlet point moisture. If condensation occurs early (when the line connecting the inlet point at the point of the saturation curve at the surface temperature intersects the saturation curve at two points), outlet point is sought on the saturation curve, the enthalpy being calculated from the humidifier effectiveness;
* Calculate the outlet point to get the desired humidity. When there is early condensation (see above), the outlet point is searched on the saturation curve for the specific humidity desired. The effectiveness of the humidifier is then calculated.

Dehumidification of a mixture by desiccation

Principle and Equations

In order to dehumidify a moist mixture by desiccation, it passes over a bed of solid hygrophilic adsorbent, which extracts moisture from the gas by exothermic physical (adsorption) or physicochemical (chemisorption) effect. The regeneration of the desiccant is made by the opposite effect, warming. Among the systems that accomplish desiccation, one of the most used takes the form of a rotary regenerative heat exchanger which simultaneously dehumidifies an air stream and regenerates part of the desiccant (Figure 12.18).

The same equations govern the two effects.

The equation of adsorptive dehumidification or desiccant regeneration is provided by the enthalpy balance, which is written here, L_s being the heat of sorption:

$$(1+w_2)(h_{mm}(t_2)-h_{mm}(0°C))-(1+w_1)(h_{mm}(t_1)-h_{mm}(0°C))+L_s(w_2-w_1)=0 \qquad (12.21)$$

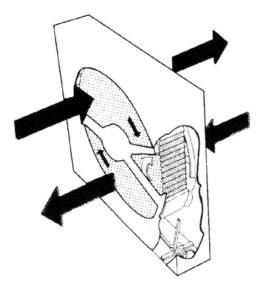

FIGURE 12.18

Rotary regenerative heat exchanger. (Courtesy Techniques de l'Ingénieur.)

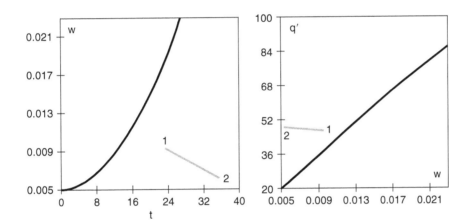

FIGURE 12.19

Desiccation process in Carrier and Mollier charts.

Knowing w_2, simply invert this expression in t_2, taking care that the composition of the moist mixture varies between points 1 and 2.

Knowing t_2, w_2 is obtained by expressing the above equation as a function of dry gas, which leads to a simple first order polynomial equation in w_2.

Dehumidification by adsorption results in the Carrier chart in a line segment tilted to the right and on the Mollier chart in a segment close to the horizontal, q' increasing slightly (Figure 12.19). That the operation is carried out at substantially constant enthalpy does not mean it is free in terms of energy, since the desiccant must be regenerated.

Calculation of a desiccation in thermoptim

This process (Figure 12.20) allows the study of dehumidification by desiccation or regeneration of the desiccant. Different types of desiccant materials are proposed, their name and their heat of sorption being displayed in the right of the screen in kJ/kg. If the user wishes he can change the latter.

Two actions are possible here:

- The outlet point temperature being known, calculate its moisture;
- Calculate the outlet point temperature to obtain the desired humidity.

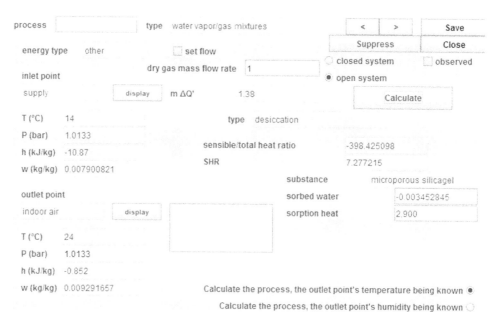

FIGURE 12.20
Desiccation process.

Note that the same process calculates both desiccation and the desiccant regeneration, according to the relative values of the inlet point and outlet point. For regeneration, the software does not check, however, if the temperature is sufficient.

Determination of supply conditions

Principle and equations

In air conditioning a space, we often experience the following problem: we want to maintain the atmosphere inside a building at a given dry bulb temperature and relative humidity. External climatic conditions are known: dry bulb temperature and relative humidity. We must evacuate the internal and external thermal loads of known magnitude, and a quantity of water corresponding to the internal gains. Depending on countries, calculations are slightly differently expressed: in the U.S., for example, sensible $\dot{Q}_s$ and latent $\dot{Q}_l$ loads are considered (Agami Reddy, 2001), while in France, we talk of enthalpy $\dot{Q}_s$ and water $\dot{m}_{water}$ loads (AICVF, 1999). This simply means that in the first case, water to be extracted is directly converted in energy terms (BTU/hr or kW), while in the second, it is expressed in kg/s.

L_{water} being the latent heat of evaporation at indoor air temperature, we have

$$\dot{Q}_l = \dot{m}_{water} L_{water}$$

In addition, for reasons of hygiene and comfort, the supply temperature should generally not be less than a given value, and the proportion of recirculated air should not exceed a limit. You have to then determine the supply conditions, that is to say the flow rate $\dot{m}_{air}$ and state of the air blown, so that the atmosphere obtained is the desired one.

1. If the air flow is known, equations are as follows:

$$\text{balance on water}: w_{su} = w_1 + \frac{\dot{m}_{water}}{\dot{m}_{air}} \tag{12.22}$$

$$\text{enthalpy balance}: q'(t_{su}, w_{su}) = q'_1 + \frac{\dot{Q}_s + \dot{m}_{water} h_{water}(t_1)}{\dot{m}_{air}} \tag{12.23}$$

$h_{water}(t_1)$ being the enthalpy of water at temperature t_1.

or (U.S. notations) $q'\left(t_{su},w_{su}\right)=q_1'+\dfrac{\dot{Q}_s+\dot{Q}_1}{\dot{m}_{air}}$

You therefore determine w_{su}, then seek t_{su} such that $q' = q'(t_{su}, w_{su})$.

2. If the supply temperature is known, the method is as follows:

$$\text{balance on water: } \dot{m}_{air}=\frac{w_{su}-w_1}{\dot{m}_{water}} \tag{12.24}$$

$$\text{enthalpy balance: } q'\left(t_{su},w_{su}\right)=q_1'+\frac{\dot{Q}_s+\dot{m}_{water}h_{water}\left(t_1\right)}{\dot{m}_{air}}\left(w_{su}-w_1\right) \tag{12.25}$$

t_{su} being known, w_{su} can be calculated by

$$q'\left(t_{su},w_{su}\right)=h_{dg}\left(t_{su}\right)-h_{dg}\left(0°C\right)+w_{su}\left(h_{H_2O}\left(t_{su}\right)-h_{H_2O}\left(0°C\right)\right)+w_{su}L_{0water}$$

$$w_{su}=\frac{q_1'-w_1\left(\dfrac{\dot{Q}_s}{\dot{m}_{water}}-h_{lwater}-h_{dg}\left(t\right)+h_{dg}\left(0°C\right)\right)}{h_{H_2O}\left(t_{su}\right)-h_{H_2O}\left(0°C\right)+L_{0water}-\dfrac{\dot{Q}_s}{\dot{m}_{water}}-h_{lwater}\left(T_1\right)} \tag{12.26}$$

Note that eliminating from Equations (12.22) and (12.24) yields

$$\theta=\frac{q_{su}'-q_1'}{w_{su}-w_1}=\frac{\dot{Q}_s+\dot{m}_{water}h_{water}\left(t_1\right)}{\dot{m}_{water}}$$

θ is called the enthalpy-moisture ratio. It expresses that variations of specific enthalpy and humidity are proportional whatever the air flow rate value $\dot{m}_{air}$. All points lying on a straight line called the condition line or load line satisfy the sensible and latent loads, i.e., are solutions of Equations (12.22) and (12.24).

The position of point "su" on the line determines the value of $\dot{m}_{air}$. It depends on various factors, such as the maximum allowable temperature difference to avoid any inconvenience (usually 6°C–12°C according to the technique used).

Calculation of supply conditions in thermoptim

This process connects the point describing the known indoor conditions (inlet point) to the point corresponding to the supply conditions sought (outlet point). Its screen is given at the beginning of this section (Figure 12.14).

Let us point out that the determination of supply conditions does not really correspond to what we called a Thermoptim process until now: this is strictly speaking a misnomer. The problem is finding the thermodynamic state of air to be blown into the space to be cooled in order to compensate for thermal and water loads.

There are two possible options in this case:

* Determine the temperature and the specific humidity of the supply point when the dry gas flow rate is known;
* Set the supply temperature, so that Thermoptim determines the specific humidity of the supply point, as well as the dry gas flow rate required.

Fields in the right of the screen allow you to enter the water (rate of moisture gain) and thermal (sensible rate of heat gain) loads to be removed, respectively, in kg/s and kW, which defines the supply line. It is assumed that the water load must be dissipated at the inside point

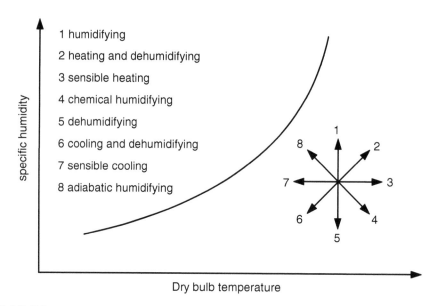

FIGURE 12.21

Air conditioning processes in a psychrometric chart.

temperature. Note that units here are not exactly those generally used in the US where the water load is usually directly expressed in kW.

Thermoptim then calculates the value of the sensible/total heat ratio (also known as sensible heat factor SHF or sensible heat ratio SHR) which is the ratio of the rate of sensible heat gain $\dot{Q}_s$ (in kW) for the space to the rate of total energy gain for the space $\dot{Q}_s + \dot{Q}_l$ (in kW). As this ratio is proportional to the slope of the supply or condition line, the ASHRAE psychrometric chart includes a protractor which facilitates drawing the condition line.

$$SHR = \frac{\dot{Q}_s}{\dot{Q}_s + \dot{Q}_l} = \frac{\Delta h_s}{\Delta h_{tot}}$$

In France, the SHR is replaced by the ratio, directly expressed in the psychrometric chart units, between changes in enthalpy and specific humidity. Known as the "slope ratio γ", it is equal to the enthalpy-moisture ratio θ previously introduced.

$$\gamma = \theta = \frac{q'_{su} - q'_1}{w_{su} - w_1}$$

Air conditioning processes in a psychrometric chart

Figure 12.21 summarizes how the different air conditioning processes can be represented in a psychrometric chart.

Air conditioning

To size an air conditioning system, we must first be able to evaluate the thermal and water loads to import or evacuate. These loads depend on the climate and conditions of occupancy. They vary throughout the year, and the design of an air conditioning system can only be made by taking into account these factors.

In this section, we will only deal superficially with these issues. The interested reader should refer to the literature listed in the bibliography, including ASHRAE publications.

Basics of an air conditioning system

Figure 12.22 shows the sketch an air conditioning system, and how it fits into a building.

The room is represented in the lower right. It has two sets of airflow veins, one for extraction and one for blowing. To cool the room, a certain air flow is "supplied" through vents. As this blown air mixes with room air, heat and water balance is achieved. Supply conditions are determined so that this equilibrium corresponds to the desired comfort conditions as indicated in the previous section.

Air extracted is partly discharged outside, and partly recycled, the recirculation rate depending on hygiene standards in force. Return air is mixed with fresh air, optionally preheated, the mixture being then "processed" so that its state corresponds to the supply conditions desired.

Figure 12.23 shows the cutaway of an air treatment vein set including various boxes including a return fan, mixer, filters, humidifier, coils and a supply blower.

The design of an air conditioning system comprises a series of distinct steps:

- Climatic conditions of reference that will be used to calculate the enthalpy and water loads must be first determined. In practice, it is based on climatic data published by national meteorological services, which are eventually corrected as indicated below. The values to take into account are not the extreme conditions but those likely to be reached or exceeded a few days per year on average. For winter, corrections must be made to temperatures at high altitude sites (–1°C every 200 m) and cities (+1 to 2°C depending on the size of the

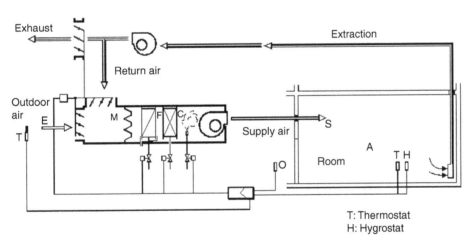

FIGURE 12.22

Air conditioning system.

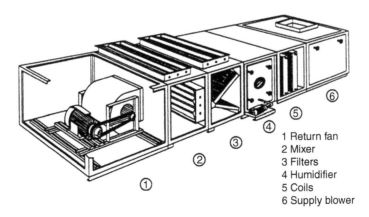

FIGURE 12.23

Cutaway of an air treatment vein.

metropolitan area), and an estimate of relative humidities can be obtained simply (100% along the coast or a lake, 90% elsewhere);

- Environmental conditions must then be defined (Agami Reddy, 2001). For specific industrial applications, refer to the information of AICVF Guides in France or ASHRAE in the U.S. For comfort cooling, it should be noted that most healthy people do not feel any noticeable difference as long as the relative humidity is between 30% and 60% and the temperature is below 25°C. As a first approximation, we can therefore choose for summer a temperature of 25°C and a relative humidity of 60% and for winter, temperatures of 19°C or 20°C and a relative humidity of 30%;

- Once these values are chosen, it becomes possible to calculate loads. Depending on countries, calculations are slightly differently expressed: in the U.S. for example, sensible and latent loads are considered while in France, we talk of enthalpy and water loads (AICVF, 1999). This simply means that in the first case, water to be extracted is directly converted in energy terms, while in the second, it is expressed in kg/s. Detailed calculation of heat loss from a building is outside the scope of this book: you should refer to the AICVF Guides (1999) or methods proposed by CSTB or ASHRAE (Rabl &Curtiss, 2001). Sensible (or enthalpy) load Q_s is given by Equation (12.27), in which ϕ_p represents losses through the walls, ϕ_i losses by air infiltration (which should certainly not be confused with those due to air renewal that are implicitly taken into account by the calculation method), ϕ_s solar gains and P the set of internal gains due to occupants, lighting, appliances, machinery, office equipment, etc.

$$Q_s = \phi_p + \phi_i + \phi_s + P \qquad (12.27)$$

- The supply or **condition line** can then be determined as explained in the previous section:
  With French notations:

$$\gamma = \frac{\Delta q'}{\Delta w} = \frac{\dot{Q}_s + \dot{m}_{water} h_{water}(t_1)}{\dot{m}_{water}} \qquad (12.28)$$

With US notations:

$$SHR = \frac{\dot{Q}_s}{\dot{Q}_s + \dot{Q}_l}$$

In the psychrometric chart, the condition line is the line of slope SHR or γ passing through the point representing the desired comfort conditions. Each point of this line corresponds to a different supply flow rate.

The actual flow rate value depends on various factors, such as the maximum allowable temperature difference to avoid any inconvenience (usually 6°C–12°C according to the technique used) or the rate of mixing required (generally between 3 and 20 volumes/hour) to ensure good uniformity without drafts;

- Once the supply point is determined, it remains to choose an air treatment for bringing a mixture of outdoor air and indoor air in this state. The recirculation rate depends on hygiene constraints. The more important it is, the higher the energy expenditure will be. The following examples show how basic treatments can be combined to form a proper air conditioning unit. Note that in first approximation, the fan heats pumped air by about 1°C. In winter, this means less heating need and in summer, greater cooling need.

Examples of cycles

There are many possible air conditioning cycles, and their presentation is beyond the scope of this book. We will, therefore, present only two examples of typical installations for summer cooling and winter heating.

Note that the Thermoptim diagram editor does not have components representing moist processes. Modeling cannot be done graphically (guidance on how to describe a project are given in the Getting Started guide on air conditioning and in guided explorations CLIM 1 and CLIM 2).

Summer air conditioning

The facility that we are going to study corresponds to the cooling of a large building like an airport located in a hot and humid climate. We simply summarize here the main results.

The problem data are as follows: we seek to maintain the internal ambience of the building at a temperature of 24°C and a relative humidity equal to 50%. External climatic conditions are temperature equal 30°C and relative humidity of 80%. It is necessary to remove external and internal thermal loads of 162.6 kW, as well as a quantity of water equal to 60 kg/h, i.e., 0.01667 kg/s or 41.7 kW.

Knowing that, for sanitary and comfort reasons, the supply temperature must not be less than 14°C, and that the recycled air proportion must not exceed 70%, the purpose of the exercise is to determine the following:

* Supply conditions;
* A way of processing of the outdoor air/recycled air mix.

A possible treatment of the mixed air is to cool it, condensing water in excess to obtain specific humidity corresponding to the supply conditions and then warm it to supply temperature. There are others, but we will present this one here (Figure 12.24).

The first step is to determine the supply conditions, which Thermoptim can do. The calculation leads to the following: a flow rate of 12 kg/s, specific humidity w = 0.0079 and temperature t = 14°C.

The second step calculates the state of the mixed air. We know that, in a coordinate system (w, q′), and while there is no supersaturation, the representative point of the mixture of indoor and outdoor air is the barycenter of the two points representative of the moist mixtures, the coefficients being equal to their mass flow rates. It follows that the three points are aligned, and that the determination of the mixture can be simply done graphically in a moist mixture chart using this system of axes. This yields specific humidity w = 0.013 and temperature t = 25.8°C (point 1).

The air conditioning unit chosen requires cooling mixed air in the dehumidifier to w = 0.0079 and then heating it at t = 14°C.

The cooling coil chosen has a surface temperature of 7°C. A perfect theoretical cooling in an infinite surface coil would lead to cool the moist mixture at the temperature of the coil in the saturated state (we have added a fictitious point 0 to represent it). The actual process is characterized taking as reference the theoretical cooling to point 0 and introducing either the cooling coil effectiveness α (here 75%) or its bypass factor b such as b = 1−α (here 25%). Its calculation shows that the mixture is then cooled at 11.7°C (point 2).

The heater allows it to be brought to the desired supply temperature (point S). Table 12.2 gives the properties of the various points of interest.

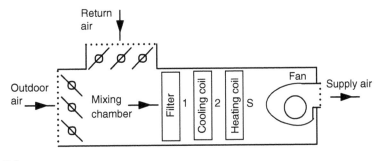

FIGURE 12.24
Summer air conditioning system.

TABLE 12.2

SUMMER AIR CONDITIONING POINT PROPERTIES

Point Name	Dry Bulb Temp. (°C)	Specific Humidity (kg/kg)	Relative Humid. (ε or RH)	Spec. Enth. (kJ/kg)	Wet Bulb Temp. (°C)	Spec. Volume (m³/kg)
Outdoor air	30	0.02155	0.8	85.1836	27.09	0.8887
Indoor air	24	0.00929	0.5	47.7032	17.06	0.8546
Mixed air (1)	25.8	0.013	0.6248	58.998	20.55	0.8648
0	7	0.00621	1	22.623	7	0.8017
Cooled air (2)	11.7	0.0079	0.923	31.6851	11.04	0.8175
Supply (S)	14	0.0079	0.7956	33.9941	12	0.824

BOX 12.1

Guided educational exploration

Summer air conditioning cycle (exploration CLIM 1)

The objective of this guided exploration is to guide you through your first steps in using Thermoptim to study a building air conditioning cycle.

The plot of the cycle on the psychrometric charts is given in Figures 12.25 and 12.26.

Winter air conditioning

The facility that we will study corresponds to the heating of a large building like a bank, located in a cool, moist climate. For this, we have a ventilation system that allows air to blow in different parts of the building. For reasons of hygiene, it is necessary to renew the air, but some can be recycled, however, which reduces heating needs.

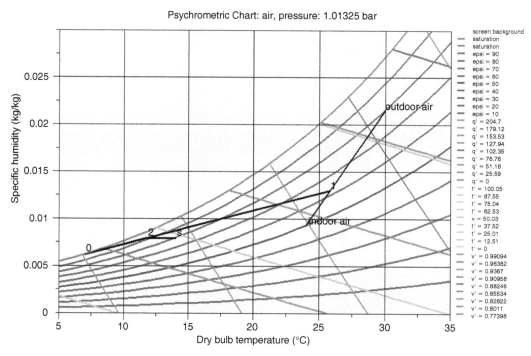

FIGURE 12.25

Summer air conditioning in the Carrier psychrometric chart.

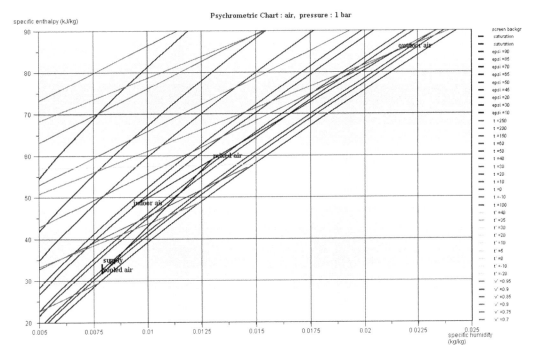

FIGURE 12.26

Summer air conditioning in the Mollier psychrometric chart.

So we recycle some of the indoor air that is mixed with outside air, previously preheated to prevent condensation on the ducts or jamming of registers. This mixture must be treated before being injected into the ventilation system so that its state corresponds to the supply conditions. These are calculated so that the external thermal loads are compensated by taking into account internal inputs. It is assumed that the water load is zero, and air temperature is 27°C.

A possible treatment of the mixed air is to humidify it adiabatically until its specific humidity corresponds to the desired supply conditions and then to warm it up at the desired air temperature (Figure 12.27). There are others, but this will be presented here.

The problem data are as follows: we seek to maintain the atmosphere inside the building at a temperature of 20°C and a relative humidity of 30%. External climatic conditions are temperature equal to –10°C and relative humidity of 90%. We must provide a heat input of 100 kW but no water. To avoid parasite condensation, fresh air is preheated at 14°C (point 1).

Knowing that, for reasons of hygiene and comfort, the proportion of return air should not exceed 70%, the purpose of the exercise is to determine the following:

• Supply conditions;
• A way of processing of the outdoor air/recycled air mix.

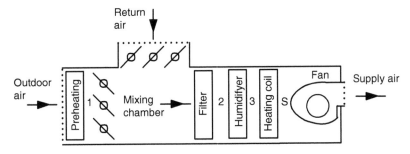

FIGURE 12.27

Winter air conditioning system.

We have neglected here the blowing fan air warming (about 1°C).

The first two steps are similar to those of the previous example. The calculation of supply conditions leads to the following: a flow rate of 14.1 kg/s, specific humidity w = 0.0433 and temperature t = 27°C.

The specific humidity of outside air is w = 0.0014. It is preheated at 14°C (point 1). For indoor recycled air, w = 0.0043. Given the rate of recirculation, mixture moisture is here w = 0.0346 and temperature 18.2°C (point 2).

It is therefore necessary to moisten the mixture at the supply humidity, for example, in an adiabatic humidifier. Its effectiveness is calculated: 23.7%, the temperature of the moist mixture being 16.2°C (point 3). Recall that an adiabatic humidification is achieved by spraying water to form a rain sprinkling the air, the heat needed to vaporize the water being supplied by air. As is done to cool a moist mixture, the reference is a theoretical humidifying leading the air to saturation, the real wetting being characterized by its effectiveness α.

A heat addition for heating the mixture at 27°C is necessary (point S). Table 12.3 gives the properties of the various points of interest.

The plot of the cycle on the psychrometric charts is given in Figures 12.28 and 12.29.

TABLE 12.3
WINTER AIR CONDITIONING POINT PROPERTIES

Point Name	Dry Bulb Temp. (°C)	Specific Humidity (kg/kg)	Relative Humid. (ε or RH)	Spec. Enth. (kJ/kg)	Wet Bulb Temp. (°C)	Spec. Volume (m³/kg)
Outdoor air	−10	0.001415	0.9	−6.50	−10.32	0.7473
Preheat. air (1)	14	0.001415	0.144	17.6	4.509	0.8155
Indoor air	20	0.004334	0.3	31.04	10.84	0.8364
Mixed air (2)	18.2	0.003461	0.268	27	9.034	0.83
Humid. air (3)	16	0.004334	0.385	27	9.08	0.825
Supply (S)	27	0.004334	0.197	38.1	13.71	0.8564

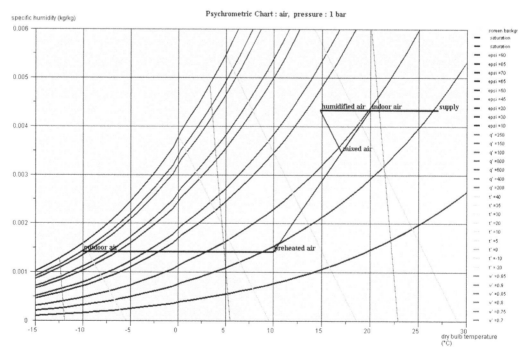

FIGURE 12.28
Winter air conditioning in the Carrier psychrometric chart.

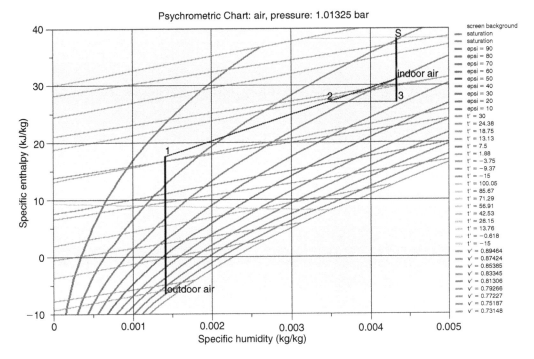

FIGURE 12.29

Winter air conditioning in the Mollier psychrometric chart.

Guided educational exploration

Winter air conditioning cycle (exploration CLIM 2)

The objective of this guided exploration is to guide you through your first steps in using Thermoptim to study a building air conditioning cycle.

Bibliography

T. Agami Reddy, *Psychrometrics and comfort, handbook of heating, ventilation, and air conditioning*, (Edited by J. F. Kreider), CRC Press, Boca Raton, FL, 2001, ISBN 0-8493-9584-4.

AICVF, Conception des installations de climatisation, et de conditionnement de l'air, Collection des guides de l'AICVF, PYC Edition, Paris, octobre 1999.

A. Bensafi, Air humide: Traitement et conditionnement de l'air, Techniques de l'Ingénieur, Traité Génie énergétique, BE 8 026.

R. Casari, Cahier Technique, données théoriques et technologiques, conduite de projets, mallette pédagogique Conditionnement d'air, Paris, Documentation interne, École des Mines de Paris, septembre 1992.

M. Duminil, Air Humide, Techniques de l'Ingénieur, Traité Mécanique et chaleur, B 2 220.

J. W. Mitchell, J. E. Braun, *Principles of HVAC in buildings*, John Wiley & Sons, Inc, New York, January 2012.

A. Rabl, P. Curtiss, *Energy calculations – Building loads, handbook of heating, ventilation, and air conditioning*, (Edited by J. F. Kreider), CRC Press, Boca Raton, FL, 2001, ISBN 0-8493-9584-4.

T.A.H. Ratlamwala, I. Dincer, Efficiency assessment of key psychometric processes. *International Journal of Refrigeration*, 36, 1142–1153, May 2013.

S. K. Wang, Z. Lavan, P. Norton, *Air conditioning and refrigeration engineering*, CRC Press, Boca Raton, 2000, ISBN 0-8493-0057-6.

W. J. Wepfer, R. A. Gaggioli, E. F. Obert, Proper evaluation of available energy for HVAC. *ASHRAE Transactions*, 85(1), 214–230, 1979.

Liquid absorption refrigeration cycles

Introduction

In the previous chapter, we considered moist mixtures for which the dry gas could be assumed to behave as an ideal gas, which allowed us to simplify the modeling. This chapter deals with real fluid mixtures which cannot be modeled as ideal gases.

Because of interactions between the molecules of the various constituents, the estimation of thermodynamic properties of real fluid mixtures is much more difficult than that of pure ones. In the general case, it relates to chemical engineering and is outside the scope of this book. However, we give below some guidance on how to proceed, so that the reader can get an idea of the problems posed.

The representations of (NH_3-H_2O) and ($LiBr$-H_2O) absorption cycles with absorber, desorber, rectifier and solution exchanger in Oldham and Merkel diagrams will be explained.

The main interest of the liquid absorption refrigeration cycles is that they require only low power compared to their counterparts in mechanical vapor compression (less than 1%). Using a three-temperature thermodynamic cycle, they allow direct use of medium or high-temperature heat to produce cooling, requiring little or zero mechanical energy input. As such, their theoretically total efficiency in terms of primary energy is greater than that of vapor compression cycles.

Moreover, they involve fluids whose impact on the ozone layer and the greenhouse effect is zero: $ODP = GWP = 0$. However, they require an input of heat at intermediate or high temperature, so that their indirect impact is not necessarily zero: it depends on the energy source used.

Another advantage of liquid absorption cycles is that they can be used in integrated energy facilities producing both mechanical power, heat at intermediate temperature and cooling. This is known as trigeneration, and total efficiencies obtained are extremely high (see Chapter 10).

Real fluid mixtures

Physical phenomena brought into play

The behavior of mixtures in liquid-vapor equilibrium is generally different from that of pure substances, the molar or mass fraction of each component moving between limits that depend on pressure and temperature, due to the distillation that takes place. In the presence of several components, the phase change is more complex than for a pure substance. Its graphical representation in a chart is simple if the mixture is binary, the case to which we will restrict ourselves in this introduction.

DOI: 10.1201/9781003175629-13

To understand the phenomena that occur, consider the behavior of a binary mixture of propane and butane. Some properties of pure components (critical pressure and temperature, boiling temperature at 1 and 2 bar) are shown in Table 13.1. They show that propane is much more volatile than butane, as at 1 bar, respectively, they evaporate at −42.4°C and −0.7°C.

Consider a mixture of these two substances, and examine how it behaves at constant pressure when its temperature changes. Figure 13.1, called "equilibrium isobaric lens", shows the phase chart of this mixture at a pressure of 2 bar.

The vapor zone is at the upper right of the chart, the liquid zone at the bottom left and the liquid-vapor equilibrium in the lens delimited by the dew and bubble curves. The horizontal axis represents the mole fraction of propane, which means that the value 0 corresponds to pure butane (T_{sat} = 18.8°C) and the value 1 pure propane (T_{sat} = −25.4°C). As is usual when studying vapor mixtures, x represents the mole fractions in liquid phase, y the mole fractions in vapor phase and z, the overall mole fractions. The vertical axis is temperature, in °C.

Consider a mixture of given composition (here z_1 = 0.6 for propane, z_2 = 1−z_1 = 0.4 for butane) in the vapor state, corresponding to point A at temperature 10°C. If the mixture is cooled at constant pressure (e.g., in a heat exchanger), the characteristic point moves along the vertical segment AB as it remains in the vapor state.

In B, at around 0°C, distillation begins on the dew curve (or drop curve), so named because the first drop of liquid appears. The two components are liquefied together, but since butane is less volatile than propane, it condenses more, so that the first drop composition differs from average z. The mole fraction x in the liquid state of component 1 (here propane) is given by the abscissa of point E on the bubble curve, so called because in a reverse process of temperature change, it is on this curve that the first bubbles appear (here x_1 = 0.28), and the mole fraction of liquid butane is x_2 = 1−x_1 = 0.72.

Gradually, as the cooling continues, the composition of the liquid phase evolves, the mole fraction of propane being given by the abscissa of the intersection of the bubble curve and

TABLE 13.1
PURE COMPONENT PROPERTIES

	P_c (bar)	T_c (°C)	T_{sat} (1 bar) (°C)	T_{sat} (2 bar) (°C)
Propane	42.48	96.7	−42.4	−25.4
Butane	37.98	152.01	−0.7	18.8

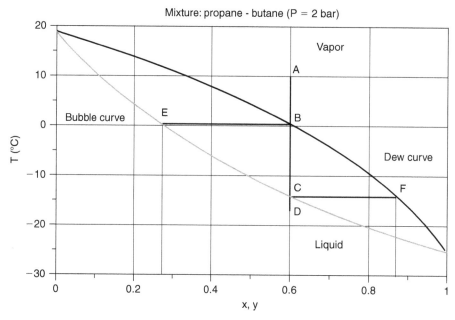

FIGURE 13.1
Isobaric equilibrium lens.

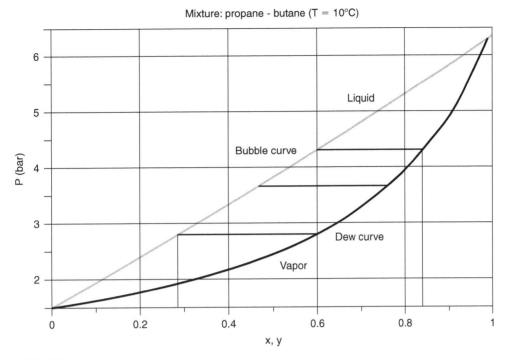

FIGURE 13.2

Isotherm equilibrium lens.

the corresponding isotherm, the figurative point moving from E to C, while that of the gas phase moves from B to F. At point C, there is only liquid, except for one last bubble of gas, whose composition is $y_1 = 0.85$ in propane and $y_2 = 1 - y_1 = 0.15$ in butane. Beyond, the mixture remains in liquid state, with its original composition.

In other words, at a given pressure, the state change of the mixture is not at constant temperature, but with a temperature "glide" (distillation range) that can be more or less important depending on the mixture. For example, for the above mixture, and the pressure of 2 bar, the temperature glide is 14°C.

Now consider an evolution of the mixture at constant temperature and variable pressure. On the chart of Figure 13.2, called equilibrium isotherm lens, the abscissa represents the propane mass fraction and the ordinate the pressure in bar. Let us still assume that the mixture has as overall composition ($z_1 = 0.6$ and $z_2 = 0.4$). If the pressure is lower than 2.78 bar, the mixture is entirely in the vapor state, and if it exceeds 4.3 bar, it is entirely liquid. Between these two pressures, it is in liquid-vapor equilibrium, and the mixture composition changes continuously in the liquid and vapor phases. For P = 2.78 bar, the liquid phase has a composition ($x_1 = 0.28$, $x_2 = 0.72$). For P = 3.7 bar, it is ($x_1 = 0.48$, $x_2 = 0.52$), and for P = 4.3 bar, it corresponds to the overall composition. Vapor phase compositions are as follows: ($y_1 = 0.6$, $y_2 = 0.4$) for P = 2.78 bar, ($y_1 = 0.77$, $y_2 = 0.33$) for P = 3.7 bar and ($y_1 = 0.85$, $y_2 = 0.15$) for P = 4.3 bar.

A special case is the so-called azeotropic mixtures where the bubble and dew curves meet for a given overall composition. One can show that an azeotropic mixture is a pressure extremum at constant temperature or temperature extremum at constant pressure. Such a mixture behaves in practice as a pure substance. Finally, when the temperature glide is small (<1°C), the error being committed by neglecting the distillation is very low. This is called quasi-azeotropic mixture. We chose here the propane/butane system because the glide is important and helps to demonstrate the phenomena peculiar to mixtures.

We introduced above two charts, binary mixture equilibrium isothermal and isobaric lenses, very useful for qualitatively understanding the phenomena involved. However, quantitatively, they are only valid for a single pressure or temperature, and above all, they contain no information on the enthalpies put in. They are not of much interest for applications in power engineering.

Principles of calculation of vapor mixtures

The calculation of vapor mixtures is made using, for each of the pure components, a well-chosen representation, such as an equation of state (EOS), and appropriately modeling interactions between molecules of different components, often through empirical correlations. The equation of the mixture is thus distinct from that of pure substances by the existence of "mixing rules" allowing one to calculate the coefficients.

You will find in the Thermoptim-UNIT portal a special section on this subject and thermodynamic software packages available.[1]

To illustrate, we will use the Peng-Robinson equation of state for a pure substance, which in molar units writes:

$$P = \frac{RT}{V-b} - \frac{a}{V^2 + 2bV - b^2} \tag{13.1}$$

with $b = 0.0778 \frac{RT_c}{P_c}$

$$a(T) = 0.45724 \frac{R^2 T_c^2}{P_c} \left[1 + \left(0.37464 + 1.54226\omega - 0.26992\omega^2 \right) \left(1 - T_r^{0.5} \right) \right]^2$$

With this equation, four data are sufficient to characterize a pure substance: the critical coordinates T_c, P_c, the acentric factor ω, and the molar mass M to move to mass quantities.

For Equation (13.1) to be representative of a mixture, it is necessary to estimate the values of the mixture coefficients $a_m(T)$ and b_m, which is possible by using the following type of relations, indices i and j corresponding the various constituents:

$$a_m(T) = \sum_{i=1}^{n} \sum_{j=1}^{n} \left(x_i x_j a_{ij}(T) \right)$$

$$\text{with } a_{ij} = \left(1 - k_{ij} \right) \sqrt{a_i(T) a_j(T)} \quad \text{if } i \neq j \tag{13.2}$$

$$b_m = \sum_{i=1}^{n} x_i b_i \tag{13.3}$$

The coefficients k_{ij} must be either determined experimentally or estimated. It is generally assumed that the binary coefficients can be used even for mixtures of order higher than two, thus implicitly assuming that the interactions of ternary and higher order are negligible.

For example, an equation giving k_{ij} is the following, V_c being the critical volume:

$$k_{ij} = 1 - \frac{\sqrt[8]{V_{ci} V_{cj}}}{\left(\sqrt[3]{V_{ci}} + \sqrt[3]{V_{cj}} \right)^3} \tag{13.4}$$

It is important to know that an error in the determination of these interaction coefficients can lead to completely wrong results. It is strongly recommended to validate the models on the basis of reference values, obtained experimentally if possible, and if not by comparison with validated software. The great danger in building models of fluids from equations of state and mixing rules is that one almost always gets results. In the absence of reference values, we may be tempted to have too much confidence in them... It is in this case recommended to try and build different models based on data from different sources and compare their results to remove the doubt.

---

[1] https://direns.mines-paristech.fr/Sites/Thopt/en/co/Progiciels_de_thermodynamique.html.

When the equation of state of the mixture is determined, the calculation of vapor-liquid equilibrium is done by writing the equality of fugacities of each species in each phase, that is to say, for each component:

$$f_{li}(P,T,x) = f_{vi}(P,T,y) \tag{13.5}$$

Without going into detail, let us recall that fugacities are related to the departure of the chemical potential of a real fluid from that of the ideal gas. Equation (13.5) is thus derived from the Gibbs free enthalpy minimum at equilibrium.

This system of equations generally results in complicated expressions, the resolution of which can be quite difficult numerically. In addition, the accuracy of the model obtained depends of course on that of the equation of state retained. For example, the main interest of the Peng-Robinson is that it is a cubic equation whose solutions can be calculated formally, but it is generally not very good in the liquid phase, for which methods based on activity coefficients are more accurate.

In practice, therefore, it is generally necessary to retain much more complicated models for pure components, and the estimation of the mixture properties becomes even more difficult.

In conclusion, the calculation of mixtures is often difficult to achieve and does not lend itself to simple graphical representations.

However, it is usually possible for applications in power engineering to simplify things by introducing a number of assumptions chosen wisely and well justified experimentally. The solutions adopted depend on the applications as we shall see.

As already mentioned in Chapter 5, we will use in the rest of this book the CTP Lib library, to model various fluids not present in the core of Thermoptim. This library can be downloaded from the Thermoptim-UNIT portal (www.thermoptim.org). It allows in particular to model many fluids using a Peng Robinson EOS and also has very precise specific equations for the pair (NH_3-H_2O) which will be used later in this chapter for the study of absorption machines and in Chapter 14 for the Kalina cycle.

Note that the Peng Robinson EOS accuracy can be improved by using corrective factors such as Mathias Copeman's Alpha function, implemented in CTP Lib, which uses three coefficients that can be fitted numerically, either against experimental data or against a very precise reference equation.

BOX 13.1
Key issue

CTP lib thermodynamic properties server

CTP Lib is the library for calculating thermodynamic properties of the Center for Process Thermodynamics of Mines ParisTech: https://direns.mines-paristech.fr/Sites/Thopt/en/co/CTPLib.html.

Coupled to Thermoptim, it can calculate the properties of mixtures of real fluids, which the software cannot do itself.

Charts for refrigerants with temperature glide

For refrigeration applications, the composition of a refrigerant blend remains constant in the vapor and liquid phases (if we neglect the influence of lubricating oil). In the liquid-vapor equilibrium zone, it is not strictly true, but almost always one does not require a precise calculation of the exact composition of the gas and liquid phases at the outlet of the expansion valve and the evaporator. By analogy with pure substances, we generalize the concept of vapor quality by that of mean quality, equal to the mass of the vapor phase, all components combined, relative to the total mass of liquid and vapor phases. We can then content ourselves with a fixed composition mixture model, much simpler. An equation numerically adjusted to the experimental values of a given mixture will thus be at the same time more accurate and easier to calculate than the general formulation we have just seen. Furthermore, equations of state having been developed specifically for these applications, the reader will find references in the literature.

We give below the charts of two blends, R404A and R407C (Figures 13.5 and 13.6). In what follows, we shall only point out the differences introduced by the temperature glide, lower for R404A than for R407C. The plot of a refrigeration cycle in the R407C charts has been given in Chapter 11.

Refrigerant blend charts

Essentially, at a given pressure, the differences between the component saturation temperatures introduce a temperature glide in the liquid-vapor equilibrium zone: when boiling starts, the temperature (bubble temperature) is lower than when boiling ends (dew point), while they are equal for pure substances. Thus, the pressure and temperature do not remain constant during boiling or condensation. In entropy (T, s) and (h, ln(P)) charts, isobars or isotherms are not horizontal in the vapor-liquid equilibrium zone (Figures 13.3 and 13.4).

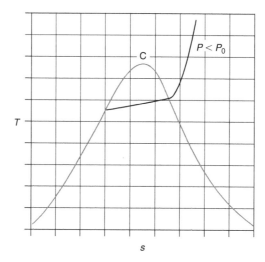

FIGURE 13.3
Entropy chart.

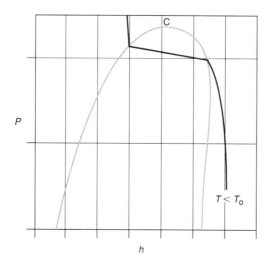

FIGURE 13.4
(h, ln(P)) chart.

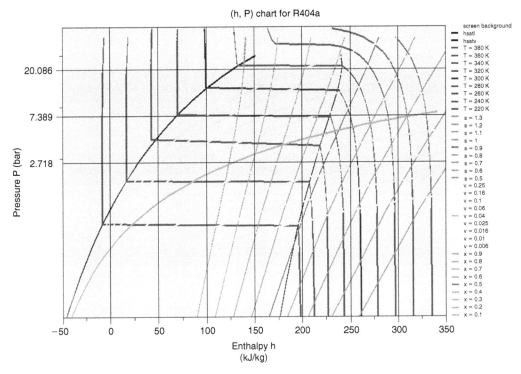

FIGURE 13.5

(h, ln(P)) chart for R404a.

As shown in Figures 13.3–13.6, an isobar is now represented in this area by an ascending line segment in the entropy chart and an isotherm by a straight down slope line in the (h, ln(P)) chart.

The existence of a temperature glide has a direct impact on the design of heat exchangers used in refrigeration cycles. In first approximation, these can be considered as isobaric, so that differences in temperature between the coolant and other fluids are not the same as that observed in machinery using a pure fluid. Generally speaking, for blends with temperature glide, heat exchangers should be of the counter-flow type, which was not imperative with pure fluids.

Figures 13.5 and 13.6 correspond to (h, ln(P)) chart for R404A and entropy (T, s) chart for R407C.

Charts used in absorption refrigeration cycles

The main interest of liquid absorption refrigeration cycles, which will be presented below, is that they require only low power compared to their counterparts in vapor compression cycles (less than 1%). Using a tri-thermal thermodynamic cycle, they can directly make use of heat at medium or high temperatures to produce cooling, requiring no or little mechanical energy input. As such, they theoretically have total efficiencies in terms of primary energy greater than vapor compression cycles.

Liquid absorption cycles involve at least two fluids: a solvent and a solute (the coolant). While other pairs are being studied, the only ones used in practice for almost all applications are the two LiBr-H$_2$O and H$_2$O-NH$_3$ pairs. Among the requirements for the prospective pair to be appropriate, the solvent should firstly have a high affinity toward the solute, and secondly, the latter should be much more volatile than the solvent so that the separation of the two components is optimal. The curve in Figure 13.7 shows, for water-ammonia pair, the relationship between the ammonia liquid and vapor mass fractions at different pressures (1, 2, 5, 10 and 20 bar). It clearly shows that, as soon as the liquid mass fraction of solute x exceeds 0.4, the ammonia content of the vapor exceeds 0.95. For lithium bromide-water mixture, the separation is even stronger, so that one can legitimately assume that water vapor is pure.

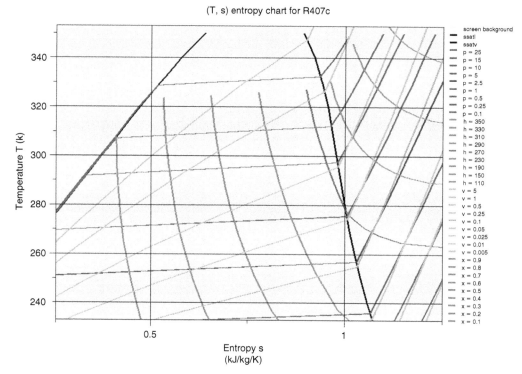

FIGURE 13.6

Entropy chart for R407c.

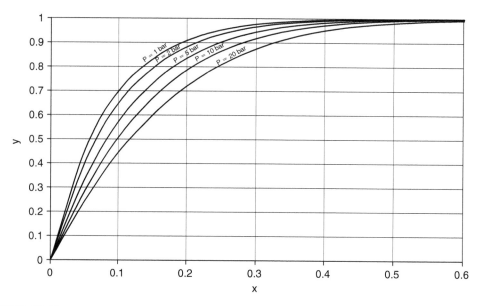

FIGURE 13.7

Vapor composition function of liquid NH_3-H_2O composition.

With the LiBr-H_2O pair, water is the refrigerant, which imposes two constraints: first working pressures are very low given the saturation pressure law of water, and secondly the minimum cycle temperature must be greater than 0°C. Machines using the lithium bromide-water pair are only used for air conditioning.

Oldham chart

The Oldham chart can view the saturation pressure curves of the mixture in question, for different values of its composition (mass fraction of solute or solvent, as appropriate). Pressure appears as the ordinate, with a logarithmic scale, while the abscissa corresponds to the bubble

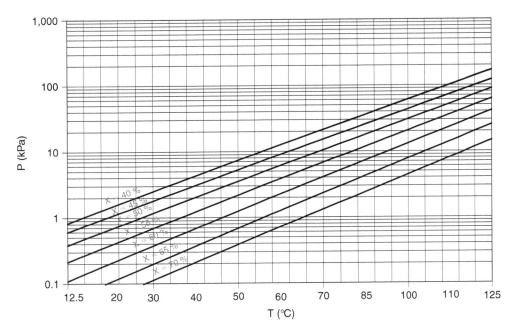

FIGURE 13.8

Oldham chart for LiBr-H$_2$O.

temperature T_b (Figure 13.8). It is also possible to show the curves corresponding to the dew point, but this is of less interest.

If one chooses ($-1/T_b$) as temperature scale (T_b in K), the iso-quality curves are very close to straight lines. The extremes correspond to the vapor-liquid equilibrium of pure substances. An evolution of the rich or weak solution corresponds to a line segment tilted parallel to the nearest iso-quality curve. As the general shape of cycles in this chart is close to a parallelogram, scientists are used to illustrating the architectures of the machines in the form of diamond (Figure 13.9).

In machines using LiBr-H$_2$O mixture, the difference in vapor pressure of the solvent (LiBr) and solute (H$_2$O) is such that we can neglect the mass fraction of solvent vapor, thereby simplifying calculations: the equilibrium chart can then be directly calibrated as a function

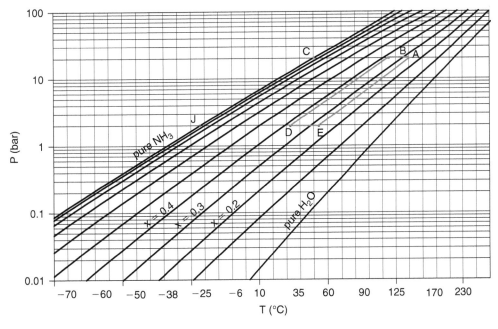

FIGURE 13.9

Cycle representation in the Oldham NH$_3$-H$_2$O chart.

TABLE 13.2

COEFFICIENTS OF EQUATIONS (13.6) AND (13.7)

A_0	−2.00755	B_0	124.937	C	7.05
A_1	0.16976	B_1	−7.71649	D	−1,596.49
A_2	−3.13E-03	B_2	0.152286	E	−104,095.5
A_3	1.98E-05	B_3	−7.95E-04		

of the solution temperature. Incidentally, it is customary to set the LiBr-H$_2$O pair chart as a function of solvent (LiBr) mass fraction and not of solute. Water being likely to crystallize at low temperatures, the crystallization curve of the mixture often appears on the chart, which corresponds to the lower operating limit of machines.

For this pair, ASHRAE (2001) provides Equations (13.6) and (13.7), established by generalizing the mixture saturation law of the coolant (water) pressure, in which the water temperature t' (°C) is replaced by a linear function of the solution temperature T (°C). P, expressed from the decimal logarithm, is the pressure in kPa, and X the mass fraction of LiBr mixture (Table 13.2).

These equations, which were used to prepare the chart in Figure 13.8, are valid in the ranges of values:

$$-15 < t' < 110°C, 5 < t < 175°C, 45 < X < 70\%$$

$$\log(P) = C + \frac{D}{t' + 273.15} + \frac{E}{(t' + 273.15)^2} \tag{13.6}$$

$$t' = \frac{t - \sum_{i=0}^{3} B_i X^i}{\sum_{i=0}^{3} A_i X^i} \tag{13.7}$$

BOX 13.2
Key issues

Modeling of the pair LiBr-H$_2$O

The ASHRAE equations giving the properties of the pair LiBr-H$_2$O have been implemented in Thermoptim as an external substance.
It is thus possible to model absorption refrigeration machines using this pair.
Results obtained are given in several sections of this book, and the associated models are available on the portal:

* Trigeneration by micro-turbine and absorption cycle, Chapter 10, whose absorber is presented below;
* Single effect LiBr-H$_2$O absorption cycle, below.

Figures 13.8 and 13.9 provide Oldham charts for LiBr-H$_2$O and H$_2$O-NH$_3$ pairs.

Merkel chart

The Oldham chart is easy to understand and use as discussed below, but, like isobar or isotherm equilibrium lenses, it has the disadvantage of not providing any information about the energies put into play.

The Merkel chart (or enthalpy-concentration chart) is constructed with the enthalpy as ordinate, and the mass fraction of solute (or often the solvent for the LiBr-H$_2$O pair) as

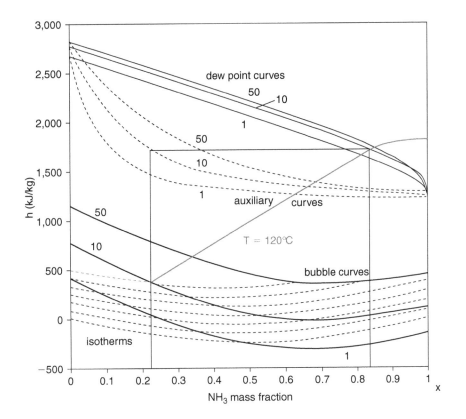

FIGURE 13.10

Isotherms in the Merkel chart.

abscissa. It allows us to superimpose information from multiple isobar equilibrium lenses while showing enthalpies. Figure 13.10 gives the ammonia-water mixture Merkel chart.

There are five types of curves:

- A set of bubble curves at P = Const, continuous lines with the concavity turned upwards;
- A set of dew curves for the same pressures, solid lines with the concavity turned downwards
- A set of liquid zone isotherms, dashed in the lower part of the chart, which is valid whatever the pressure, the influence of pressure on the enthalpy of the liquid being negligible in first approximation (for example, the difference in enthalpy of the liquid solution and 30% ammonia by weight is only 5 kJ/kg at 0°C when pressure increases from 0.5 to 50 bar, while the useful part of the chart covers more than 1,000 kJ/kg in the liquid zone);
- A set of curves called auxiliary or construction, dashed in the upper part of the chart, which, for a given pressure, knowing the solute liquid mass fraction x, determine the solute vapor mass fraction y;
- A set of refrigerant iso-quality curves in the vapor phase, which, for a given pressure, provide the mass fraction of liquid corresponding to the chosen vapor iso-quality (the latter set of curves is sometimes omitted from the Merkel chart, as in this example).

However, isotherms in the vapor zone do not appear in the Merkel chart, because they depend on pressure, and the chart would become unreadable.

Notice that this chart superimposes two kinds of curves: liquid-vapor equilibrium and isotherm curves in the liquid state. There may be difficulties in interpreting the cycle points if you do not specify whether they are or not in liquid-vapor equilibrium.

In the NH_3-H_2O pair chart of Figure 13.10, the saturation pressure for x = 0.22 and T = 120°C is 10 bar. To know the mass fraction value y of the vapor, we draw the vertical segment passing by x = 0.22 to the auxiliary curve corresponding to 10 bar, then the horizontal

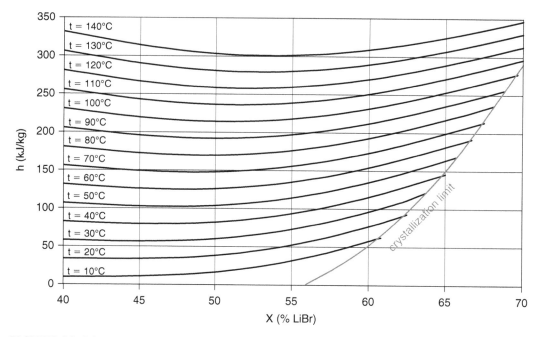

FIGURE 13.11

Merkel chart of LiBr-H$_2$O mixture.

segment from that point until the dew isobar for 10 bar and then we draw the vertical to the horizontal axis, which gives y = 0.83.

We have already reported that for LiBr-H$_2$O pair, we can neglect the mass fraction of solvent vapor, that is to say, consider that the refrigerant vapor behaves like pure water. The Merkel chart is thus simplified, and we generally content ourselves to build it by making only the liquid zone isotherms appear, bounded on the lower right by the crystallization curve (Figure 13.11).

Given this simplification, it is possible to determine the LiBr-H$_2$O mixture enthalpy through a simple equation, as proposed by ASHRAE (13.8) (Table 13.3, ASHRAE, 2001), where t is the temperature of the solution in °C, h its specific enthalpy in kJ/kg and X the LiBr mass fraction of the mixture. This equation, which was used to establish the chart of Figure 13.11, is valid within the ranges of values: 5 < T < 165°C, 40 < X < 70%. If this equation is chosen, the origin of the water vapor enthalpy is set at 0°C in the liquid state (which corresponds to the usual conventions for pure water charts).

$$h = \sum_{i=0}^{4} A_i X^i + t \sum_{i=0}^{4} B_i X^i + t^2 \sum_{i=0}^{4} C_i X^i \tag{13.8}$$

Principle of the absorption machine

An absorption machine exchanges heat with at least three heat sources at three different temperature levels (Figure 13.12).

TABLE 13.3

COEFFICIENTS FOR EQUATION (13.8)

A_0	−2.024.33	B_0	18.2829	C_0	−0.037008214
A_1	163.309	B_1	−1.1691757	C_1	2.89E-03
A_2	−4.88161	B_2	0.03248041	C_2	−8.13E-05
A_3	0.06302948	B_3	−4.03E-04	C_3	9.91E-07
A_4	−2.91E-04	B_4	1.85E-06	C_4	−4.44E-09

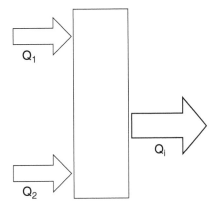

FIGURE 13.12
Absorption machine principle.

Let us mark by index 1 the hot source, 2 the cold one and i the intermediate temperature source and call W the work put into play in the machine and ΔS the entropy generation. Writing the two laws of thermodynamics leads to Equations (13.9) and (13.10).

$$W + Q_1 + Q_i + Q_2 = 0 \qquad (13.9)$$

$$\frac{Q_1}{T_1} + \frac{Q_i}{T_i} + \frac{Q_2}{T_2} = \Delta S \qquad (13.10)$$

Two cases can be distinguished according to whether the temperature of the environment is or is not above the temperature of the cold source.

In the first case, which corresponds to the refrigeration cycle, the machine cools an enclosure at low temperature and rejects heat into the environment, T_i being greater than or equal to T_0.

In the second case, which corresponds to the heat pump cycle, the machine extracts heat from the cold source (which may be the environment) and uses it to heat a chamber at intermediate temperature T_i.

Let us consider the first case and assume that the intermediate temperature source is the environment, T_i being equal to T_0.

Combining Equations (13.9) and (13.10), we get

$$W = -Q_c\left(1 - \frac{T_0}{T_c}\right) - Q_f\left(1 - \frac{T_0}{T_f}\right) + T_0\Delta S \qquad (13.11)$$

The first and the third term being negative and the second positive, this equation shows that work W may be zero, that is to say that it is possible to design a three-temperature machine for producing refrigeration cooling without consuming work. In practice, however, many absorption chillers include circulating pumps.

Study of a NH₃-H₂O absorption cycle

An ammonia absorption cycle machine has eight main components (Figure 13.13):

* A desorber-rectifier, which receives heat flux from the heat source, and wherein enters the strong solution at high pressure, preheated in solution exchanger (B). The exiting fluids are on the one hand, the almost pure refrigerant vapor (NH₃) (C), and secondly the depleted solution (A);

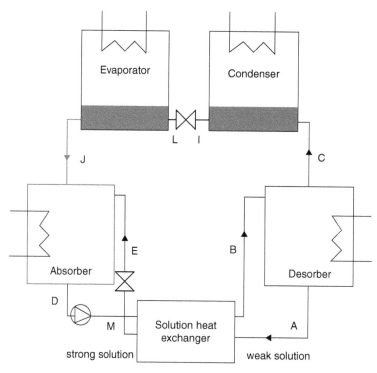

FIGURE 13.13

Sketch of a NH_3-H_2O absorption cycle.

- The condenser, from which exits the vapor condensed and possibly sub-cooled (I), the heat removed being rejected in the surroundings;

- A refrigerant expansion valve, which reduces the pressure of the refrigerant, exiting in the two-phase state at low temperature (L);

- An evaporator where the refrigerant at low temperature and pressure is vaporized and possibly slightly superheated (J), extracting useful heat (refrigeration effect) from the cold source;

- An absorber, in which enters the vaporized refrigerant and the weak solution preheated in solution exchanger, and out of which flows the strong solution (D), the heat removed being rejected in the surroundings;

- A solution heat exchanger, which allows for internal regeneration between the strong solution and the weak solution;

- A pump which is needed to pressurize the strong solution before it enters the heat exchanger;

- An expansion valve at the outlet of the heat exchanger which reduces the weak solution pressure to that of the absorber.

Note that the condenser, the refrigerant expansion valve and the evaporator work the same way as in a vapor compression refrigeration cycle.

We discuss below such a cycle and represent it both in isobaric equilibrium lenses and in the Merkel chart, and then we plot it in Oldham chart. The numerical values for constructing these charts can be obtained from equations proposed by Ziegler and Trepp implemented in CTP Lib.

The purpose of this section being simply to illustrate the use of thermodynamic charts, the cycle used is a theoretical cycle, in which some actual cycle irreversibilities are neglected. Readers interested in a very detailed study of these cycles may refer to the literature, especially to Techniques de l'Ingénieur articles No. BE 9735 and BE 9736 by Maxime Duminil. They will find there detailed explanations on how to calculate a rectification.

In the desorber, which receives a heat supply, the strong solution is partially vaporized (B–C) at 120°C and partially depleted and heated (B–A) to 140°C. B–C is a horizontal segment on the isobaric equilibrium lens of Figure 13.14 and an oblique segment on the Merkel

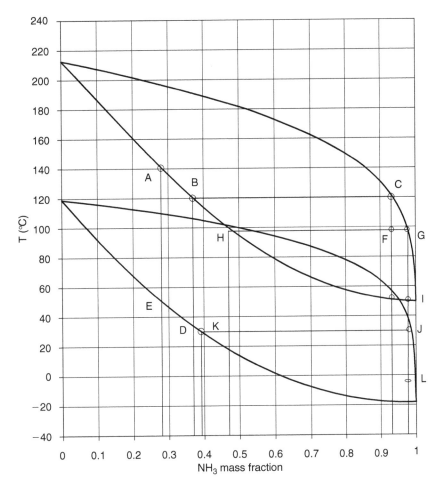

FIGURE 13.14

Isobaric equilibrium lenses at 2 and 20 bar.

chart of Figure 13.15, which is easily constructed using auxiliary curves. B–A follows in both cases the saturated isobaric line. Point A determination assumes that we are given a hypothesis about the operation of the desorber: either the temperature T_A, or the liquid mass fraction of weak solution x_p or the heat provided. Note that the depletion of the solution is made possible by a special arrangement of the apparatus, e.g., counter-flow, otherwise $T_A = T_C$. It is important to emphasize this point because very often in the literature, the implicit or explicit assumption is made that the desorber is isothermal, while presenting points A, B and C positioned in the Merkel chart as in our example, which is physically impossible.

A: $T = 140°C$, $P = 20$ bar, $x = 0.28$ weak liquid exiting counter-flow desorber

B: $T = 120°C$, $P = 20$ bar, $x = 0.37$ strong liquid beginning vaporization

C: $T = 120°C$, $P = 20$ bar, $y = 0.93$ vapor stemming from the strong liquid

The vapor exiting the desorber in C is then rectified, that is to say slightly cooled (here at 97°C) because its quality in refrigerant is insufficient. This operation (not described here because it is outside the scope of this book) results in almost pure vapor in G ($y = 0.995$). Determining point G (on the dew point curve) assumes that its temperature or vapor quality is given. A very small amount of liquid (H) also leaves the rectifier. It generally falls into the absorber where it is mixed with the weak liquid. Point H is deduced simply from G on both charts: it is on the bubble curve at the temperature and pressure of G.

G: $T = 97°C$, $P = 20$ bar, $y = 0.995$ steam rectifier output

H: $T = 97°C$, $P = 20$ bar, $x = 0.47$ liquid rectifier output

F: $T = 97°C$, $P = 20$ bar, $z = 0.93$ average rectifier state

The refrigerant vapor is directed to the condenser, which it exits in I, at high pressure and 54°C. It is then expanded isenthalpically (point L in two-phase zone at low pressure).

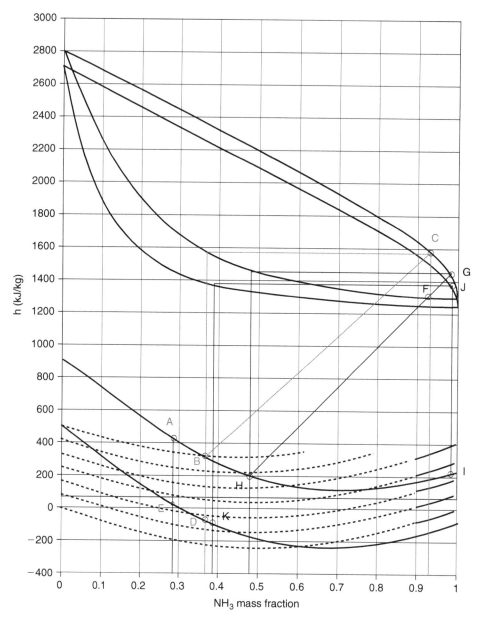

FIGURE 13.15

Merkel chart.

The refrigerant is then vaporized in the evaporator at point J. To determine points I and L, we generally assume that the refrigerant is pure, and we use the ammonia (h, P) or (T, s) chart.

I: $T = 54°C$, $P = 20$ bar, $y = 0.995$ condensed liquid condenser outlet;

L: $T = -15°C$, $P = 2$ bar, $z = 0.995$ two-phase throttling outlet, evaporator inlet.

The weak solution heats the strong solution in a liquid-liquid regenerator heat exchanger. Mass fractions x_r and x_p of both solutions being constant, these two processes are substantially vertical segments (AE) and (DB) in both charts that we consider. Note that in the Merkel chart, enthalpies put into play are comparable and temperature levels consistent. As the flow rates of the two solutions are slightly different, heat exchange can take place. Strictly speaking, points E and B are not located exactly on the bubble curves corresponding to their pressure: their exact state depends on the characteristics of the solution heat exchanger. For simplicity of analysis, we will not consider this discrepancy in what follows: we simply estimate *a posteriori* the effect of this assumption.

D: $T = 30°C$, $P = 2$ bar, $z = 0.37$ strong liquid absorber exit;

E: $T = 55°C$, $P = 2$ bar, $z = 0.28$ weak liquid solution heat exchanger exit.

In the absorber, the vapor (assumed pure) coming out of J is condensed in K and then mixed with weak liquid to give the strong solution that exits in D, the operation involving heat extraction. The determination of point D assumes that we are given an additional assumption on the operation of the absorber: either the value of the quality of the strong solution or its temperature or heat extracted. As mentioned above, point B is obtained from D, knowing its pressure and x_r.

J: $T = 45°C$, $P = 2$ bar, $y = 0.995$ steam evaporator outlet;

K: $T = 45°C$, $P = 2$ bar, $x = 0.395$ steam condensed in the absorber.

To be precise, we should represent point M at the outlet of the strong solution pump. However, as the compression in the liquid state is almost isothermal, and involves a small amount of work, M is almost coincident with D, knowing that it is not in the liquid-vapor equilibrium state but in the liquid sub-cooled state.

M: $T = 30°C$, $P = 20$ bar, $z = 0.37$ strong solution pump outlet.

Calling $\dot{m}$ the refrigerant flow, $\dot{m}_p$ the weak solution flow, $\dot{m}_r$ the strong solution flow, the desorber material balance (total mass flow rate and mass flow of refrigerant) provides

$$\dot{m}_r = \dot{m} + \dot{m}_p \tag{13.12}$$

$$\dot{m}_r\left(1 - x_r\right) = \dot{m} + \dot{m}_p\left(1 - x_p\right) \tag{13.13}$$

This system of equations allows us to express $\dot{m}_p$ and $\dot{m}_r$ in terms of $\dot{m}$, x_p and x_r:

$$m_r = m\frac{1 - x_p}{x_r - x_p} \tag{13.14}$$

$$m_p = m\frac{1 - x_r}{x_r - x_p} \tag{13.15}$$

Enthalpies put into play in the processes are (under the simplifying assumptions that we used, including assuming points B and E on their respective bubble curves) as follows:

Condenser: $Q_{cond} = \dot{m}(h_I - h_G)$
Evaporator: $Q_{evap} = \dot{m}(h_J - h_I)$
Desorber: $Q_{des} = \dot{m}h_C + \dot{m}_p h_A - \dot{m}_r h_B$
Absorber: $Q_{abs} = -\dot{m}h_J - \dot{m}_p h_E + \dot{m}_r h_D$
Solution heat exchanger: $Q_{sol} = \dot{m}_r(h_B - h_D) = \dot{m}_p(h_A - h_E)$
Work of the strong solution pump: $W = \dot{m}_r(h_M - h_D) \neq 0$, which we can estimate equal to $v\Delta P$.

The COP is defined as the ratio of useful energy Q_{evap} to purchased energy $(Q_{des} + W)$.

By neglecting work W, we get

$$COP = \frac{\dot{m}\left(h_J - h_I\right)}{\dot{m}h_C + \dot{m}_p h_A - \dot{m}_r h_B} \tag{13.16}$$

$$COP = \frac{\left(h_J - h_I\right)}{h_C + \dfrac{1 - x_p}{x_r - x_p}h_A - \dfrac{1 - x_r}{x_r - x_p}h_B} \tag{13.17}$$

This way of working includes the implicit assumption that the solution exchanger is balanced, which is not totally verified as shown in Tables 13.4 and 13.5, valid for a cycle without subcooling at the condenser outlet.

However, the lowest temperature in the absorber being equal to 30°C, we can easily use the cold source to sub-cool the liquid refrigerant at that temperature, which lowers the enthalpy of points I and L, and thus the vapor quality after expansion. The enthalpy change in the evaporator increases, and the COP increases slightly (about 10%, from 0.4 to 0.44).

TABLE 13.4

STATE OF ABSORPTION CYCLE POINTS

	T (°C)	P (Bar)	z	x	y	h (kJ/kg)	m (kg/s)
A	140	20		0.28		450	8
B	120	20		0.37		325	7
C	120	20			0.93	1,575	7
D	30	2	0.37			−65	7
E	55	2		0.28		50	8
F	97	20	0.93			1,300	
G	97	20	0.995			1,440	1
H	97	20				200	
I	54	20			0.995	232	1
J	45	2			0.995	1,375	1
K	45	2		0.395		−60	1
L	−15	2	0.995			232	1
M	45	20				−60	7

TABLE 13.5

PERFORMANCE OF THE ABSORPTION CYCLE

Condenser	−1,208
Evaporator	1,143
Desorber	2,900
Absorber	−2,230
Strong solution exchanger	2,730
Weak solution exchanger	3,200
Pump work	5 (estimate)
COP	0.394

It is possible to represent the points of this cycle at the vapor-liquid equilibrium in the Oldham chart (Figure 13.16). We call this representation the solution cycle. It has the characteristic shape of a diamond. Note that the analyses that this type of chart allows are much more cursory than those provided by the Merkel chart: the only information available are the pressures, temperatures and mass fractions of the solution.

The absorption cycle that we have presented operates through a process called single effect, leading to relatively limited performance. If you have a heat source hot enough, you can use more efficient cycles, known as multiple effect or cascade. The interested reader can refer to the literature.

Modeling LIBR-H$_2$O absorption cycle in thermoptim

Modeling a cycle using the pair LiBr-H$_2$O is easier than what we have just seen since we can consider that water vapor in the desorber outlet is almost pure, so that the rectifier of Figure 13.13 becomes useless. The diagram of the machine is a little simplified. As also the properties of the pair can be easily modeled using Equations (13.6–13.8), the cycle can be modeled with Thermoptim without too much difficulty, using three external classes[2] to represent the mixture properties, the absorber and the desorber, since, as we said above, condenser, refrigerant expansion valve and evaporator operate the same way as in a vapor compression refrigeration cycle.

---

[2] These classes are available in Thermoptim model library: https://direns.mines-paristech.fr/Sites/Thopt/en/co/module_Logiciel_7.html.

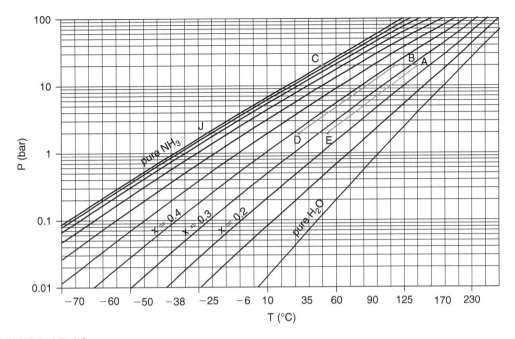

FIGURE 13.16

Representation in the NH_3-H_2O Oldham chart.

The absorber and desorber are modeled by two external nodes which exchange heat with the outside, as a high temperature input in the desorber and an intermediate temperature cooling in the absorber.

Representation of thermal coupling is possible using two thermocouplers, called "heat input" and "cooling" on the synoptic view in Figure 13.17. Each node calculates the external heat energy that must be exchanged, and each thermocoupler recalculates the "exchange" process to which it is connected.

On this synoptic view, we recognize in the central part the absorption machine, with the three sources with which it exchanges heat: above the cooling at medium temperature, in the lower left part chilled water and at the bottom right high temperature steam. The effectiveness of the solution heat exchanger has been set equal to 0.8.

Cooling energy is considered here as useful energy and thermal energy supplied to desorber as purchased energy, which leads to a single effect absorption cycle COP equal to 0.72.

Design of the external component used in the trigeneration example

In Chapter 10, we presented a trigeneration facility using a microturbine and an absorption cycle. The COP of this cycle is also equal to 0.72. Trigeneration indicator values are given at the end of Chapter 10.

In this section, we explain how the external component representing the absorption unit has been designed.

The difference with the model of Figure 13.17 is that the four elements comprising the absorption machine, the absorber, the desorber and the solution exchanger are included in a single external component.

To be easily comprehensible, the model presented here is relatively simple thermodynamically. It thus presents several limitations: in particular, as it is implemented, it allows for a direct calculation of the component knowing the temperature of the absorber and the desorber and the solution exchanger effectiveness but not an inverse calculation where we seek to determine, for given operating conditions, one of these quantities. However, it has the great advantage of being already sufficiently complicated to show clearly problems that arise during the creation of an external component and solutions that help solve them.

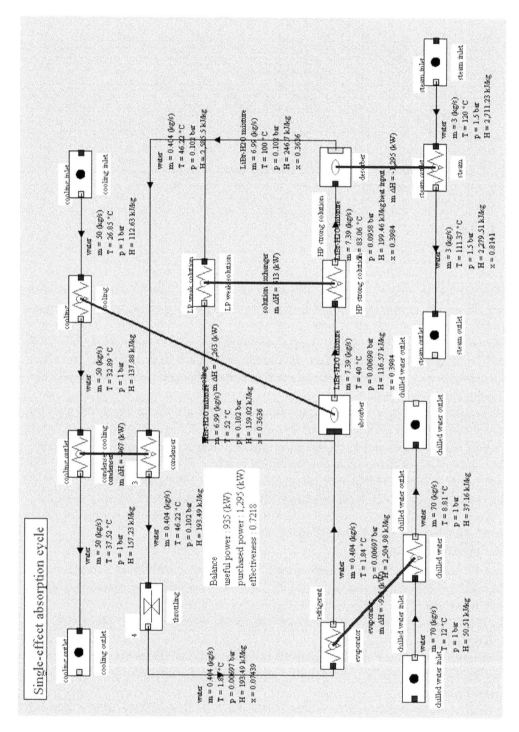

FIGURE 13.17

Synoptic view of a single-effect absorption cycle.

The missing component is the external subsystem (absorber–solution exchanger–desorber) of the absorption cycle using the pair LiBr-H_2O (Figure 10.21). The model used is that of a simple effect machine with the following assumptions (note that in what follows, for reasons of consistency of notations, strength or weakness of the solution is expressed relative to the refrigerant (water) and not to the solvent (lithium bromide), unlike the usual convention in the U.S., which is that adopted by ASHRAE):

* The desorber and condenser are at the same pressure;
* The absorber and evaporator are at the same pressure;
* The refrigerant vapor leaving the evaporator is saturated;
* The liquid refrigerant leaving the condenser is saturated;
* The refrigerant vapor leaving the desorber is saturated at the equilibrium temperature of the weak solution in refrigerant at desorber pressure;
* The weak solution is boiling at the exit of the desorber, assumed isothermal;
* The strong solution leaves saturated the absorber, assumed isothermal.

Model parameters are as follows:

* The absorber and desorber temperatures;
* The heat exchanger solution effectiveness.

The input data of the model are as follows (provided by other system components):

* The absorber and desorber pressures;
* The refrigerant flow rate.

The outputs are as follows:

* The absorber and desorber thermal loads;
* The weak and strong solution concentrations;
* The weak and strong solution flow rates.

A component graphical interface can be deduced (Figure 13.18). The bottom right of the screen has to be created, the rest being defined as Thermoptim standard.

Parameters correspond to the first three lines added, the results of calculations to others. Below are listed the two thermocouplers that the component defines and sets.

The input data are provided by other components of the global model: the refrigerant flow set by the upstream "cooling effect" process is here 0.055 kg/s, the absorber pressure (point 1) is equal to 0.00697 bar, corresponding to an evaporation temperature of 1.93°C and desorber pressure (point 2) is equal to 0.102 bar, corresponding to a condensing temperature of 46.22°C.

A peculiarity of this component is that it does not change its downstream point 2, whose state is considered set. It would of course be possible to approach the problem from another angle, but as we said earlier, we do not want to overcomplicate things in this example.

Thermodynamic model

The model equations are obtained as follows.

LiBr-H_2O mixture

The LiBr-H_2O mixture is modeled using the Equations (13.6–13.8).

Absorber

The weak refrigerant solution is sprayed into rain and washes the refrigerant vapor at low pressure, which is absorbed, releasing its heat of condensation and heat of dilution.

This heat Q_{abs} is extracted by cooling water which then cools the condenser.

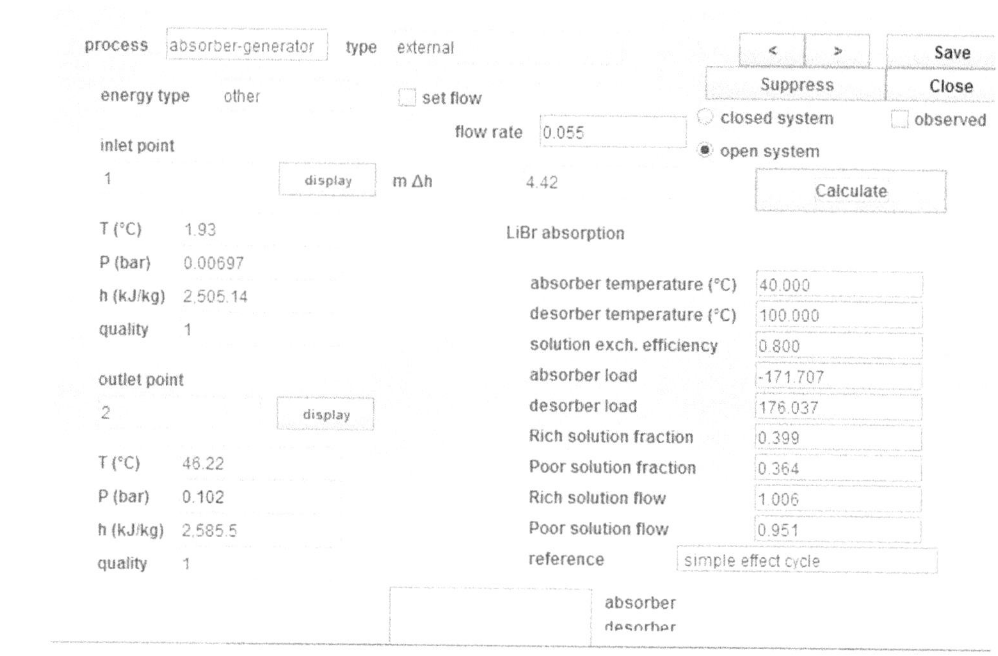

FIGURE 13.18

Screen of the external component.

With the assumption that the absorber is at a constant temperature T_{abs} and the strong solution is saturated, the absorber equations are the following:

The equation of solution saturated vapor pressure $P_{abs} = P(x_{sr}, T_{abs})$ provides the concentration of the strong saturated solution and therefore its enthalpy $h_{srD} = h(x_{sr}, T_{abs})$.

Known data: m_r, h_{r1}, T_{abs}, P_{abs}

Unknown: m_{sr}, m_{sp}, x_{sp}, h_{spE}

Conservation of mass: $m_r + m_{sp} = m_{sr}$

Conservation of mass of solution: $(1 - x_{sp}) m_{sp} = (1 - x_{sr}) m_{sr}$

These two equations provide m_{sp} and m_{sr} if x_{sr}, x_{sp} and m_r are known:

$$m_{sr} = m_r \frac{1 - x_p}{x_r - x_p}$$

$$m_{sp} = m_f \frac{1 - x_r}{x_r - x_p}$$

Conservation of enthalpy: $m_r h_{r1} + m_{sp} h_{spE} = m_{sr} h_{srD} + Q_{abs}$

There are eight variables, four known and three equations.

Generator/desorber

The strong solution in refrigerant is introduced at high pressure in the high-temperature generator where it boils by contact with tubes heated either directly by a fuel or by steam. Vapor produced is almost pure refrigerant, due to the saturation pressure difference between the two fluids. It is then directed to the condenser. The depleted solution is extracted to be recycled.

With the assumption that the generator is at a constant temperature T_{gen} and the weak solution is saturated, the equations are given below.

The inversion of the equation of the solution saturated vapor pressure $P_{gen} = P(x_{sp}, T_{gen})$ provides concentration x_{sp}, and enthalpy h_{spA}

New data: h_{r2}, T_{gen}, P_{gen}

New unknowns: h_{srB}, h_{spA}

Conservation of enthalpy: $m_r h_{r2} + m_{sp} h_{spA} = m_{sr} h_{srB} + Q_{gen}$

There are five variables, among which three are known and one equation.

Solution exchanger

At this stage, two equations are missing to solve the model. They correspond to the solution heat exchanger:

$$\varepsilon = f\left(T_A, T_B, T_D, T_E\right)$$

$$m_{sp}\left(h_{spA} - h_{spE}\right) = m_{sr}\left(h_{srB} - h_{srD}\right)$$

Sequence of calculations

Specifically, the sequence of calculations is as follows:

1. Update the component before calculation by loading the values of m_r, h_{r1} and P_{abs} from the upstream point;
2. Reading T_{abs}, T_{gen} and ε on the external component screen;
3. Inversion of $P_{abs} = P(x_{sr}, T_{abs})$ for x_{sr} then h_{srD};
4. Loading values h_{r2} and P_{gen} from the downstream point;
5. Reverse $P_{gen} = P(x_{sp}, T_{gen})$ for x_{sp} and then h_{spA};
6. Computing of m_{sp} and m_{sr};
7. Calculation of T_B and T_E, thanks to solution exchanger equations;
8. Calculation of h_{srB} and h_{spE} assuming that these points are in equilibrium at their respective temperature and concentration;
9. Calculation of thermal loads Q_{abs} and Q_{gen};
10. Update of the external component display;
11. Update and calculation of the associated thermocouplers.

Bibliography

ASHRAE, Fundamentals handbook (SI), Thermophysical properties of refrigerants, 2001.

Chase et al., Janaf thermochemical tables. *Journal of Physical and Chemical Reference Data*, 14(Suppl. 1), 1985.

M. Duminil, Machines thermofrigorifiques, Techniques de l'Ingénieur, Génie énergétique, BE 9. 730 to 9 736.

R. C. Reid, J. M. Prausnitz, B. E. Poling, *The properties of gases and liquids*, 4th edition, Mc Graw-Hill, s, 1987.

S. Sandler, *Chemical and engineering thermodynamics*, 3rd edition, J. Wiley and Sons, New York, 1999.

B. Ziegler, C. Trepp, Equation of state for ammonia-water mixtures, *International Journal of Refrigeration*, 7 (2): 101–106, March 1984.

4

Innovative cycles including low environmental impact

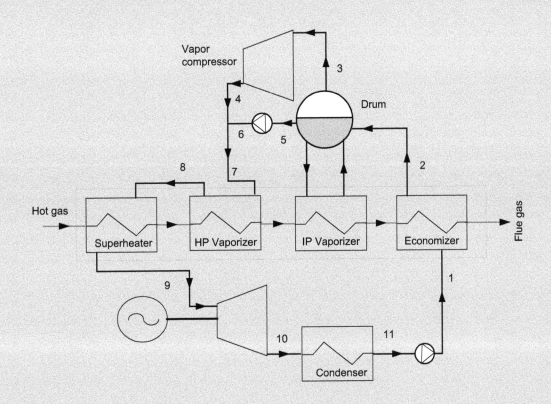

DOI: 10.1201/9781003175629-4

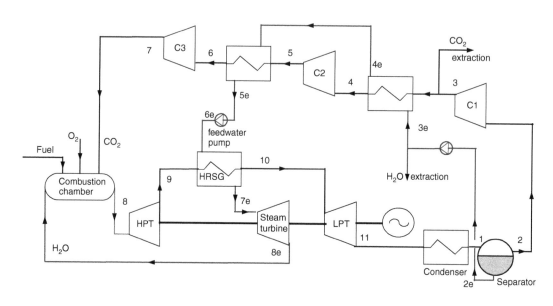

Part 4 completes Part 3 by addressing innovative advanced cycles, including low environmental impact ones: advanced gas turbine cycles, Stirling engines, future nuclear reactors, oxy-combustion cycles, new and renewable energy thermodynamic cycles, evaporation, mechanical and thermal vapor compression, desalination, drying by hot gas and electrochemical converters.

Advanced gas turbine cycles

Introduction

In this chapter, which completes Chapters 2, 9 and 10, we present three families of advanced cycles based on the use of gas turbines. In the first, the working fluid is no longer an ideal gas but a mixture a little more complex: the humid air gas turbine (HAT) cycle takes advantage of the variation of humidity.

The second family corresponds to the supercritical CO_2 cycle whose interest is to benefit from a compression work in supercritical liquid state much lower than when the working fluid remains in the gaseous state. Its main use considered today is electricity production from high-temperature nuclear reactors (HTR, cf. Chapter 15).

The third family is four advanced combined cycles. The first three are variants of conventional combined cycles, with flash and recompression, while the Kalina cycle involves the ammonia-water mixture whose properties have been studied in Chapter 13. The Kalina cycle has the advantage of reducing thermal irreversibilities between the gas stream and the working fluid.

Humid air gas turbine

HAT is as shown in Figure 14.1. It uses as working fluid system "water–air", which can significantly improve the capacity and efficiency compared to the simple gas turbine cycle

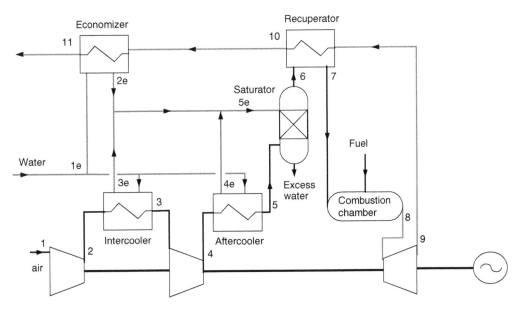

FIGURE 14.1

HAT cycle.

DOI: 10.1201/9781003175629-14

(Mori et al., 1983; Chiesa et al., 1995).

The intake air is compressed (1–4) at about 25 bar with intercooling and postcompression cooling (4–5) to reduce work involved while recovering heat to preheat water.

Air is then introduced into a saturator which it leaves saturated with water. The humid air is preheated in a recuperator (6–7) by exchange with the expanded fumes (9–10) and then headed to the combustion chamber, where it serves as an oxidizer to a fuel. Burnt gases are expanded in the turbine (8–9) and then cooled in the recuperator prior to preheat water in the economizer before entering the saturator (10–11).

Water that comes in (1e) is heated in the various heat exchangers before being introduced into the saturator (5e).

As in a steam injection gas turbine cycle, water intake increases the mass flow passing through the turbine, which participates in the performance gain.

By varying humidity of air leaving the saturator, it is possible to modulate the system capacity. The cycle is much simpler than a combined cycle, and does not require expensive components, so that its capital cost is relatively low.

Westinghouse (Nakhamkin et al., 1996) has proposed a cycle variant called cascaded humidified advanced turbine cycle (CHAT) and announced very high efficiencies (55%–65% with a compression ratio of 80 and a turbine inlet temperature of 1,500°C).

The cycle diagram in Thermoptim is given in Figure 14.2. This cycle uses a saturator. Since the amount of makeup water needed is not *a priori* known, we have provided a flow of water supply larger than required. At the saturator outlet, a divider allows to separate excess water, which is recycled.

Settings are as follows: cycle pressure 25 bar and end of combustion temperature 1,250°C. Heat exchanger effectiveness is equal to 0.9, except that of the after-cooler, which requires a special setting described below.

The overall efficiency is then 58%. These values are relatively close to those provided in the literature: air entering the saturator at 227°C and water entering at 264°C. Exit conditions are here somewhat higher (150°C instead of 120°C). To avoid any temperature crossover in the saturator, the water temperature must be high.

The need to take as much heat as possible on compressed air leads to define a rather complex heat exchanger architecture, especially for the after-cooler (Figure 14.3). A counter-flow

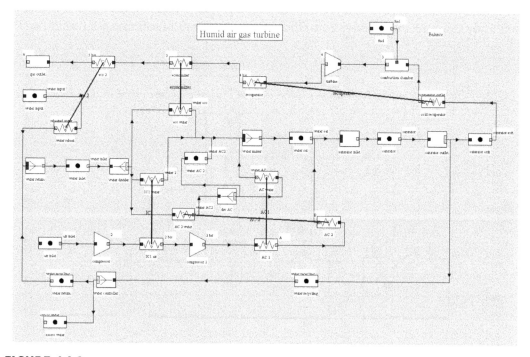

FIGURE 14.2

Diagram of the cycle in Thermoptim.

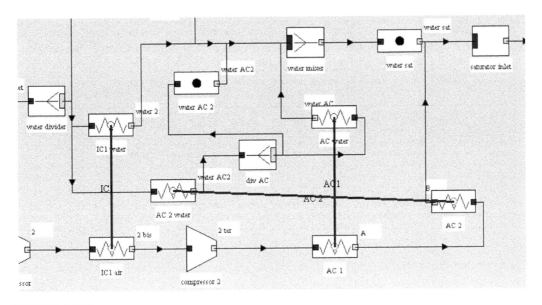

FIGURE 14.3

Internal recovery exchangers.

arrangement is chosen, so that water leaving the divider exchanges with air already cooled. At the outlet of "AC 2 water", the stream of water is divided into two parts, one providing the vaporization of a portion of the flow and the other being by-passed, otherwise the setting of the heat exchanger oscillates. The choice of the flow of water to vaporize is a bit tricky and must be done by successive approximations, but we thus succeed in cooling appropriately the compressed air.

Settings of the initial water divider also influence the results (Figure 14.4). Here, the inter-cooler was set to cool air well before recompression.

The saturator (Figure 14.5) behaves like a moist mixer and is calculated as such.

A simplified model involving just twenty points and twenty processes can be constructed (Figures 14.6 and 14.7). The efficiency is slightly lower (53.4%), but the main elements are there.

There is no recycling of water in excess, and we must therefore seek the minimum flow of water input, given the parameter values.

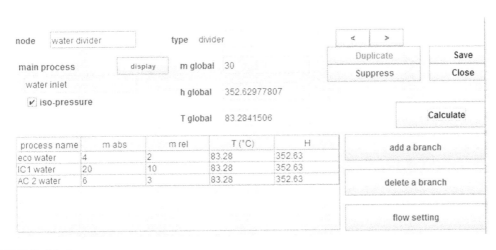

FIGURE 14.4

Water inlet divider.

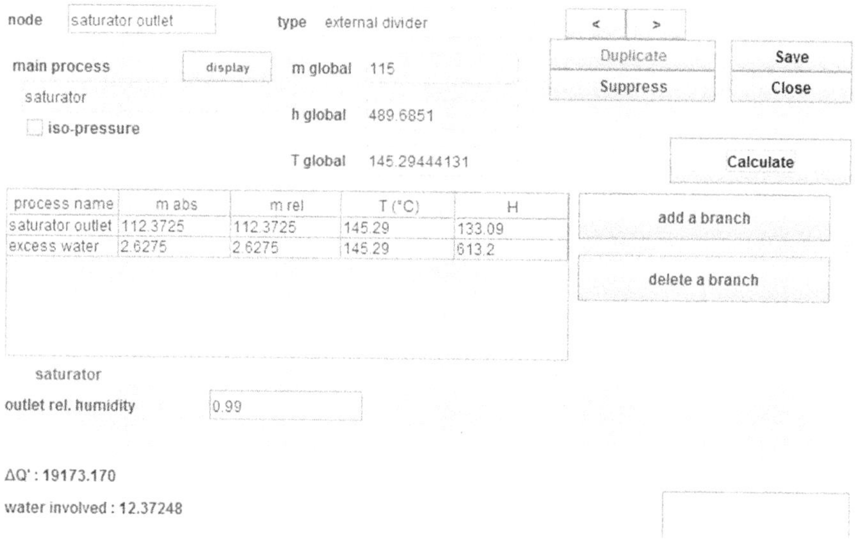

node
saturator outlet
type
external divider
<
>
Duplicate
Save
Suppress
Close
main process
display
m global
115
saturator
h global
489.6851
iso-pressure
T global
145.29444131
Calculate
process name
m abs
m rel
T (°C)
H
saturator outlet
112.3725
112.3725
145.29
133.09
excess water
2.6275
2.6275
145.29
613.2
add a branch
delete a branch
saturator
outlet rel. humidity
0.99
ΔQ' : 19173.170
water involved : 12.37248

FIGURE 14.5
Saturator screen.

Humid air gas turbine (simplified)
gas outlet
economizer
recuperator
turbine
combustion chamber
fuel
Balance
cold recuperator
water eco
eco water
water sat
water inlet
div water
IC1 water
AC water
saturator inlet
saturator
saturator outlet
air inlet
compressor
IC1 air
compressor 2
AC1
recyclage water
excess water

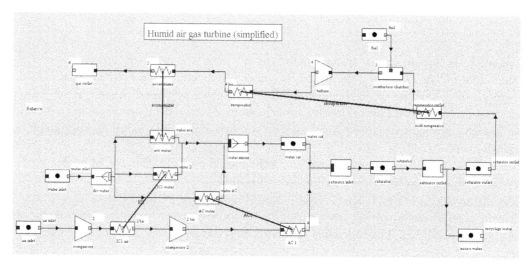

FIGURE 14.6
Diagram of the simplified cycle in Thermoptim.

Humid air gas turbine (simplified)
Balance
useful power 64,390 (kW)
purchased power 120,652 (kW)
effectiveness 0.5337

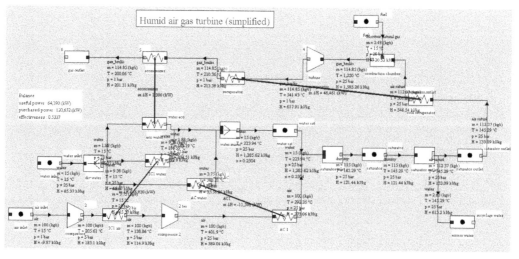

FIGURE 14.7
Synoptic view of the HAT simplified cycle in Thermoptim.

Supercritical CO_2 cycles

MIT has recently worked on cycles using supercritical carbon dioxide, which, it considers, lead to better performance than others to produce electricity from nuclear reactors at moderate temperatures between 650°C and 800°C (Dostal et al., 2003).

Proponents of these cycles argue that efficiencies are, for this temperature range, higher than those of steam cycles and that machines are much more compact.

Let us detail a little, to fix ideas, the magnitudes of the respective sizes of these machines:

- Expansion ratio in the turbine is much lower (about 2.6) than that in steam cycles (about 7,300). Mass flow rate is however very high (about 3 t/s). Volume flow at the turbine inlet is equal to about 27 and 55 m³/s at the outlet. At outlet conditions, assumed to be half the speed of sound m/s, this corresponds to a flow area of 0.27 m²;
- For a regenerative Rankine steam cycle with extraction and reheats of equivalent capacity, mass flow is 170 kg/s, with, at the LP turbine outlet, a specific volume of 55 m³/kg or a volume flow 170 times greater than for the supercritical CO_2 cycle. At outlet conditions, assumed to be half the speed of sound m/s, this corresponds to a flow area of 45 m²;
- In a 600 MW flame power plant, of which the inlet conditions are 565°C and 163 bar, the main flow is about 500 kg/s. The three LP turbine sections have an exhaust section of 150 m²;
- 1,300 MW PWR nuclear power plants have three double flow LP turbine sections with a total exhaust area of 112 m², last blades having a height of 1.45 m and a wheel diameter of 5.55 m. To limit the peripheral velocity, their rotation speed is 1,500 rpm. The flow through them is only 45% of steam flow at admission because of the different extractions. The turbine length is 56 m (73.5 m with the generator). The HP rotor mass is 83 tons, and that of each LP rotor is 218 tons.
- However, if the steam engine turbine is much larger than that of the supercritical CO_2 machine, heat exchangers of the latter are of significant size. On the basis of an exchange coefficient h close to 100 W/m²/K for this gas, the cycle modeled in Thermoptim (Figure 14.9) would involve a 210,000 m² HT regenerator, a 180,000 m² LT regenerator, and an 80,000 m² precooler, not to mention the IHX, while the steam cycle condenser would have an area of tens of thousands of m² only.

Several types of supercritical CO_2 cycles are considered: the simplest is a Brayton cycle with regenerator (Figure 14.8), the principal variants involving partial cooling, precompression or recompression.

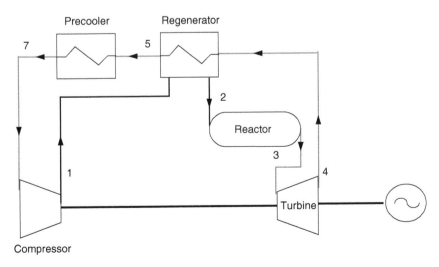

FIGURE 14.8

Supercritical CO_2 regenerative cycle.

In thermodynamics, the value of using such a cycle is to benefit from a compression work in supercritical liquid much lower than if the working fluid remains in the gaseous state as in a classic Brayton cycle.

In the following sections, we compare the performance of four variants of these cycles, by considering examples similarly set. In all cases, the carbon dioxide properties are modeled, thanks to the CTP Lib external mixture library, which implements a very precise CO_2 equation model published by Span and Wagner (1996).

Simple regeneration cycle

In a 350-MWe supercritical CO_2 **simple regenerative** cycle (Figure 14.9), a flow of 3 t/s of CO_2 at 200 bar enters the reactor at a temperature of 364°C and leaves it at 650°C. It is then expanded at the pressure of 77 bar in a turbine. Regeneration takes place between expanded CO_2 and that which enters the reactor. CO_2 then enters the cooler, which it leaves at 32.5°C, before being sucked into the compressor, which brings it at 200 bar.

The efficiency of this cycle remains fairly low, close to 34% with a regenerator of effectiveness 0.9 and turbomachinery polytropic efficiency equal to 0.9.

Precompression cycle

A first improvement is to use **precompression**. In such a cycle (Figure 14.10), compression is two-stage with intercooling. After passing through the high-temperature regenerator (4–5), a first compressor compresses CO_2 at about 100 bar (5–10). Precompressed CO_2 is cooled in the second low-temperature regenerator (10–9), followed by exchange with the cold source (9–7).

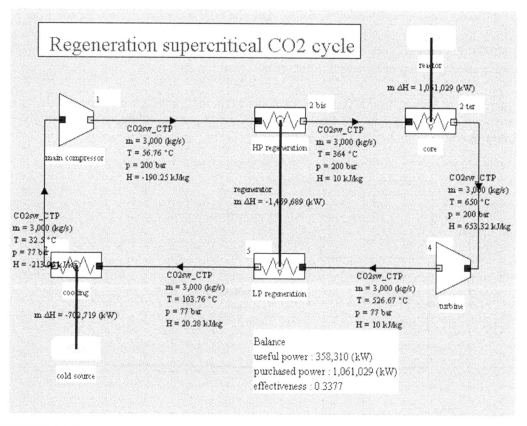

FIGURE 14.9

Synoptic view of the supercritical CO_2 regenerative cycle.

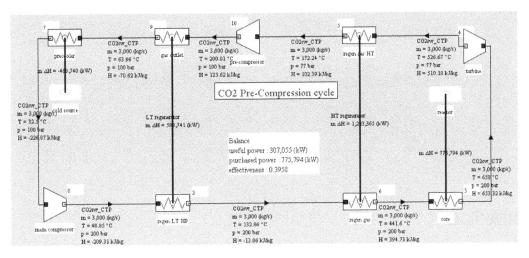

FIGURE 14.10

Synoptic view of the supercritical CO_2 precompression cycle.

FIGURE 14.11

LT regenerator screen.

It is then compressed (7–8) in the main compressor and then heated (8–2) by exchange with the flow exiting the precompressor (10–9). A second exchange (2–6) with the flow exiting the turbine allows heating of CO_2 to be continued before entering the reactor.

As shown in Figure 14.10, with the same assumptions as before on heat exchanger effectiveness and turbomachinery polytropic efficiency, the capacity of this cycle drops to 307 MW, but its efficiency reaches 39.6%.

The advantage of this cycle is that when the temperature difference between the two fluids is reduced in the regenerator, the insertion of a precompression allows one to increase it, thus promoting regeneration. A problem remains however, as shown in Figure 14.11: CO_2 heat capacities of the two fluids becoming very different, the heat exchanger internal irreversibilities increase.

Figure 14.12 shows the plot of the precompression cycle in the CO_2 (T, s) and (h, ln(P)) charts. These are not the usual Thermoptim charts, as the CO_2 is modeled using the CTP Lib library. The backgrounds for them were generated using the CreateMixtureCharts external class.[1]

---

[1] https://direns.mines-paristech.fr/Sites/Thopt/en/co/createur-fond-diag.html.

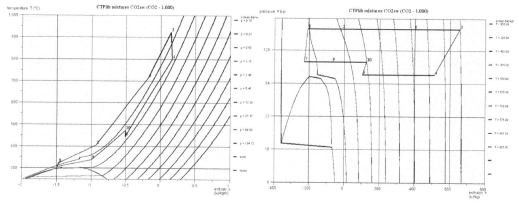

FIGURE 14.12

CO_2 precompression cycle in the (T, s) and (h, ln(P)) charts.

Recompression cycle

In a **recompression** cycle (Figure 14.13), compression is two-stage with intermediate cooling of only a portion of the fluid, which allows recycling of a larger amount of heat.

The main flow of CO_2 (3 t/s) leaving the turbine passes through a high-temperature regenerator (4–5) and then in a low temperature one (5–6), before being split into two.

Sixty percent of the flow is cooled in a precooler (6–7), then compressed in the main compressor from 77 to 200 bar (7–8) and heated in the LT regenerator (8–9).

The remaining CO_2 flow is compressed at 200 bar in the recompressor (6–8b), then mixed with another stream exiting the LT regenerator (1) and the total flow is heated in the HT regenerator (1–2) before entering the reactor.

CO_2 exiting the reactor is then expanded in the turbine (3–4), and the cycle is thus closed.

In this case, the reactor inlet temperature is higher, and with the same assumptions as before on the polytropic efficiency of turbomachinery (0.9) and exchanger effectiveness (0.9), the capacity remains close to 300 MWe, and the efficiency exceeds 45% (Figure 14.14). Even higher values are hoped for, the order of 50%, with exchanger effectiveness around 95%.

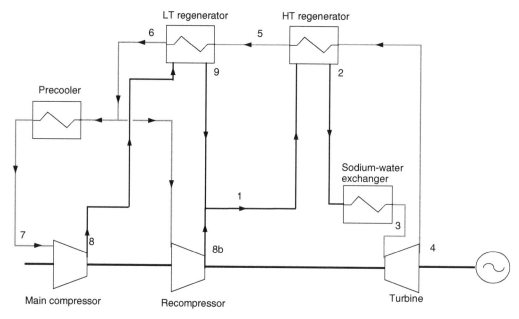

FIGURE 14.13

CO_2 supercritical recompression cycle.

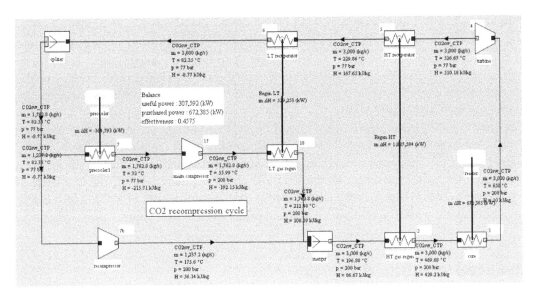

FIGURE 14.14

Synoptic view of the CO_2 supercritical recompression cycle.

The performance improvement is that the LT regenerator is much better balanced in terms of enthalpy than in precompression cycle because only a fraction of the flow traverses the high-pressure part, thus compensating for the increased thermal capacity of CO_2 in the supercritical region at low temperature.

Partial cooling cycle

The cycle we present here is a simplified version of the **partial cooling** cycle (Dostal, 2004), where compression is still two-stage, intercooled with only a portion of the fluid.

After passing through a high-temperature regenerator, then in a second low-temperature one and in a cooler (Figure 14.15), a first compressor compresses CO_2 from 30 to 77 bar. The precompressed CO_2 flow rate is then split into two, one part being cooled by exchange with the cold source, and then compressed in the main compressor and heated in a LT regenerator.

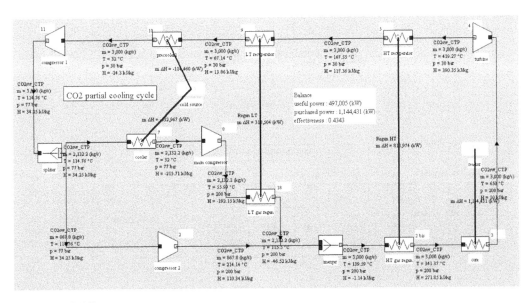

FIGURE 14.15

Synoptic view of the partial cooling CO_2 supercritical cycle.

The remaining flow passes through a third compressor to be compressed at 200 bar before being remixed with the stream exiting the LT regenerator, to be heated in the HT regenerator before returning to the reactor compartment. With the same assumptions as before on heat exchanger effectiveness and turbomachinery polytropic efficiency, the efficiency of this cycle is 43.4%, slightly less than the recompression cycle, which is also simpler. Its capacity, however, is much higher (497 MW).

Supercritical CO_2 cycles appear very interesting in thermodynamic terms. The two main technological constraints are the realization of regenerators in the circum-critical zone in particular to avoid any temperature crossing and that of turbomachinery effective for CO_2, an area where there is no reference. Moreover, these cycles can only work if the CO_2 state at the precooler output is supercritical, which implies a temperature limit of 32°C, which can be difficult to achieve when the cold source is outside air, a river or sea water.

Advanced combined cycles

Air combined cycle

In a conventional combined cycle, exhaust gases of a high-temperature gas turbine are used as a heat source for a steam cycle. In an air combined cycle (Figure 14.16), the steam cycle is replaced by a second air gas turbine cycle operating with adequate compression ratio and air flow (Weston, 1993; Najjar & Zammout, 1996). In English, we talk about an air bottoming cycle.

The advantage of such a cycle is not to require water cooling and to be less expensive than a conventional combined cycle, and it can be considered if regeneration is impossible.

The hot air leaving the high-temperature second cycle can be used in cogeneration.

Low compression ratio and low-temperature cycle

Let us consider a gas turbine which sucks 0.78 kg/ s of air at 15°C and 1 bar. Its compression ratio is equal to 5, and the compression isentropic efficiency is 0.875. At the combustor outlet, turbine inlet temperature is 950°C, and expansion isentropic efficiency is 0.885. It is assumed that fuel is natural gas.

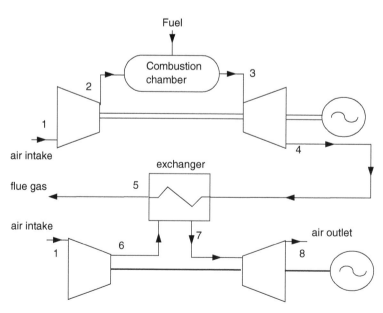

FIGURE 14.16

Air combined cycle.

For the second cycle, we seek the pair (air flow/compression ratio) which leads to the best performance, taking 0.95 as intercooler effectiveness and compressor and turbine polytropic efficiencies of 0.9. The cycle synoptic view is given in Figure 14.17. Its efficiency is 38.5%.

If we also use intermediate cooling, its efficiency becomes 41.6% (Figure 4.18). In this case, where the compression ratio is low, a simple regeneration cycle leads to slightly better

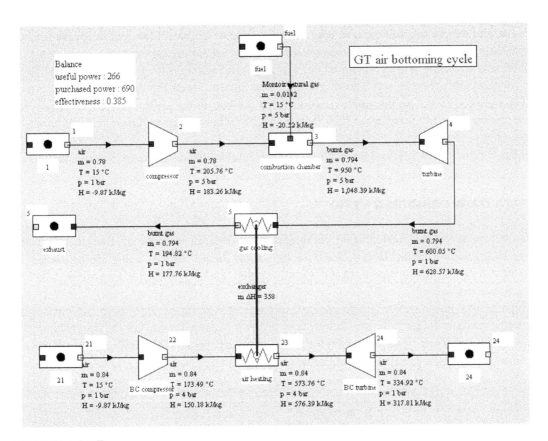

FIGURE 14.17

Synoptic view of the air combined cycle.

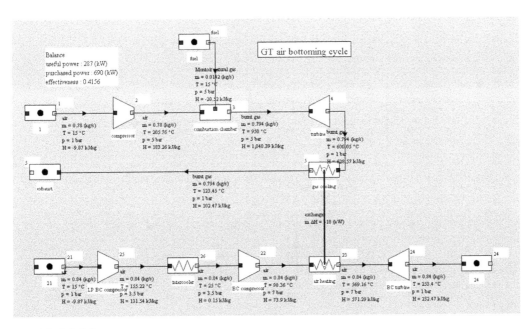

FIGURE 14.18

Air combined cycle with intercooler.

performance (45%), as you can easily verify starting from the example of Guided exploration C-M2-V2 and modifying its settings as indicated above.

High compression ratio and high-temperature cycle

For the remainder of this section, we will consider as base case the single pressure combined cycle studied in Chapter 10, whose gas turbine has a compression ratio equal to 16, turbine inlet temperature of 1,120°C and polytropic efficiency equal approximately to 0.85 for compression and expansion, and we will take pressure drop of 0.3 bar in the heat recovery steam regenerator and the exchanger between the air topping and bottoming cycles.

The synoptic view of this combined cycle is given in Figure 14.19. Its efficiency is 47.8%, and the temperature of the gases leaving the stack is equal to about 165°C.

Taking into account an intermediate cooling for the bottoming cycle, one gets for this new compression ratio an efficiency of 42% (Figure 14.20), whereas regeneration becomes almost impossible.

The efficiency of this cycle is lower than the reference, but its architecture is much simpler.

Steam flash combined cycle

The performance of a conventional combined cycle depends directly on that of the heat recovery steam generator (HRSG). We have shown in Chapter 10 that a single pressure level does not allow the gas turbine to be sufficiently cooled, and it is desirable to provide two or three pressure levels.

Such technology, however, is restricted to machines of high capacity, for both technical and economic reasons. Technically, it is indeed very difficult to properly control the distribution of the total flow of water or steam between the different circuits. In fact, small combined cycles are limited to a single pressure without reheat.

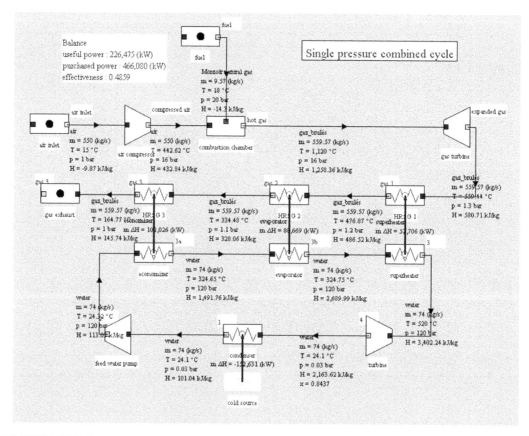

FIGURE 14.19

Reference combined cycle.

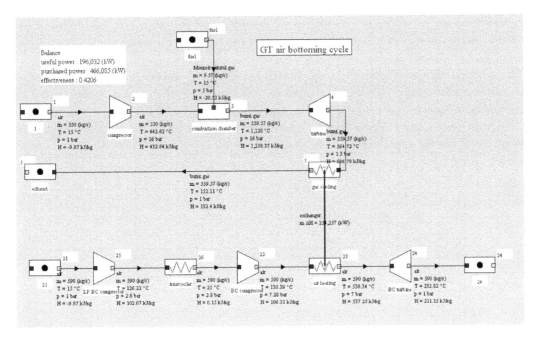

FIGURE 14.20

High pressure air combined cycle with intercooler.

To overcome this difficulty, a new steam flash cycle has been proposed (Dechamps, 1994) (Figure 14.21). In English, we talk about *water flashing bottoming cycle* or WFBC. The advantage of such a cycle is to reduce irreversibilities in the HRSG and be less costly than a combined cycle with several pressure levels.

The total flow of liquid water at HP pressure is vaporized at the corresponding saturation temperature. Much of this flow is then superheated and expanded in a turbine at LP pressure, while the rest of the flow is expanded in two-phase state at this LP pressure, and then the liquid and vapor phases are separated. The two LP steam streams are then remixed before

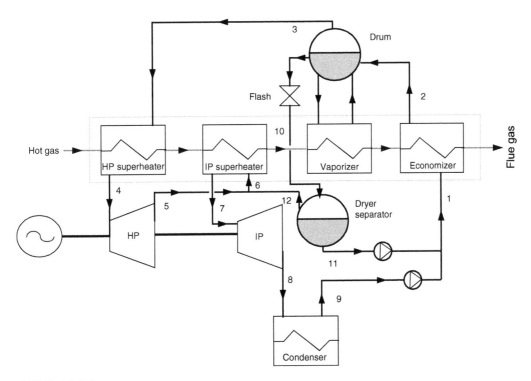

FIGURE 14.21

Steam flash combined cycle.

being reheated and expanded in a turbine at the condensation pressure. The liquid flow at LP pressure is recompressed at HP pressure by a pump and mixed with water at the outlet of the recirculation pump located downstream the condenser.

The synoptic view of Figure 14.22 shows that the efficiency obtained is 48.3%, as exhaust gases can be cooled at 120°C, which is a small improvement over the reference single pressure combined cycle for which these figures were, respectively, 47.8% and 165°C.

Composites (see Chapter 7) for this cycle are given in Figure 14.23. They show that the hot fluid is cooled, but the superheat could be increased.

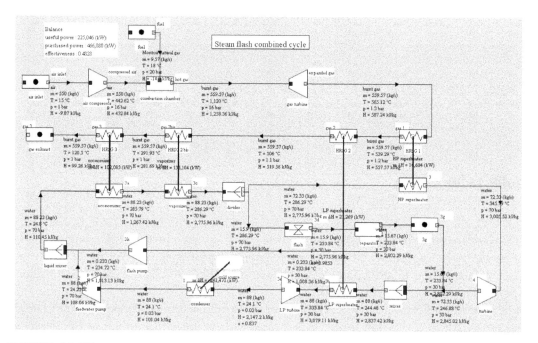

FIGURE 14.22

Synoptic view of a steam flash combined cycle.

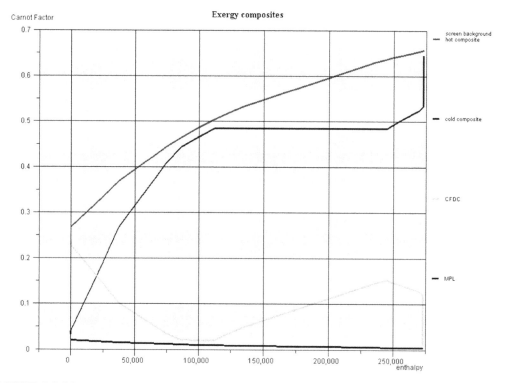

FIGURE 14.23

Steam flash combined cycle composites.

Steam recompression combined cycle

A patent was filed in 1978 by Cheng on a new steam recompression cycle (see Figure 14.24). In English, we speak of *steam recompression bottoming cycle* or SRBC. The advantage of such a cycle is to reduce irreversibilities in the HRSG and be less costly than a combined cycle with several pressure levels.

The total flow of liquid water leaving the condenser is compressed at LP pressure and then heated at the corresponding saturation temperature. A fraction of this flow is vaporized and then compressed at HP pressure by a compressor, while the rest of the flow is compressed in the liquid state at the same pressure by a pump. Both streams are then remixed before feeding the high-pressure evaporator, the superheater and the turbine.

In searching, for steam recompression combined cycle, a set of values (superheat temperature, primary and secondary steam flow rates, IP and HP pressures) that leads to good performance, keeping the same values for HRSG pinches, we get the synoptic view of Figure 14.25.

The efficiency is 49%, as the exhaust gas can be cooled to 160°C, which is still a small improvement over the reference single pressure combined cycle.

Composites for this cycle are given in Figure 14.26. They show that the hot fluid is cooled less than with the flash steam cycle, but the superheat is better, so that, overall, HRSG irreversibilities are lower.

Kalina cycle

As we discussed in Chapter 7, the optimization of a combined cycle is a particularly complex problem that leads to try to minimize irreversibilities stemming from temperature heterogeneity between the two cycles. When the recovery cycle uses a pure substance like water, the existence of a vaporization plateau induces losses that can be reduced only with difficulty by multiplying pressure levels. To overcome this constraint, Kalina has proposed to use as working fluid water-ammonia mixture, which has a large temperature glide (Kalina, 1983, 1984a, 1984b).

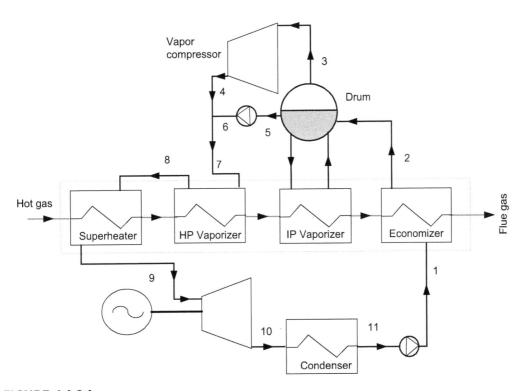

FIGURE 14.24

Steam recompression combined cycle.

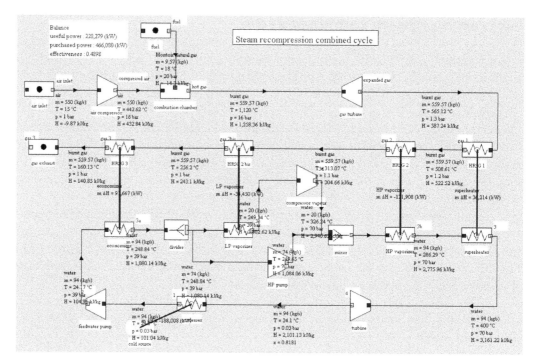

FIGURE 14.25

Synoptic view of the steam recompression combined cycle.

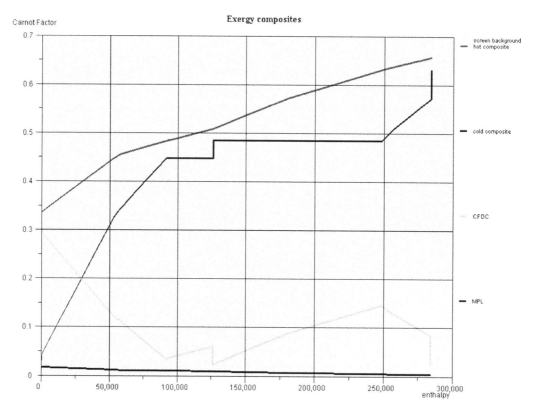

FIGURE 14.26

Steam recompression combined cycle composites.

An elementary Kalina cycle (Figure 14.27) consists of three main elements:

* A recovery steam generator (HRSG);
* A turbine;
* A distillation and condensation system (DCS).

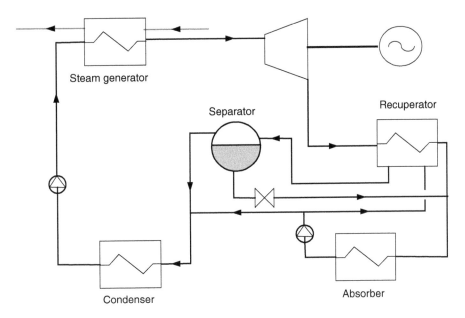

FIGURE 14.27

Kalina cycle.

The complexity of the cycle results from two causes:

- To benefit from a significant temperature glide in the temperature range corresponding to the hot gas cooling, the ammonia quality of the mixture should be quite high (typically 50% NH_3, 50% H_2O);
- To be condensed at a temperature above that of the surroundings and at low pressure, the ammonia quality of the mixture should be low (typically 20% NH_3, 80% H_2O).

The structure of a Kalina cycle is given in Figure 14.27. It can produce mechanical power from a heat source at variable temperatures, such as effluent or exhaust gas from a gas turbine.

As shown in Figure 14.28b, the existence of this temperature glide reduces the average difference in temperature between the gas stream cooling in the HRSG and the working fluid as compared to what is happening in a Rankine cycle (Figure 14.28a). Internal irreversibilities due to the temperature gradient are reduced.

One is therefore led to change the mixture composition between the hot zone (HRSG) and cold zone of the cycle, which is achieved in the DCS. Let us recall that you will find explanations on the behavior of the water-ammonia mixture in Chapter 13.

In the DCS, the rich mixture leaving the turbine is cooled in a recuperator and then mixed with a weak solution of NH_3 to raise the condensation temperature. The mixture (basic solution) is condensed in the absorber, recompressed and directed to the recuperator. A portion of the flow is used to dilute the rich solution leaving the separator, while the main stream

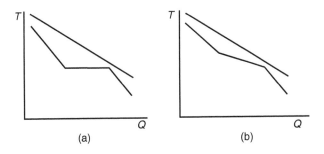

FIGURE 14.28

Internal irreversibilities in a Hirn (a) and Kalina (b) cycle.

is distilled at the recuperator outlet. The vapor is mixed with the extraction on the basic solution, condensed and compressed before entering the HRSG.

There are many Kalina cycle variants. The one presented here (Figure 14.29) is relatively simple and easy to model, as long as we have functions for calculating the ammonia-water mixture properties, which can be done by Thermoptim connected to an appropriate external thermodynamic property server. In the examples shown here, the NH_3-H_2O mixture properties are modeled, thanks to the CTP Lib external mixture library, which implements a very precise equation model of the NH_3-H_2O system (Ziegler & Trepp, 1984).

Mixer and separator nodes have been modified to enable them to correctly calculate flow mixtures of different compositions for the first and distillation of a mixture of known pressure and temperature for the second.

Note that the cycle parameter initialization is not trivial, mainly because the ammonia-water system compositions vary depending on the cycle points, coupled with the flow, heavily dependent on the pressure and temperature values. The values shown here are tentative and may change slightly from one setting to another.

At point 1, at the rich vapor condenser outlet, the working fluid is condensed at 2.1 bar and 15°C. It is a NH_3-H_2O mixture approximately equimolar (Figure 14.30). It is pressurized in a pump and then vaporized in a boiler. At pressure of 82 bar, the bubble point temperature of the mixture is about 195°C. Gradually, as boiling continues, the liquid phase ammonia mass fraction decreases, the bubble point temperature increases to about 250°C and then steam is superheated.

This steam is raised to 520°C and then expanded in a turbine at the pressure of 0.6 bar and the temperature of 88°C. At this pressure and for the composition of Figure 14.29, the dew

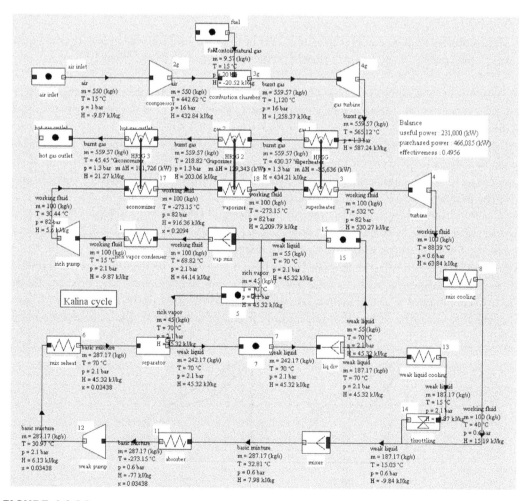

FIGURE 14.29

Synoptic view of a Kalina cycle.

component name	mass fraction	molar fraction
ammonia	0.5055938	0.5198732
water	0.4944062	0.4801268

FIGURE 14.30
Composition of the working fluid.

point is very low, and we cannot condense the expanded working fluid with a cold source at the surroundings' temperature.

To circumvent this difficulty, we proceed as follows.

The expanded working fluid (m = 100 kg/s) is cooled down at 15°C and mixed with a high flow rate (m = 350 kg/s) of weak liquid (Figure 14.31), thus forming the basic mixture (Figure 14.32), which is condensed, compressed to 2.1 bar and heated at 70°C;

- The basic mixture is then distilled, thereby separating the rich vapor (Figure 14.33) and weak liquid;
- A fraction of the weak liquid is remixed with the rich vapor to form the working fluid entering the turbine, which should not be too rich in ammonia if we want it to be condensed at room temperature;
- The working fluid is condensed and then pressurized and led to the boiler; the rest of the weak liquid is cooled and then expanded before being mixed with the working fluid.

Note that, since the pressure and temperature at the separator inlet are set, the compositions of the weak liquid and rich vapor are independent of the basic mixture, which only affects the distribution of these two fluid flows.

The cycle efficiency approaches 50%, exhaust gases being cooled at 55°C.

The representation of certain points on the ammonia-water mixture isobaric lenses allows one to understand the changes in composition in the Kalina cycle (Figure 14.34).

Plotted in this diagram are the following:

- From bottom to top, respectively, the three lenses at 0.6, 2.1 and 82 bar;
- In the form of vertical segments from left to right, respectively, the ammonia compositions of the weak liquid (7-P), basic mixture (6-B), working fluid (18-17-16-T) and rich vapor (5-R).

component name	mass fraction	molar fraction
ammonia	0.190206	0.1991662
water	0.809794	0.8008338

FIGURE 14.31
Composition of weak liquid.

component name	mass fraction	molar fraction
ammonia	0.3013353	0.3135038
water	0.6986647	0.6864962

FIGURE 14.32
Composition of the basic mixture.

component name	mass fraction	molar fraction
ammonia	0.8762739	0.8823388
water	0.1237261	0.1176612

FIGURE 14.33
Composition of rich vapor before mixing.

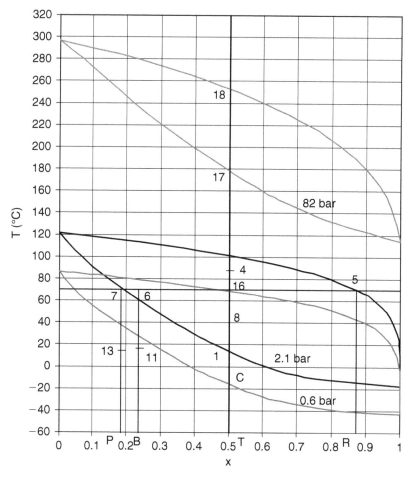

FIGURE 14.34

Water-ammonia isobaric lenses.

The working fluid leaves the turbine at point 4, at 0.6 bar. To condense it at this pressure, it should be cooled to point C, i.e., at about −18°C, which is impossible by simple exchange with the surroundings.

The solution is to cool it to point 8, at 40°C and mix it in an absorber with weak liquid subcooled at 15°C (point 13). The mixture leaves the absorber at 19°C with the basic mixture composition, in a slightly subcooled liquid state. It can then be compressed to 2.1 bar at the cost of reduced work.

The heated basic mixture enters in state 6, the separator which operates distillation, separating weak liquid (7) and rich vapor (5). A fraction of the weak liquid flow is remixed with the rich vapor to form the working fluid at 2.1 bar (16), which is then condensed in the liquid state (1), thereby allowing to compress it up to 82 bar with a reduced work.

In the steam generator, the working fluid is heated in the liquid state to point 17 and then vaporized with temperature glide at point 18 and superheated at 532°C (point outside diagram). It is then expanded in the turbine to point 4, closing the cycle.

Figures 14.35 and 14.36 show the shapes of composites for the Kalina cycle just described. They show that the system can be balanced in terms of enthalpy: heat recovered during the cooling of the expanded steam and the weak liquid is enough to preheat the basic mixture before entering the separator.

Thermal integration of the Kalina cycle is thus relatively simple.

The importance of the temperature glide and its impact on minimizing internal irreversibilities is very clear in these two curves. The Kalina cycle can recover much better the enthalpy available in the exhaust fumes than a conventional steam cycle (El Sayed & Tribus, 1985; Johnsson & Yan, 2001).

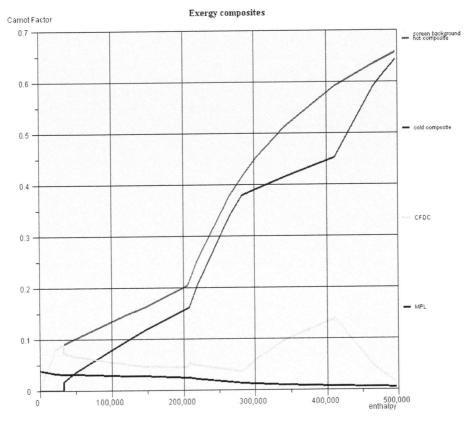

FIGURE 14.35

Exergy composites of the Kalina cycle.

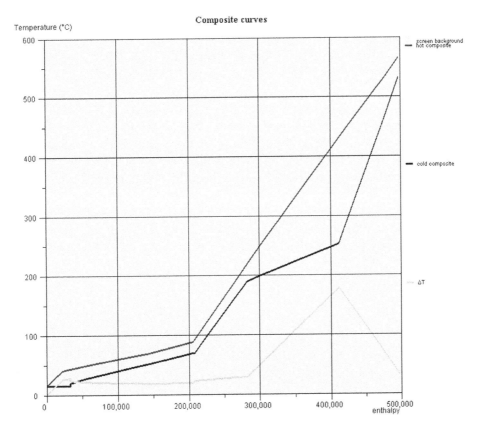

FIGURE 14.36

Composites of the Kalina cycle.

Bibliography

D. Y. Cheng, Regenerative parallel compound dual-fluid cycle heat engine, U.S. Patent 4,128,994, 1978.

P. Chiesa, G. Lozza, E. Macchi, S. Consonni, An assessment of the thermodynamic performance of mixed gas-steam cycles: Part B —water injected and HAT cycles. *Journal of Engineering for Gas Turbines and Power*, 117, 499–508, July 1995.

P. J. Dechamps, A study of simplified combined cycle schemes with water flashing, *Proceedings of the Florence World Energy Research Symposium*, FLOWERS'94, Florence, Italy, July 6–8, 1994.

V. Dostal, A supercritical carbon dioxide cycle for next-generation nuclear reactors, PhD thesis, MIT, January 2004.

V. Dostal, M. J. Driscoll, P. Hejzlar, N. E. Todreas, A supercritical CO_2 gas turbine power cycle for next-generation nuclear reactors, *Proc. ICONE-10*, Arlington, VA, April 14–18, 2003.

Y. M. El-Sayed, M. Tribus, A theoretical comparison of the Rankine and Kalina cycle, 1985, ASME publication AES-Vol. 1.

M. Johnsson, J. Yan, Ammonia-water bottoming cycles: A comparison between gas engines and gas diesel engines as prime movers. *Energy*, 26, 31–44, 2001.

A. I. Kalina, Combined cycle and waste heat recovery power systems based on a novel thermodynamic energy cycle utilising low temperature heat for power generation, ASME Paper 83-JPGC-GT-3, 1983.

A. I. Kalina, Combined cycle system with novel bottoming cycle. *Journal of Engineering for Gas Turbines and Power*, 106, 737–742, 1984a.

A. I. Kalina, Combined cycle system with novel bottoming cycle. *ASME Journal of Engineering for Power*, 106(4), 737–742 or ASME 84-GT-135, Amsterdam, Oct. 1984b.

T. R. Mori, H. Nakamura, T. Takahashi, K. Yamamoto, *A Highly Efficient Regenerative Gas Turbine System by New Method of Heat Recovery with Water Injection Proceedings*, 1983 Tokyo International Gas Turbine Congress, Vol. 1, 297–303.

Y. S. H. Najjar, M. S. Zaamout, Performance analysis of gas turbine air-bottoming combined system. *Energy Conversion and Management*, 37, 399–403, 1996.

M. Nakhamkin, E. C. Swensen, J. M. Wilson, G. Gaul, M. Polsky, The Cascaded humified advanced turbine (CHAT). *Journal of Engineering for Gas Turbines and Power*, 118, 565–571, July 1996.

R. Span, W. Wagner, A new equation of state for carbon dioxide covering the fluid region from the triple point temperature to 1100 K at pressures up to 800 MPa. *Journal of Physical and Chemical Reference Data*, 25, 1509–1596, 1996.

K. C. Weston, Dual gas turbine combined cycles, *Proceedings of the 28th Intersociety Energy Conversion Engineering Conference IECEC'93*, Boston, 1993, Vol. 1, pp. 955–958.

F. Wicks, C. Wagner, Synthesis and evaluation of a combined cycle with no steam nor cooling water requirements, *Proceedings of the 28th Intersociety Energy Conversion Engineering Conference IECEC'93*, Boston, 1993, Vol. 2, 105–110.

B. Ziegler, C. Trepp, Equation of state for ammonia water mixtures. *The International Journal of Refrigeration*, 101–106, 1984.

Stirling, future nuclear reactor and oxyfuel cycles

Introduction

In this chapter, we begin by studying the Stirling engine with displacer and regenerator, whose technical fluid was originally air. Its theoretical efficiency is equal to that of the Carnot cycle, but actual Stirling cycles performance is much lower.

The second topic which is presented is future nuclear reactors. We expand our previous analyses of thermodynamic nuclear energy conversion by presenting the main types of cycles that are considered to be used in the future.

To conduct an oxyfuel combustion is to replace by pure oxygen the usual oxidizer, namely air. Oxy-fuel technology allows at the same time to get fumes composed almost exclusively of water and carbon dioxide and to drastically reduce emissions of nitrogen oxides.

The separation of CO_2 and H_2O is then easily done by simple water condensation.

Stirling engines

The first patent for a Stirling engine type was introduced in 1816 by Scotsman Robert Stirling who sought to develop a device safer than steam engines. As its technical fluid was air in the nineteenth century, it is known as hot-air engine which quickly became a huge success. Its applications were numerous, such as industrial water pumping as well as domestic or space ventilation. Since the early twentieth century, internal combustion reciprocating engines and the development of electric motors gradually dethroned Stirling engines, which are now much less used.

The Stirling engine is an external combustion reciprocating gas engine which operates under the closed regenerative cycle shown in the entropy diagram in Figure 15.1. As we shall see later, its theoretical efficiency is equal to that of Carnot, which explains the fascination it exerts on many researchers.

By the mid-30s, the Dutch company Philips invested heavily in this technology, its aim being to have small generators for autonomous power supply for isolated radio stations or relays, with the aim of developing a silent motor, operating with a minimum maintenance, and good efficiency.

The Philips work yielded many significant breakthroughs and has enabled the power density to increase by 50, completely renewing the development prospects of the Stirling engine. In 1957, General Motors teamed up with Philips to develop a motor for traction, but in spite of technical successes, the partnership was not renewed ten years later.

Today, many experts agree in considering that the Stirling engine could experience a significant rise in the future considering first its benefits, including protection of the environment, and secondly the many hot sources that could be used. It has particular strengths for uses such

DOI: 10.1201/9781003175629-15

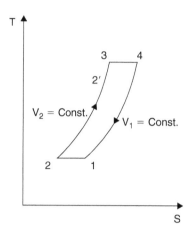

FIGURE 15.1

Stirling cycle in the entropy chart.

as marine propulsion, especially submarines, for small power electricity generation in isolated areas from different energy sources (solar, biomass, fossil fuels) and for small cogeneration.

Principle of operation

The basic principle of a Stirling engine is very simple: a gas enclosed in a cylinder closed with a piston is successively heated and then cooled. The pressure difference which is established in the enclosure can be transformed into motor work. The difficulty is that it is virtually impossible, given the thermal inertia of the cylinder to heat and cool it quickly enough so that the system operates under optimum conditions. Robert Stirling's idea was to put the gas in motion by introducing a special device whose operation we will analyze, the displacer. The engine structure is shown in Figure 15.2.

In a main vessel can move on the one hand a working piston and on the other hand a displacer, whose role is to transfer the working fluid from the compression volume to the expansion volume and *vice versa*.

During displacement, the fluid passes successively in one direction or another, in the boiler at temperature T_3, the regenerator, and the cooler at temperature T_1.

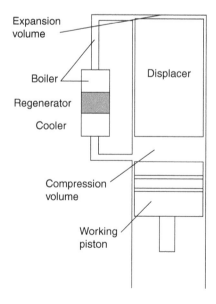

FIGURE 15.2

Sketch of a Stirling cycle.

The cycle includes four phases illustrated by the diagrams of Figures 15.3–15.6 and the entropy diagram of Figure 15.1.

During the compression phase a), the displacer is in top position, and the fluid confined in the cold zone is compressed by the working piston in its upward stroke, which requires the provision of work W_{12}.

At point 2, the piston is at top dead center, and the displacer is moved in the bottom position, which has the effect of transferring the compressed fluid, which passes during phase b) from the cold zone to the hot zone, starting with warming up in the regenerator and then receiving heat from the boiler. This process takes place at constant volume, the pressure increasing in the ratio of temperatures T_3 and T_1.

During expansion phase c), the fluid expands into the expansion volume, where it continues to be heated by the boiler tubes. This expansion has the effect of pushing the working piston down and provides useful work W_{34}. During this phase, the displacer and the piston move together.

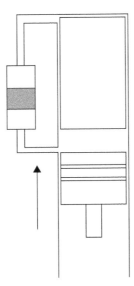

FIGURE 15.3

Compression phase.

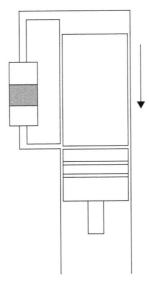

FIGURE 15.4

Constant volume heating.

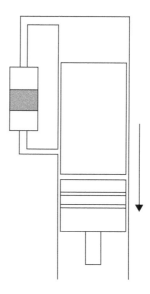

FIGURE 15.5
Expansion phase.

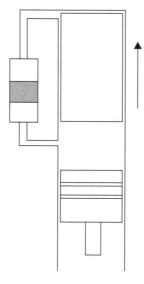

FIGURE 15.6
Constant volume cooling.

During phase d), after the working piston reaches bottom dead center, the displacer is moved in the upper position, which has the effect of transferring the fluid from the hot zone (expansion volume) to the cold zone (compression volume). During the transfer at constant volume, the fluid begins to yield its heat to the regenerator, before being cooled by the cooler.

Note that the displacer does not produce work. Pressure on its two opposite faces is always the same, if we neglect the pressure drop in the three heat exchangers (heater, regenerator and cooler).

In practice, the relative movements of the displacer and piston are obtained in various ways, especially by mechanical devices made from rods and crankshafts. Movement, which was described above as being discontinuous, is in reality very nearly sinusoidal, and a phase shift of about 90° between the displacer and working piston is generally accepted.

Piston drive

The main considerations that guide the choice of a coupling mechanism between the movements of the two pistons are as follows:

- Systems are simple and therefore inexpensive to manufacture and maintain;
- Dynamic aspects are essential, both in terms of stress repartition and vibration;
- The Stirling engine is a closed cycle engine, quiet by nature, and this quality should not be affected by the drive mechanisms;
- Finally, we search systems as tight as possible because, as discussed below, one of the Stirling engine characteristics is to operate at relatively high pressures.

The most used mechanisms are connecting rods, rhombic drive (Figure 15.7), cylinder rods and gliding rods. The interested reader will find illustrations in Walker (1973) and Reader and Hooper (1983).

Several geometric configurations have been proposed for producing Stirling engines. The procedure we present is that of single-acting piston, where piston and displacer are two separate bodies.

There is another category of engines, known as double-acting piston, in which the working piston also plays the role of displacer. This is made possible by coupling together a number of cylinders (between three and six) and adjusting the relative movements of the pistons in the cylinders so that the expansion phase of one of them corresponds to the displacement phase of another. The fluid is thus transferred successively between two cylinders. By placing the cylinders around an axis, it is possible to produce compact engines that share the same boiler burner and have a very efficient drive system.

Free-piston Stirling engines have also been made, in which the working piston and displacer are not coupled mechanically but through the working fluid. This type of arrangement makes it possible assembling a linear alternator around the working piston, to realize hermetic motors and thus solve sealing problems mentioned above.

The plan in Figure 15.8 shows part of the elements that make up a reduced model Stirling engine produced by MR. J. Maiwald, from Bavaria, Germany.

We can clearly see, at the bottom right, the cylinder containing the displacer, the boiler being located on the far right and the cooler more in the center. The working piston is placed vertically, and a set of connecting rods connected to the cranckshaft sited on the left makes it possible to coordinate the movements of the displacer and the working piston.

Figures 15.9 and 15.10 show two Stirling engines, the first corresponding to the plan we have just seen and the other to a two-cylinder variant engine.

In both cases, the flywheels are necessary to regulate the movement of the machine since it has only one working piston.

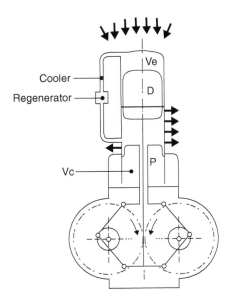

FIGURE 15.7

Rhombic drive. (Courtesy Techniques de l'Ingénieur.)

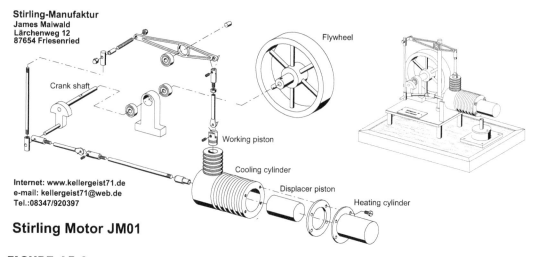

Stirling-Manufaktur
James Maiwald
Lärchenweg 12
87654 Friesenried

Internet: www.kellergeist71.de
e-mail: kellergeist71@web.de
Tel.:08347/920397

Stirling Motor JM01

FIGURE 15.8

Sketch of a Stirling motor. (Courtesy J. Maywald.)

FIGURE 15.9

Photo of the Stirling motor. (Courtesy J. Maywald.)

Thermodynamic analysis of stirling engines

In what follows, having quickly established the theoretical cycle efficiency, we consider two cycles representative of Stirling engines: the ideal cycle with imperfect regenerator and the para-isothermal Stirling cycle, in which isotherms are replaced by an adiabatic followed by an isovolume.

Theoretical cycle

The theoretical Stirling cycle (Figure 15.1) decomposes as shown below.

Between points 1 and 2, the fluid is compressed at constant temperature. We must provide work W_{12} given by the integration of equation $\delta W = -Pdv = -r\,T_1\,dv/v$ where $T_1 = $ Const.

$$W_{12} = rT_1\ln(\rho),$$

ρ being the volumetric compression ratio

FIGURE 15.10
Photo of a Stirling motor (2). (Courtesy J. Maywald.)

As on an isotherm, $du = \delta W + \delta Q = C_v\, dT = 0$, it is necessary to extract heat $Q_{12} = -W_{12}$.

Between points 2 and 3, heat Q_{23} is supplied to the fluid at constant volume.

Between points 3 and 4, the fluid is expanded at constant temperature and provides work W_{34}:

$$W_{34} = -rT_3 \ln(\rho)$$

We should also provide the fluid with heat $Q_{34} = -W_{34}$.

Between points 4 and 1, heat Q_{41} is extracted from fluid at constant volume. As the heat transfers (2–3) and (4–1) take place at the same temperature levels, and as the fluid is assumed to be perfect, both isovolumes can be deduced by translation and it is possible, assuming the existence of a perfect regenerator, to ensure that

$$Q_{23} + Q_{41} = 0$$

The amount of heat to provide to the cycle is then Q_{34} and the useful work $(-W_{12} - W_{34})$.

The cycle efficiency is

$$\eta = \frac{-W_{12} - W_{34}}{Q_{34}} = \frac{r\ln(\rho)(T_3 - T_1)}{rT_3\ln(\rho)} = 1 - \frac{T_1}{T_3} \tag{15.1}$$

The efficiency of the theoretical Stirling cycle is equal to that of Carnot.

This result largely explains the interest in the Stirling engine: its theoretical cycle has the best efficiency that allows the second law of thermodynamics. However, it is extremely difficult to make isotherm compression or expansion, as the compression or expansion machines are compact machines rotating at high speed. Furthermore, achieving heat exchange close to isothermal would require a large exchange surface and low transfer rates.

In practice, Stirling engines deviate significantly from the theoretical cycle consisting of two isotherms and two isovolumes, and their efficiency is far below the Carnot efficiency. There are several other reasons:

- First, the regenerator is not perfect, and only a fraction of the available thermal energy is actually recovered between phases (2–3) and (4–1);
- Second, heating and cooling do not occur only during the expansion and compression phases. These are not perfect isotherms, and in some cases, may be close to adiabatic, or, more generally, heated or cooled polytropic processes.

As we will show, it is possible to somehow refine Stirling engine models by considering more complex cycles, but the exercise has its limits because the continuous movement of the piston and displacer creates losses of aerodynamic nature (pressure drops, establishment of pressure waves caused by the pulsating flow) that are extremely difficult to model, especially analytically. The interested reader will find in (Walker, 1973) an analysis proposed by Schmidt in 1861, taking into account a harmonic motion of the moving parts, or may refer to the literature for more advanced developments. However, even today, the modeling of Stirling engines is still unsatisfactory, particularly as regards the simulation of instantaneous temperature and pressure fields prevailing in the various heat exchangers that make up these engines (Stouffs, 1999).

In this book, we limit ourselves to two simple models: the ideal cycle and the para-isothermal cycle. We present only one model made with Thermoptim. Note that some basic assumptions of the software (steady-state components in particular) are not met here, so that modeling with this tool finds its limits in this case, even if the working fluid is helium, which can legitimately be regarded as perfect.

Ideal stirling cycle

In the ideal cycle (see Figure 15.1), the fluid still follows two isotherms and two isovolumes, but effectiveness ϵ of the regenerator is less than 1:

$$\epsilon = \frac{T_{2'} - T_1}{T_3 - T_1}$$

In addition to heat Q_{34}, we must therefore provide the cycle with heat $Q_{2'3}$:

$$\eta = \frac{-W_{12} - W_{34}}{Q_{34} + Q'_{23}} = \frac{r \ln(\rho)(T_3 - T_1)}{r T_3 \ln(\rho) + Q'_{23}}$$

Introducing the ratio $\zeta = \dfrac{T_1}{T_3}$ we get

$$Q_{2'3} = C_v\left(T_3 - T_{2'}\right) = C_v\left(T_3 - T_1 - \epsilon\left(T_3 - T_1\right)\right) = C_v T_3 (1 - \zeta)(1 - \epsilon)$$

$$Q_{2'3} = T_3(1 - \zeta)(1 - \epsilon)$$

All developments made, we find

$$\eta = \frac{(\gamma - 1)(1 - \zeta) r \ln(\rho)(T_3 - T_1)}{(\gamma - 1)\ln(\rho) + (1 - \zeta)(1 - \epsilon)} \tag{15.2}$$

The curve in Figure 15.11 gives the appearance of efficiency depending on volume ratio ρ for $\xi = 0.3$ and $\epsilon = 0.75$.

The importance of the γ value can be seen on this graph. It seems preferable to use monatomic or diatomic gases. In practice, fluids used until now were air, hydrogen or helium.

Moreover, the efficiency of this cycle increases with the volumetric compression ratio.

We can also calculate a dimensionless work W_0, ratio of useful work to the product of pressure P_1 by the swept volume $(V_1 - V_2)$:

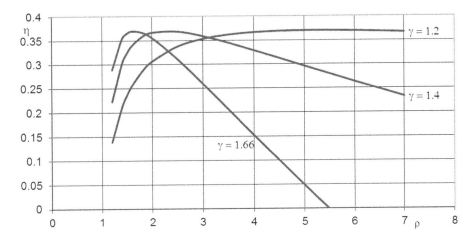

FIGURE 15.11

Stirling engine efficiency.

$$W_0 = \frac{-W_{12} - W_{34}}{P_1(V_1 - V_2)}$$

The swept volume is

$$V_1 - V_2 = V_1 \frac{\rho - 1}{\rho}$$

All developments made, we find

$$\eta = \frac{\rho \ln(\rho)(1-\zeta)}{\zeta(\rho-1)} \tag{15.3}$$

The ideal Stirling cycle is not fully satisfactory because the expressions of efficiency it provides are in contradiction with an experimental fact: there is an optimum volumetric ratio (a value between 2 and 3) for which efficiency is maximum.

Paraisothermal stirling cycle

The para-isothermal Stirling cycle seeks to remedy the inadequacies of the ideal cycle, better approaching real processes undergone by the fluid. If one refers to the principle of operation outlined previously, we note that a significant portion of heat exchange between the fluid and the surroundings takes place not during the phases of compression and expansion, but during the transfer of fluid, when it goes into the cooler or boiler.

In the para-isothermal cycle (Figure 15.12), it is assumed that isotherm (1–2) is replaced by adiabatic compression (1–2″), followed by cooling (2″–2), and that isotherm (3–4) is replaced by adiabatic expansion (3–4′), followed by heating (4′–4).

Under these conditions, the heat to provide to the cycle becomes

$$Q = Q_{2'3} + Q_{4'4}$$

and expression of the compression and expansion work is no longer that of the isotherm but that of the adiabatic (calculated in closed system).

$$\eta = \frac{-W_{12''} - W_{34'}}{Q_{2'3} + Q'_{44'}}$$

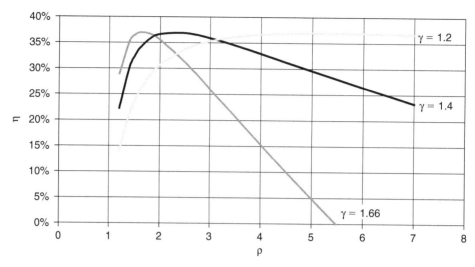

FIGURE 15.12

Para-isothermal cycle.

All developments made, we find

$$\eta = \frac{\left(1-\rho^{1-\gamma}\right)-\zeta\left(\rho^{\gamma-1}\right)}{\left(1-\zeta\right)\left(1-\varepsilon\right)+1-\rho^{1-\gamma}}$$

(15.4)

The shape of the efficiency function of the volumetric ratio is given in Figure 15.13. It shows a maximum, which is between 1.8 and 5 depending on the fluid used. These results are much closer to experimental reality of engines built to date.

If we show θ, reverse of ζ, representative of the heating, the efficiency becomes (Figure 15.14):

$$\eta = \frac{\theta\left(1-\rho^{1-\gamma}\right)-\left(\rho^{\gamma-1}\right)}{\left(\theta-1\right)\left(1-\varepsilon\right)+\theta\left(1-\rho^{1-\gamma}\right)}$$

(15.5)

The power factor W_0 can also be determined analytically:

$$W_0 = \frac{-W_{12''}-W_{34''}}{P_1\left(V_1-V_2\right)}$$

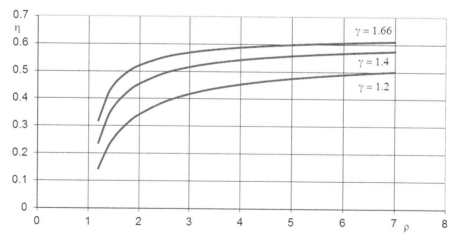

FIGURE 15.13

Stirling engine efficiency.

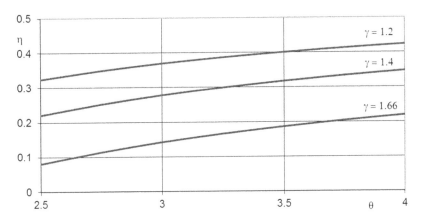

FIGURE 15.14

Stirling engine efficiency function of θ.

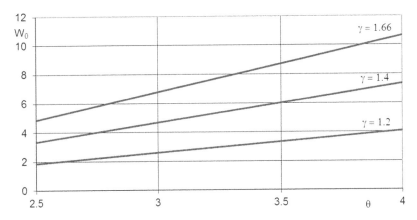

FIGURE 15.15

Stirling engine power factor function of θ.

All developments made, we find

$$W_0 = \frac{\rho\left(1-\rho^{1-\gamma}\right)\left(1-\zeta\rho^{\gamma-1}\right)}{\zeta(\rho-1)(\gamma-1)} \tag{15.6}$$

W_0 is a linear function of θ (Figure 15.12)

$$W_0 = \frac{\rho\left(1-\rho^{1-\gamma}\right)}{(\rho-1)(\gamma-1)}\left(\theta-\rho^{\gamma-1}\right) \tag{15.7}$$

To fix ideas, cooling is often provided with water at about 40°C, 313 K, and the maximum temperature is around 750°C, 1,023 K, which gives a value of θ around 3.3. The curves of Figures 15.14 and 15.15 give the look of efficiency and power factor as a function of θ around this value.

Influence of pressure

By definition of W_0, the work W that may be provided by an engine of displacement V_s is equal to $W = W_0 P_1 V_s$.

In the case of the Stirling engine, theory suggests therefore, which is verified by experience, that the power output is roughly proportional to the fluid pressure: for a given capacity, it is possible to have a machine either compact (small V_s) and under high pressure or larger with lower working pressure.

One generally seeks to use very high pressures (about 100–200 bar), and the limit is determined by the particular sealing problems at the joints between the piston and the casing of the machine. The difficulties encountered at this level partly explain why in the past, the Stirling engine development has been slower than that of its competitors.

Experience also shows that the efficiency itself is an increasing function of pressure, probably because of improved heat transfer within the machine.

Finally, the almost proportional relationship existing between the work done by the engine and the pressure therein allows control of the machine by varying the pressure, which is generally done.

In all cases, the risks of leakage are important, and there should be a mechanism for resetting the pressure of the working fluid, especially when it is a light gas, like hydrogen.

Choice of the working fluid

The study of the para-isothermal cycle suggests that the same maximum efficiency can be obtained with different values of γ, but for different volumetric ratios, and experience does not invalidate this conclusion.

The choice of the fluid depends more on thermal and fluid mechanics considerations than thermodynamics. From this point of view, hydrogen is very attractive, with a heat capacity fourteen times greater than air and nearly three times larger than that of helium. However, it poses specific containment problems.

Ultimately, the choice of the working fluid is strongly dependent on the type of engine and its operating conditions. A careful optimization study must be done.

Heat exchangers

The various heat exchangers that appear in a Stirling engine are the cooler, the regenerator and the boiler. We have already seen how heavily the engine efficiency depends on the effectiveness of these very specific exchangers.

Cooler

The design of the cooler is less problematic than the other two heat exchangers, insofar as the cold source is most often water. This is a relatively conventional exchanger.

Note however that in a Stirling engine, the heat removed by cooling water is a much larger percentage of the heat of combustion than in a reciprocating internal combustion engine.

Table 15.1 gives an indication of the amount of heat removed by cooling water and exhaust for different types of engines.

Regenerator

The regenerator is a key component of the Stirling engine. It generally consists of a porous material of high heat capacity, to ensure the best heat exchange with the fluid. Ideally, its axial conductivity should be as low as possible so that it retains a good stratification, but its radial conductivity should be high so that its temperature is uniform.

As in the case of the boiler, there is a contradiction between seeking a reduction in pressure drop and an increase in heat transfer between the fluid and the regenerator.

TABLE 15.1

HEAT REJECTION PER KW OF USEFUL POWER

	Cooling Water	Flue Gas
Gasoline engine	1.00	1.33
Diesel engine	0.56	1.22
Stirling engine	1.39	0.39

The regenerator is by far the exchanger which transfers most energy in the Stirling engine: its capacity should generally be between five and ten times that of the engine or even more.

Despite all these difficulties, we are now able to make very efficient regenerators, of effectiveness about 0.9.

Boiler

The boiler is used to transfer heat from the heat source to the working fluid in the hot zone. As the Stirling engine follows a closed cycle, it is possible to use a wide variety of hot sources, classically as burners, or even more particular as heat pipes or the furnace of a solar concentrator.

In the classic case of a burner, the boiler is subjected on one side (outside the tubes) to the action of combustion gases, possibly corrosive, but at low pressure, and on the other side (inner tubes) to the internal pressure of the engine (above 100 bar). Once the steady state reached, variations in temperature are of relatively low amplitude, but at startup, the boiler is subject to strong dilatation. These operating conditions generate high stresses, and only very strong materials can be used.

In addition, we seek to optimize the design of the boiler so that pressure drops will be as low as possible, which requires large diameters, but also so that the dead volume is reduced, and the heat exchange coefficients are high, which requires small diameters. A compromise must therefore be found between all these requirements, which ask for careful analysis.

Characteristics of a stirling engine

The shape of the specific consumption curve of a Stirling engine is much flatter than that of a reciprocating internal combustion engine. Indeed, one can consider that the work required to run the motor with no load is substantially proportional to the compression work with no load, itself proportional to $(\rho^{\gamma-1}-1)$. The compression ratio being much lower in a Stirling engine, the no-load losses are lower too, and experience confirms this analysis.

We have already indicated that it is possible to modify the work provided by a Stirling engine by acting on the working fluid pressure. It is this way that we can at best regulate the engine. In fact, acting on the amount of heat injected into the boiler induces too high delays and carries the risk of overheating which can weaken the tube mechanical strength, already strained because of the pressure difference that exists between inside and outside.

The control principle is therefore, firstly to vary the average pressure of the working fluid depending on the desired capacity, using a compressor and an annex fluid reserve, which are needed to compensate for leaks, and secondly, to pilot the amount of fuel injected into the combustion chamber through a temperature control of tubes at a value possibly modified by the user.

In addition, a Stirling engine runs much quieter than diesel or gasoline because the combustion is continuous, and torque is more regular.

Simplified stirling engine thermoptim model

The Thermoptim model that we present corresponds to a Stirling engine placed in the focal region of a solar dish (see Chapter 16 for explanations on solar energy collectors). Although rather simple, this model allows us, however, to illustrate how to operate. It could be improved by developing well-designed external classes, especially to represent the regenerator (Figure 15.16).

This model implements six components:

- A compressor operating in closed system, set compression ratio (here taken equal to 7);
- An exchange process representing the hot part of the regenerator;
- An external process corresponding to the solar concentrator. The latter is a variant of class SolarConcentrator that performs calculations in a closed system;

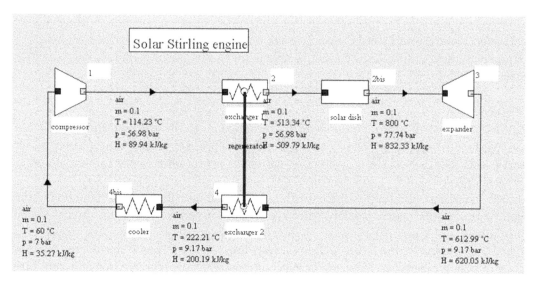

FIGURE 15.16

Synoptic view of a solar Stirling engine.

* An expansion process operating in a closed system of set expansion ratio (here taken equal to 7);
* An exchange process representing the cold part of the regenerator and coupled by a heat exchanger representing the hot part;
* An exchange process to model heat transferred to the cold source.

We retained as hypothesis on the one hand that compression is partially cooled and expansion partially heated, their polytropic efficiencies being equal to 0.9, and secondly that the effectiveness of the regenerator is equal to 0.8.

The results provided by the model are summarized in Figure 15.17, which shows the screen of the external driver developed to build up its energy balance.

The power output is 26.8 kW and the efficiency 41%.

This model is not entirely satisfactory, since the model mixes components operating in a closed system and others in open system, which has the effect that the overall balance is not exact.

FIGURE 15.17

Stirling engine driver screen.

Future nuclear reactors

Introduction

Although PWR technology is today the most developed on the industrial side, future nuclear reactor projects mainly concern other systems and correspond to concepts already studied in the years 1960–1970, often abandoned because of development difficulties or disappointing economic performance.

The aim is twofold: first to overcome technological obstacles encountered at that time and secondly to achieve viable economic and technical performance, with better adaptation to sustainable development criteria.

For this, a number of qualitative leaps are to be made:

* The first is security: it is to develop inherently safe reactors, i.e., for which there is no possibility of core fusion due to chain reaction and loss of control;
* Second, these reactors should not produce highly active and long-term radioactive waste or produce only a limited amount relative to reactors currently in operation, so as to reduce or eliminate the end of cycle problems. An alternative would be the development of reactors burning waste from other systems;
* The third qualitative leap concerns size and modularity: the capacity of new reactors should be about a hundred MWe to be adaptable to all types of networks. Note however that this target is not new but has been postponed, economies of scale favoring high capacities;
* Fourth, the innovative reactors should be versatile and lend themselves to different applications, such as cogeneration of electricity and heat at medium or high temperature, desalination of seawater or production of hydrogen from water;
* Finally, higher efficiencies, including conversion of nuclear fuel, is a prime criterion, both to save on cost per kWh and also to improve all ratios (waste/kWh, water consumption/kWh, etc.): we should be able to produce fifty times more electricity with the same amount of uranium.

Since the early 1990s, different types of projects have emerged, taking into account to varying degrees the objectives outlined above. Among these projects, we will describe on the one hand, those for high temperature gas-cooled reactors, entitled PBMR (pebble bed modular reactor) and GT-MHR (gas turbine modular helium-cooled reactor), and secondly those put up from 2001 following the initiative of the United States Department of Energy (DOE), called *Generation IV Nuclear Energy Systems Initiative*, which has managed to unite around it a group of nuclear research organizations from ten countries in a loose cooperation called the GIF (*Generation IV International Forum*). It is to study systems that could be the fourth generation of reactors, after those in use or planned in the short term (EPR) and deployable by 2030.

The notion of generations of reactors has been introduced to distinguish the main developments that marked the history of civilian nuclear reactors:

* First generation corresponds to the first prototypes built mainly in the U.S. until the late 50s. It saw in France the development of natural uranium graphite gas systems (GCR), with completion of nine reactors, efficiency close to 29%;
* Second generation, from 1960 to 1995, marks the first phase of commercialization of nuclear power plants, with three main systems: light water reactors (BWR Boiling Water Reactors), pressurized water reactors (PWR, efficiency about 33%), and heavy water (in Canada). Most operational plants are currently second generation;
* Third generation has the main objective to increase plant safety, accidents at Three Mile Island in the United States, Chernobyl in the USSR and Fukushima in Japan having shown the major risks associated with certain units of second generation. The EPR

(European pressurized reactor) is a third-generation reactor. Its performance is slightly higher: 35%;

- It is increased efficiencies which primarily justifies the value of the work on the fourth-generation reactors, including the great innovation which is to design the reactor with the cycle that goes with it and optimize the whole, which allows sustainability issues to be properly handled.

After studying in detail a hundred reactor types, the GIF has selected six concepts on which efforts of its members will focus:

- Supercritical water reactors (SCWR);
- Very high-temperature reactors (HTRs and VHTRs);
- Gas-cooled fast neutron reactors (GFR);
- Sodium-cooled fast neutron reactors (SFR);
- Lead-cooled fast neutron reactors (LFR);
- Molten salt reactors (MSR).

Among these six options, a major interest is given to gas-cooled systems.

Let us recall that there are two main categories of reactors, depending on the neutron energy put into play: fast neutron reactors and slow neutron reactors.

- We call "fast neutrons" neutrons produced by fission reactions before they are slowed by a large number of shocks. Their energy is about 0.1 MeV to 2 or 3 MeV. These neutrons are able to split, thus destroying not only nuclei known to be fissile but also actinides, which are nuclei heavier than uranium accumulating in the reactor fuel. If you wish to burn effectively this radioactive waste, fast neutrons are required. We say that fast reactors operate as breeders, which means they produce more plutonium-239 fuel than they consume: fuel becomes inexhaustible;
- We call "slow neutrons" or "thermal neutrons" neutrons slowed by a large number of shocks, usually in a medium called a moderator. Their energy is of the order of one electron volt or a fraction of electron volt, i.e., six orders of magnitude smaller than that of fast neutrons. They can only split a small number of nuclei: uranium 235 (the only one existing in nature), plutonium 239 and uranium-233 produced in the reactors. Such reactors are almost all the existing fleet.

Let us specify, in addition to what has been said, those which we call "epithermal neutrons" neutrons with an energy intermediate between the two previous modes. This type of spectrum occurs in the MSR which will be discussed later.

Reactors coupled to Rankine cycles

Sodium cooled fast neutron reactors

Two variants of SFR are considered, differing according to capacity (150–500 MW for one, 500–1,500 MWe for the other) and fuel. In both cases, a primary cooling loop is provided, the working fluid differing from the coolant. The core outlet temperature is about 550°C (Figure 15.18).

These reactors are designed to valorize high-level waste, particularly plutonium and other actinides, but they can also use fissile and fertile fuel, with a much higher efficiency than PWRs (see Chapter 8).

Supercritical water reactors

The principle of supercritical cycles has been exposed in Chapter 8. The cycle used by the GIF reaches an efficiency of 45%, with a turbine inlet temperature from 510°C to 550°C and

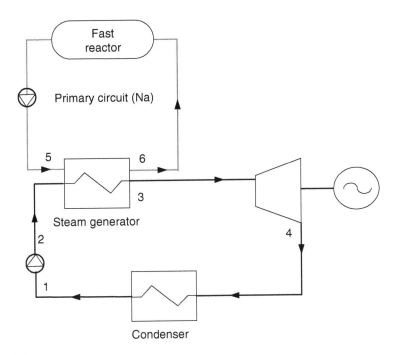

FIGURE 15.18

Sodium cooled fast neutron reactor.

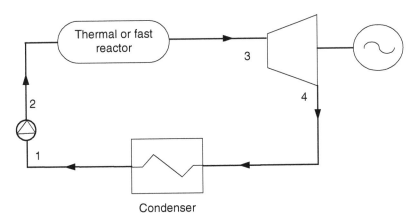

FIGURE 15.19

Supercritical water reactor.

a pressure of 250 bar. Two versions of the reactor are considered, one for thermal neutrons and one for fast neutrons.

While in PWRs and SFR, it is necessary to provide a primary cooling circuit mainly due to the steam change of state, it could be deleted in this type of supercritical cycle, the working fluid being also the reactor coolant, which would significantly simplify the system (Figure 15.19).

The reference capacity of such a reactor would be 1,700 MWe. Note that opinions on the actual feasibility of such a reactor are fairly mixed.

Figure 15.20 shows the synoptic view of a supercritical cycle with two reheats at 550°C, maximum pressure 280 bar. Its efficiency is about 44%.

Reactors coupled to brayton cycles

HTR that were studied in the years 1960–1970, including Germany and the United States, had the following characteristics:

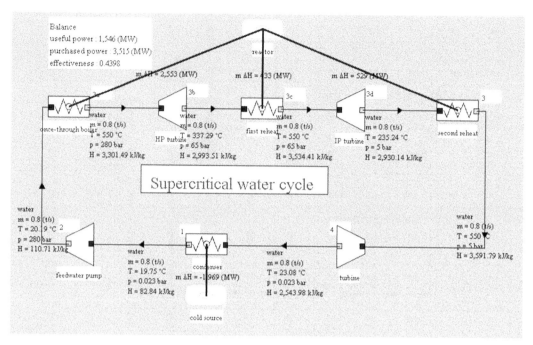

FIGURE 15.20

Synoptic view of a supercritical water cycle.

- Fuel was packaged in the form of nuclei of oxide or carbide of uranium, plutonium or thorium, coated with pyrolytic carbon and silicon carbide supposed to prevent fission products from escaping from fuel spheres;
- Helium was considered as gas coolant in order to reach temperatures of around 800°C for a high thermodynamic efficiency of 40%;
- Reactors were generally of the slow neutron type with graphite as moderator.

One of the advantages of these reactors is to significantly exceed temperature levels to which those using water are limited, which allows high conversion efficiencies to be considered.

Their design concept is shown in Figure 15.21, in case the reactor cooling is provided by the working fluid.

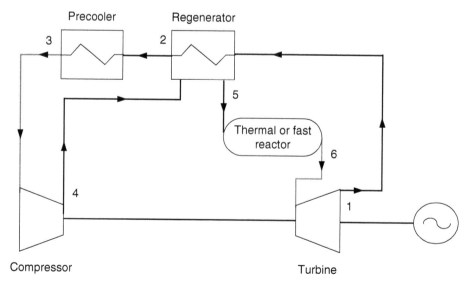

FIGURE 15.21

Gas-cooled reactor.

Closed cycle gas turbines and variants

Consider a helium Brayton cycle coupled to a nuclear reactor at 900°C, using pebbles of low-enriched uranium coated in carbon which acts as a moderator. In this cycle, helium plays both the role of heat transfer fluid in the nuclear reactor and of working fluid.

In the example shown in Figure 15.22, a rate of 140 kg/s of helium is compressed to 70 bar (1), enters a regenerator, which it exits at 514°C (2) before cooling the reactor core which it exits at 900°C (3) and at 67.4 bar.

This flow is expanded in a HP turbine (3–5) used to drive the compressor, before entering a second turbine (5–6), which produces the useful power.

In Thermoptim, balancing a turbine with a compressor can be done by choosing the "mechanically balanced with" option located in the lower right part of the turbine screen (Figure 15.23).

This option allows, for the "expansion" processes only, to determine the downstream pressure of the process such that the absolute value of the expansion work involved is equal to the work of the compression process chosen.

To choose the compression, just double-click in the field on the right of the check box, and choose from the list of proposed compression processes.

The cycle is closed by cooling the helium in two stages: regeneration (6–7) then cooling (7–8) to around 28°C before it is compressed to 70 bar.

In this example, the polytropic efficiencies of the turbomachines are assumed to be 0.9, as well as the efficiency of the regenerator.

The cycle efficiency is 44%,

Small capacity modular reactor PBMR

PBMRs (Figure 15.24) are high-temperature thermal neutron reactors with an output of 100 MW using pebbles of weakly enriched uranium fuel embedded in carbon (which plays the role of moderator) and using helium as coolant.

In the example shown in Figure 15.25, a rate of 140 kg/s of helium is compressed at 70 bar (1), enters a regenerator, which it exits at 514°C (2) before cooling the reactor core which it exits at 900°C (3) and at 67.4 bar. This flow is expanded in a HP turbine (3–4) used to drive

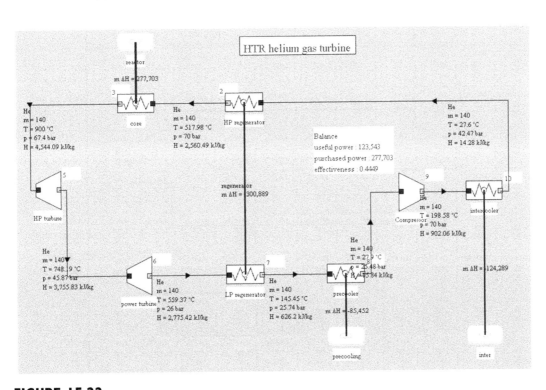

FIGURE 15.22

Helium gas turbine.

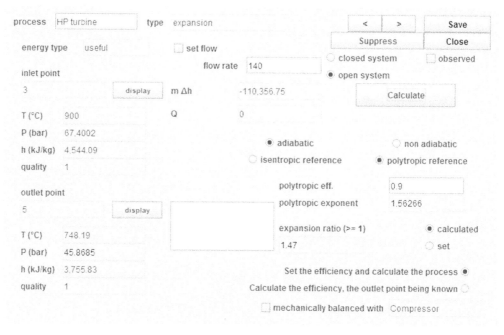

FIGURE 15.23

Screen of the turbine coupled to a compressor.

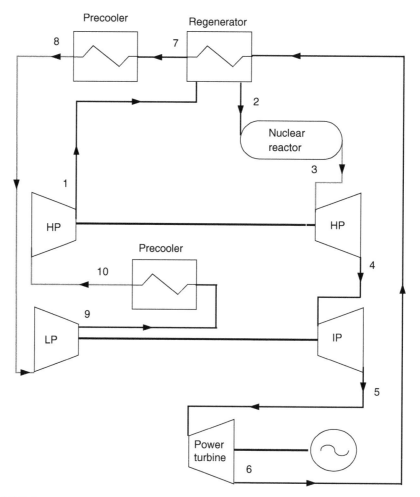

FIGURE 15.24

PBMR reactor.

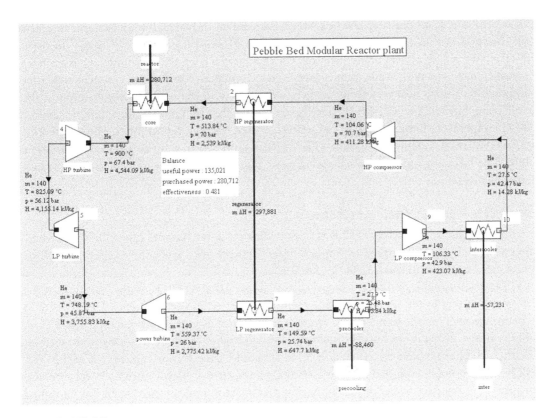

FIGURE 15.25

Small capacity PBMR modular reactor.

the HP compressor, before entering a second IP turbine (4–5), equilibrated with the LP compressor.

Helium is then expanded at 26 bar and 559°C in the LP turbine (5–6), which produces useful power. The cycle is closed by two-step helium cooling (regeneration (6–7) and then cooling at about 28°C (7–8)), followed by a two-stage compression (8–9) and (9–1) with intermediate cooling (9–10). In this example, turbomachinery polytropic efficiency and regenerator effectiveness are assumed equal to 0.9.

PBMR cycles theoretically have an excellent efficiency, close to 48%, as shown in Figure 15.25. However, their industrial production is hampered by various technological difficulties, particularly in terms of on-site review of the fuel, which means that actual efficiencies are actually much lower.

In addition, there is currently no high-efficiency helium turbomachinery. Launched in South Africa in the early 1990s, this concept was abandoned in 2002.

GT-MHR reactors

Studied by an international consortium led by General Atomics (USA), GT-MHR is a modular thermal neutron reactor (300 MW units), helium-cooled, high-temperature, which can use various fuels (plutonium or enriched natural uranium, thorium) packed in silicon carbide beads. Differences between GT-MHR and PBMR are numerous, although they are both high-temperature helium cooled reactors.

The GT-MHR capacity is 300 MW, a compromise between the requirement of intrinsic safety, which involves reduced capacity, and economic competitiveness which requires economies of scale. As fuel, GT-MHR uses the same principle as PBMR, beads coated with millimeter fuel refractory beads by TRISO process. However, instead of being agglomerated in the form of pebbles, the balls are in the form of small cylinders of a few centimeters in height, the cylinders themselves being formed into fuel elements in graphite prismatic shape. This method allows a better knowledge of the location of the fuel and yields a higher unit power than for the PBMR. It also reduces by a factor of five to ten the pressure drop in the reactor.

To control the chain reaction, the GT-MHR has control rods which may be supplemented by an injection system by gravity of boron particles. Moreover, in case of damage on these two active systems, designers made sure that the reactor can cool itself through a heat transfer by conduction to the tank walls, themselves water circulated. Finally, the concrete structure is supposed to absorb heat and transmit it by conduction in the surrounding basement, the reactor being buried. The GT-MHR developers ensure that fuel temperature remains below 1,600°C temperature limit for stability of materials used for making millimeter beads, including in case of reactor depressurization. GT-MHR is supposed to achieve a thermal efficiency of 48%. The refueling outages are scheduled every eighteen months, with replacement of half the fuel elements.

Very high-temperature reactors

Very high temperature reactor VHTR is in line with modular reactors type GT-MHR (thermal neutrons). It is distinguished by a much higher temperature, since the temperature of the coolant gas is expected to reach 1,000°C–1,100°C.

VHTR reactor fuel is designed along the same lines as that of HTR, with a package in the form of millimeter beads agglomerated as cylinders inserted in fuel elements. VHTR should use helium as coolant, the temperature of gases leaving the reactor vessel reaching 1,000°C. A VHTR priority is that it can burn its fuel with a much higher efficiency than current reactors.

With high thermodynamic efficiency, VHTR would have a unit capacity of 600 MWe. VHTR was originally intended to burn essentially a mixture of highly enriched uranium and thorium. The objective today is explicitly that this reactor can not only burn low-enriched uranium but also incinerate plutonium and plutonium mixed with some minor actinides. Contested by some experts, this capability would allow a resumption of PWR waste.

The goal of the proponents of VHTR is not only to cover the needs of power generation but also to advance the technology of refractories, which will be useful for the development of other systems and finally to open new markets for nuclear power. For electricity production, efficiencies achieved with as high operating temperatures would be above 50%, which is significantly higher than those of existing PWRs (33%), which should lead to competitive production costs.

However, to successfully make them work, many technological problems must be solved, particularly the development of materials capable of withstanding very high temperatures. The know-how for the development of HTR would then be useful to that of gas-cooled fast reactor GFR.

New markets opened by VHTR should be multiple. Many industrial processes are indeed performed at high temperature: production of cement, glass, steel, coal gasification and thermochemistry.

However, given the inertia of industrial processes and hopes for developing alternative fuels for transportation, the main VHTR application would be hydrogen production.

Gas-cooled fast neutron reactors

GFR are somehow variants of GT-MHR or VHTR, where the thermal neutron core is replaced by a fast neutron core. The reference capacity is 288 MWe, with a turbine inlet temperature of 850°C. The cooling gas is typically helium, and many thermodynamic cycles are possible, from direct cycles where helium that cools the core is directly expanded in the turbine, which avoids the intermediate heat exchanger IHX and improves performance up to combined cycles, through innovative cycle-type supercritical CO_2.

Lead-cooled fast reactors

Fast neutron reactors cooled by lead (or lead-bismuth eutectic) are variations from Russian submarine reactors (Figure 15.26). Their reference capacity is thus small, between 120 and 400 MWe, and the tank outlet temperature is between 550°C and 800°C, allowing us to consider a gas thermodynamic cycle (e.g., He or CO_2).

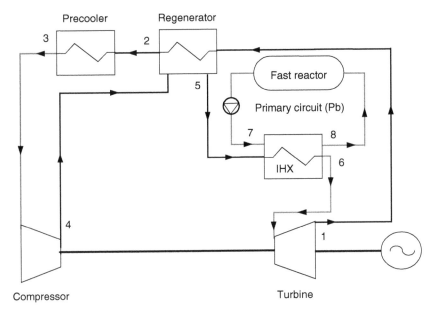

FIGURE 15.26

Lead cooled fast reactor.

Molten salt reactors

Although regarded as potentially very attractive in the long term, molten salt epithermal neutron reactors are not currently a priority for the GIF.

One of their characteristics is to have a homogeneous core and a fuel flow in the cooling circuit, fuel acting also as thermal fluid. If corrosion problems are solved one day, their value is to minimize radioactive waste and mitigate proliferation risks. They are indeed characterized by the possibility of an online processing of spent fuel, with a small chemical unit, which processes all the salt every ten days.

The proposed reference capacity is 1,000 MWe, with a core outlet temperature between 700°C and 800°C, allowing use of high temperature gas thermodynamic cycles. Their diagram is similar to Figure 15.26, with a two-stage primary circuit to prevent molten salts from directly interacting with the working fluid.

Thermodynamic cycles of high-temperature reactors

In the previous sections, we briefly introduced the various types of future nuclear reactors envisaged, especially in the Generation IV project. The six systems studied by the GIF differ more by the type of neutron involved (slow or fast) or by the type of core cooling envisaged than by the temperature level and thus the thermodynamic cycles that can be associated with them. In this section, we summarize what has been said of gas cycles in the previous pages, and we introduce two new types of cycles that have good potential for HTR.

Brayton cycles

The simple helium Brayton cycle used for the PBMR reactor leads to excellent performance (Figure 15.27) but has the drawback that there are currently no suitable industrial turbomachines and that it will take many years to develop them.

An alternative solution consists in using helium as coolant for the reactor core, and another gas as thermodynamic fluid, in particular a mixture of nitrogen and helium (79% N_2, 21% He by volume), whose thermodynamic properties are relatively close to those of air for which industrial technology is well mastered.

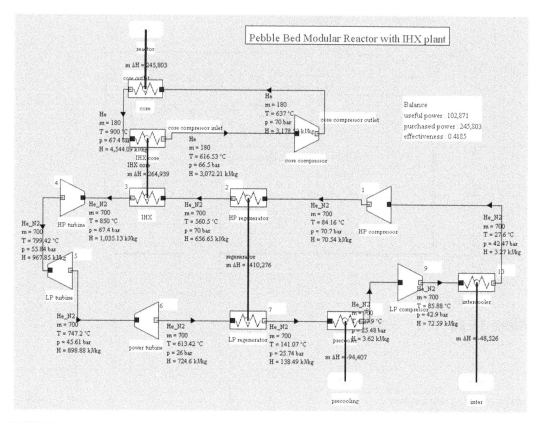

FIGURE 15.27

Synoptic view of the PBMR cycle with IHX.

This solution however requires the introduction of an intermediate exchanger called IHX for Intermediate Heat eXchanger (we then speak of an indirect Brayton cycle), the realization of which is not simple, even if it is more or less controlled, and which introduces irreversibilities which have the effect of lowering the overall efficiency.

Figure 15.27 shows the synoptic view of such a cycle, which is the subject of guided exploration (C-M4-V4).

The drop in efficiency is significant. It is mainly due to pressure drops in the intermediate exchanger, which is a gas-gas exchanger whose heat exchange coefficients are low, which requires large exchange surfaces and small sections.

We assumed an average temperature difference of around 50°C in this exchanger, but the price to pay is a consumption of more than 19 MW in the helium compressor.

BOX 15.1

Guided educational exploration

High-temperature nuclear cycle (exploration C-M4-V4)

This exploration presents the model of a high-temperature nuclear cycle using gas turbines operating in closed system, and not open system like those studied in previous explorations. It shows in particular how to balance a turbine with a compressor in Thermoptim.

Supercritical CO$_2$ cycles

Another type of cycle envisaged for future nuclear reactors would use supercritical carbon dioxide as working fluid.

As indicated in the previous chapter, cycles using supercritical carbon dioxide could lead to better performance than others at moderate reactor temperatures between 650°C and 800°C (Dostal et al., 2003).

Proponents of these cycles argue that efficiencies are, for this temperature range, higher than those of steam cycles and that machines are much more compact.

In thermodynamics, the value of using such a cycle is to benefit from a compression work in supercritical phase much lower than if the working fluid remains in the gaseous state as a Brayton classic cycle.

Several types of supercritical CO_2 cycles were presented in Chapter 14, to which you should refer for details. Figure 15.28 shows a diagram of a recompression cycle and Figure 15.29 its synoptic view.

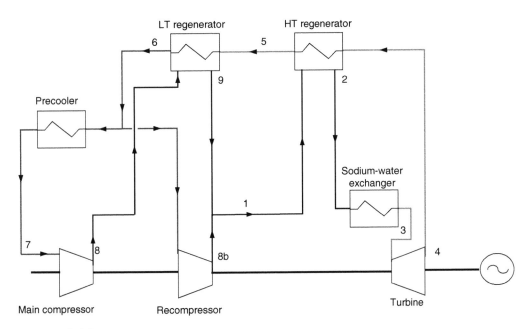

FIGURE 15.28

Recompression supercritical CO_2 cycle.

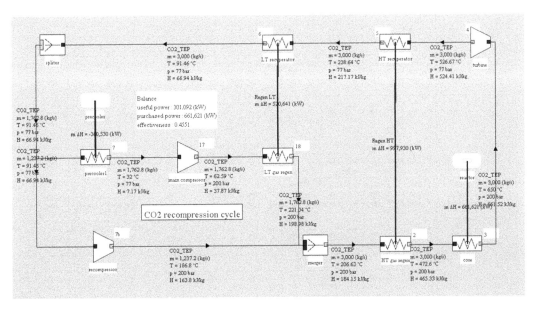

FIGURE 15.29

Synoptic view of the supercritical recompression CO_2 cycle.

Supercritical CO_2 cycles appear very interesting in thermodynamics terms. The two main technological constraints are the realization of regenerators in the circum-critical zone in particular to avoid any temperature crossover and that of efficient turbomachinery for CO_2, an area where there is no reference. Moreover, these cycles can only work if the CO_2 state at the precooler outlet is supercritical, which implies a temperature limit of 32°C, which can be difficult to achieve when the cold source is outside air, a river or sea water.

HTR combined cycles

The GT-MHR reactor is the result of optimization work performed by General Atomics in 1985. Since 2000, Areva, which is involved in this work, changed the basic concept (helium Brayton cycle) so as to make the design easier and the adaptation to cogeneration more immediate (Gauthier et al., 2005; Gosset et al., 2005).

Considerations governing the selection of cycles are numerous. Obviously, intrinsic cycle efficiency is fundamental, but technological feasibility is not less. This has led Areva to choose for HTR-VHTR a combined cycle using a helium-nitrogen mixture (20–80 wt%) instead of pure helium, because of the considerable experience gained over several decades in air gas turbines.

In 2002, Areva optimized with Thermoptim a combined cycle associated with a HTR (Figure 15.30). Variations of this cycle can provide superheated steam at 110 bar using water at 120°C for some cogeneration applications, for capacities from 50 to 300 MW. The efficiency announced by Areva in electricity production is only 47%, but the objective is actually 50%, without any particular technological development except the IHX intercooler, the Mitsubishi company committing itself of making all turbomachinery (gas and steam) on the basis of current knowledge.

The gas circuit (a mixture of helium and nitrogen) is composed of a heat exchanger IHX that allows the transfer of energy from the coolant fluid (helium) warmed in the nuclear reactor core. A temperature difference of about 50°C between the hot fluid inlet and of cold fluid outlet is desirable to maintain the surface of this exchanger in a range of reasonable values.

Before passing through the exchanger IHX, the gas is first compressed by a compressor at a pressure between 55 and 70 bar, the latter value of 70 bar being a maximum value for reasons of strength of materials. At the exchanger outlet, the gas is expanded at about 40 bar in a turbine and exits at still relatively high temperature (about 600°C). It is then cooled in a heat exchanger and redirected to the compressor. The thermodynamic cycle selected utilizes the enthalpy available in this exchanger to operate a steam cycle and thereby obtain a combined cycle.

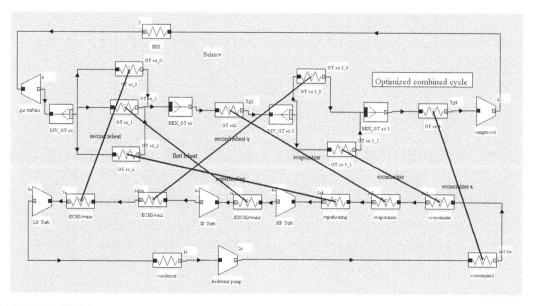

FIGURE 15.30

Optimized combined cycle.

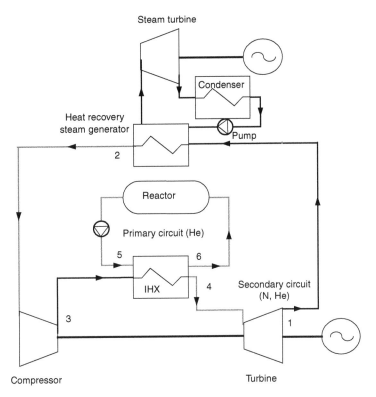

FIGURE 15.31

Combined cycle optimized by Areva with Thermoptim.

Gas temperature at the IHX heat exchanger inlet (thus at the compressor outlet) is 300°C and pressure 55 bar. Gas temperature at the IHX exchanger outlet is 800°C and pressure 55 bar. This results from a temperature differential of 50°C selected on the IHX exchanger.

The proposed steam circuit consists of three successive turbines (high, intermediate and low pressure). Superheated steam available at the entrance of the HP turbine goes through an economizer that heats the liquid, an evaporator that vaporizes it and a superheater. Two reheats are provided at the entrance of the IP and LP turbines.

Variations of this cycle adapted to CHP have also been optimized by Areva with Thermoptim (Figure 15.31).

This section has shown that thermodynamic cycles to be used in future nuclear reactors should have much higher efficiencies than PWR, thanks to the increased working fluid temperature. It is too early to know, among the major cycles considered (supercritical water, Brayton helium, supercritical CO_2, combined cycle), those which will prove the best. The answer depends on many parameters and primarily the type of reactor that will be the safest and most economical.

Table 15.2 summarizes the main options that exist today with regard to future nuclear reactors (HTR here is essentially GT-MHR, close to PBMR as we have seen).

TABLE 15.2

OPTIONS FOR THE FUTURE NUCLEAR REACTORS

	T (°C)	P (MWe)	Cooling	Neutrons	Thermo Cycle
SFR	550	500–1,500	Indirect	Fast	Steam/super-crit.
SCWR	550	1,700	Direct	Th. or fast	Supercritical steam
HTR	900	100–600	Dir./indir.	Thermal	Gas, CO_2, CC
VHTR	1,100	600	Dir./indir.	Thermal	Gas, CO_2, CC
GFR	850	300	Dir./indir.	Fast	Gas, CO_2, CC
LFR	550–800	100–400	Indirect	Fast	Gas, CO_2, CC
MSR	700–800	1,000	Indirect	Thermal	Gas, CO_2, CC

Oxy-combustion cycles

Introduction

To conduct an oxyfuel combustion is to replace by pure oxygen the usual oxidizer, namely air, a mixture mainly of oxygen and nitrogen (respectively 21% and 78% by volume). Oxy-fuel technology allows at the same time to get fumes composed almost exclusively of water and carbon dioxide and to drastically reduce emissions of nitrogen oxides. These are technologies already used in industry, including glass and steel.

The separation of CO_2 and H_2O is then easily done by simple water condensation, and the absence of nitrogen allows in addition NO_x emissions and the volume of smoke to be greatly reduced.

The CO_2 can then be stored, and the net greenhouse gas emissions of the cycle are zero.

Despite these advantages, these technologies have so far hardly been used for electricity generation, given the difficulties and costs involved in the production of oxygen. Several technical solutions exist, but the more developed today, called the air separation unit, is to separate oxygen and nitrogen from air by cryogenic operation both costly and energy consuming.

Four oxycombustion plants using this separation technique have been recently proposed: cycles called water cycle, Matiant, oxy-fuel and Graz.

Their principle is to use as oxidizer a gas containing on the one hand oxygen and secondly a mixture of CO_2 and H_2O in different proportions depending on the case.

Under these conditions, flue gas thermophysical properties are similar to those encountered in a gas turbine whose oxidizer is air, which allows the use of existing high performance expansion turbines without the need to make further technological developments.

The four cycles are distinguished by their configurations, close to regenerative GT for the first two, a combined cycle for the third and more original for the Graz one, whose gas and steam cycles are integrated directly rather than simply juxtaposed.

This process based on separation of the external air into oxygen and nitrogen is not the only one, and innovations are being considered in the form of internal separation cycles.

All of these cycles but two (water cycle and Matiant) involve both gas turbines and steam turbines.

This process based on separation of the external air into oxygen and nitrogen is not the only one, and innovations are being considered in the form of internal separation cycles such as those based on use of

- Membranes permeable to oxygen (AZEP cycle);
- Or transport of oxygen by chemical carrier by making, in the presence of air, a metal oxide which is then reduced before combustion (this is called chemical looping combustion (CLC)).

Their diagrams are given in the following sections and synoptic views in Figures 15.41–15.46. Water extraction is modeled by class DehumidifyingCoil.

Oxy-fuel cycle

In this combined cycle the oxidizer is a mixture of O_2 and CO_2.

In an Oxy-fuel cycle (Figure 15.32), at the condenser inlet (6), we get fumes containing CO_2 and H_2O at a pressure of 1 bar. They are cooled by an external heat sink at a temperature low enough for almost all water to be condensed (7). The condensed water is extracted from the cycle.

At the condenser outlet, a fraction of the gas composed mainly of CO_2 is extracted (2), the remaining flow, equal to 35 kg/s in this example being compressed at the pressure of 40 bar in a compressor (2–3) before being mixed with pure oxygen (about 4.4 kg/s). This mixture is

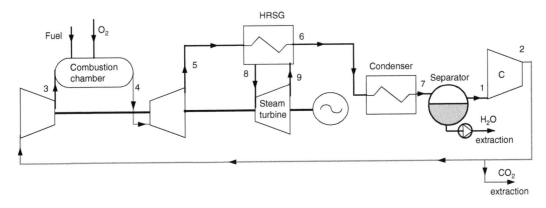

FIGURE 15.32

Oxy-fuel cycle.

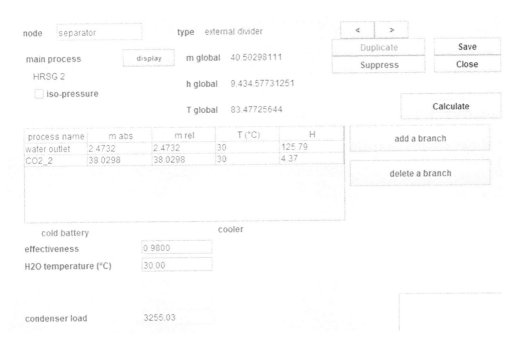

FIGURE 15.33

Steam condensation component screen.

then used as an oxidizer in a stoichiometric combustion chamber (3–4) whose fuel is methane, the temperature at the end of combustion being about 1,350°C.

Fumes are expanded at 1 bar in a turbine (4–5), then cooled in a recovery steam generator (5–6) before entering the condenser, thus closing the cycle.

A steam cycle, working between pressures of 0.03 and 150 bar and 530°C, involving a water flow of 10 kg/s, is coupled to the previous cycle to form a combined cycle. In the diagram of Figure 15.32, a simplified steam cycle is represented as a HRSG, and the turbine pump and condenser are not shown nor oxygen and fuel compressors.

The modeling of this cycle uses an external divider to represent the condensation of mixed water vapor: the external class ColdBattery,[1] coupled to a thermocoupler (Figure 15.33).

The thermocoupler extracts the heat of condensation of the water (Figure 15.34).

This cycle is the subject of guided exploration (C-M4-V5). It leads to the following results: power output 35 MW and gross efficiency 60.6%, but these excellent performances must be tempered by the "cost" of separation of oxygen from the air. The synoptic view of this cycle is given in Figure 15.41.

---

[1] http://direns.mines-paristech.fr/Sites/Thopt/fr/co/modele-cold-battery.html.

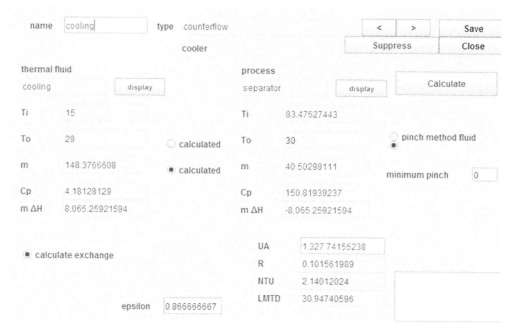

FIGURE 15.34

Thermocoupler screen.

BOX 15.2
Guided educational exploration

Oxycombustion cycle (exploration C-M3-V5)

This exploration presents the model of an oxyfuel-type oxycombustion cycle.

Water cycle

In this regeneration cycle (Figure 15.35), the oxidizer of the first combustion chamber is a mixture of O_2 and H_2O, and expansion is sequential with intermediate reheat.

At the condenser inlet (9), we get fumes containing O_2 and H_2O at a pressure of 0.1 bar. They are cooled by an external heat sink at a temperature low enough for almost all water to be condensed. The condensed water (1) is compressed at 4 bar (2), and the surplus to 125 kg/s is extracted from the cycle. At the separator outlet, the gas is composed mainly of extracted CO_2.

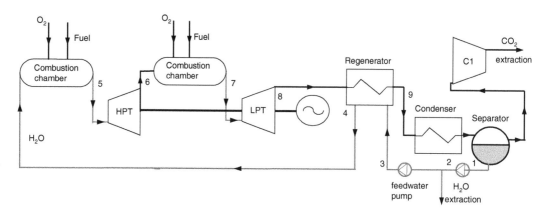

FIGURE 15.35

Water cycle.

Liquid water is compressed at 83 bar (3) before it enters the regenerator (actually a steam generator), which it exits in the state of saturated liquid at 300°C (4).

Water is mixed with pure oxygen (about 33 kg/s), the mixture being used as oxidizer in a first combustion chamber (stoichiometric) whose fuel is methane, the end of combustion temperature being at about 900°C–1,000°C (5).

In modeling, one must be careful that it is impossible to mix oxygen with liquid water to form steam, because we do not have enough enthalpy in oxygen for the water vaporization. In reality, this energy is drawn from the available heat in the combustion chamber. We may either build a thermocoupler between the combustion chamber and the "evaporator" process or model the whole as a combustion process followed by an exchange process designed to vaporize the water before mixing with oxygen.

The burnt gases are expanded in a high-pressure turbine at the pressure of 8.3 bar (6) and then mixed with pure oxygen (about 25 kg/s). This mixture is then used as an oxidizer in a second combustion chamber, also stoichiometric, whose fuel is methane, the end of combustion temperature being 1,200°C–1,300°C (7). The fumes are expanded in a low-pressure turbine at the pressure of 0.1 bar (8), then cooled in the steam generator, thus closing the cycle.

It leads to the following performance: capacity 437 MW and 59.5% gross efficiency. The synoptic view of this cycle is given in Figure 15.42.

Matiant cycle

In this regeneration cycle (Figure 15.36), whose architecture is very similar to the previous one (Mathieu & Nihart, 1999), the oxidizer of the first combustion chamber is a mixture of O_2 and CO_2, and expansion is sequential with intermediate reheat.

At point 1, a gas composed mainly of CO_2, at a pressure we assume initially equal to 1 bar, is compressed at the pressure of 110 bar (2) in a four-stage compressor with intermediate cooling at 30°C.

A fraction of CO_2 is then extracted, the remaining flow, equal to 50 kg/s, being heated in a regenerator (3), before being mixed with pure oxygen (about 2.5 kg/s). This mixture is then used as an oxidizer in a first combustion whose fuel is methane, the temperature at the end of combustion being equal to 1,300°C (4).

The burnt gases are expanded in a high-pressure turbine at a pressure of 40 bar, then mixed with oxygen (about 5 kg/s). This mixture (5) is then used as an oxidizer in a second combustion chamber where the fuel is methane, the temperature at the end of combustion being equal to 1,300°C (6). The fumes are expanded in a low-pressure turbine at the pressure of 1 bar (point 7) and then cooled in the regenerator to point 8.

They are then cooled by an external heat sink at a temperature low enough for almost all water to be condensed. The condensed water is extracted from the cycle, while the residual gas, whose mass flow rate is greater than 1 kg/s, enters the multi-stage compressor, closing the cycle.

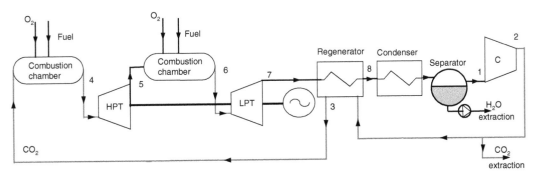

FIGURE 15.36

Matiant cycle.

It leads to the following performance: 47.8 MW capacity and gross efficiency 56.5%. The synoptic view of this cycle is given in Figure 15.43.

Graz cycle

In this cycle (Figure 15.37), the oxidizer is a mixture of O_2, CO_2 and H_2O, the latter two gases being mixed in the combustion chamber inlet and then separated after expansion, each following a particular subcycle (Heitmeir & Jericha, 2003).

At the condenser inlet (11), we get fumes containing CO_2 and H_2O. They are cooled by an external heat sink at a temperature low enough for almost all water to be condensed. The condensed water is compressed at 12 bar (2e–3e), and the surplus to 110 kg/s is extracted from the cycle.

At the outlet of the separator (2), gas composed mainly of CO_2 at a pressure of 0.25 bar, is compressed (2–7) at the pressure of 40 bar in a three-stage compressor with intermediate cooling (C1, C2 and C3). A fraction of CO_2 is extracted from C2, the remaining flow being equal to 200 kg/s. The intermediate cooling is by exchange with liquid water, which is compressed at 180 bar (5e–6e) before it enters the recovery steam generator (HRSG on the diagram), which it exits superheated at 560°C (7e).

The gas at the outlet of C3 (7) is mixed with pure oxygen (about 0.5 kg/s) and steam from turbine "Steam turbine", the mixture being used as oxidizer in the combustion chamber (stoichiometric) whose fuel is methane, the end of combustion temperature being equal to 1,328°C (8).

Burnt gases are expanded in a high-pressure turbine (8–9) at the pressure of 1 bar and then cooled in the recovery steam generator (9–10). They are finally expanded in a low-pressure turbine (10–11) at the pressure of 0.25 bar and then cooled in the condenser, thus closing the cycle.

It leads to the following performance: 396 MW capacity and 56.4% gross efficiency. The synoptic view of this cycle is given in Figure 15.44.

AZEP cycle

The AZEP cycle (Sundkvist) (Advanced Zero Emission Process) is a combined cycle involving internal separation of air through a membrane (Figures 15.38 and 15.39).

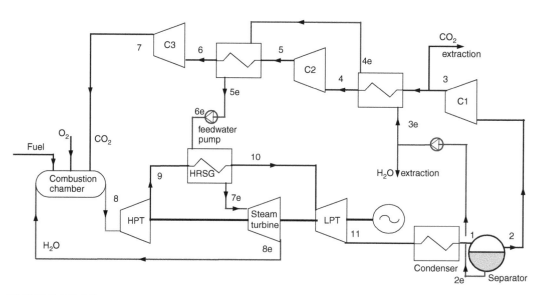

FIGURE 15.37

Graz cycle.

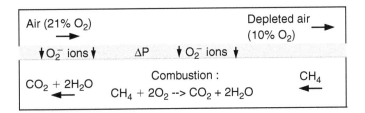

FIGURE 15.38

Schematic of the MIEC membrane.

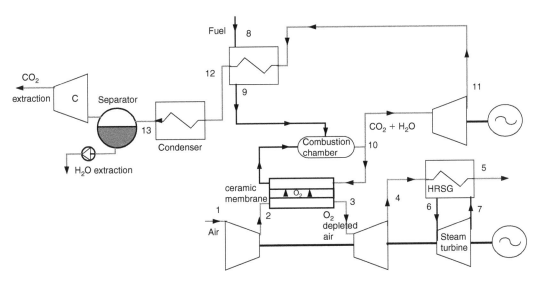

FIGURE 15.39

AZEP cycle.

Air is sucked into the compressor of a modified gas turbine, where the combustion chamber is replaced by a chamber with two compartments separated by a ceramic membrane permeable to oxygen (mixed ionic-electronic conducting membrane (MIEC)). In one of the chambers, air is depleted in oxygen because of the difference in oxygen partial pressures between the two media. In the other, this oxygen acts as oxidizer for combustion of fuel in the presence of an inert gas that is but a recirculation of exhaust gases, composed mainly of CO_2 and H_2O, plus inert gases if the fuel contains some.

At high temperatures (above 700°C), the ceramic membrane (Figure 15.38) is a mixed ionic and electronic conductor, through which pass simultaneously O^{2-} ions and electrons, oxygen being adsorbed on the surface.

Depleted air is expanded in a turbine (3–4) and then used as a heat source (4–5) for a steam cycle (see Figure 15.40, the pump and condenser not being shown in figure so as not to overload). The part of the flue gas not recirculated is expanded in a turbine (10–11), then cooled (11–12) either to preheat the fuel (8–9) or as a hot source for the steam cycle, before being condensed for water extraction (12–13). Remaining CO_2 can then be captured. The part of flue gas that is recirculated is first cooled slightly by preheating air.

Much of the gas leaving the combustion chamber is recycled, in particular so that the average molar fraction of oxygen in the oxidizer is low enough that oxygen can pass through the membrane. Exchangers around the ceramic membrane have not been shown.

Although this is not the only possible solution, the steam generator can be split into two parts, one heated by the depleted air and the other by the flue gases.

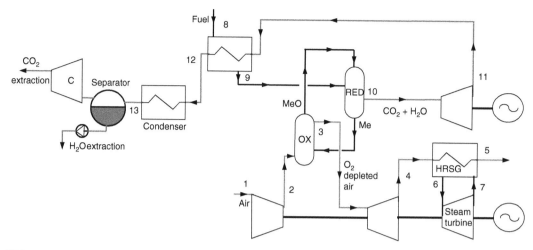

FIGURE 15.40

CLC cycle.

It leads to the following performance: 403 MW capacity and gross efficiency 59%. The synoptic view of this cycle is given in Figure 15.45.

CLC cycle

The general principle of the CLC (Chemical Looping Combustion) cycle (Wolf, 2004) is close to the AZEP cycle: this is still a combined cycle, but air separation is accomplished through a chemical reaction allowing oxygen to be preferentially attached (Figure 15.40).

Air (about 50 kg/s) is compressed (1–2) between 15 and 20 bar by the compressor of a modified gas turbine, whose combustion chamber is replaced by a chamber with two compartments between which circulates a metal oxide such as NiO, thanks to a circulating fluidized bed. In one of the chambers, air is depleted in oxygen (2–3) due to oxidation of the metal. In the other, the oxide is reduced, and the oxygen released burns with a fuel (9–10).

The depleted air (mole fraction of oxygen of about 0.14) is expanded in a turbine (3–4), and then used as a heat source (4–5) for a steam cycle (6–7). The exhaust gases are also expanded in a turbine (10–11) and then used as heat source (11–12) on the one hand to preheat the fuel (8–9) and on the other hand (not shown in figure so as not to overload it) as hot source for the steam cycle, before being condensed for water extraction. The remaining CO_2 can then be captured.

Although this is not the only possible solution, the steam generator can be divided into two parts, one heated by the depleted air and the other by the flue gases with a reheat.

It leads to the following performance: 378 MW capacity and gross efficiency 53.28%. The synoptic view of this cycle is given in Figure 15.46.

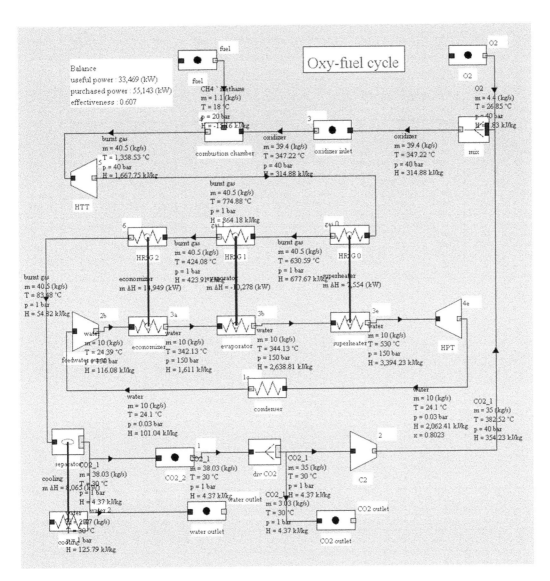

FIGURE 15.41

Synoptic view of the Oxyfuel cycle.

Estimate of oxygen separation work

Many oxyfuel cycles require oxygen as oxidizer. As mentioned, the excellent performance of these cycles has to be tempered by the "cost" of separating oxygen from air.

To estimate this cost, we will make the simplifying assumption that the power required to produce oxygen is equal to that required to condense the gas, which is calculated as the theoretical power corresponding to the condensation of the flow of pure substance, divided by a carefully chosen exergy efficiency. This method is of course debatable, but it is much closer to reality than the consideration of the minimum separation power alone, which is simply equal to exergy lost during the mixing of air constituents.

We proceed on the basis of a rate of 1 kg/s of oxygen. To separate the air, we must operate in three phases:

* We begin by cooling it to the boiling point of oxygen at atmospheric pressure, 85 K;
* Oxygen is condensed and separated from nitrogen;
* Oxygen and nitrogen are separately heated to room temperature.

For simplicity, we can assume that the thermal power required to cool air is almost equal to that corresponding to heat separated gases, so we can perform internal regeneration between

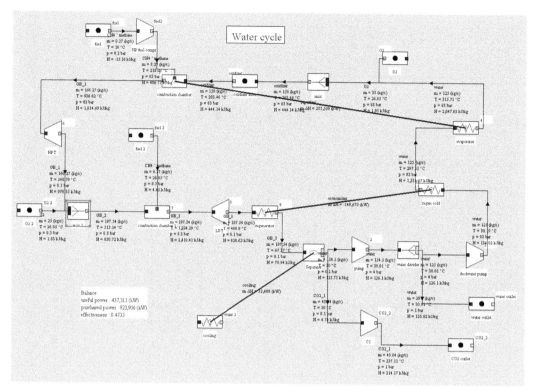

FIGURE 15.42

Synoptic view of the WaterCycle.

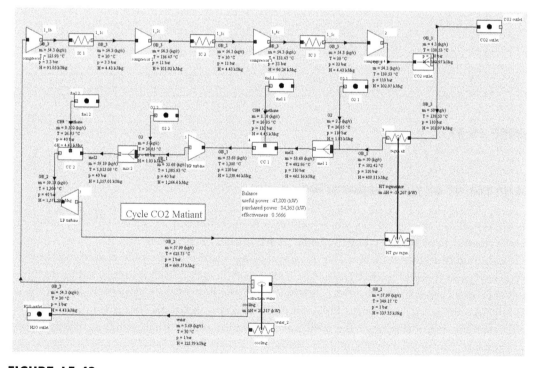

FIGURE 15.43

Synoptic view of the Matiant cycle.

these two gas streams. The only power input is then that corresponding to the condensation of oxygen, i.e., 211 kW.

If this thermal power was produced in a perfect reverse Carnot engine, its efficiency would be equal to 85/(300–85) = 0.395. The mechanical power of a perfect machine allowing 1 kg/s of oxygen at 85 K to condense is then equal to 211/0.395 = 534 kW.

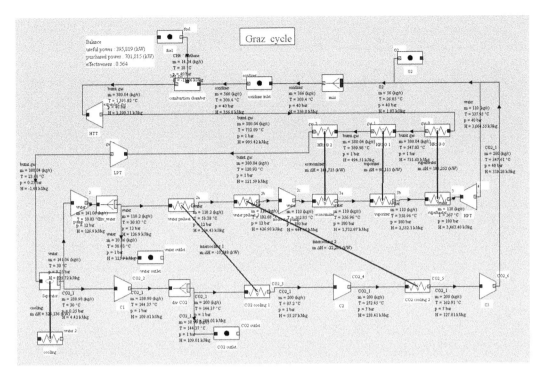

FIGURE 15.44

Synoptic view of the Graz cycle.

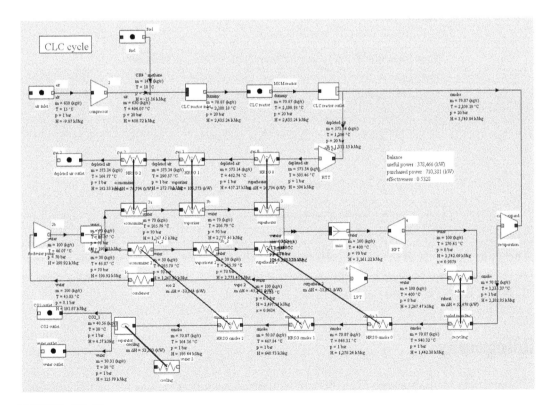

FIGURE 15.45

Synoptic view of the AZEP cycle.

The various examples of refrigeration machines discussed in Chapter 11 have exergy efficiencies between 25% and 42%. We use here an exergy efficiency value of 40% for our simplified air separation unit. Under these conditions, the power required to separate 1 kg/s of oxygen is equal to 534/0.4 = 1,335 kW or a work of 1,335 kJ/kg of O_2.

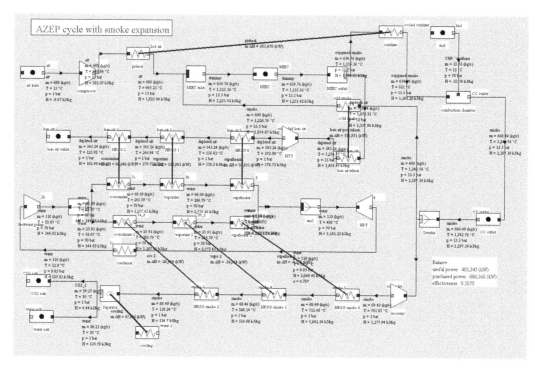

FIGURE 15.46

Synoptic view of the CLC cycle.

TABLE 15.3

EFFICIENCY REDUCTIONS OF OXY-COMBUSTION CYCLES

Capacities in kW	OxyFuel	Water Cycle	Graz	Matiant
Gross capacity	33,469	437,311	395,819	47,800
Purchased capacity	55,143	734,449	701,815	84,363
Gross efficiency	60.7%	59.54%	56.40%	56.66%
O₂ flow rate (kg/s)	4.4	58	56.1	7.5
O₂ power	5,874	77,430	74,894	10,013
Net capacity	27,595	359,881	320,926	37,788
Net efficiency	50.04%	49%	45.73%	44.79%
Efficiency reduction	10.65%	10.54%	10.67%	11.87%

This estimate allows us to correct performance of the cycles we studied previously to determine their net capacities and efficiencies, given in Table 15.3. The reduction in efficiency is from 10.5 to about twelve points.

Bibliography

O. Bolland, H. M. Kvamsdal, J. C. Boden, A thermodynamic comparison of the oxy-fuel power cycles, water-cycle, graz-cycle and matiant-cycle, Power generation and sustainable development, 293–298, (Liège, 8–9 October 2001).

G. Descombes, J. L. Magnet, *Moteurs non conventionnels*, Techniques de l'Ingénieur, Traité Génie Mécanique, BM 2593.

V. Dostal, A supercritical carbon dioxide cycle for next-generation nuclear reactors, PhD thesis, MIT, January 2004.

V. Dostal, M. J. Driscoll, P. Hejzlar, N. E Todreas, *A supercritical CO₂ gas turbine power cycle for next-generation nuclear reactors*, Proc. ICONE-10, Arlington, Virginia, April 14–18, 2003.

J. C. Gauthier, M. Lecomte, P. H. Billot, The Framatome-anp near term HTR concept and its longer term development perspective, *13th International Conference on Nuclear Engineering*, Beijing, China, May 16–20, 2005.

J. Gosset, R. Gicquel, M. Lecomte, D. Queiros-Conde, Optimal design of the structure and settings of nuclear HTR thermodynamic cycles. *International Journal of Thermal Sciences*, 44, 1169–1179, 2005.

F. Heitmeir, H. Jericha, Graz cycle # An optimized power plant concept for CO_2 retention, *First International Conference on Industrial Gas Turbine Technologies*, Brussels # 10/11, July 2003.

H. M. Kvamsdal, O. Maurstad, K. Jordal, O. Bolland, Benchmarking of gas-turbine cycles with O_2 capture, GHGT-7, Vancouver, 2004.

P. Mathieu, R. Nihart, Zero-emission MATIANT cycle, *International Gas Turbine and Aeroengine Congress and Exhibition*, Stockholm, Sweden, vol. 121(1), 116–120, 1999.

P. F. Peterson, Multiple-reheat Brayton cycles for nuclear power conversion with molten coolants. *Nuclear Technology*, 144(Numbre 3), 279–288, Dec. 2003.

P. Pradel, *La R&D sur les filières nucléaires actuelles et futures : enjeux et perspectives*, Réalités Industrielles, Annales des Mines, ISSN 1148.7941, 23–30, Fev. 2007.

G. T. Reader, C. Hooper, *Stirling engines*, Spon Editors, London, 1983.

T. Schulenberg, H. Wider, M. A. Fütterer, Electricity production in nuclear power plants – Rankine vs. Brayton Cycles, *ANS/ENS International Winter Meeting* (Global 2003), New Orleans, LA, Nov. 2003.

W. B. Stine, *A compendium of solar dish/Stirling technology, Sandia national laboratories*, SAND93-7026 UC-236, 1994.

P. Stouffs, *Machines thermiques non conventionnelles, état de l'art, applications, problèmes à résoudre*, Société Française des Thermiciens, Thermodynamique des machines thermiques non conventionnelles, Comptes rendus de la journée du, 14 octobre 1999.

S. G. Sundkvist, Å. Klang, M. Sjödin, K. Wilhelmsen, K. Åsen, A. Tintinelli, S. Mccahey, H. Ye, AZEP, gas turbine combined cycle power plants # thermal optimisation and lca analysis, [Online] Available from: http://uregina.ca/ghgt7/PDF/papers/peer/079.pdf

G. Walker, *Stirling-cycle machines*, Clarendon Press, Oxford, 1973.

J. Wolf, CO_2 mitigation in advanced power cycles – Chemical looping combustion and steam-based gasification, Doctoral Thesis 2004, KTH # Royal Institute of Technology, Department of Chemical Engineering and Technology Energy Processes, SE-100 44 Stockholm, Sweden.

New and renewable thermal energy cycles

Introduction

This chapter deals with the thermodynamic conversion of new and renewable energies (NRE) into electricity. Renewable energy sources, as the name suggests, are distinguished from other energy sources in that they are in the form of flow rather than stock.

It begins by introducing technical and economic problems, focusing on the specific terms that they induce thermodynamically and on available technologies for solar radiation collection, including concentration: flat plate, parabolic trough (PT), parabolic dish (PD), Fresnel and power tower.

We will then successively discuss ocean thermal energy conversion (OTEC), geothermal energy cycles: direct-steam, flash, binary, combined cycle, and biomass combustion: pyrolysis, gasification, fluidized bed, downdraft and updraft.

Solar thermodynamic cycles

Introduction

Presumably promising to be a significant development in the medium to long term, most solar energy conversion technologies are not yet mature enough to compete with conventional energy on a large scale, although this situation could evolve fairly rapidly in the future.

Technical constraints are mainly threefold:

* Power density available is relatively low, which means large areas of collectors and high material costs;
* Source variability is high: solar energy fluctuates a lot, which requires often complex control systems;
* Need to store: being an energy flow, storage is required for most applications, which poses a problem because today we do not know how to store energy in good conditions.

Economic constraints are twofold:

* High capital cost. Even when properly managed, technologies are relatively expensive in investment, while their operating costs are generally low;
* Need for a back-up. In case of source unavailability, another energy is often required, imposing additional costs, sometimes significant.

DOI: 10.1201/9781003175629-16

By contrast, emissions of pollutants are zero in operation, which is an undeniable asset for sustainable development.

Solar energy comes from thermonuclear reactions that occur within the sun, causing the emission of high power electromagnetic radiation, appearing much like a blackbody at 5,800 K.

Outside the atmosphere, the radiation received by the earth varies depending on time of the year between 1,350 and 1,450 W/m^2. It is then partially reflected and absorbed by the atmosphere, so that the radiation received at ground level has a direct part and a diffuse part, the total ranging from 200 W/m^2 (overcast) to about 1,000 W/m^2 (zenith clear sky). The energy received by a given surface depends on its tilt and orientation and local climatic conditions. Solar radiation atlases at ground level are issued by national and international meteorological services in the form of maps and charts, on paper or digital.

Direct conversion of solar radiation is in three main ways:

* Thermal;
* Photoelectric effect;
* Photosynthesis.

Thermal conversion of solar energy

Thermal conversion of solar energy is to intercept the incident photons on an absorbing material, whose temperature increases.

Several methods of capture are possible:

* **Passive solar housing**. For heating and space cooling applications, it is possible to design the architecture of buildings so that they optimize naturally (or passively) the use of the solar resource, without using fluid flow and capture and storage auxiliary devices. The advantage of passive solar design of buildings is that it can lead to substantial energy savings with low incremental costs;
* **Flat plate collectors** generally use the greenhouse effect to minimize heat loss from the absorber. Indeed, the glass is transparent to visible radiation, and therefore lets through the incident solar energy, but opaque to infrared radiation, which has the effect of trapping heat. According to the technology used, the operating temperatures of flat plate collectors vary from 40°C to 120°C (vacuum collectors). Figure 16.1 shows the sectional view of a collector. The absorber consists of a metal plate on which are welded pipes in which circulates the thermal fluid. Heat losses to the front of the collector are reduced by one or more glazings (2 on the figure) and those to the rear by insulation;
* **Concentration collectors**. To reach temperatures above about 120°C, it is necessary to concentrate the solar radiation by suitable sets of reflective elements (mirrors) or lenses (usually Fresnel). The main constraint, apart from the higher device cost, is the tracking system intended to follow the sun in its course. A series of concentrators have been proposed and developed (Figures 16.2–16.5).

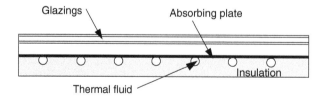

FIGURE 16.1

Flat plate collector section.

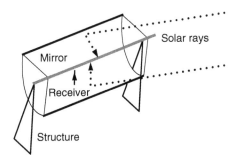

FIGURE 16.2

PT.

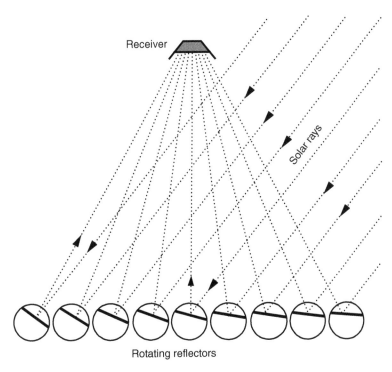

FIGURE 16.3

CLFR.

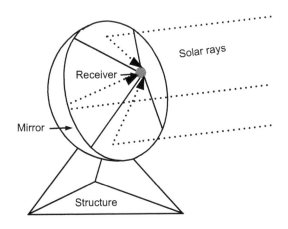

FIGURE 16.4

Dish collector.

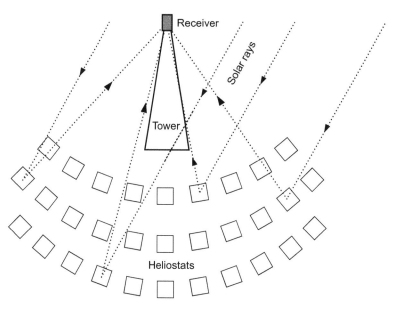

FIGURE 16.5

Central or power tower.

To thermodynamically convert solar radiation into electricity with a good efficiency, it is generally necessary to concentrate it, otherwise heat losses of absorptive surfaces are very high and the collector effectiveness low.

It is customary to characterize concentration collectors by their concentration factor C, ratio of the collector surface to that receiving the concentrated solar flux.

The experience of the last thirty years shows that four main technologies are used in practice to concentrate solar radiation in technical and economically viable conditions:

* Parabolic trough (PT) (C≈40–80, Figure 16.2), which are cylinders of parabolic cross section, that concentrate sunlight onto a straight tube;
* Concentrating linear Fresnel reflectors (CLFR) (C≈30, Figure 16.3): these use narrow rectangular plate mirrors to concentrate sunlight onto a fixed absorber consisting of a series of parallel tubes;
* Parabolic dish (PD) (C≈1,000–2,500, Figure 16.4), where the reflector is a paraboloid of revolution;
* Power towers (PT) or central tower plants (C≈200–700, Figure 16.5), in which thousands of tracking reflectors, called heliostats, redirect incoming solar radiation to an absorber at the top of a tower, allowing achievement of both high concentrations and high radiation fluxes.

The two first technologies only require to track the sun in one direction, but concentration, and thus collector temperature is lower (400°C). The other two require a double tracking movement but can reach much higher temperatures (750°C–1,000°C).

Thermodynamic cycles considered

Electro-power solar energy systems are relatively complex and include at least one collection circuit, an energy distribution circuit, storage and various control systems (Figure 16.6). In these, both the load (demand) and availabilities vary continuously, leading to methodological difficulties.

Four thermodynamic cycles are now mostly considered for conversion to electricity:

* Hirn (or Rankine) cycle for PT, CLFR and CT;
* Brayton cycle for PD (micro GT) and CT (hot air cycles);

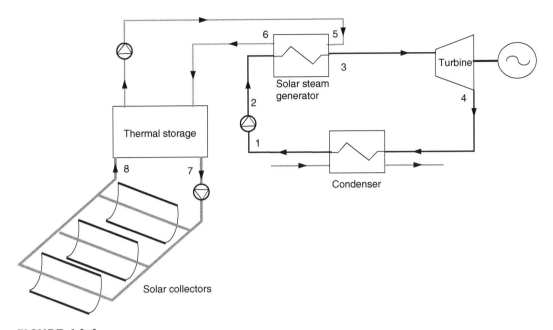

FIGURE 16.6

Schematic diagram of an electro-solar plant.

- Stirling cycle for PD;
- Combined cycle, combining the first two, for PT or CLFR (ISCCS).

We discuss them below, after presenting solar collector models we will use.

Performance of solar collectors

Flat plate solar collectors differ depending on the type of coverage: unglazed (swimming pools), single or double glazing, vacuum, antiemissivity coatings. Their transmittivity τ depends on this coverage. Because the effectiveness of solar collectors drops with temperature, it is preferable to operate at temperatures as low as possible given the intended use.

Low-temperature solar collectors

For space heating, a very good solution (but one limited to new housing) is to use a direct solar floor, in which hot water circulates through a solar heating device at about 23°C in winter and is used to make hot water in summer. This system can provide 40%–70% of heating needs. It requires a specific design of the floor (heavy, up to 30 cm thick, high inertia, which allows a temperature shift from 5 to 10 hours). Per 100 m² housing, the heating floor is 45–70 m², and the collector surface 8–15 m².

There are three main types of solar water heaters:

- Integrated storage, standard-size piece, factory assembled, driven by hot water, the tank being placed above the collector;
- Compact, the tank being close to the collector, but distinct, with a primary circuit generally antifreeze, operating by thermosyphon, sized (collector area, volume of tank) as needed;
- Separate elements; the collector is far from storage, primary circuit antifreeze, operating by thermosyphon or forced circulation.

Low-temperature flat plate solar collector model

The solar flux received by the collector goes generally through a glazing intended to isolate the absorbing surface. Reflection, transmission through the glazing and absorption result in optical losses, characterized by an overall transmittivity τ.

The absorber heats up and loses heat to the outside essentially by radiation and convection. This heat exchange can be characterized by a thermal loss coefficient U. A thermal fluid cools the absorber, taking useful heat that is then transferred or converted for different uses.

Based on this brief analysis, it is possible to produce a model whose parameters are as follows:

* Glazing transmittivity τ;
* Thermal loss coefficient U (W/m^2/K);
* Incident solar flux G (W/m^2);
* Collector surface A (m^2);
* Outside temperature Tout (°C).

The heat balance of a small slice of collector length dx is the power carried out by the fluid $\dot{m}C_p dT$ equal to the radiation received reduced by optical losses τG A/L dx, less the collector heat loss U $(T-T_{out})$ A/L dx, which is also written (Figure 16.7):

$$\dot{m}C_p dT = \left[\tau G - U\left(T - T_{out}\right)\right]\frac{A}{L}dx \tag{16.1}$$

This is a differential equation that integrates seamlessly and leads to

$$Q = \dot{m}C_p\left(T_0 - T_i\right) = \dot{m}C_p\left[\tau\frac{G}{U} - T_i + T_{out}\right]\left[1 - \exp\left(-\frac{UA}{\left(\dot{m}C_p\right)}\right)\right] \tag{16.2}$$

Equation (16.2) is valid whatever the value of the flow. However, it can be replaced by a simplified formula when the flow is high enough that it can be assumed that the temperature varies almost linearly between the inlet and outlet of the collector. We write then that the losses are proportional to the average temperature T_m, itself equal to half the sum of inlet and outlet temperatures:

$$Q = \dot{m}C_p A\left[\tau G - U\left(T_i - T_{out}\right)\right] \tag{16.3}$$

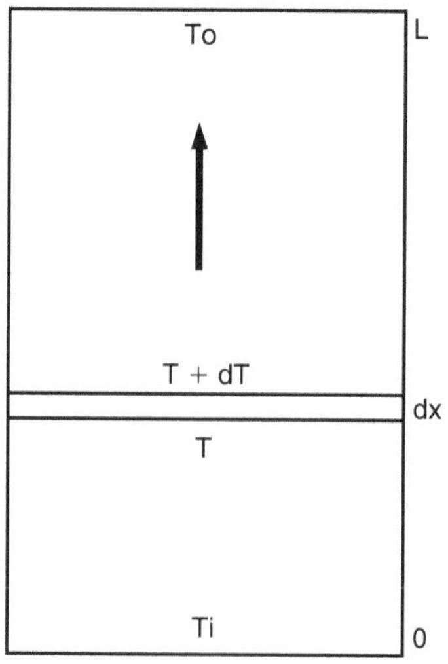

FIGURE 16.7
Flat plate solar collector model.

These are the models proposed in Thermoptim model library.

Two calculation methods are possible: to determine the outlet point state knowing the collector surface or determine this surface knowing the outlet point state.

The model input data are as follows (provided by other system components):

* The thermal fluid temperature at the collector inlet T_i (°C);
* The flow $\dot{m}$ of the thermal fluid (kg/s).

The outputs are as follows:

* The thermal fluid temperature at the collector outlet T_o (°C);
* The thermal power received by the thermal fluid Q (W/m^2);
* The collector effectiveness.

High-temperature solar collectors

The low-temperature flat plate collectors we just presented do not allow electricity to be produced with acceptable efficiencies. It is desirable to approach or exceed 100°C to consider thermodynamic conversion of solar energy.

It is therefore possible to use flat plate solar collectors provided that their absorber is very well thermally insulated, which implies both an antiemissivity treatment and good convective insulation.

Vacuum tube collectors are now seriously considered a solution for that. These collectors are usually made up of modules comprising a U-shaped pin of small diameter (of the order of 1 cm) provided with fins and inserted into a glass tube of large diameter (about 1 dm). The coolant circulates inside the pin. Such modules are placed to form parallel flat plate collectors (Figure 16.8).

Modeling high-temperature concentration collectors

The solar collector model presented above is based on two assumptions that we will refine in this section:

* First, we assumed that losses are proportional to the temperature difference between the thermal fluid and the surroundings;
* Second the absorber section is assumed to be equal to that of the collector.

The new model is a variation of the above, adapted to represent high-temperature concentration collectors.

The solar flux received by the collector is first reflected on the concentrator mirrors and goes then through the glazing material thermally insulating the receiver where it is absorbed by a suitable surface. In high-concentration collectors, only the direct component of solar radiation can be directed to the receiver, as the diffuse component cannot be concentrated.

The loss coefficient U can often be decomposed into a constant term and a term proportional to the temperature difference between absorber and ambient air: $U = U_0 + U_1 (T_m - T_{out})$. τ is a

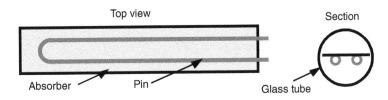

Top view   Section

Absorber   Pin   Glass tube

FIGURE 16.8

Vacuum collector module.

function of radiation angle of incidence, mirror reflectivity, absorber absorptivity and transmittance of the glazing protecting the absorber.

With the previous notations, and assuming a linear distribution of temperatures in the collector (the assumption is only valid if the flow is not too small, which is often the case in practice), the model equation is as follows, T_m being the average absorber temperature, and S_c and S_a being, respectively, the collector and absorber surfaces:

$$Q = \dot{m}C_p\left(T_o - T_i\right) = \tau E_s S_c - S_a\left(U_0 + U_1\left(T_m - T_{out}\right)\right)\left(T_m - T_{out}\right) \tag{16.4}$$

It would be possible to generalize relation (16.1) to obtain the differential equation of the high-temperature collector, but its integration would be a bit more difficult.

Table 16.1 gives values of coefficients valid for three types of concentrating collectors, among which are the two PTs used in Luz SEGS power plants presented below.

The characteristics of the Luz 3 PT are as follows: area $235\,m^2$, width $5\,m$, length $48\,m$, receiver consisting of a steel tube $70\,mm$ in diameter covered with a cermet selective coating and surrounded by a vacuum glass tube (with pressure of the order of 0.013 Pa). The absorption coefficient of the selective coating is equal to 0.96 with respect to direct radiation, and its emissivity at $350°C$ is 0.19. Mirrors are made of glass panels, hot shaped, silvered on their posterior surface and protected against external damage. Collectors track the sun by rotating around a north-south axis, thanks to hydraulic control.

Figure 16.9 compares the effectiveness of different solar collectors, depending on the temperature difference between the absorber and the surroundings for an irradiation of $1,000\,W/m^2$. It illustrates the value of concentration but also shows that vacuum tube collectors, which do not require device tracking, remain attractive to just above $100°C$.

TABLE 16.1

CONCENTRATION OF COLLECTOR PARAMETERS

	Luz 2	Luz 3	Dish	Fresnel
T	0.737	0.8	0.7	0.66
S_c/S_a	22.6	26.1	500	20
U_0	−0.0223	−0.0725	0.21	−0.031
U_1	0.000803	0.00089	0.000134	0.00061

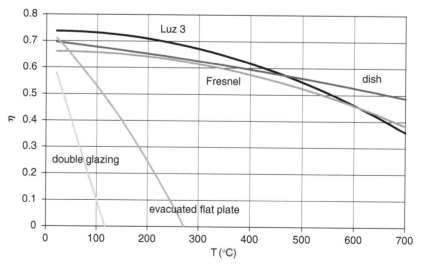

FIGURE 16.9

High-temperature collector effectiveness.

PT plants

Technology today considered the most mature is PTs, thanks to the experience in California's Mojave Desert, where nine plants have been operational since 1984. Their capacity ranges between 14 and 80 MW, totaling 354 MW. These plants, called Solar Electric Generating Systems (SEGS) were built by the Luz company under an agreement with Southern California Edison Company benefiting from the Public Utility Regulatory Policies Act (PURPA). The particularly appealing electricity sale financial conditions of this agreement being no longer available for new plants, their development stopped since the late 80s.

In SEGS plants (Figure 16.4), the steam gets its heat from a thermal oil heated (at a maximum temperature of 393°C) by a field of PT solar collectors. Superheating temperature under these conditions is limited to 371°C and steam pressure at 100 bar. Moreover, these plants being used in hot, sunny areas, the condensing temperature is relatively high (42°C, a pressure close of 0.082 bar).

Note that CLFRs operate at temperatures close to PTs, so that they could also be used in SEGS plants.

Optimization of the collector temperature

To optimize the cycle, a compromise must be found between the superheat temperature and effectiveness of solar collection. The latter is a decreasing function of the average temperature in the collector, while the maximum exergy of water vapor is an increasing function of steam temperature (Figure 16.10).

Under the conditions of the SEGS plants, a simple model can show that the optimum temperature of operation is about 400°C and 100 bar, which corresponds precisely to the choice made for these plants (Figure 16.11).

The temperature level is relatively low in comparison with conventional power plants, and efficiency without feedwater heaters is about 33% (Figure 16.12).

Plant model

To increase the plant capacity, and especially to enable it to function in the absence of sunshine, a back-up is often expected. When the auxiliary boiler runs continuously, the plant is also known as a hybrid system.

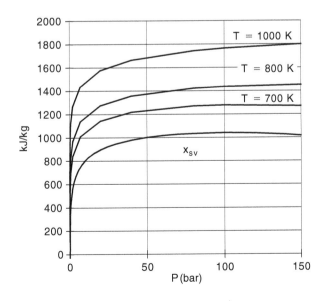

FIGURE 16.10

Exergy of steam.

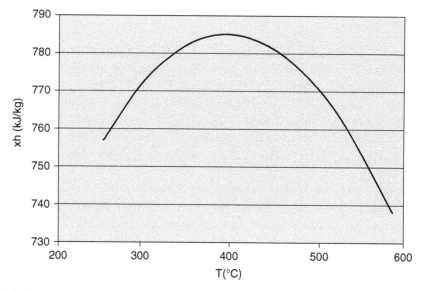

FIGURE 16.11

Optimal operation temperature.

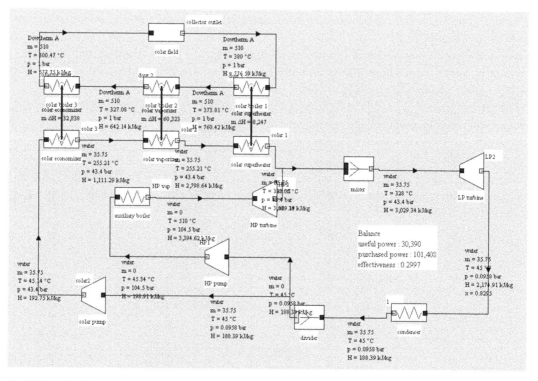

FIGURE 16.12

SEGS 3 plant with collector field and back-up.

We have seen in Chapter 8 that it is often possible to improve the efficiency of a steam cycle by conducting an internal regeneration, thanks to a series of feedwater heaters. The gain in efficiency is obtained by reducing irreversibilities taking place when heat at low temperature in the economizer is provided by the source at high temperature. The more efficient the basic cycle, the less important is the relative effect of regeneration. In the case of SEGS plants, optimization of feedwater heaters achieves an efficiency of about 37.5%, an improvement of 11% (see Figure 16.13).

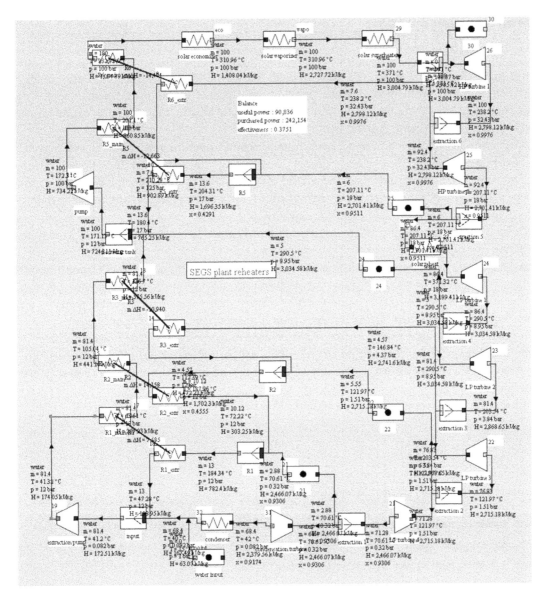

FIGURE 16.13

Detailed model of the SEGS IX plant.

BOX 16.1

Worked example

Modeling of a SEGS solar plant

The modeling of a SEGS solar power plant is presented in a guidance page of the Thermoptim-UNIT portal (https://direns.mines-paristech.fr/Sites/Thopt/en/co/fiche-sujet-fg1.html).

It allows you to study the operation of solar power plants and show how they can be realistically modeled with Thermoptim. The solar collector is type SEGS developed by the company Luz. The cycle is as a simple variant of a Rankine cycle, where the boiler is replaced by a steam generator in which the thermal fluid is heated by the field of collectors

The model uses two external classes, "solar concentrator" and "Dowtherm A".

Parabolic dish systems

Parabolic dish systems are those that achieve the highest concentrations and thus allow the use of high temperature cycles, the most efficient. However, the constraints of accurate tracking of the sun and mechanical resistance, including effects of wind, just limit their size, and thus their capacity. Technologically, few machines are available, the pair high temperature and low capacity occurring only rarely.

Two technologies are competing here: micro-turbines operating with the Brayton cycle, and Stirling engines.

The latter, which lead to the best theoretical efficiency, but often to much lower actual efficiencies, have the advantage of being mainly suited for small capacities, unlike gas turbines, whose speed increases with inverse of the capacity.

In both cases, a back-up is normally used to compensate for the momentary lack of solar radiation due to a passing cloud.

As we showed in Chapter 15, it is possible to model a Stirling cycle in Thermoptim, but the gap between theoretical and practical performance being important, the confidence one can have in this model is relatively limited. For a Stirling engine using a heated expansion, we must include heat supplied during the expansion in the energy balance (Figure 16.14).

Figure 16.15 shows a micro-turbine solar concentrator cycle: the solar hot-air receiver is located upstream of the combustion chamber of a micro-turbine, thereby reducing fuel consumption and leading to correct efficiency (31% by taking into account the solar heat in the purchased energy).

This is the subject of guided exploration (C-M4-V1).

BOX 16.2
Guided educational exploration

Micro-turbine solar concentrator (exploration C-M4-V1)

This exploration presents the cycle of a parabolic solar concentrator with regenerative gas micro-turbine: the hot air solar receiver is placed upstream of the combustion chamber, thus reducing consumption of fuel.

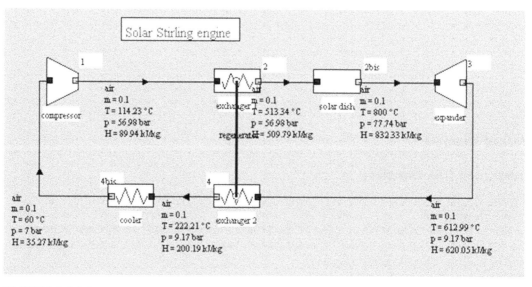

FIGURE 16.14
Solar Stirling engine.

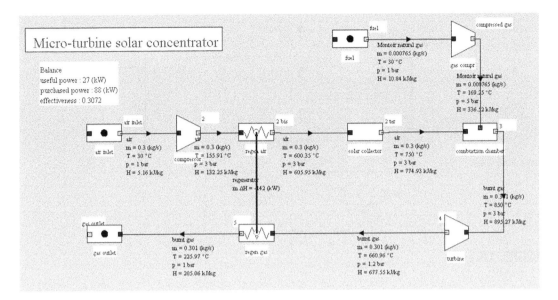

FIGURE 16.15

Micro-turbine solar concentrator.

Power towers

Two types of thermodynamic cycles are used in power towers: either a steam cycle at about 520°C and 100 bar, with a reheat, or a hot air Brayton cycle. A number of experimental plants were constructed over the last thirty years (Themis in France, Solar One and Two at Brastow California, CESA in Almeria, Spain, Eurelios).

Given the very high temperatures reached in the absorber, it is difficult to find an organic fluid resistant enough, and molten salt are used, which can optionally be stored to decouple the capture of solar radiation from its use, allowing thus the plant to operate during cloudy periods or at night.

OTEC cycles

Introduction

We introduced OTEC cycles when we discussed ORC cycles in Chapter 8.

OTEC means Ocean Thermal Energy Conversion. OTEC cycles are designed to generate electricity in warm tropical waters using the temperature difference between water at the surface (26°C–28°C) and in depth (4°C–6°C), from 1,000 m (Figure 8.29) which gives the seawater temperature profile in different locations.

In all cases, the need to convey very high water flow rates and pump cold water at great depth induces significant auxiliary consumption. Optimization of an OTEC cycle is imperative to take into account those values.

Two main types of cycles are used: closed cycles and open cycles.

Although technically valid, OTEC cycles are not yet economically viable. Prototypes of various capacities have been realized or are being considered, including in Hawaii and Tahiti.

OTEC closed cycle

Closed cycles use hot water at about 27°C to evaporate a liquid that boils at a very low temperature, such as ammonia or an organic fluid (Figure 16.16). The vapor produced drives a

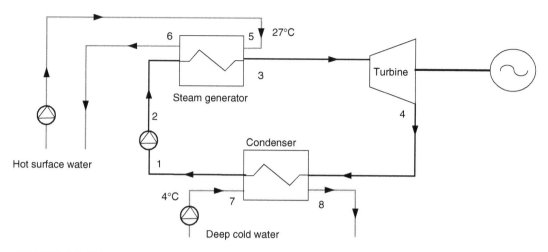

FIGURE 16.16

OTEC closed cycle.

turbine, then is condensed by heat exchange with cold water at about 4°C from deeper layers of the ocean.

The thermodynamic cycle is similar to the one we have studied in Chapter 2. The sizing of heat exchangers is of course even more critical given the very small temperature difference between hot and cold sources. Pinch values should be as low as possible while remaining realistic.

This cycle is the subject of a guided exploration (C-M1-V9). Refer to Chapter 8 for further explanation.

OTEC open cycle

In open cycles, warm water at about 26°C is expanded in a low-pressure chamber (called flash), which allows evaporating a small fraction (around 5%). The steam produced drives a turbine and is condensed in a low pressure chamber by heat exchange with cold water at about 4°C from deeper layers of the ocean. The condensate is virtually pure water, which can be used for food.

The open cycle thus has the advantage of producing both electricity and fresh water, but the very low expansion ratio involves using very large turbines.

Figure 16.17 shows the cycle operation. It involves five elements: a flash evaporator, a turbine, a condenser, a basin for collecting used sea water and a vacuum pump.

Hot water is pumped at the surface and brought to a certain height (1), and then it is injected into the evaporator in which there is a slight depression determined by the height of the water column between the evaporator and the collection basin. Because of the pressure difference, water undergoes an isenthalpic throttling (flash) and a small fraction is vaporized (2) and then headed to the turbine (2–3).

In the condenser, pressure is lower than in the evaporator, thanks to the vacuum pump and the height of the water column between the condenser and the collection tank. The turbine expands the steam produced in the evaporator, producing mechanical power. The steam is then condensed (3–4) by exchange with cold water, producing fresh water.

The hot and cold water mixed in the collection basin (8) are fed back into the sea at a depth of 60 m.

It is possible to model such a cycle with Thermoptim, to calculate its efficiency, then to build its exergy balance and estimate the magnitudes of the system sizes (exchange surfaces, flow sections, etc.), not forgetting to take into account the pumping power, *a priori* not negligible.

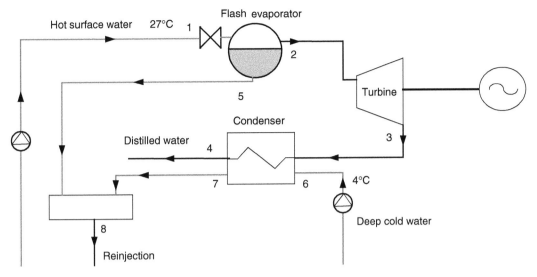

FIGURE 16.17

OTEC single flash open cycle.

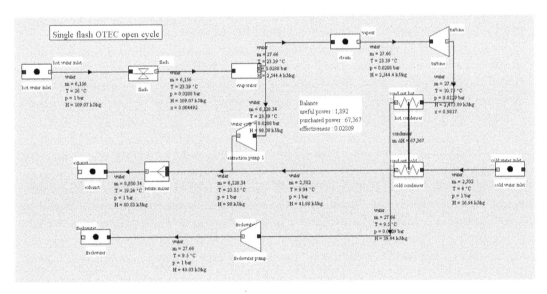

FIGURE 16.18

Synoptic view of a single flash OTEC open cycle.

The synoptic view obtained for a cycle involving a little more than 6 t/s of surface water is given in Figure 16.18. It leads to production of 28 kg/s of potable water and a capacity of 1,300 kW.

It is also possible to model a more complex cycle with double flash, which would be a little more efficient and produce more fresh water.

Geothermal cycles

Geothermal energy comes from the gradual temperature increase as one penetrates deeper into the earth's crust, either because of the natural gradient (3°C/100 m, with an average flux of 60 mW/m^2) or because of geophysical singularities (high-temperature natural geothermal reservoirs of porous rock).

It is customary to distinguish three broad categories of reservoirs, according to their temperature levels:

* High temperature (>220°C);
* Intermediate temperature (100°C–200°C);
* Low temperature (50°C–100°C).

In the first case, the geothermal fluid can be essentially composed of water or steam; in the other two, it is water, optionally under pressure. A special feature of geothermal fluid is that it is never pure water: it also includes many impurities, corrosive salts (the concentration limit for an operation to be possible is equal to 1.5 mol/kg) and noncondensable gas (NCG) in varying amounts (0.1%–10%). We shall see that this feature imposes constraints on thermodynamic cycles that can be used.

For environmental reasons, the geothermal fluid should generally be reinjected into the reservoir after use, but it is not always the case.

The thermodynamic conversion of geothermal energy uses four main techniques:

* Plants called "direct-steam" can be used if the geothermal fluid is superheated steam that can be directly expanded in a turbine. Historically, this type of plant was first implemented in Larderello in Italy since 1904;
* Flash vaporization power plants can exploit sites where geothermal fluid is in the form of pressurized liquid or liquid-vapor mixture. Today, it is the type of plant most used. Geothermal fluid begins by being expanded in a chamber at pressure lower than that of the well, thereby vaporizing a portion, which is then expanded in a turbine;
* Systems known as binary use a secondary working fluid, which follows a closed Hirn or Rankine cycle, the boiler being a heat exchanger with the geothermal fluid;
* Fluid mixture systems, such as Kalina cycle studied in Chapter 14, a variant of binary systems where the working fluid is no longer pure but consists of two fluids to achieve a temperature glide during vaporization.

Mixed or combined cycles can use both a direct or flash system and a binary system. In what follows, we present these different cycles modeled in Thermoptim.

Immediately note a small feature of some of these models: in a geothermal cycle, calculating purchased energy is not always immediate, since the geothermal fluid (which will be modeled as water) is most often distributed in several streams, reinjected or not. We can therefore rarely directly estimate the enthalpy it provides. When this happens, it is preferable not to declare in Thermoptim a process as "purchased energy" and simply compare cycles on the basis of mechanical power produced.

To estimate an efficiency on a comparable basis, we may consider as a reference a cycle that would allow the entire geothermal fluid to be reinjected at a temperature of 50°C. We will talk then of reference efficiency.

Note that temperature and pressure levels of the geothermal fluid considered in the examples that follow are not necessarily the same, leading us to temper these comparisons.

Direct-steam plants

Direct-steam cycle is very close to that of Hirn or Rankine. The main difference comes from the need to extract the NCG in order to condense water at the turbine outlet, which allows the steam to be expanded at pressure below the ambient. Depending on circumstances, the extraction is done using an ejector driven by geothermal steam or a compressor coupled to the turbine (Figure 16.19).

Generally, the condenser cooling is provided by a cooling tower whose makeup water may be taken from the condensate itself.

As mentioned above, this type of plant requires the existence of dry steam in production wells, which is exceptional: the only known sites that have this property are Larderello in Italy and the Geysers in NW California.

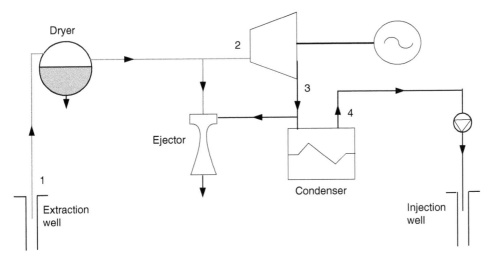

FIGURE 16.19

Direct-steam plant.

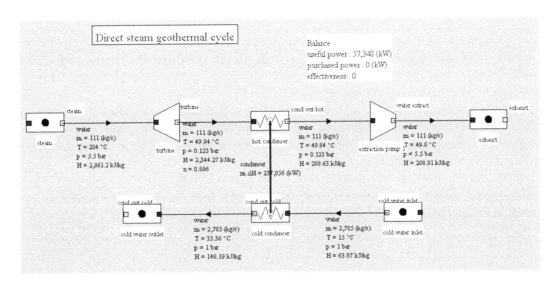

FIGURE 16.20

Synoptic view of a direct steam geothermal cycle.

The synoptic view of such a cycle is given in Figure 16.20. We considered available 111 kg/s of steam at 5.5 bar and 204°C, which represents approximately a 50°C superheating. This steam is expanded at 0.123 bar, 50°C and then condensed and recompressed before reinjection. In this case, the mechanical power produced is 57.3 MW, the cycle efficiency being 20.9%.

Simple flash plant

Generally, the well contains a low-quality (below 0.5) liquid-vapor mixture, which cannot be sent directly to the turbine.

If the initial pressure is sufficient, a solution is to partially expand the mixture in order to vaporize a portion, which is then sent to the turbine, while the liquid fraction is reinjected (Figure 16.21).

As in the case of direct-steam plant, the vapor phase typically contains a significant amount of NCG to be extracted if we want to condense water at the turbine outlet.

Note that steam through the turbine is distilled water which can sometimes be valorized notably as drinking water.

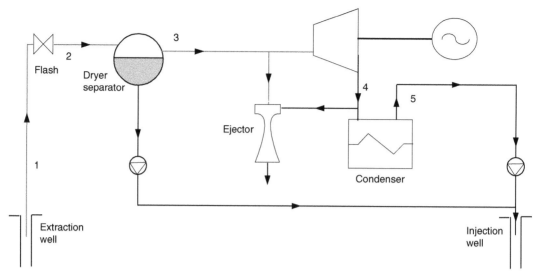

FIGURE 16.21

Simple flash plant.

Figure 16.22 shows the synoptic view of such a cycle modeled in Thermoptim. We assumed that we had 760 kg/s of hot water in the saturated liquid state at 230°C and 28 bar. This cycle is the subject of guided exploration (C-M4-V3).

This water undergoes a flash at 6 bar, leading to a 0.15 quality. The liquid and vapor phases are then separated, the first being recompressed before reinjection, whereas the latter is expanded at the pressure of 0.123 bar (50°C) and then condensed. The mechanical power produced is 57 MW and efficiency 9.6%.

Note that the pressure at which the flash is performed (6 bar) has not been optimized. The condensing pressure and temperature are relatively high because of noncondensable gases present in the geothermal fluid.

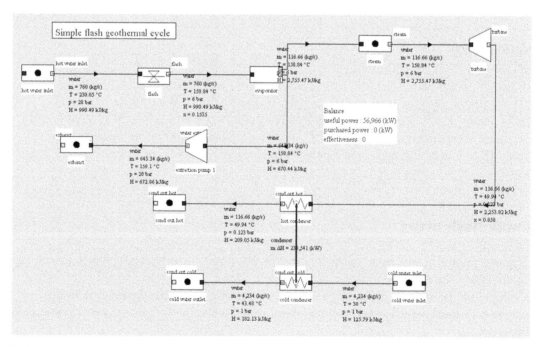

FIGURE 16.22

Synoptic view of a single flash geothermal cycle.

Double flash plant

In some cases, if the pressure at the well outlet is sufficient, it is possible to achieve a double flash, which allows steam to be obtained at two different pressure levels and increases plant performance (Figure 16.23).

Theoretically, we could thus increase the number of flashes, but technological and economic constraints limit them in practice to 2.

As shown in the synoptic view of Figure 16.24, the liquid stream, which in the previous cycle was recompressed and re-injected, undergoes this time a second flash at a pressure of 0.931 bar, which leads to a 0.115 quality.

The liquid phase is recompressed and reinjected, while the vapor is mixed with the steam flow from the first flash expanded at the same pressure. The whole is then expanded at the condenser pressure in a LP turbine. Mechanical power produced rises from 57 to 77 MW, representing an increase of 35%. Efficiency becomes 13%.

Here too, the flash pressures (6 and 0.931 bar) have not been optimized.

Binary cycle plants

When the temperature or pressure at the well outlet is low, it becomes impossible to make use of direct-steam or flash cycles. We then use a second working fluid, which follows a closed Hirn or Rankine cycle (with or without superheating) (Figure 16.25).

The geothermal fluid then transfers its heat to the fluid before being reinjected.

A cooling tower ensures condensation of the working fluid, whose choice depends on many considerations, technological, economic and environmental. Since this is often an organic fluid, it is customary to speak of Organic Rankine Cycle (ORC, Chapter 8).

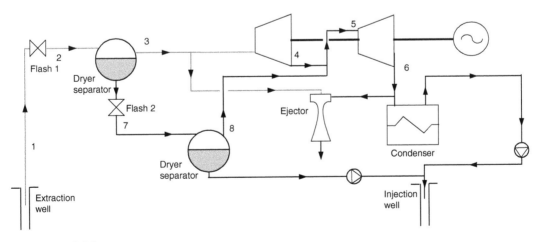

FIGURE 16.23

Double flash plant.

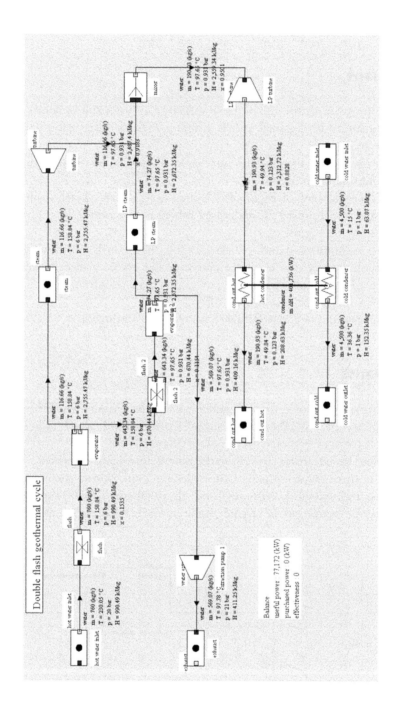

FIGURE 16.24

Synoptic view of a double flash geothermal cycle.

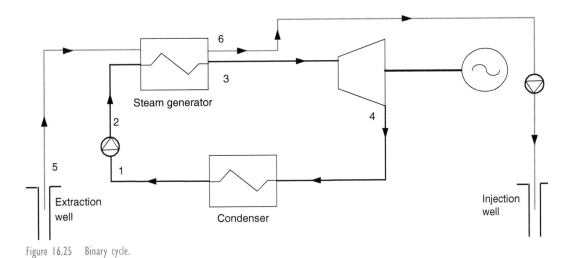

Figure 16.25   Binary cycle.

Figure 16.26 shows the synoptic view of such a cycle modeled in Thermoptim. We assumed we had 310 kg/s of hot water in the subcooled liquid state at a temperature of 169°C and pressure of 20 bar.

This water is used to vaporize butane with a very small superheat (2°C). It is then expanded in a turbine and condensed in an entirely conventional Rankine cycle. The mechanical power produced here is 18.9 MW, and the efficiency is 12%.

The condensation pressure and temperature of butane may be here lower than for the geothermal fluid in flash cycles because of the absence of noncondensable gases in the butane cycle.

Like any Rankine cycle, this cycle can be improved by judiciously introducing reheats and/or feedwater heaters.

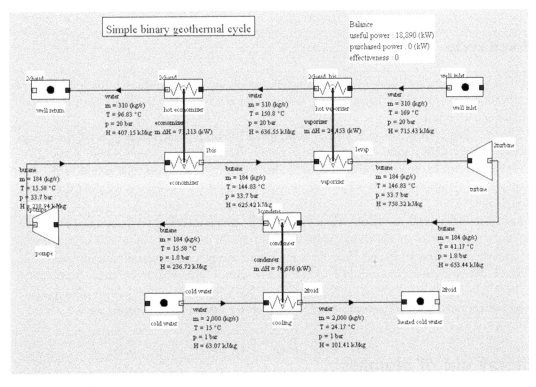

FIGURE 16.26

Synoptic view of a butane binary geothermal cycle.

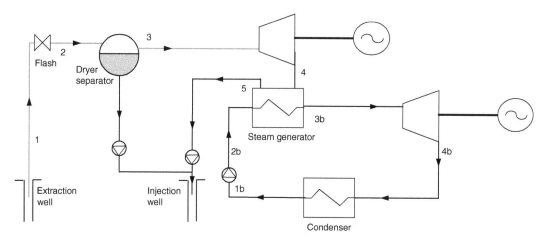

FIGURE 16.27
Combined cycle.

Combined cycles

As we have seen, one of the constraints encountered in condensation of direct-steam or flash cycles is the need to extract NCG, resulting in significant parasitic energy consumption.

An alternative is to use a combined cycle, combining direct-steam or flash cycle with an ORC, the steam leaving the turbine being at a pressure higher than atmospheric and being cooled in the boiler of the second cycle (Figure 16.27).

Let us for example consider the case of the double-flash cycle studied previously, the steam being expanded this time at only 0.9 bar instead of 0.123 bar and then cooled at 50°C in a heat exchanger used as vapor generator for a butane ORC.

This gives the combined cycle in Figure 16.28: mechanical power increases from 77 to 95 MW and efficiency from 13% to 16.1%.

Mixed cycle

One major drawback of flash cycles is that they exploit only a small share of the total flow of geothermal fluid, the one corresponding to the vapor fraction after the flash, the liquid fraction being reinjected.

Mixed cycles valorize the geothermal fluid, using the liquid fraction to provide the energy necessary for an ORC cycle.

One can thus obtain a total efficiency well above that of a flash cycle. Of course, if the topping cycle of a combined cycle is the flash type, it is possible to associate a second ORC of the mixed type.

Let us consider again the double flash cycle studied previously, and add a butane Hirn cycle, no longer on the circuit of expanded steam but this time on the circuit before reinjection of geothermal fluid. Figures 16.29 and 16.30 show the result: mechanical power output increases from 77 to 94 MW and efficiency from 13% to 15.8%. The gain is somewhat lower than that obtained in the combined cycle but already quite significant. Note that it would indeed be quite possible to add this butane cycle on the combined cycle, which would improve performance.

Energy use of biomass

In this section, after a brief presentation of the main biomass energy conversion systems, including thermochemical processes, we give some details on gasification and then introduce an external class that models different types of combustion of this fuel type.

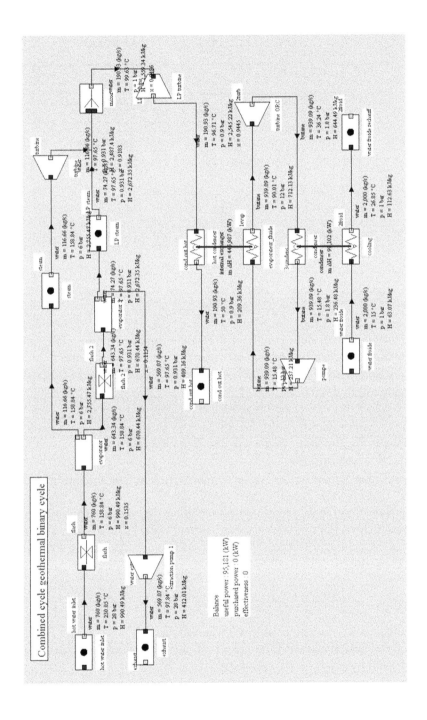

FIGURE 16.28

Synoptic view of a geothermal combined cycle.

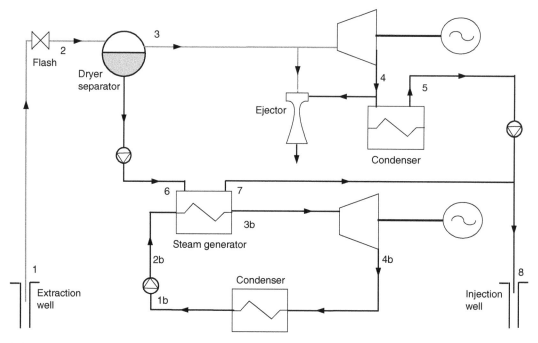

FIGURE 16.29

Mixed cycle.

Introduction

We call biomass all organic materials, mainly of vegetable origin, natural or cultivated, land or sea, produced by chlorophyll conversion of solar energy, fossil fuels excluded.

Biomass is mainly composed of lignin ($C_{40}H_{44}O_6$) (25%) and carbohydrate $C_n(H_2O)_m$ (cellulose $C_6H_{10}O_5$ and hemicellulose) (75%).

Biomass can be converted in energy by three main categories of processes:

* Biochemical conversion: digestion, hydrolysis and fermentation;
* Chemical conversion (esterification);
* Thermochemical conversion: combustion, cocombustion, pyrolysis and gasification.

Biochemical conversion comprises two main types: anaerobic digestion and aerobic digestion.

In the first case, which occurs in the absence of oxygen, there is production of a "biogas", consisting mainly of methane (50%–65%), CO_2 (30%–35%) and other gases. Reactions take place at temperatures between 20°C and 70°C. The basic reaction is

$$CH_3COOH \rightarrow CH_4 + CO_2 \qquad (16.5)$$

In the second case, there is production of more CO_2 and H_2O.

Thermochemical conversion is divided into combustion and cocombustion (excess air), gasification (lack of air) and pyrolysis (in the absence of air).

Combustion is the oldest and probably most used conversion mode for both domestic and industrial uses. Its efficiency is good insofar as the fuel is rich in structured carbohydrates (cellulose and lignin) and especially dry enough (humidity below 35%).

Cocombustion is to simultaneously burn a fossil fuel, usually coal, and biomass (up to 15%) to reduce, in an existing boiler, the amount of initial fuel.

We call the C/N ratio the ratio of the amounts of carbon and nitrogen in biomass. It varies from 10 to about 100. Pyrolysis allows conversion of relatively dry biomass (humidity below 10%) of C/N ratio greater than 30, in various high LHV fuels, storable, gaseous, liquid and solid (charcoal). It takes place at temperatures between 400°C and 800°C and can be done in

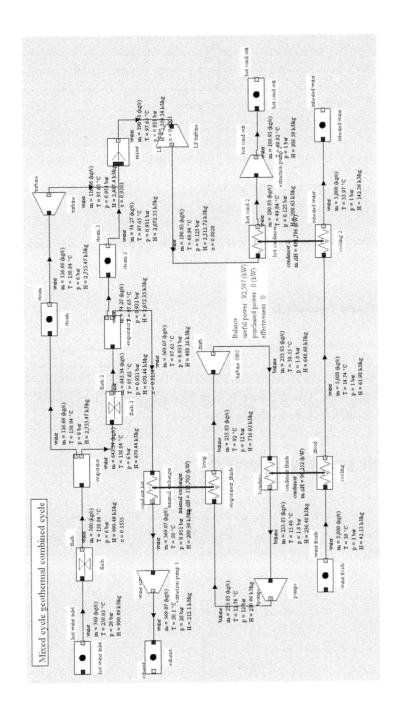

FIGURE 16.30

Synoptic view of a geothermal mixed cycle.

several modes: slow pyrolysis or char (solid product only), conventional pyrolysis at moderate temperature (600°C) (1/3 gas, 1/3 liquid, 1/3 solid), fast pyrolysis at 500°C–600°C, which takes place in about one second (70%–80% of liquid), flash pyrolysis at over 700°C and in less than one second (more than 80% of liquid).

Gasification of biomass is obtained by performing combustion with lack of air in schematically two main stages: pyrolysis producing gas, liquid and solid phases, followed by the gasification itself of the last two phases.

In a gasifier, the fuel begins by being dried and is then pyrolyzed, these two steps being endothermic. The gaseous products are then burned at high temperature, releasing heat, a part of which is used by the previous two steps. The exhaust gases are then put into contact with the solid phase after pyrolysis and with water from the drying, which causes a reduction reaction which leads to the formation of a synthesis gas rich in CO and H_2, whose LHV is approximately 70%–75% of that of the original biomass.

Several gasification processes exist:

* Fixed bed systems, which consist of two main technologies, the cocurrent gasifiers (downdraft) and the counter-current (updraft);
* Fluidized bed systems, which consist of three categories depending on the fluidization speed, dense, circulating and entrained fluidized beds.

The first match small or medium facilities and the latter large ones. The advantages and disadvantages of these systems are given below (van de Steene et al., 2003).

Cocurrent gasifiers (downdraft)

Of simple design and construction (Figure 16.31), these devices have a high conversion efficiency and are very well suited for certain fuels that they convert into a relatively clean gas. Their maintenance cost is however high (wear). The fuel should be uniform and of low humidity and significant size. The capacity is very limited (350 kWe), and ash fusion may happen in the reactor grid, which locks the device.

Counter-current gasifiers (updraft)

These gasifiers are also simple in design and construction (Figure 16.32) and have high conversion efficiency but are unsuitable for producing electricity. However, they are more flexible with respect to raw material moisture. Gases highly loaded in tar exit at low temperature with condensation risk.

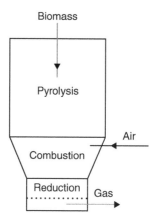

FIGURE 16.31

Cocurrent downdraft gasifier.

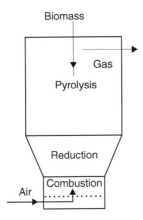

FIGURE 16.32
Counter-current updraft gasifier.

Dense fluidized bed gasifier

Since the 1970s, we looked for combustion processes which can significantly reduce pollutant emissions during combustion of low-quality fuels.

Fluidized bed technology involves blowing a gas vertically beneath a layer of solid particles of suitable size. At a certain gas velocity, the particles are raised and the layer swells, creating a suspended medium whose behavior is similar to a fluid: it is said that the bed is fluidized. If the gas velocity increases, the particles are mixed and transported: the bed becomes turbulent.

The technology of dense fluidized bed gasifiers, characterized by high reaction rates and good solid/gas contact, allows temperatures to be controlled well. Of relatively simple and operational construction, they are not limited in size, although their economically optimal capacity exceeds 20 MW. Treatment in the catalytic bed is possible, tar is moderate to high and particles are high. Entrainment losses of carbon in ash limit their performance. Fuel must be low humidity <20%, be packaged in the form of small pieces, combustion being sensitive to their distribution in the bed.

Circulating fluidized bed gasifier

These high conversion efficiency devices provide good temperature and reaction speed control. They have a high tolerance for fuel (type, size), although the best performance is achieved with small particle size and low humidity (<20%). Tar in the gases is moderate, but particles are high.

As dense fluidized beds, they are not limited in size, although their economically optimal capacity exceeds 20 MW, and entrainment losses of carbon in ash limit their efficiency.

Entrained fluidized bed gasifier

These high conversion efficiency gasifiers ensure good gas-solid contact. Their high temperature operation produces gas of quality but low LHV, clean with little tar, and leads to vitrification of ashes. However, the cost of biomass preparation is high and the types of fuels limited. The capacity of these plants is large (>50 MWe).

Modeling thermochemical conversion

This section presents a simplified model that can simulate different types of biomass combustion, and in which it is possible to vary with considerable flexibility fuel composition and moisture and combustion conditions. The external class in which it is implemented, called BiomassCombustion, can be used to simulate both a boiler and a cocurrent gasifier.

This model is quite simplified compared to those currently being studied in research laboratories, particularly in its representation of pyrolysis (Vijeu et al., 2005). Its main interest is to allow Thermoptim users to approach the study of thermochemical biomass conversion and the insertion of gasifiers or combustors into complete systems.

To simplify the writing of the model, we took advantage of using the combustion calculation functions already present in Thermoptim, just adding equations corresponding to reactions that the software does not take into account.

Explanations are presented in Volume 3 of the Thermoptim reference manual, while combustion setting is presented in Volume 2.

Combustion parameters are exactly the same as that required for combustion calculated in Thermoptim core, with the proviso that it is possible to set a heat load to be taken into account in calculating the end of combustion temperature. It is also possible to perform preliminary calculations before starting calculations in Thermoptim.

In thermodynamics, the main parameters that influence biomass combustion are as follows:

* First, of course, fuel composition;
* Second, moisture, which firstly determines enthalpy required for drying and, on the other hand, plays on gas composition, and finally influences CO_2 dissociation;
* Finally, quenching temperature and CO_2 dissociation rate.

To separate as much as possible the influence of these first two parameters, composition is that of dry fuel, and moisture is taken into account by adding water. Given that the overall moisture is not necessarily the one that governs the thermodynamic equilibrium, as part of the steam may not react, for reasons of kinetics or geometry, we introduced an additional parameter, equal to the fraction of water involved in combustion. Physically, this means that a fraction of the water is not dissociated: it must be vaporized, but its influence on the composition of the gas is the same as if it were inert.

The biomass composition varies widely depending on its origin and humidity. However, we can retain values close to 25% lignin ($C_{40}H_{44}O_6$) and 75% carbohydrates $C_n(H_2O)_m$ (cellulose $C_6H_{10}O_5$ and hemicellulose).

Estimates of the composition and the LHV value referred to the dry gas for various synthesis gases are given in Table 16.2.

Definition of fuel and progress of combustion

Fuel definition is made in Thermoptim as follows (Figure 16.33):

* Dry gas composition, excluding species not included in Thermoptim core, is estimated, for example in a spreadsheet, and entered in the form of a compound gas (process-point "dry fuel");
* Species not considered by Thermoptim core are considered separately: they are entered in the component screen and are subject to precombustion;

TABLE 16.2

COMPOSITION AND LHV VALUE FOR VARIOUS SYNTHESIS GASES

	[CO]	[H$_2$]	[CH$_4$]	[CO$_2$]	[N$_2$]	LHV (MJ/m^{3}N)
Charcoal	28–31	5–10	1–2	1–2	55–60	4.6–5.6
Moist wood 12%	17–22	16–20	2–3	10–15	50–55	5–5.85
Wheat straw	14–17	17–19		11–14		4.5
Typical estimate	19–15	15–20	3–5	10–15	40–50	4–6

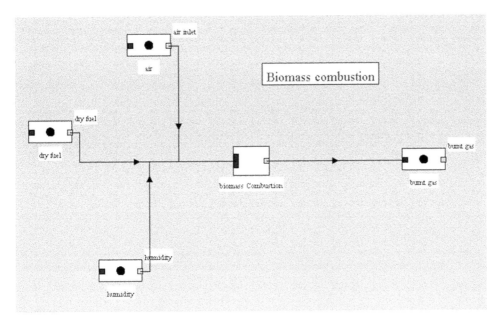

FIGURE 16.33
Diagram of the combustion structure.

* Dry gas is mixed with water (H_2O gas) to form moist fuel (process-point "humidity");
* Final combustion is calculated by core Thermoptim functions, which are emulated from the external class (external mixer "biomass combustion"), taking into account precombustion and enthalpy that would have been required to evaporate water.

The oxidizer (usually air) is itself defined in process-point "air inlet". Flows involved are also specified in each of the three process-points entering the external mixer. Their respective ratios allow the system to adjust the fuel moisture and excess air.

Example of lack of air combustion (gasifier)

Biomass gasification is partial oxidation of an organic resource, mainly composed of cellulose ($C_6H_{10}O_5$) to produce synthesis gas. Biomass being generally very wet, oxidation takes place in four stages:

* Fuel drying;
* Pyrolysis or carbonization (in the absence of oxygen), producing tars and carbon;
* Combustion of carbon and oxygen, exothermic reaction drying fuel among others;
* Reduction of CO_2, H_2 and water by carbon and tar.

By performing lack of air combustion, we can model a downdraft gasifier (Figure 16.34) using class BiomassCombustion. We assumed here that only 50% of fuel moisture takes part in combustion, the rest not participating.

Figure 16.34 shows the component screen allowing one to model biomass combustion. In the case presented, it is assumed that fuel comprises 0.634% ammonia and 10.5% carbon referred to dry mass, that the quenching temperature equals 900°C and that 50% of the moisture is involved in combustion.

As we have said, the input flow rates of fuel, water and oxidizer obviously play a fundamental role, as their ratios directly influence the synthesis gas composition (Figure 16.35).

In this example, fuel moisture is 50% by mass, the flows of the two processes "dry fuel" and "humidity" being equal. Air flow being less than that which would have ensured a stoichiometric combustion (about 7.8 kg/s), combustion occurs in lack of air and the CO_2 dissociation rate is recalculated.

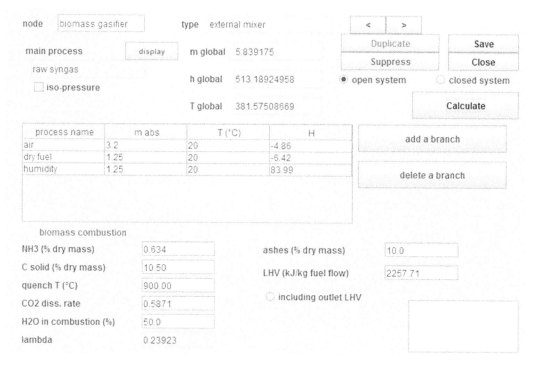

FIGURE 16.34

Biomass gasifier component.

component name	molar fraction	mass fraction
CO2	0.09095625	0.1780174
H2O	0.2800306	0.2243506
N2	0.3332404	0.4151491
CO	0.1293294	0.1611006
H2	0.1626134	0.01457812
Ar	0.003829836	0.006804206

FIGURE 16.35

Composition of raw synthesis gas (LHV: 3.4 MJ/kg).

The synthesis gas temperature depends a lot on the heat of reaction, itself a function of oxygen available: if there is very little, it is mainly CO that is produced with few CO_2.

The raw synthesis gas, high humidity, can be washed and partially dried, using class WaterQuench presented in the Thermoptim-UNIT portal model library.

Example of combustion in excess air

In this example, simulating a boiler burning biomass, combustion is performed in excess air.

Figure 16.36 shows the component screen allowing to model biomass combustion. In the case presented, we as previously assume that fuel comprises 0.634% ammonia and 10.5% carbon by dry mass and that quenching temperature is equal to 900°C. We assume however that all the moisture is involved in combustion, and CO_2 dissociation rate is 0.05, while its value was previously calculated.

A rate of 1.25 g/s of dry fuel, composition given in Figure 16.37, humidity 50% by weight (water flow of 1.25 g/s) is burned with 10 g/s of dry air, leading to air factor $\lambda = 1.23$. Exhaust gas composition is given in Figure 16.38.

In Figure 16.36, in addition to values for exhaust gases (flow, temperature and enthalpy), magnitudes displayed are air factor λ and CO_2 dissociation rate (if $\lambda > 1$, it is an entry, otherwise it is recalculated).

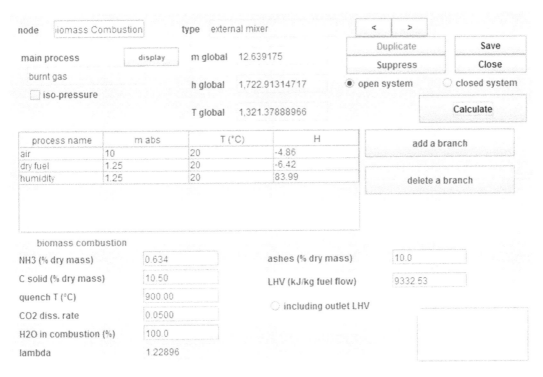

FIGURE 16.36

Biomass combustion component.

component name	molar fraction	mass fraction
O2	0.01374655	0.01641626
H2	0.1041061	0.007832275
C2H6 ' éthane	0.1096048	0.1230014
CO2	0.2733782	0.4490144
CO	0.2348086	0.2454598
CH4 ' méthane	0.2643558	0.1582759

FIGURE 16.37

Composition of dry gas (LHV: 17.2 MJ/kg).

component name	molar fraction	mass fraction
CO2	0.1180614	0.1892233
H2O	0.2398588	0.1573671
O2	0.03282001	0.03824633
N2	0.5864343	0.5982773
CO	0.006213756	0.006338563
H2	0.009859662	0.0007238424
Ar	0.006752062	0.009823597

FIGURE 16.38

Composition of smoke (LHV: 0.15 MJ/kg).

Bibliography

Bechtel Corporation, ISCCS Study Integrated Solar Combined Cycle System, Report for the National Renewable Energy Laboratory, 1998.

P. Bombarda, E. Macchi, Optimum cycles for geothermal power plants, Proceedings World Geothermal Congress 2000, Kyushu-Tohoku, Japan, June 10, 2000.

E. A. Demeo, Solar-Thermal Electric Power, 2003 Status Update, EPRI report, February 2003.

R. Dipippo, Small geothermal power plants: design, performance and economics, GHC Bulletin, June 1999.

R. Dipippo, Second law assessment of binary plants generating power from low-temperature geothermal fluids, *Geothermics* 33: 565–586, 2004.

J. A. Duffie, W. A. Beckman, *Solar engineering of thermal processes*, John Wiley and sons, New York, 1980.

R. Gicquel, Behavior of plane solar collectors under transient conditions. *International Chemical Engineering*, 19 (1), January 1979.

A. Rabl, *Active solar collectors and their applications*, Oxford University Press, New York, 1985.

K. W. Kwant, H. Knoef, Status of gasifi cation in countries participating in the IEA and GasNet activity, 2004.

K. J. Riffelmann, D. Kruge, R. Pitz-paal, Solar thermal plants, Power and process heat, DZLR.

L. Van de steene, G. Philippe, Le point sur la gazéification de la biomasse, Bois-Energie n° 1, 2003.

R. Vijeu, L. Gerun, J. Bellettre, M. Tazerout, Z. Younsi, C. Castelain, Modèle thermochimique bidimensionnel de pyrolyse de la biomasse, International Congress on the Renewable Energies and the Environment – CERE, 24–26. mars 2005, Sousse, Tunisie, 8 p.

Evaporation, mechanical vapor compression, desalination and drying by hot gas

Introduction

Concentration by evaporation facilities is widely used in industries including food as well as for seawater desalination. The idea is to evaporate the product, the solute being usually non-volatile. This produces a separation of two components, which allows the concentration to be increased. Desalination is a variant of evaporation, where the valuable output is the vapor evaporated, which is condensed to produce fresh water.

Mechanical vapor compression (MVC) is often presented as an efficient alternative to heat pump if the cold source is a gas or vapor. The two most conventional MVC applications being distillation and concentration by evaporation, we deemed it appropriate to address this issue here. Thermal vapor compression (TVC) is a variant where the compressor is replaced by an ejector.

An emerging application of the technologies studied in this chapter is zero liquid discharge (ZLD), the objective of which is to recycle all of a plant's liquid effluents. ZLD technologies combine thermal treatment of evaporation and distillation with MCV, RO and crystallization.

They have been developed both for economic reasons and so that companies can meet increasingly stringent environmental standards.

Evapoconcentration

Evapoconcentration being a major consumer of heat, it is necessary to optimize the design of concentration facilities if we want to get good performance. Thermal integration methods presented in Chapter 7 can find a scope of interest in this area.

Modeling of product thermophysical properties

To model an evapoconcentration installation, it is first of all necessary to know the properties of the product as a function of the temperature and its concentration. As the product is not a pure fluid, their modeling requires certain precautions which we will present before going further.

DOI: 10.1201/9781003175629-17

TABLE 17.1

MASS COMPOSITION OF ORANGE JUICE DRY MATTER

Protein	Fat	Carbohydrate	Fiber	Ash
5.28%	1.25%	88.00%	1.79%	3.67%

Products to be concentrated are mixtures one of the characteristics of which is that one can generally consider that they are composed on the one hand of various components which are stable on the physicochemical level and on the other hand of water liable to evaporate.

They can therefore be considered as the solute, a mixture of a dry product of possibly complex formulation, and the solvent, water. The mixture is then characterized by the concentration x of the dry product, the temperature and the pressure, which is only second order, the liquid being incompressible.

The product remaining in the liquid state, we can in Thermoptim use the quality field to enter this concentration.

Researchers who have worked on the modeling of food products have shown that one can represent them with reasonable precision by breaking them down into five types of components: protein, fat, carbohydrate, fiber and ash, for which they have determined equations giving the main thermodynamic and transport properties (Choi & Okos, 1986).

In the literature, estimates of the composition of a large number of food products can be found according to the five types of elementary components, which allows them to be modeled.

The orange juice external substance that we use in the following is an example of this. Its dry matter composition is given in Table 17.1 (ASHRAE, 2002).

The detail of its modeling is specified in the documentation of the external class.

Boiling point elevation

Another important characteristic to take into account is what is called the boiling point elevation.

The saturated vapor pressure of a mixture (solute-solvent) decreases following Raoult's law. It follows that, at given pressure, the boiling point increases slightly compared to that of pure solvent. We call the boiling point elevation this temperature difference ΔT_{eb}. Physically, the presence of the solute impedes the evaporation of the solvent, which may only be made at higher temperature.

It can be shown that ΔT_{eb} is given by the following law:

$$\Delta T_{eb} = i x \frac{K_{eb}}{\rho} \tag{17.1}$$

* i is the solute van't Hoff coefficient, representing the number of elementary particles (ions, etc.) formed in the solution;
* x is the solute mass concentration;
* K_{eb} is the pure solvent ebullioscopic constant;
* ρ is the solvent density.

Law (17.1) indicates that the boiling point elevation is proportional to x. In practice, if we can assume that ρ and K_{eb} are constant, an assumption valid for a small temperature range, it suffices to know some values of the boiling point elevation to identify it as (17.2):

$$\Delta T_{eb} = K x \tag{17.2}$$

Other laws can be found in the specialized literature, depending on the product, such as $\Delta T_{ed} = a^{bx}$ or variants.

The external classes used in the models presented in this chapter take into account the boiling point elevation.

Single-effect concentration cycle

In a conventional single-effect evaporation cycle (Figure 17.1), a product to concentrate (solute + solvent) is injected into a unit, heated by any heat input Q (steam 4–5). The concentrated product is extracted in 3, at the bottom of the unit, while the solvent vapor exits in 2 and is condensed, its enthalpy being lost.

Calling x the mass concentration of solute, equations governing the behavior of this unit are as follows:

$$\text{conservation of the total flow: } \dot{m}_1 = \dot{m}_2 + \dot{m}_3 \tag{17.3}$$

$$\text{conservation of solute: } x_1 \dot{m}_1 = x_3 \dot{m}_3 \tag{17.4}$$

$$\text{conservation of enthalpy: } h_1 \dot{m}_1 + Q = h_2 \dot{m}_2 + h_3 \dot{m}_3 \tag{17.5}$$

The calculation principle of the evapo-concentrator is as follows.

Knowing the desired concentration of rich product at the outlet, defined by the downstream point, these three equations make it possible to calculate the distillation unit.

The device is represented in Thermoptim by an external divider receiving at the inlet the flow rate of product to be concentrated, and from which two flows exit, that of the concentrated product and that of the evaporated water vapor.

Knowing the outlet concentration, Equations (17.3) and (17.4) provide the flow rates. The heat necessary to vaporize the water is deduced from this, and the enthalpy balance makes it possible to calculate the outlet temperature, and therefore the saturation pressure, taking into account the boiling point elevation.

The heat to be supplied is transmitted to the thermocoupler which calculates the process to which it is coupled.

Figure 17.2 shows the synoptic view of such an evaporator modeled with Thermoptim by external class EvapoConcentrator2. In this example, there are 1.22 kg/s of feed product (orange juice) containing 11.7% of dry matter at 1 bar, that we want to bring to a concentration of 16%. It is necessary for this to provide heat in the form of 0.487 kg/s of saturated steam at 1.08 bar, which is fully condensed.

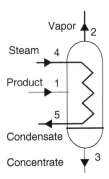

FIGURE 17.1

Single-effect evaporation cycle.

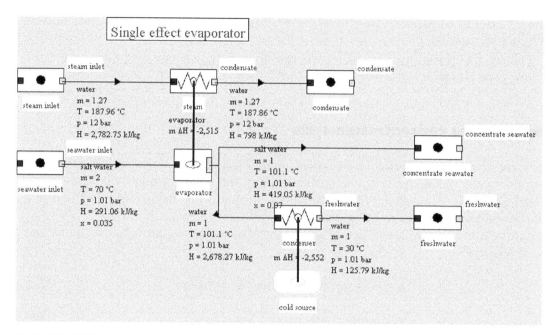

FIGURE 17.2

Single-effect evaporator.

Another way of presenting the performance is to say that it took 0.487 kg/s of steam to extract 0.328 kg/s of water from the initial orange juice. The operation specific consumption OSC, which is the ratio between these two flow rates, is equal to 1.48. The energy consumed to evaporate water, energy specific consumption ESC, is equal to 825 kWh/t.

Multieffect concentration cycle

In a multi-effect (ME) cycle (Figure 17.3), vapor produced is condensed in the evaporator of a second unit operating in series with the previous (the different types of irreversibilities by heterogeneity temperature, boiling point elevation, pressure drops, etc. make it necessary to condense the vapor at a lower pressure and therefore temperature). A multi-effect concentrator thus allows recovering a portion of the solvent vapor and reduces the operation-specific consumption.

Figure 17.4 shows the synoptic view obtained for a triple-effect evaporator operating in conditions close to the previous. The flow of live steam is here 0.47 kg/s, thanks to the use of vapor from the first effect for vaporizing part of the second.

The improvement is about 50%.

The OSC is equal to 0.48 and the ESC to 264 kWh/t. The consumption has been divided by three.

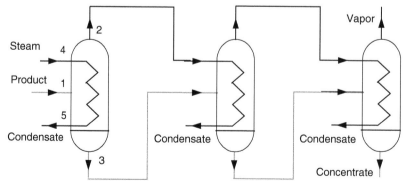

FIGURE 17.3

Triple-effect evaporator.

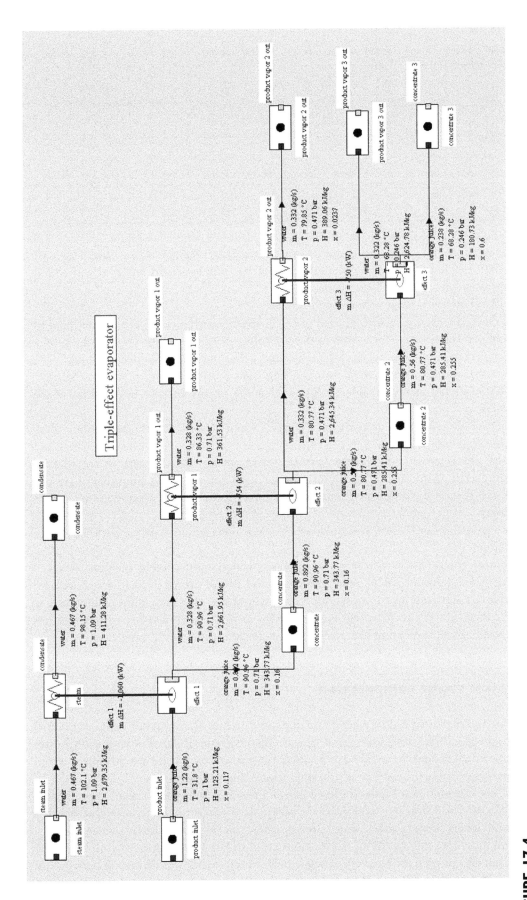

FIGURE 17.4

Triple-effect evaporator.

TABLE 17.2
EXERGY BALANCE OF THE TRIPLE-EFFECT EVAPORATION CYCLE

Steam inlet	265.6					$T_0 = 283.15$ K
Product inlet	2.4					
Concentrate 3	2.9					
Effect 1	296.3	245.0	79.1%	51.3	19.5%	Evaporator
Effect 2	229.5	207.2	89.3%	22.2	8.4%	Effect 2
Effect 3	189.5	158.6	80.5%	30.9	11.7%	Effect 3
Condensate				19.8	7.5%	Loss
Product vapor 2 out				12.2	4.6%	Loss
Product vapor 1 out				10.4	4.0%	Loss
Product vapor 3 out				116.5	44.2%	Loss
Global	267.9	3.0	1.7%	263.3	100.0%	

In this example, we were inspired by data provided by Balkan, Colak and Hepbasli, in their 2005 article given in the references, which we supplemented when not provided. The results we achieve are consistent with theirs.

Table 17.2 shows the exergy balance of this cycle.

In this table, the exergy balance of each of the evapoconcentrators has been established by considering that the resource is the sum of the incoming exergies, and the product the sum of the outgoing exergies, which explains why the exergy efficiencies of each effect are high.

On the other hand, in the overall balance, only that corresponding to the concentrated product is considered as recoverable exergy, so that the exergy balance is very low, of the order of 1.7%.

If we add as a recoverable exergy that of the water vapor at the outlet of effect 3, the overall balance rises to about 45%.

Note that the interest of an exergy balance is less obvious for this type of installation, because the exergy of the concentrated orange juice is not very different from that of the initial juice, the reverse operation, the dilution of a concentrated juice, corresponding to low irreversibility.

The exergy efficiency is from this point of view less representative of the performance of the installation.

In summary, the results of the exergy balance are strongly influenced by the possibilities of recovering the enthalpy available in the output streams. In this exergy balance, we considered that the sole concentrated juice corresponds to a recoverable exergy and all the condensates to losses. The level of concentration that is reflected by such an exergy balance is therefore less than the possibility of recovering heat from the streams at the outlet of the device.

Mechanical vapor compression

In most situations that would be suitable for the use of a heat pump (small temperature difference between a "free" cold source and heat needs), mechanical vapor compression is presented as an efficient alternative if the cold source is a gas or vapor. It is then possible to avoid the evaporator, possibly the condenser, and thus achieve high performance.

Indeed, compression of a vapor (steam or gas in a process) allows its pressure and temperature to be raised and thus its enthalpy. If the compressed vapor can be recovered either in a process, or in terms of energy, the compression operation can be very attractive economically.

The two most conventional MVC applications are distillation and concentration by evaporation. It can also be used for recovery of waste heat. However, its cycle is then akin to that of a HRSG such as we have discussed in Chapters 5 and 10.

A variant called TVC consists of replacing the compressor with an ejector. We will present an example of such a cycle below.

Evaporative mechanical vapor compression cycle

In an evaporative mechanical vapor compression cycle (Figure 17.5), the idea is to raise the solvent vapor enthalpy level so that it can be directly used to provide heat to the evaporator.

The basic scheme is as in Figure 17.5 and its cycle is plotted in the entropy diagram of Figure 17.6: in 0, the product to be concentrated is brought in two heat recovery exchangers where it is preheated (0–1) by cooling the concentrated product ((1'–7)) and condensate ((5–6)).

In the evaporator, the product to be concentrated is partially evaporated (1–2) by exchange with its own vapor, which condenses (4–5).

At the bottom of the evaporator, the concentrate is extracted in 1', while the solvent vapor exits in 2 to be recompressed in 3 and gain enough enthalpy to be able to serve as cycle hot source. To ensure large heat exchange coefficients (two-phase) in the evaporator, steam is frequently desuperheated from 3 to 4 via a separate heat exchanger, cooled by desuperheating water, even if that requires a steam addition in 4.

The energy interest of the operation is that by providing little additional enthalpy h_{23} (but a mechanical one), it is possible to recover the solvent vapor condensation enthalpy h_{45}. As an indication, if the solvent is water, which has a very high heat of vaporization (almost ten times that of oil), h_{23} is in practice between 3% and 9% of h_{45}.

Figure 17.7 presents a modification of a single-effect evaporator presented in Figure 17.1 at which was added a steam mechanical compression for much of the heat supply. The inlet steam flow is divided by a factor greater than 3, while the compressor power is only 14.4 kW. The OSC is equal to 0.44 and the ESC to 256 kWh/t.

Types of compressors used

Evaporators can be combined in series or in parallel, to form multiple-effect mechanical vapor compression evaporators. If evaporators are arranged in parallel, the compressor operates at high flow rate and low pressure ratio; if they are arranged in series, it works at low flow rate and high pressure ratio. In the first case, one uses centrifugal or screw compressors and in the second, screw and piston compressors. When the saturation temperature difference to achieve is low (5°C–6°C), the pressure difference (about 0.1 bar) can even be supplied by blowers that look like large fans. The latter configuration, which has the advantage of being the most reliable and most flexible, allows, with a speed controller, to modulate within minutes the rate of evaporation from the rated load down to 50% of its value.

In certain situations, if we do not attach too much importance to the compressor isentropic efficiency value, other compression means can be employed: either lobe or liquid ring compressors, or, if steam at high pressure compared to that of the evaporator is available, ejector

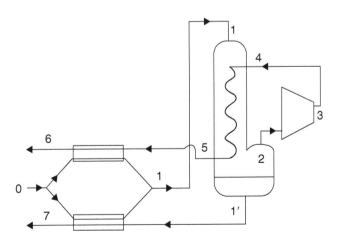

FIGURE 17.5

Evaporative mechanical vapor compression cycle.

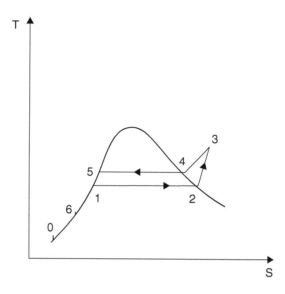

FIGURE 17.6

Mechanical vapor compression cycle.

(or ejecto-compressors), where a steam jet is used to entrain the vapor at the evaporator outlet and recompress it enough so it can be used as heat source. An example is given below.

Design parameters of a MVC

To design and size a MVC installation, one must know precise characteristics of the fluid being treated, and in particular, in addition to general thermodynamic data, the following:

- Concentrations C_i and C_o at the evaporator inlet and outlet. If the fluid flow at the entrance is $\dot{m}_{fi}$, the flow of solvent to evaporate is $\dot{m}_s = \dot{m}_{fi}\left(1 - \dfrac{C_i}{C_o}\right)$;

- The boiling point elevation (see above): depending on the nature of the solute and its final concentration, boiling occurs at a temperature different from that corresponding to the solvent saturation pressure, the gap being up to several tens of degrees Celsius. This elevation depends on the solutions considered. In the case of water, it can lead to increase the compression ratio, and therefore costs;
- Thermosensitivity: some products do not stand being processed at too high a temperature or for too long periods of time. This results in specific constraints for processing devices, which is preferable to take into account in the design phase;
- Viscosity can vary significantly from one product to another. It increases with the concentration and decreases with temperature. Its value determines the capacity of circulating pumps;
- Noncondensable gases drawn or dissolved in the fluid to concentrate reduce heat exchange coefficients. Their extraction can require vacuum pumps, of sometimes important capacity. Careful maintenance of these facilities is crucial for the proper functioning of the whole;
- Corrosion: the choice of materials to select and therefore the cost of various organs is determined by how aggressive are fluids that run through the installation, whether the fluid to treat or cleaning products.

Figure 17.8 shows a MVC evaporator manufactured by company France Evaporation, used to concentrate various products. The evaporator is of the "falling flow" type, i.e., the solution to concentrate trickles down inside tubes which are heated from the outside by steam. The energy consumption by ton of evaporated water is 15 kWhe as well as 630 kWhth for a single effect unit, 230 kWhth for a triple effect and 115 kWhth for a six-fold unit.

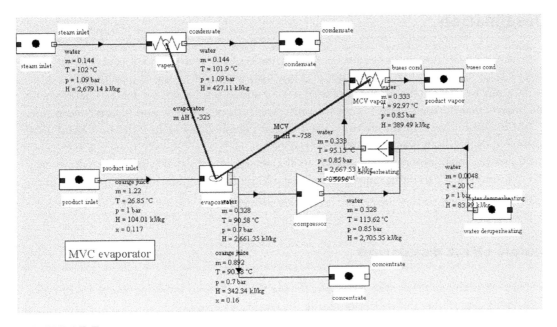

FIGURE 17.7

Mechanical vapor compression evaporator.

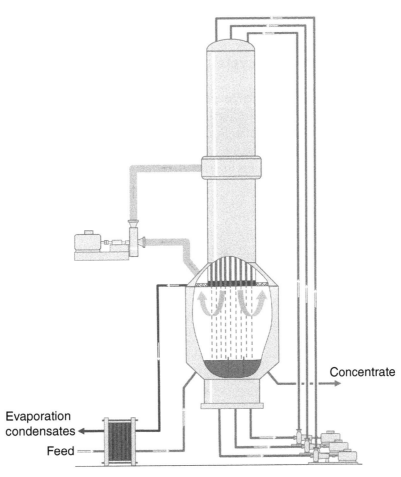

FIGURE 17.8

MVC plant. (Courtesy France Evaporation.)

Desalination

The limitation of freshwater resources in many parts of the world has for centuries led man to seek to produce water from seawater and brackish water. The issue of desalination is similar in many respects to that of the concentration of aqueous solutions we studied before, with the proviso that the useful effect is not the same: we focus on distillate and not on concentrate.

In this section, we will simply overview the main desalination techniques now being used commercially, without giving a complete description.

One of the most important parameters to characterize the treated water is its salinity usually expressed in g/l. Its value depends on the sea: 35 g/l for the Atlantic 50 g/l and more in the Persian Gulf.

Simple effect distillation

Distillation is the simplest way to desalinate sea water. It is to vaporize water by bringing high-temperature heat and then condensing the steam produced.

A single effect desalination cycle can be easily modeled in Thermoptim with class EvapoConcentrator (Figure 17.9).

In this example, 2 kg/s of seawater of salinity equal to 35 g/l, preheated at 70°C, enters the evaporator to provide 1 kg/s of freshwater. The thermal power input is 2,513 kW in the absence of losses, and 2,552 kW should be rejected at the condenser to cool the distilled water at 30°C. The OSC is equal to 1.27 and the ESC to 705 kWh/t.

Double-effect desalination cycle

In order to recover some of the available enthalpy in the distilled steam, it is possible to arrange several distillation units in cascade, as indicated previously, to obtain a multi-effect cycle. However, increasing the concentration of sea water in the evaporator outlet leads to an increase in evaporation temperature, because of the boiling point elevation, and you have to gradually reduce the pressure in the units in order to compensate it.

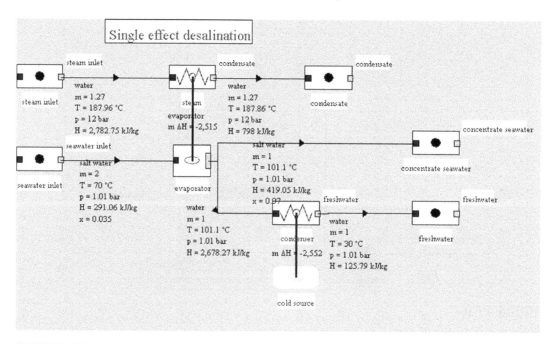

FIGURE 17.9

Single-effect desalination.

Figure 17.10 shows a double-effect cycle fed with 3 kg/s of seawater, which produces 2 kg/s of distilled water, consuming a thermal power just above the single-effect evaporator (we have neglected the losses and the compression work of the distilled water at the outlet of the second unit).

The synoptic view of Figure 17.10 shows that the steam input required for evaporating 1 kg/s of water (OSC) is here equal to 0.66 kg/s and the ESC to 705 kWh/t. By adding other units, we gradually improve the performance of the facility.

Mechanical vapor compression desalination cycle

Figure 17.11 presents a modification of the simple effect evaporator of Figure 17.10 with added mechanical compression of steam distilled to ensure most of the heat supply. Heat input is divided by ten, while the compressor power is only 35 kW. Ratio (steam evaporated)/(heating steam) is 9.

Thermal vapor compression

Another fairly general solution is to utilize the vapor available as motive fluid in an ejector, to entrain the steam distilled exiting the last unit of a multiple-effect cycle, allowing to recompress it and to use it to provide the evaporator with heat needs. The presentation of ejectors is made in Chapter 11.

Figure 17.12 shows the architecture of such a desalination ejector cycle. The motive steam flow is significantly greater than for a MVC facility, but the cost of the ejector is much lower than that of a compressor.

MSF desalination cycle

In a MSF cycle, seawater is preheated by heat exchange with distilled steam. To reduce the temperature differences between the two fluids, a counter-flow arrangement is used, with several chambers (stages) in series as shown in Figure 17.13 (Lior, 2000). In each stage, salt water is expanded by flash, vaporizing it partially.

Seawater at low temperature warms in the lowest pressure chamber, then passes into the next stage and so on. A high-temperature heat input is supplied by live steam.

The warm seawater is expanded in the first stage, the distilled steam generated being condensed as described above. The salt-concentrated residue is then routed to the next stage at lower pressure and so on.

The OSC is equal to 0.54 and the ESC to 300 kWh/t.

BOX 17.1
Key issues

External class FlashBrine

External class FlashBrine has been set up in order to allow you to make multi-stage flash desalination models. It is presented in Thermoptim portal model library (https://direns.mines-paristech.fr/Sites/Thopt/en/co/modele-chbre-flash.html).

The flash chamber acts as a divider receiving as input salt water, from which exit two fluids, water vapor and the concentrated solution. The chamber being adiabatic, the enthalpy of vaporization is taken from the aqueous solution, whose temperature drops.

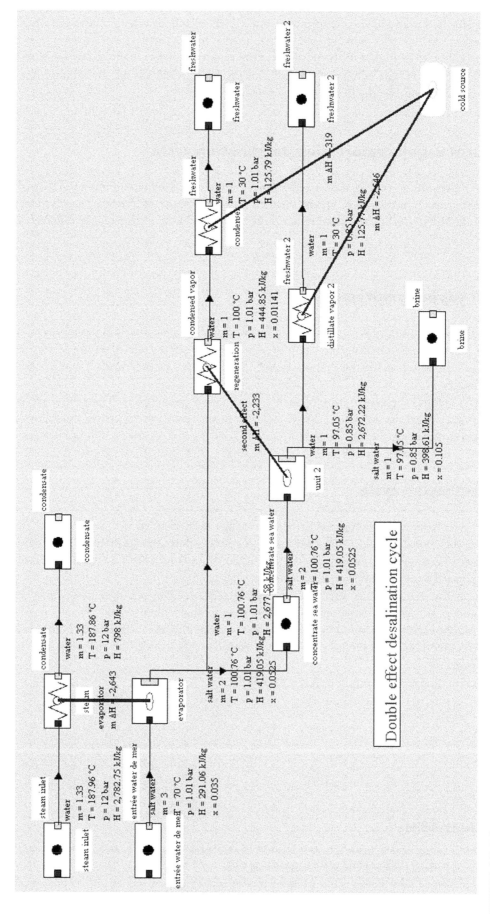

FIGURE 17.10

Double-effect desalination cycle.

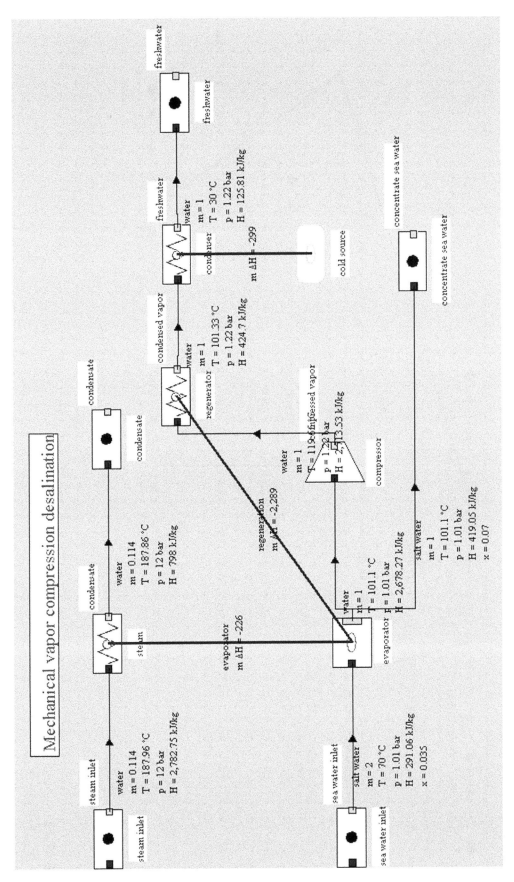

FIGURE 17.11

Mechanical vapor compression desalination.

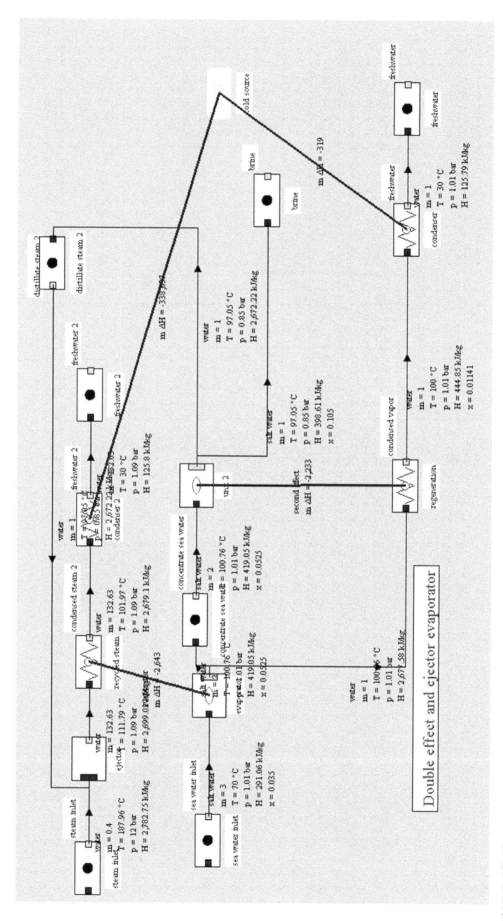

FIGURE 17.12

Double-effect ejector desalination cycle.

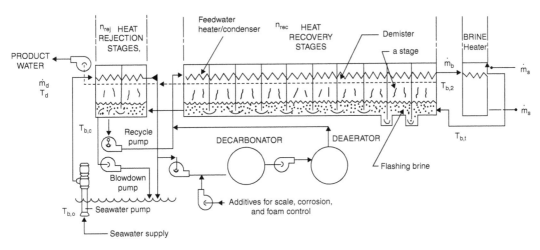

FIGURE 17.13

MSF desalination system.

Exergy balance

Table 17.3 gives the exergy balance of this system.

The synoptic view of Figure 17.14 shows that the steam input required for evaporating 1 kg/s of water is here equal to 0.54 kg/s.

The remarks we made for the exergy balance of Table 17.2 also apply here. In this case, the only valued exergy is that of the produced fresh water.

Reverse osmosis desalination

Let us consider two media consisting of two mixtures of the same pair solvent-solute but of different concentration and separated by a semipermeable membrane. Experience shows that

- If the two media are at the same pressure, a solvent transfer called **osmosis** takes place;
- If you apply some pressure, called osmotic pressure π, to the most concentrated medium, this transfer can be canceled;
- If pressure is greater than π, the solvent migrates from the more concentrated solution to the other: this is called **reverse osmosis** (RO).

TABLE 17.3

EXERGY BALANCE OF THE MSF CYCLE

Component	Resource	Product	Exergy Efficiency (%)	Irreversibilities	% Total (%)	$T_0 = 283.15$ K
Steam inlet	515.658					
Seawater inlet	24.055					
Water produced	2.778	2.778				
Collecteur	2.781	2.778	99.9	0.003	0.0	
Condensation 1	112.505			28.095	6.0	Condenser 1
Condensation 2	139.081			45.330	9.7	Condenser 2
Condensation 3	71.107			0.530	0.1	Condenser 3
Steam	415.099			195.172	41.8	Heat input
Condensate	0			94.293	20.4	Loss
Brine	0			28.346	6.1	Loss
Condenser	3.015			3.015	0.7	$T_k = 10.00°C$
Condenser 2	5.523			5.523	1.2	$T_k = 10.00°C$
Condenser 3	65.816			65.816	14.2	$T_k = 10.00°C$
Global	545.236	82.77	15.2	466.123	100.2	

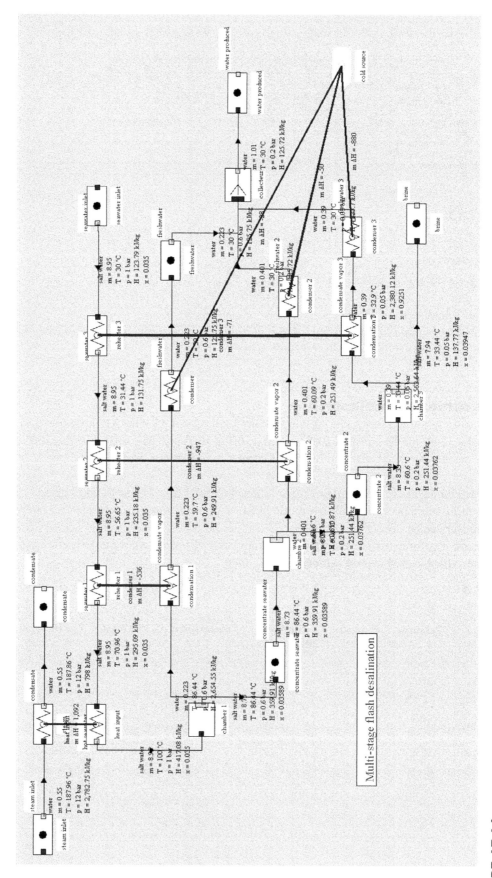

FIGURE 17.14

Synoptic view of the three-stage flash desalination system.

Desalination plants using RO operate on this principle (Figure 17.15): by applying a pressure higher than osmotic pressure of salt water in a chamber equipped with a semipermeable membrane, it is possible to collect on the other side of the membrane a solution of very low salinity, called the permeate.

Typically, a spiral wound-type RO module appears externally as a cylinder of about 1 m long and 20 cm in diameter: the flat semipermeable membrane is spirally wound around the tube-manifold, enclosing a fine polyester mesh through which passes permeate, forming a stack sealed by glue on three sides. Outside this envelope, water to desalinate flows through a spacer.

The energy consumed is only the compression work of the initial solution and is therefore much lower than that brought into play (as heat) in most devices that we have studied so far.

The van't Hoff law (17.6) states that the osmotic pressure exerted by the solute is equal to the pressure it would have exercised in the perfect gas state in the same volume and at the same temperature. If the solute is dissociated as ions, osmotic pressure is multiplied by the number of ions present.

$$\pi = n_i XRT \tag{17.6}$$

For seawater, the magnitude of π is 25–30 bar.

ΔP being the pressure difference across the membrane, we can show that the flow of solvent through the membrane J_e is given by (17.7).

$$J_e = A(\Delta P - \Delta\pi) \tag{17.7}$$

A is called water permeability of the membrane. It is a characteristic parameter of the membrane, which depends on temperature according to an Arrhenius-type law (17.8).

$$A = A_0 \exp\left(\frac{E}{R}\left(\frac{1}{298} - \frac{1}{T}\right)\right) \tag{17.8}$$

Even if there is a preferential transfer of the solvent, a small fraction of the solute also crosses the membrane. The flow of solute J_s is given by (17.9). It is proportional to the concentration difference.

$$J_s = B\Delta X \tag{17.9}$$

B is called the salt permeability of the membrane. This parameter depends on the membrane but not the temperature.

We call conversion rate the ratio of flow through the membrane to the feed flow rate and retention rate the ratio of the concentration difference between the initial solution and permeate to the initial solution concentration.

Figures 17.16 and 17.17 show the Thermoptim diagram and component screen allowing one to model a Dow Chemical FilmTech SW30–4040 membrane, used to desalinate sea water. The values of A and B were estimated from the manufacturer's documentation. The component is modeled by external class ReverseOsmosis.

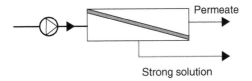

FIGURE 17.15

RO unit.

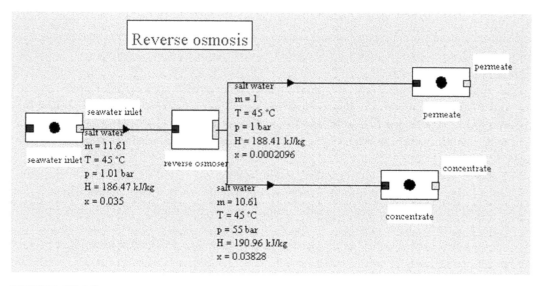

FIGURE 17.16

Thermoptim diagram of a RO unit.

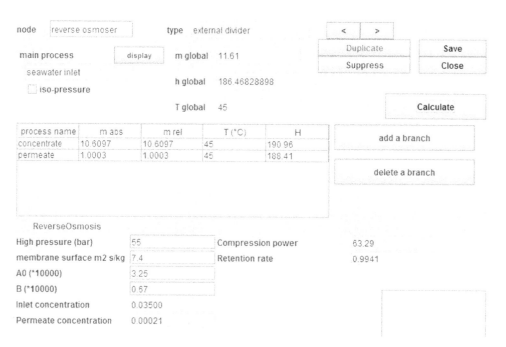

FIGURE 17.17

RO component screen.

The compression power to produce 1 kg/s of fresh water is equal to 63.3 kW, or nearly twice that required by mechanical vapor compression. However, it is possible to recover some of that power by expanding the concentrate in a turbine connected to the compressor.

Note that the salt water must be very clean in order to prevent the membrane from clogging.

Drying by hot gas

Working principle

At the beginning of this chapter, we introduced evaporation concentration facilities. Drying by hot gas is to evaporate the product by increasing the absolute humidity of the gas. There is thus also a separation of the solvent and solute which increases the concentration.

The difference with evapo-concentration techniques is that the final product is in solid form, with very low humidity (less than 3% approximately).

There are other methods of drying products than by hot gas, but we will not study them here. These include, for example, drying using rotating drums heated from the inside, onto which the products to be dried are poured.

The hot gas must be clean enough to be placed in direct contact with the product to concentrate. It is generally air or gas turbine exhaust gases.

Several types of hot air-drying apparatus exist, such as tray or cabinet, tunnel and conveyor, etc. In what follows, we will give an example of a spray dryer, in which the product to be dried, initially in liquid form, is atomized into a current of hot air in the form of small droplets, which dry on contact with air, whose humidity increases, and temperature drops accordingly.

The dried product, a powder of fine particles, falls to the bottom of the apparatus and is collected. Since a small part is carried away by the air flow, cyclones recover it at the outlet.

Air inlet temperatures range from 120°C to 250°C, depending on the product. Due to evaporation, the air temperature drops to less than 100°C, the product temperature being 20°C–30°C lower.

Such devices are used in the food industry, and in particular for the manufacture of powdered milk, downstream of evapo-concentrators which perform a first concentration of the milk.

Modeling a spray dryer in thermoptim

The model presented here is used to spray dry skimmed milk previously brought to the concentration of 39.42% (the initial concentration is 9.2%). On leaving the dryer, the product concentration is 94.2%.

The settings correspond to data from Bimbenet et al., (2002), with the caveat that we have expressed the flow rates in g/s and not in kg/h, which means that capacities are multiplied by a factor of 3.6. The performance of the dryer is completely consistent with those they indicate, which correspond to measured values. Our modeling therefore seems quite satisfactory, although this is a particularly simple model built in order to allow us to get an idea of how such a facility works.

It confines itself to establishing the mass and energy balances, assuming firstly that the concentration of the product at the outlet is known, and secondly, that only the hot gas provides the enthalpy corresponding to the evaporated water, taking into account heat loss from the component as a percentage of that value. Flow rates of both inlet streams are set by conditions upstream of the component and not recalculated. If the gas flow is insufficient for allowing its cooling to saturation to evaporate the water involved, a message warns the user.

The principle of our model is as follows:

* The only parameters are the concentration of the product at the outlet, as entered in the point screen, and the percentage of losses;
* We first calculate the absolute humidity of the incoming gas and determine the mass flow of dry gas from that of moist gas;
* Absolute humidity w at the outlet is determined, and we iterate on the outlet air temperature, which allows us to calculate the total specific enthalpy change $\Delta Q'$. When the variation of air specific enthalpy is equal to that required to evaporate water, the solution is found (we neglect here the variation of sensible heat of the dry product);
* The composition of the outlet moist gas is changed;
* Values downstream of the node are updated.

Such a component can be easily implemented in Thermoptim as an external class, using moist mixture properties' calculation equations which are presented in Chapter 12.

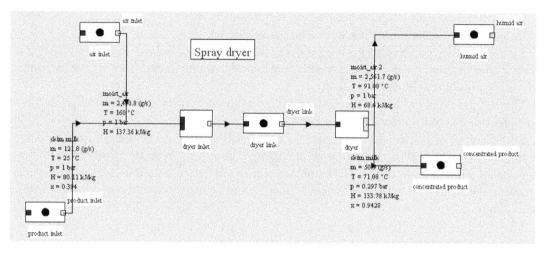

FIGURE 17.18

Spray dryer unit.

Figure 17.18 shows the Thermoptim synoptic view of such a device (external classes SprayDryerInlet and SprayDryerOutlet): a rate of 2,471 g/s dry air of hot air (w = 0.008 kg water/kg air) at 160°C is blown on a stream of 121.1 g/s of skim milk at 25°C. 96.5% of the water contained in the product evaporates increasing air moisture. Note that on this synoptic view, the air flow values are related to dry air. At the component outlet, air thermodynamic state is temperature T = 91.7°C, relative humidity ε = 7.44%.

This air can then optionally be cooled and condensed and then recycled if necessary.

$\Delta Q'$ represents the change in the total specific enthalpy of the air in the spray dryer. Its value (15.9 kW) is low and very far from the thermal power that has to be supplied to the product to evaporate it. This is because humidification can in reality be broken down into two successive fictitious processes: cooling at constant specific humidity of the air, down to the outlet temperature of the dryer, which provides thermal power ΔH_t (rate of sensible heat gain), followed by isothermal humidification to the output state, which requires a thermal power ΔH_w (rate of latent heat gain).

$$\Delta Q' = \Delta H_w - \Delta H_t$$

In the screen of Figure 17.19, we have displayed the rate of latent heat gain which is an estimate of the heat exchanged with the heating air to ensure drying: ΔH_t = 189.5 kW, as well as a performance indicator, the specific consumption of the device, equal to the ratio of this value to the amount of water evaporated. As can be seen, the variation in the specific enthalpy of the air $\Delta Q'$ is very far from the thermal power actually put into play.

The specific consumption is 2,673 kJ/kg of extracted water, while the theoretical one is around 2,500. The efficiency is therefore quite good.

The sensible heat ratio SHR is also calculated: $SHR = \dfrac{\Delta H_t}{\Delta Q'}$.

The model for skim milk is analogous to the one we presented for orange juice (class SkimMilk). The composition of the corresponding dry matter is given in Table 17.4.

Figures 17.20 and 17.21 show the plot in the Carrier and Mollier psychrometric charts of the humidification experienced by the drying air. As we can see, it is very close to adiabatic humidification. Recall that the properties of moist gases and the psychrometric charts are presented in Chapter 12.

In Figure 17.20, the breakdown of the humidification in two processes is illustrated by the horizontal cooling of the hot air at constant humidity (air inlet–fictitious point) followed by the vertical humidification at constant temperature (fictitious point–humid air).

In Figure 17.21, the breakdown of the humidification in two processes is illustrated by the vertical cooling of the hot air at constant humidity (air inlet–fictitious point) followed by the

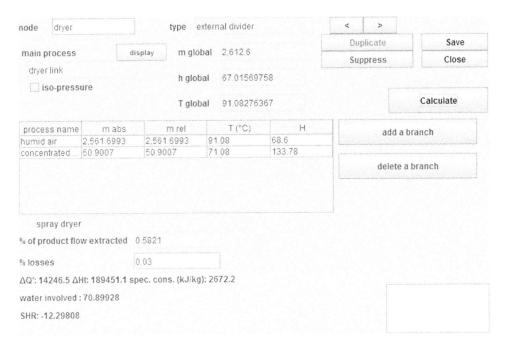

FIGURE 17.19

Spray dryer component screen.

TABLE 17.4

MASS COMPOSITION OF SKIM MILK DRY MATTER

Protein	Fat	Carbohydrate	Ash
37.07%	1.96%	52.72%	8.26%

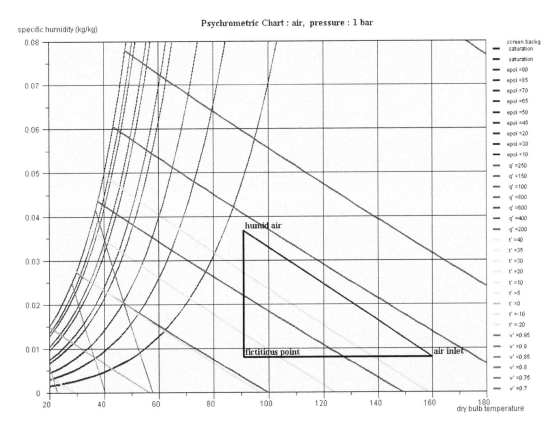

FIGURE 17.20

Air humidification in the Carrier psychrometric chart.

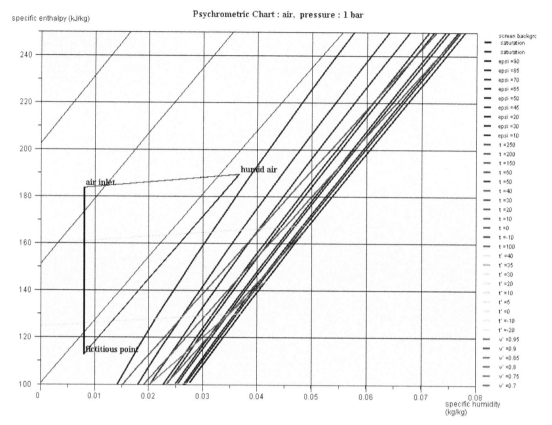

FIGURE 17.21

Air humidification in the Mollier psychrometric chart.

humidification at constant temperature (fictitious point–humid air) represented by a slightly inclined segment.

ΔH_t, ΔH_w and $\Delta Q'$ can be directly read on the y-axis.

The initial heating of the drying air is often provided by the combustion of a clean fuel such as natural gas.

The following model introduces a combustion chamber upstream of the spray dryer (Figure 17.22).

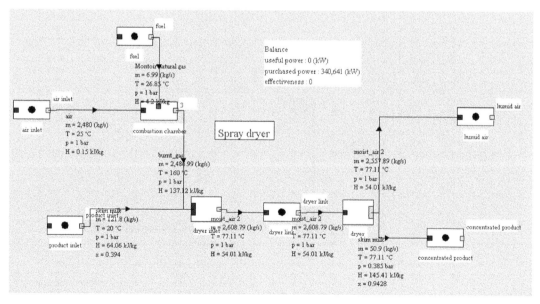

FIGURE 17.22

Synoptic view of a spray dryer unit heated by a combustion chamber.

Thermoptim can calculate this model without difficulty, even if the hot gas is now the burnt gases and not air. It provides an interesting information: the heat input in the combustion chamber is worth 340.6 kW. This value, about twice that of ΔH_w, means that the specific consumption is 4,811 kJ/kg of extracted water, and not 2,673 as the performance of the sole spray dryer might suggest.

This difference is at least partially explained by the losses of this cycle, in which the air and the concentrated product leave the device at a high temperature (93.4°C) while they were at 25°C at the inlet.

It is possible to reduce them at least partially by cooling the outlet streams by exchange with the inlet ones.

For example, in the cycle of Figure 17.23, an exchanger between the incoming and outgoing air streams makes it possible to recover 135.2 kW, and therefore to reduce the contribution of the combustion chamber, which goes from 340 to 204 kW. Specific consumption is thus greatly reduced.

Table 17.5 gives the exergy balance of the cycle with regeneration. Most of the irreversibilities stem from the combustion chamber. We have here considered that the concentrated product corresponds to a valuable exergy, while the humid air at the outlet is rejected in the atmosphere.

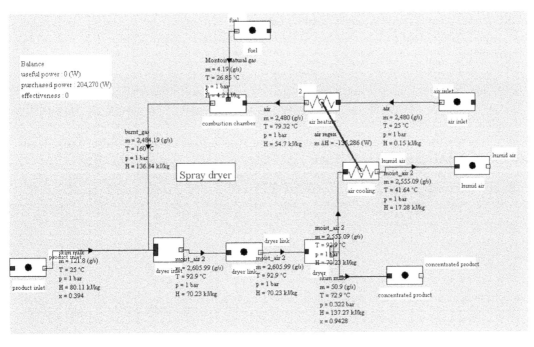

FIGURE 17.23

Spray dryer cycle with regeneration.

TABLE 17.5

EXERGY BALANCE OF THE SPRAY DRYER WITH COMBUSTION AND REGENERATION

Component	Resource	Product	Exergy efficiency (%)	Irreversibilities	% total	T₀=298.15 K
Fuel	225,784	0	0	0	0	
Combustion chamber	0	57,784	25.6	179,454	79	
Dryer	73,780	36,652	49.7	37,128	16	
Concentrated product	470	470	0	0	0	
Product inlet	132	0	0.0	0	0	
Air heating	16,581	11,013	0.0	5,568	2	Air regen
Humid air	0	0	0.0	4,524	2	Loss
Global	226,355	470	0.2	225,885	100	

The exergy efficiency of the dryer itself is about 50%, but the overall efficiency of the whole system is very low, as the only valuable exergy is that of the concentrated product, while the exergy spent is a little more than that of the fuel consumed.

Bibliography

ASHRAE, Thermal properties of foods. *Refrigeration*, Chapter 8, 8.1–8.30 (SI), 2002.

F. Balkan, N. Colak, A. Hepbasli, Performance evaluation of a triple effect evaporator with forward feed using exergy analysis. *International Journal of Energy Research*, 29, 455–470, 2005, http://dx.doi.org/10.1002/er.1074 Exergy balance.

J.-J. Bimbenet, P. Schuk, M. Roignant, G. Brulé, S. Méjean, Heat balance of a multistage spray-dryer: Principles and example of application. *Le Lait*, 82(4), 541–551, 2002, ISSN 0023-7302.

Y. Choi, M. R. Okos, Effects of temperature and composition on the thermal properties of foods, *Food engineering and process applications*, (Edited by M. LeMaguer, P. Jelen), Elsevier Applied, Science Publishers, London, 1986, 1, 93–101.

P. Danis, Dessalement de l'eau de mer, Article J2700, Techniques de l'Ingénieur, Paris.

R. Leleu, Evaporation, Article J2320, Techniques de l'Ingénieur, Paris.

N. Lior, Water desalination, *The CRC handbook of thermal engineering*, (Edited by F. Kreith), CRC Press, Boca Raton, FL, 2000, ISBN 0-8493-9581-X.

A. Maurel, Dessalement de l'eau de mer et des eaux saumâtres, 2ème édition, Ed. Tec et Doc, Paris, 2006.

F. B. Petlyuk, *Distillation theory and its application to optimal design of separation units*, Cambridge Series in Chemical Engineering, Cambridge University Press, 2004, ISBN 0-521-82092-8.

J. F. Reynaud, Concentration par évaporation et recompression mécanique de vapeur, Ed. Eyrolles, Paris, 1984.

M. E. Williams, A review of reverse osmosis theory, [Online] Available from: http://www.eetcorp.com/heepm/RO_TheoryE.pdf

Electrochemical converters: fuel cells and electrolyzers

Introduction

Electrolysis is an old and well-known process: an electrolyzer performs an electrochemical reaction that produces hydrogen and oxygen from water through electricity consumption by an endothermic reaction.

A fuel cell performs the reverse electrochemical reaction: hydrogen and oxygen react to produce electricity and water while releasing heat. The main technologies being considered are solid oxide fuel cells (SOFCs), proton exchange membrane fuel cells (PEMFCs), alkaline fuel cells (AFCs), phosphoric acid fuel cells (PAFCs) and molten carbonate fuel cells (MCFCs).

Fuel cells

A fuel cell performs the reverse electrochemical reaction of electrolysis according to reaction:

$$H_2 + \tfrac{1}{2}O_2 \rightarrow H_2O$$

Thus, we see that if the fuel used is pure hydrogen, the fuel cell's only byproduct is water: this is a particularly clean generator.

The heart of the cell consists of two electrodes, the anode and cathode, separated by an electrolyte.

In some cells, such as SOFCs, oxide ions O^{2-} migrate from the cathode to the anode where water is produced, and in others like PEMFCs, cations H_3O^+ (hydrated protons H^+) migrate from the anode to the cathode.

The stack behaves as a quadrupole: in Figure 18.1, which represents a cell stack, hydrogen enters in the top left of the stack, combines at the anode with ions O^{2-} to form water, and exits in the lower left, enriched in water, while air enters in the top right and exits in the bottom right depleted in oxygen.

A SOFC works at very high temperatures (between 600°C and 1,000°C), so that water at the outlet is in gaseous form.

Both reactions that take place are as follows:

At the anode: $H_2 + O^{2-} \rightarrow H_2O + 2e^-$
At the cathode: $O_2 + 2e^- \rightarrow O^{2-}$

DOI: 10.1201/9781003175629-18

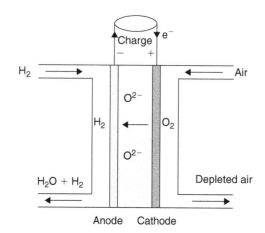

FIGURE 18.1

Solid oxide fuel cell.

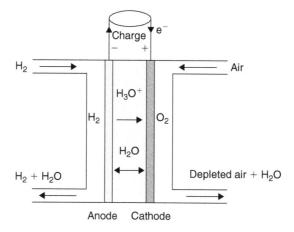

FIGURE 18.2

Proton exchange membrane fuel cell.

In a PEMFC, hydrated protons H_3O^+ migrate from the anode to the cathode, where water is produced.

In Figure 18.2, which represents a cell stack, hydrogen enters in the top left of the stack, combines at the anode with water to form cations and exits in the lower left, while the air enters in the top right and exits at the bottom right, oxygen-depleted and water-enriched.

A PEMFC works at low temperature (between 80°C and 120°C), so that water is in liquid form at the outlet.

Both reactions that take place are as follows:

At the anode: $H_2 + 2H_2O \rightarrow 2H_3O^+ + 2e^-$
At the cathode: $O_2 + 2H_3O^+ + 2e^- \rightarrow 3H_2O$

In reality, this fuel cell works only if the membrane is wetted on both sides and is permeable to water. Water balance is governed by two phenomena: a portion of the water goes through the membrane from the anode to the cathode under the effect of electro-osmosis, driven by protons, while a part crosses in the other direction by diffusion due to concentration difference between the two sides of the membrane.

The operation of a fuel cell can be characterized by two parameters:

* The rate of fuel use τ, which represents the fraction of fuel that reacts;
* The energy efficiency ε, ratio of electric output to the flow of enthalpy put into play by reaction $H_2 + \frac{1}{2}O_2 \rightarrow +H_2O$, under standard temperature and pressure conditions (25°C, 1 atm). ε is close to 50% in practice.

Each cell theoretically implements a 1.23 V voltage, but in practice, the open circuit voltage barely exceeds 1 V, and the operating voltage is around 0.7 V. The difference is due to various losses which will be analyzed below.

Figure 18.3 shows the internal content of a single cell PEMFC test bench allowing one to assess fuel cells of 50 cm^2 active surface (7.1 cm × 7.1 cm).

In the middle is the membrane electrode assembly (MEA), with graphite bipolar plates on both sides through which flow hydrogen and air.

The plate width is about 1 cm. Electric collectors are placed on the side opposite the MEA. Structural end plates allow one to maintain the various parts in place.

To get high voltages, it is necessary to set many cells in series. Commonly known as a stack of cells, it is in practice limited to 100 or 200.

Current densities are generally between 0.1 and 1 A/cm^2.

For large intensities, it is necessary to increase the cell surface, but there are limits, because it is difficult to supply gas uniformly on large areas.

There are several types of fuel cells, usually classified according to the nature of their electrolyte:

- AFC uses potash as electrolyte. Developed for space, this type of cell is used in all inhabited spaceflight at NASA, is entirely satisfactory and is cheap but suffers from the need to operate in the absence of CO_2 produced today by reforming. Its operating temperature is 80°C;
- PEMFC uses, as we have seen, a proton exchange membrane electrolyte. It seems today one of the most promising and is the subject of the largest development efforts and thereby supplants AFC. It has great potential for many applications in all capacity ranges from watts to megawatts. Its main drawback is cost, but cost reductions are hoped for in the short term. Its operating temperature is 80°C;
- PAFC uses phosphoric acid as electrolyte. Marketed by U.S. company ONSI Corp. for several years, it is especially interesting in CHP in light of its operating temperature of 200°C;
- MCFC uses molten carbonate as electrolyte. Operating at high temperature (650°C), ionic transport is performed by ions CO_3^{2-}. It has the advantage of being directly fed with a synthesis gas and can be integrated in high performance complex cycles in combination with gas turbines;
- SOFC uses a solid oxide (ZrO_2) doped with small amounts of calcium oxide and yttrium oxide. Its main advantage is being able to consume carbon monoxide. Operating at high temperature (600°C–1,000°C), it can also be integrated with combined cycle gas turbines, as we shall see below.

FIGURE 18.3

Internal view of a single cell PEMFC. (Courtesy CEP Mines-ParisTech.)

SOFC modeling

In Thermoptim, the model structure is given in Figure 18.4.

Cell model

The first model we develop is very simple: rate of hydrogen τ and energy efficiency ε are assumed known.

At the anode, only a fraction of τ is transformed in the cell, the rest emerging from it. Typically, $\tau = 0.5$. The overall reaction giving the outlet species is written:

For the part used:

$$\tau\left(H_2 + \tfrac{1}{2}O_2\right) \rightarrow \tau H_2O \; \left(\tau\Delta H_0 = -241,820\tau \, kJ/kmol\right) \tag{18.1}$$

For the unused part:

$$(1-\tau)H_2 \tag{18.2}$$

We get therefore at the anode outlet:

$$(1-\tau)H_2 + \tau H_2O \tag{18.3}$$

$\tau\Delta H_0$ represents the theoretical energy put into play by the conversion of the part used. A fraction ε of this energy is converted directly into electricity, and $(1-\varepsilon)$ is transformed into heat.

Moreover, oxygen is withdrawn from the oxidizer air at the cathode. λ representing the flow of incoming air, we get the following:

$$\text{Intake air} : \lambda\left(O_2 + 3.76N_2\right) \tag{18.4}$$

$$\text{depleted air exiting} : \left(\lambda - \tau/2\right)O_2 + 3.76\lambda N_2 \tag{18.5}$$

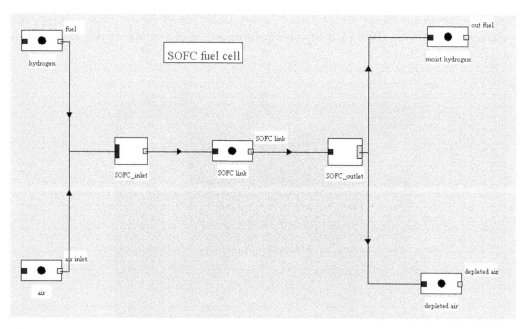

FIGURE 18.4
SOFC structure.

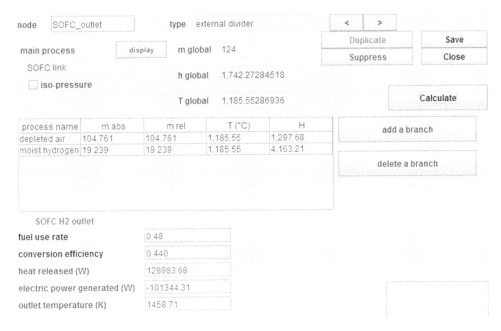

FIGURE 18.5

SOFC component screen.

The model retained is the following:

- Species composition is given by solving Equations (18.1–18.5): one determines hydrogen and air inlet molar flows, which provides the value of λ, from which we deduce molar flows at the outlet, τ and ε values being known;
- Heat released by fraction $\tau(1-\varepsilon)$ of the fuel is used to provide energy to heat gases.

$\dot{m}_{mol}$ being the molar flow of hydrogen, enthalpy released is equal to $\dot{m}_{mol}\tau\Delta H_0$. It is split between electricity $\dot{m}_{mol}\varepsilon\tau\Delta H_0$ and energy to heat gases $\dot{m}_{mol}\tau(1-\varepsilon)\Delta H_0$.

In Thermoptim, the cell is represented by an external mixer connected to an external divider (classes SOFCinlet and SOFCoutlet), the calculations being made by the latter.

The SOFC component screen is given in Figure 18.5. It is assumed that gases enter the cell at 500°C.

It is a cylindrical stack technology such as that developed by Westinghouse, corresponding to a 100 kWe system, where $\tau = 0.48$ and $\varepsilon = 0.44$.

Results of the simple model

With the following settings corresponding to Figures 18.5 and 18.9 (4 g/s of hydrogen and 120 g/s of air entering the cell at 1 bar and 500°C), gas compositions obtained are given in Figures 18.6–18.8.

Improving the cell model

We have said above that each cell theoretically implements a 1.23 V voltage, but in practice, the open circuit voltage barely exceeds 1 V, and the operating voltage is around 0.7 V. The difference is due to various losses which stem from the following causes.

- Activation losses: caused by the slowness of the reaction taking place on the surface of the electrodes. A proportion of the voltage generated is lost in driving the chemical reaction that transfers the electrons.

component name	molar fraction	mass fraction
N2	0.781	0.7555302
Ar	0.009	0.01241636
O2	0.21	0.2320534

FIGURE 18.6

Air entering the cell, flow rate 120 g/s.

component name	molar fraction	mass fraction
H2	0.52	0.1081169
H2O	0.48	0.8918831

FIGURE 18.7

Humidified hydrogen at the outlet, flow rate 19.24 g/s, LHV: 11,230 kJ/kg.

component name	molar fraction	mass fraction
N2	0.8824047	0.8654283
Ar	0.01016856	0.01422242
O2	0.1074267	0.1203493

FIGURE 18.8

Air depleted in O_2 flow rate 104.76 g/s.

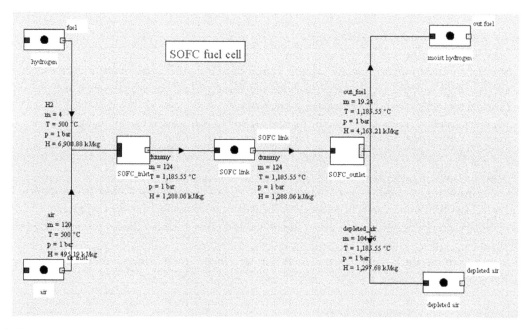

FIGURE 18.9

Synoptic view of the SOFC.

- Ohmic losses: voltage drop due to the resistance to the flow of electrons through the material of the electrodes. This loss varies linearly with current density.
- Concentration losses: result from the change in concentration of the reactants at the surface of the electrodes as the fuel is used.
- Fuel crossover losses: result from the waste of fuel passing through the electrolyte and electron conduction through the electrolyte. This loss is typically small but can be more important in low-temperature cells.

The result is a decline of the voltage when the current drawn by the receiving circuit increases. The shape of this variation is represented by what is called the polarization curve of the cell (Figure 18.10 where the voltage is the ordinate and the current density the abscissa).

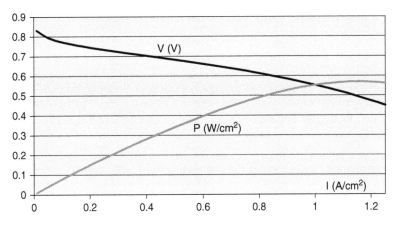

FIGURE 18.10

Polarization curve of the cell.

We will now present a second model taking into account such a curve.

We assume that rate of hydrogen use τ and energy efficiency ε are determined from an electric model (classes SOFCH2ElecInlet and SOFCH2ElecOutlet).

To calculate the cell voltage V_{cell} knowing the intensity and the active surface (Figure 18.11), various models have been proposed. We retain one developed at the Center for Energy and Processes of École des Mines de Paris (Hubert, 2005). Its equation is as follows:

$$V_{cell}(J) = E + \frac{b}{\ln\left(\dfrac{J}{J_d}\right) - 2} + \left(\frac{b}{4J_d} - \Delta\right)J \tag{18.6}$$

where $J = I/A$ is the current density (intensity divided by the active area A), in A/cm^2.

E, J_d, b and Δ are four parameters *a priori* dependent on the cell temperature and the hydrogen pressure. They can be considered constant as a first approximation, but the influence of temperature and partial pressure of H_2 on each parameter can also be taken into account if necessary.

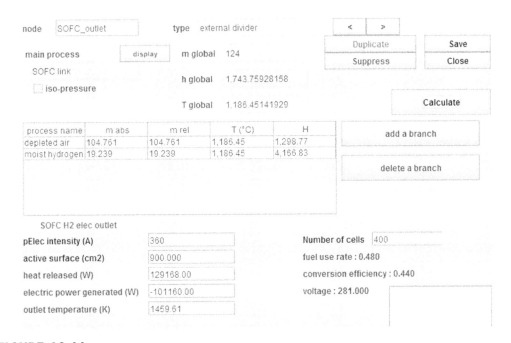

FIGURE 18.11

SOFC component screen.

Knowing V_{cell}, it is possible to simply calculate the cell power output:

$$P_{elec} = V_{cell}IN_{cell}$$

The rate of hydrogen use is assumed to be proportional to the current density J, which corresponds in first approximation to what happens in practice: the fraction of hydrogen consumed is zero in a physically open circuit and increases when the current output increases. We choose the following law: $\tau = 1.2$ J.

New model parameters are number of cells N_{cell}, current intensity sought I and active surface of a cell A. The model calculates τ and ε as well as output voltage V.

The SOFC component screen is given in Figure 18.11. It is assumed that the gases enter the cell at 500°C.

It is a cylindrical stack technology such as that developed by Westinghouse, each cell having a diameter of 19 mm and a length of 1.5 m, an area of 900 cm², leading to an intensity of 360 mA with a current density of 400 mA/cm².

Calculation results for a set of 400 cells, corresponding to a 100 kWe system are given below: $\tau = 0.48$, $\varepsilon = 0.44$, V = 281 V. In terms of electrical power and heat, this model leads to the same values as the simple model developed previously.

Model with a thermocoupler

The third model is an extension of the previous one, which assumes that the fuel cell is cooled by water at 40 bar, its temperature rising from 200°C (liquid) to 450°C (steam).

Stack cooling is modeled by a thermocoupler connected to the outlet divider.

We assume in this model that one third of the heat in the stack is recovered by steam cooling, the rest being used to heat gases that enter the fuel cell. The thermocoupler setting is explained below.

Figures 18.12 and 18.13 show model changes (classes SOFCH2ThermoInlet and SOFCH2ThermoOutlet).

With the new assumptions, the thermocoupler allows us to heat 17.4 g/s of water at 40 bar from 200°C to 450°C.

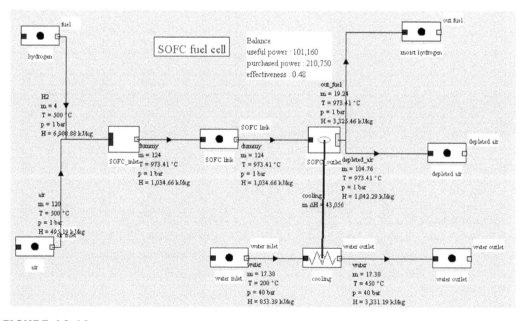

FIGURE 18.12
Synoptic view of the SOFC with thermocoupler.

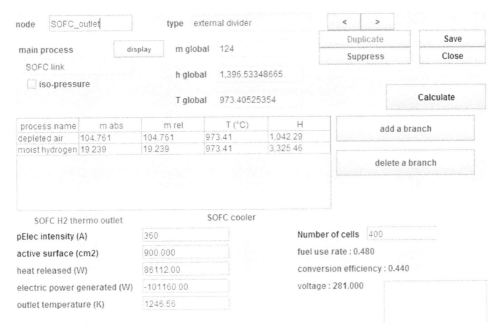

FIGURE 18.13

SOFC component screen.

The overall performance is remarkable since the electrical efficiency is already equal to 48%. Note, however, that everything depends on the way in which the incoming energy is counted, because here we have only taken into account the hydrogen used by the fuel cell.

If we consider that the energy purchased is the totality of the hydrogen entering the cell and that the moist hydrogen exiting is lost, the efficiency drops significantly. It becomes equal to only 23%.

Exergy balance

Table 18.1 shows the exergy balance of the fuel cell. In the fuel cell line (SOFC_outlet), the product (101.16 kW) is the electricity produced, and 59.74 kW are accounted as irreversibility. They correspond to the exergy loss in the thermocoupler.

As the moist hydrogen and the water outlet (steam) are considered as valuable products downstream of the cell, the overall exergy efficiency is high (79.8%).

Here we see the big difference with the combustion chambers which are always the seat of very important exergy losses because the chemical exergy of the fuel is converted into heat and not into electricity.

TABLE 18.1

EXERGY BALANCE OF THE SOFC FUEL CELL WITH THERMOCOUPLER

Component	Resource	Product	Exergy Efficiency (%)	Irreversibilities	% Total (%)	$T_0 = 283.15$ K
Air	25,945					
Hydrogen	578,544					
Moist hydrogen	340,060	340,060				
Water inlet	3,391					
Water outlet	23,755	23,755				
SOFC_outlet		101,160	21.08	59,470	48.36	
Depleted air				63,506	51.6	Loss
Global	607,881	484,872	79.8	123,009	100.0	

TABLE 18.2

EXERGY BALANCE OF THE SOFC FUEL CELL WITH THERMOCOUPLER, MOIST HYDROGEN BEING LOST

Component	Resource	Product	Exergy Efficiency (%)	Irreversibilities	% Total (%)	$T_0 = 283.15$ K
Air	25,945					
Hydrogen	578,544					
Water inlet	3,391					
Water outlet	23,755	23,755				
SOFC_outlet		101,160	21.1	59,470	12.8	
Depleted air				63,506	13.7	Loss
Moist hydrogen				340,060	73.4	Loss
Global	607,881	144,812	23.82	463,069	100.0	

If we consider that the moist hydrogen is lost, the exergy balance is given in Table 18.2. The exergy efficiency drops to 23.8%. Hence the importance of using the hydrogen which has not been converted in the cell.

Variation of the performance as a function of the current intensity

Figure 18.14 shows the variation of the cell performance as a function of the current intensity: electric power output, conversion efficiency and exergy efficiency.

The electric power output is an increasing function of the current intensity, but the counterpart is a drop in the values of the two other indicators.

Coupling SOFC fuel cell with a gas turbine

It is possible to model the coupling of a SOFC fuel cell with a gas turbine. An example is given in the diagram of Figure 18.15.

Air and hydrogen are initially at 15°C and 1 bar. Their respective flow rates are 300 and 4 g/s (we have greatly increased the airflow firstly to have enough oxygen in the depleted air to burn all the residual hydrogen at the stack output, and secondly to limit the turbine inlet temperature at a reasonable value.

Air and hydrogen are compressed at 20 bar by compressors of isentropic efficiency 0.85 and then preheated before entering the cell by exchange with gas leaving the anode and cathode. 43 kW of heat is provided by the cell to produce steam at 40 bar and 500°C.

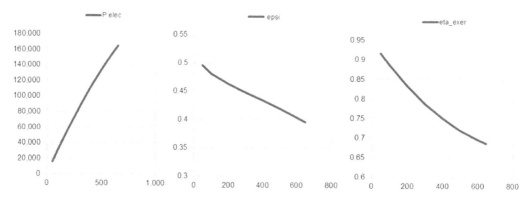

FIGURE 18.14

Performance as a function of the current intensity.

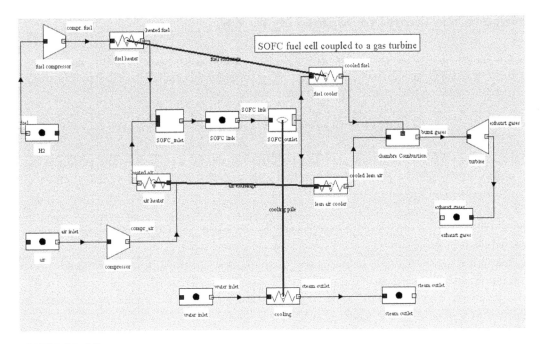

FIGURE 18.15

Thermoptim diagram of the facility.

Depleted air and humidified hydrogen are then burnt in the combustion chamber and flue gases expanded in the turbine of isentropic efficiency 0.9.

Figure 18.16 shows the synoptic view of the facility for such a simulation.

While the cell efficiency is only equal to 44%, the electrical efficiency of this facility is already equal to 45.4%, and in addition, 43 kW of steam and approximately 180 kW thermal are available in the exhaust gas exiting the turbine at about 600°C.

The available cogeneration power output is equal to 230 kWe, plus 223 kW thermal energy at a cost of 506 kW. The overall efficiency is therefore equal to 89.5%.

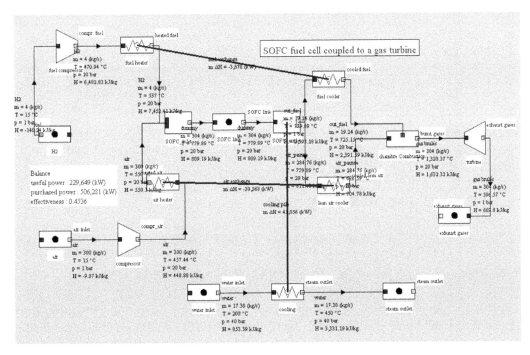

FIGURE 18.16

Synoptic view of the plant.

Change in the model to replace H$_2$ by CH$_4$

The models we have developed so far only allowed for pure hydrogen as fuel. It is actually very rare that one has hydrogen, which must be produced from another fuel. This is called reforming.

The basic equation for reforming a fuel such as methane is

$$CH_4 + H_2O \rightarrow CO + 3H_2 \left(\Delta H = 206,140\,kJ/kmol\right) \tag{18.7}$$

Reforming converts fuel into hydrogen but requires water and heat supply and produces carbon monoxide, which is a poison for some fuel cells such as PEMFC, where it is then necessary to also convert CO.

Note that one advantage of SOFC technology is that it tolerates the presence of CO well, while in other cases, a shift reaction is necessary to convert it into CO_2. The energy efficiency of the whole is improved.

Let us call α the molar ratio of water and methane flows (α must be greater than 1 so that all methane can be processed).

At the anode, given the high operating temperature of SOFC, we can consider that the whole fuel is converted by the cracking reaction

$$CH_4 + \alpha H_2O \rightarrow CO + 3H_2 + (\alpha-1)H_2O \left(\Delta H_v = 206,140\,kJ/kmol\right) \tag{18.8}$$

Then, only a fraction τ (rate of fuel use) is transformed in the cell, the rest coming out of it. The overall reaction giving the output species is written:

For the part used:

$$\tau\left(CO + 3H_2 + 2O_2 + (\alpha-1)H_2O\right)$$

$$\rightarrow \tau\left(CO_2 + (\alpha+2)H_2O\right) \left(\Delta H_0 = -1,008,450\tau\,kJ/kmol\right) \tag{18.9}$$

Heat of reaction ΔH_0 is calculated here by considering that water remains in the gaseous state due to the temperature (LHV).

For the unused part:

$$(1-\tau)\left(CO + 3H_2 + (\alpha-1)H_2O\right) \tag{18.10}$$

We therefore get at the anode outlet:

$$(1-\tau)CO + \tau CO_2 + 3(1-\tau)H_2 + (3\tau+\alpha-1)H_2O \tag{18.11}$$

Of the amount used, a fraction ε is directly converted into electricity, and $(1-\varepsilon)$ is transformed into heat (some of which is used for steam cracking).

Moreover, λ being a parameter representing the incoming air, oxygen is removed from the oxidizer air at the cathode:

$$Intake\,air: \lambda\left(O_2 + 3.76N_2\right) \tag{18.12}$$

$$depleted\,air\,exiting: \left(\lambda - 2\tau\right)O_2 + 3.76\lambda N_2 \tag{18.13}$$

The model retained is the following (classes SOFCCH4inlet and SOFCCH4outlet):

* Species composition is given by solving the equations above: we determine the molar flow rates of fuel and humidified air at the inlet, which provides values for α and λ, we deduce the molar flow at the output, the values of τ and ε being read on the screen;
* Heat released by fraction $\tau(1-\varepsilon)$ of the fuel is used to provide the energy needed to heat gas and by steam cracking.

The enthalpy released is equal to $\tau\Delta H_0$. It is divided into electricity ($\varepsilon\tau\Delta H_0$) and heat required for steam cracking (ΔH_v) and heating of gas ($\tau(1-\varepsilon)\Delta H_0 - \Delta H_v$).

Figure 18.17 shows the synoptic view of the fuel cell. The settings used are similar to that of the hydrogen-powered cell model: inlet gas temperature 500°C, flow rate 10 g/s for fuel, and 80 g/s for air.

Figure 18.18 shows the upstream mixer screen where appear the settings of the electric model, the fuel utilization rate and the fraction of the thermal power extracted by the thermocoupler.

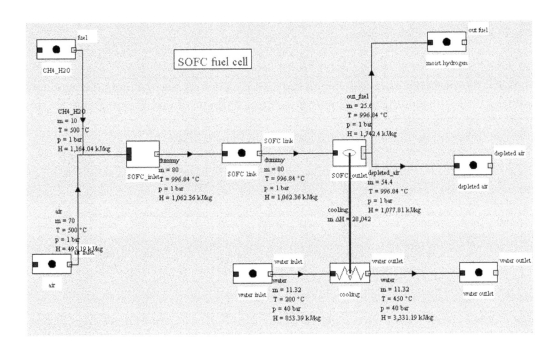

FIGURE 18.17

Synoptic view of the SOFC cell.

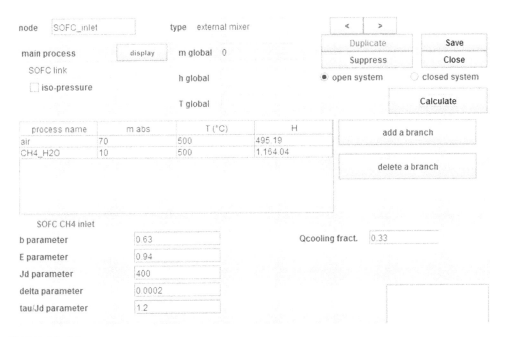

FIGURE 18.18

Screen of the SOFC upstream mixer.

BOX 18.1

Worked example

Diapason sessions on fuel cells

A series of sessions (S61–65) has been prepared to enable students to become familiar with the operation and modeling of fuel cells.

In session S61, we study a SOFC fuel cell, fueled by pure hydrogen, using the simple two-parameter model. In session S62, the previous model is progressively refined, first by taking into account the equation of polarization of the cell and then introducing a cooling of the stack. Finally, an exercise shows how to couple the battery to a gas turbine to form a high-efficiency cogeneration installation.

In session S63, we see how to modify the previously established models to replace hydrogen with a fuel such as methane. Session S64 deals with reforming, and session S65 models a PEMFC fuel cell.

The SOFC component screen is shown in Figure 18.19. We have taken a fuel utilization rate of almost 85%, corresponding to the figure announced by Siemens and Westinghouse for this type of cell operating with natural gas.

The gas compositions that are obtained are given in Figures 18.20–18.22.

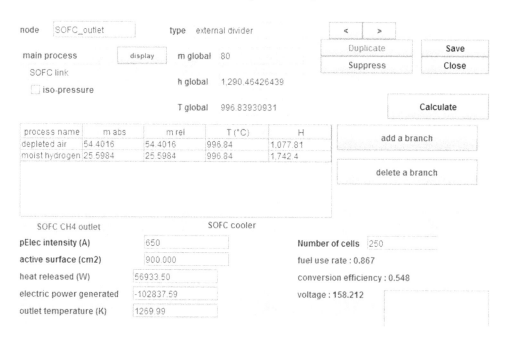

FIGURE 18.19

SOFC component screen.

component name	molar fraction	mass fraction
H2O	0.52	0.5488451
CH4 ` methane	0.48	0.4511549

FIGURE 18.20

Fuel, flow rate 10 g/s, LHV: 22,562 kJ/kg.

component name	molar fraction	mass fraction
CO	0.03265306	0.0410301
CO2	0.2122449	0.4190309
H2	0.09795918	0.008858684
H2O	0.6571429	0.5310803

FIGURE 18.21

Fuel at the outlet, flow rate 25.6 g/s, LHV: 1,477 kJ/kg.

component name	molar fraction	mass fraction
N2	0.9782662	0.9721497
Ar	0.01127323	0.01597628
O2	0.01046053	0.01187401

FIGURE 18.22

O_2-depleted air, flow rate 64.4 g/s.

Reforming

In the SOFC, we have taken into account reforming by a simple equation, but for other cases such as PEMFC, it is necessary to perform the reforming through a specific component before entering the cell. That is what we study in this section.

Modeling of a reformer in thermoptim

For simplicity, we assume in this example that natural gas contains only methane, ethane, carbon dioxide and nitrogen. Taking into account other reactive or inert components poses no particular problem, if one assumes that the other reactants react completely as ethane.

The reformer receives humidified natural gas, which reacts according to the three reactions:

$$CH_4 + H_2O \rightarrow CO + 3H_2 \ \left(\Delta H = 206,140 \, kJ/kmol \right) \tag{18.14}$$

$$C_2H_6 + 2H_2O \rightarrow 2CO + 5H_2 \ \left(\Delta H = 347,000 \, kJ/kmol \right) \tag{18.15}$$

$$CO + H_2O \leftrightarrow CO_2 + H_2 \ \left(\Delta H = -41,200 \, kJ/kmol \right) \tag{18.16}$$

Ethane reaction can be considered complete, while the other two are in equilibrium, their constants being given by

$$\ln\left(K_{P_{1a}}\right) = 31.0152 - \frac{28,357.7}{T} + \frac{610,573}{T^2} \tag{18.17}$$

$$\ln\left(K_{P_2}\right) = -3.57414 - \frac{3,642.48}{T} + \frac{292,593}{T^2} \tag{18.18}$$

Calling $F°$ the molar flow of natural gas, x the molar flow of CO from reaction (18.14), y the molar flow rate of CO_2 from (18.16), w/2 the molar flow of CO from reaction (18.15), and a and d the mole fractions of CH_4 and CO_2 in the natural gas and SC the number of moles of H_2O per mole of natural gas, we can write

$$F_{CO} = x + 2w - y$$

$$F_{CO_2} = y + d.F°$$

$$F_{H_2} = 3x + 5w + y$$

$$F_{CH_4} = a.F° - x$$

$$F_{H_2O} = SC.F° - x - 2w - y$$

$$F_{tot} = (1 + SC).F° + 2x + 4w$$

$$K_{1a}(T) = P^2 . \left(F_{H_2}{}^3 . F_{CO} \right) \Big/ \left(F_{CH_4} . F_{H_2O} . F_{tot}{}^2 \right) \tag{18.19}$$

$$K_2(T) = \left(F_{CO_2}.F_{H_2}\right) \Big/ \left(F_{CO}.F_{H_2O}\right) \tag{18.20}$$

The two equilibrium Equations (18.19) and (18.20) form a nonlinear system difficult to solve. However, it can be noted that reaction (18.20) is a quadratic form in x and y, which allows one of these two variables to be formally expressed as the solution of an equation of the second degree of the other.

If x is expressed in terms of y this way and that the value obtained is reinjected into the other equation, we obtain an implicit formulation in y, which can easily be solved numerically. Only one of the two solutions of the quadratic equation must be chosen, and equation giving y has only one real positive solution. The search range of its value is not easy to find, especially because of the existence of a singularity in the function, corresponding to the zero of F_{CH4}.

x being an increasing function of y, one solution is to seek an upper interval bound y_{max} by solving equation $y = f(aF^\circ)$ solution of (18.20). The lower bound can be chosen proportionally to y_{max}: $y_{min} = 0.9 y_{max}$. The solution of (18.19) is then easily obtained by dichotomy between these two bounds.

The model that can be retained under these conditions is as follows:

* Species composition is given by solving the equations above: from the molar flow of humidified fuel input, we determine the compositions and molar flow rates of different species at the outlet, the only parameter to consider being the reactor temperature;
* The heat required by the reaction can then be calculated and set by a thermocoupler;
* This heat is supplied by a combustion chamber located directly upstream of the thermocoupler.

In Thermoptim, the reformer is represented by an external process connected to a thermocoupler. Its class is Reformer (Figure 18.23).

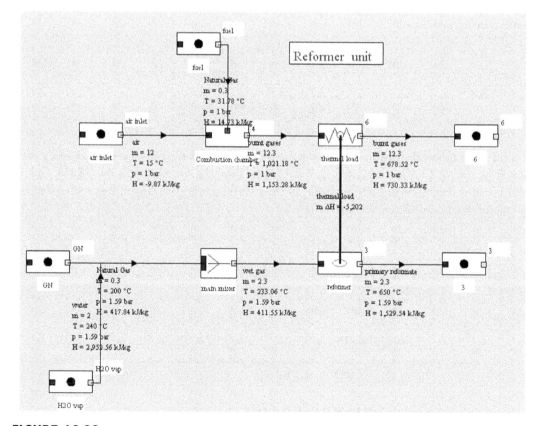

FIGURE 18.23
Diagram of the reformer.

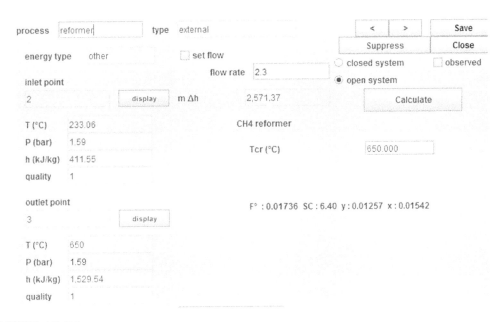

process reformer   type external

energy type   other   ☐ set flow

inlet point   flow rate 2.3

2   display   m Δh   2,571.37

T (°C)   233.06   CH4 reformer

P (bar)   1.59

h (kJ/kg) 411.55   Tcr (°C)   650.000

quality   1

outlet point   F° : 0.01736 SC : 6.40 y : 0.01257 x : 0.01542

3   display

T (°C)   650

P (bar)   1.59

h (kJ/kg) 1,529.54

quality   1

< > Save
Suppress Close
○ closed system   ☐ observed
● open system
Calculate

FIGURE 18.24

Screen of the reformer component.

name thermal load   type counterflow

thermal load

thermal fluid   process

thermal load   display   reformer   display   Calculate

Ti   1,021.17771318   Ti   233.06492863

To   678.51990757 ● calculated   To   650   ○ pinch method fluid
                                            ●

m   12.3   ○ calculated   m   2.3   minimum pinch   10

Cp   1.23432943   Cp   12.47752291

m ΔH -5,202.31714381   m ΔH 5,202.31714381

                            UA   9.96622693
● calculate exchange   R   0.529029613
                            NTU 0.656439305
                            LMTD 407.18786307
epsilon 0.434782702

< > Save
Suppress Close

FIGURE 18.25

Thermocoupler screen.

The screens of the reformer component and its thermocoupler are given in Figures 18.24 and 18.25.

Inlet flows are determined by components located upstream of the process and the combustion chamber. As they are expressed in mass units, the molar flow rate F°, the value of SC and the values of x and y are displayed on the process screen.

The thermal power supplied by the thermocoupler is determined by the process.

Results

Gas compositions that are obtained are given in Figures 18.26 and 18.27.

component name	molar fraction	mass fraction
CH4 ` méthane	0.1243921	0.1113837
C2H6 ` éthane	0.008112529	0.01361569
CO2	0.001352088	0.003321271
N2	0.001352088	0.002114077
H2O	0.8647912	0.8695652

FIGURE 18.26

Humidified natural gas, LHV: 6,217 kJ/kg.

component name	molar fraction	mass fraction
CO	0.03018695	0.06006241
CO2	0.07799918	0.2438392
H2	0.3919335	0.05612304
CH4 ` méthane	0.003366579	0.003836472
H2O	0.4954514	0.6340248
N2	0.001062407	0.002114077

FIGURE 18.27

Primary reformate, LHV: 7,531 kJ/kg.

Electrolyzers

An electrolyzer performs an electrochemical reaction that produces hydrogen and oxygen from water through electricity consumption by the endothermic reaction:

$$H_2O \rightarrow H_2 + \tfrac{1}{2}O_2$$

An electrolyzer is composed of two electrodes, the anode and cathode, separated by an electrolyte.

Technologies

Electrolyzers are currently used to produce pure hydrogen for a number of industries such as electronics, pharmaceutical and food. They are however expensive as they require electricity. Electrolysis also produces commercially valuable oxygen gas which has many industrial applications. Technologies vary in the operating temperature and the electrolyte.

Alkaline process was first developed and is the most mature method of hydrogen production with the caveat of its poor performance due to low energy conversion efficiency and high energy consumption.

Alkaline cells operate with aqueous electrolytes containing approximately 30% KOH, at operating temperatures between 70°C–90°C and pressure of approximately 30 bar.

Recently, research and development has focused on polymer electrolyte membrane (PEM) cells using a solid polymer electrolyte which is more compact and provides higher efficiency and current density.

However, at low temperature, the performance of electrolyzers is quite low due to thermodynamic barriers. The interest of carrying out electrolysis at high temperature is twofold: firstly, the total energy to be supplied to dissociate the water in the gaseous state is lower (246 kJ/mol) than in the liquid (285 kJ/mol), and secondly, a fraction of this energy can be supplied in the form of heat, which is less expensive than electricity.

The production of hydrogen by electrolysis at high temperature thus appears to be a promising path, so that much research is underway on this subject. Current interest in the solid oxide electrolysis cell (SOEC) electrolyzers are based on solid oxide ionic conductor technology, such as that used for SOFCs presented above.

Modeling a high-temperature electrolyzer in thermoptim

We will content ourselves here with presenting a fairly simple model, corresponding to a technology presented in Rodriguez & Pinteaux (2003).

High-temperature electrolysis (HTE) is an advanced process which performs the electrolysis reaction in the gas phase, i.e., from steam, at a temperature between 750°C and 950°C.

On leaving the electrolyzer, the hydrogen mixed with steam is collected. The separation does not pose any particular problem: it suffices to condense the water to obtain hydrogen, as pure as the steam is.

A high-temperature electrolyzer receives a water and hydrogen mixture (80%–20% molar composition) at 30 bar and 800°C. It releases two fluids: oxygen and the initial mixture enriched in hydrogen.

The Thermoptim model involves a conventional mixer upstream used to create the mixture and an external divider downstream, the two being connected by a process-point playing a passive role. The model structure is given in Figures 18.28 and 18.29.

The model we develop (class HTE) is very simple: it takes as parameters the molar fraction of hydrogen α at the outlet or the electricity supplied ΔH, and the conversion ratio in kWh/Nm³, which expresses the amount of electricity to provide for electrolyzing 1 Nm³ of hydrogen. This conversion ratio varies from 2.6 kWh/Nm³ in HT electrolyzers to 5 kWh/Nm³ in low-temperature ones.

The balance of species is easy to establish. Electricity provided electrolyzes water and produces hydrogen and oxygen and is partly converted into heat, which raises the temperature of the device.

Two calculation methods are possible:

- Either determine ΔH knowing α and the conversion ratio in kWh/Nm³;
- Or determine α knowing ΔH and the conversion ratio in kWh/Nm³.

In the first mode, the calculation is as follows: the model determines the inlet molar fraction and molar flow rate from the external divider main flow, deduces the molar flow of hydrogen electrolyzed and establishes its balance.

Let us call β the inlet mole fraction of hydrogen and $\dot{m}_{mol}$ the inlet total molar flow rate.

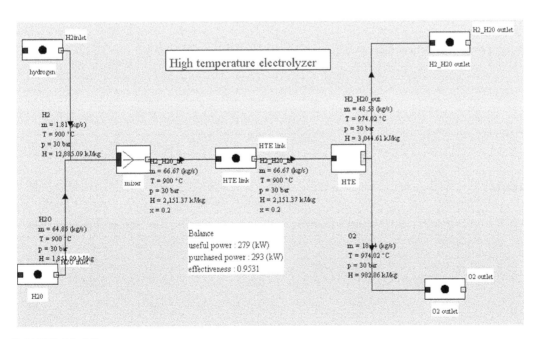

FIGURE 18.28
Diagram of HTE component with upstream and downstream connections.

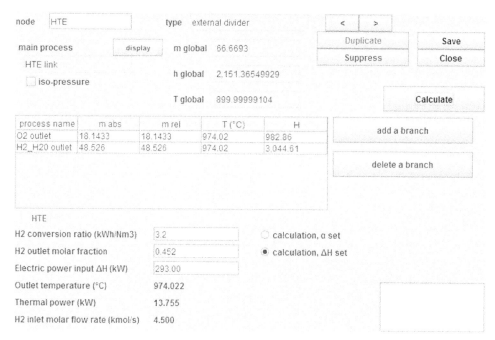

FIGURE 18.29

Screen of HTE component.

The molar flow of hydrogen electrolyzed $\dot{m}_{H_2}$ is

$$\dot{m}_{H_2} = (\alpha - \beta)\dot{m}_{mol}$$

The composition of the mixture $[H_2]/[H_2O]$ being known, its molecular weight can be determined. As its molar flow rate is equal to the inlet molar flow rate, the mass flow is known.

The molar flow rate of oxygen is $\dot{m}_{H_2}/2$. Its mass flow rate is deduced directly.

The conversion ratio in kWh/Nm^3 allows us to calculate the amount of electricity to be supplied, ΔH, by

$$\Delta H = \dot{m}_{H_2} M_{H_2} \frac{3,600}{1,000 \cdot 0.08988}$$

ΔH_0 being the theoretical heat of reaction (246 kJ/mol), heat released is

$$Q = \Delta H - \dot{m}_{mol}\Delta H_0$$

In the second mode, the sequence of calculations is slightly modified: knowing ΔH, $\dot{m}_{H_2}$ can be determined. We deduce α. Q is determined in the same way.

Results

With the settings above (0.9 kmol/s of hydrogen and 3.6 kmol/s of water at the electrolyzer inlet), gas compositions obtained are given in Figures 18.30 and 18.31:

component name	molar fraction	mass fraction
H2	0.2	0.02721331
H2O	0.8	0.9727867

FIGURE 18.30

Humidified hydrogen at inlet, flow rate 66.67 kg/s, LHV: 3,264 kJ/kg.

component name	molar fraction	mass fraction
H2	0.4520003	0.08449709
H2O	0.5479997	0.9155029

FIGURE 18.31

Humidified hydrogen at outlet, flow rate 48.53 kg/s, LHV: 10,136 kJ.

TABLE 18.3

EXERGY BALANCE OF THE HT ELECTROLYZER

Component	Resource	Product	Exergy Efficiency (%)	Irreversibilities (%)	% Total (%)	$T_0 = 283.15$ K
H_2O	97,248					
Hydrogen	277,265					
the	293,000		95.3	13,755	47.9	
$H_2\_H_2O$ outlet	744,938	744,938				
Mixer	97,248	374,513	100.0			
O_2 outlet				14,952	52.1	Loss
Global	667,513	638,806	92.6	28,707	100.0	

This model can be used to represent an electrolyzer integrated into a system, and particularly to study the thermal integration of the process, in order to minimize the external heat input.

Rodriguez and Pinteaux present the coupling of this electrolyzer with a GT-MHR cycle of the type studied in Chapter 15. Overall efficiencies ranging from 51% to 56% are expected.

Table 18.3 gives the exergy balance of this electrolyzer. It receives in exergy resources 97.2 MW of steam (H_2O) and 293 kW of electricity. The only exergy destroyed corresponds to the heat released by the reaction (13.8 kW), and we considered that the flow of oxygen is not valued (15 kW).

The exergy efficiency is quite high, but the model is very simple. In particular, it does not take into account the polarization curve of the electrolyzer nor the current density. To refine the model, it would be necessary to operate in a manner analogous to what we did for fuel cells.

Bibliography

D. Candusso, R. Glises, D. Hissel, J.-M. Kauffmann, M.-C., Pera Piles à combustible PEMFC et SOFC : Transferts de chaleur et de masse, Techniques de l'Ingénieur, Traité Génie énergétique, BE 8 596.

D. Candusso, R. Glises, D. Hissel, J.-M. Kauffmann, M.-C. Pera, Piles à combustible PEMFC et SOFC : Description et gestion du système, Techniques de l'Ingénieur, Traité Génie énergétique, BE 8 595.

G. Hoogers, Ed., *Fuel cell technology handbook*, 2nd edition, CRC Press, Boca Raton, FL, 2002, ISBN 9780849308772.

C.-E. Hubert, Etude du fonctionnement et optimisation de la conception d'un système pile à combustible PEM exploité en cogénération dans le bâtiment, Thèse de Doctorat, École des Mines de Paris, 2005.

S. Kuai, J. Meng, Eds., *Electrolysis: Theory, types and applications, Chemistry Research and Applications*, Nova Science Publishers, 2010, ISBN 978-1-60876-619-2.

G. Rodriguez, T. Pinteaux, Studies and design of several scenarios for large production of hydrogen by coupling a high-temperature reactor with steam electrolysers, *Proceedings of the first EHEC*, 2003, Grenoble.

H. Struchtrup, *Thermodynamics and energy conversion*, Springer, Berlin, Heidelberg, 2014, ISBN 978-3-662-43714-8.

B. Viswanathan, M. Aulice Scibioh, *Fuel cells: Principles and applications*, CRC Press, Boca Raton, FL, 2008, ISBN 9781420060287.

F. Werkoff, A. Marechal, F. Pra, Technico-economic study on the production of hydrogen by high-temperature steam electrolysis. CEA *Nouvelles technologies de l'énergie*. Rapport d'activités, 2001–2003. Document Internet: www.cea.fr

General conclusion

We hope this book has met your expectations and given you a better understanding of how energy systems work and how they can be modeled.

By way of conclusion, you will find in this brief section a reminder of the approach we have adopted and some avenues for further study.

Summation of the approach presented

This book offers an approach to thermodynamics applied to energy systems very different from those generally adopted and much better suited for learners who do not have a significant background in mathematics and physics.

Theoretical developments are deliberately limited to the strict minimum, in favor of technological explanations and studies of realistic models solved by using a software environment that is both powerful and relatively easy to use, which allows extremely precise results to be obtained.

The examples are treated with the Thermoptim software package, due to the fact that it is available and that its free demonstration version is sufficient to model them, but similar models could be carried out with other tools without the relevance of our approach being called into question.

Throughout the book, we have given detailed technological explanations on the functioning of machines, because it seems fundamental to us that learners understand the existing technical constraints, otherwise they risk reasoning in a way disconnected from reality.

In the first part, we sought to make thermodynamics accessible to as many people as possible by introducing only a very small number of abstract concepts and by showing how very different systems can be analyzed in a similar way once the few elementary functions provided by their components and the corresponding reference processes are identified.

This minimalist approach in terms of cognitive load has the double advantage of being at the same time generic and therefore generalizable and very simple to implement, which reassures those who are discouraged by theoretical developments.

Once they realize that they can easily move on to practice, they feel confident and are more willing to invest in the discipline.

The thirty-five self-assessment exercises offered allow everyone to test their understanding of the key concepts introduced in this part.

Learners can furthermore practice through guided explorations using a dedicated browser capable of emulating the software package.

The second part then made it possible to introduce various theoretical and technological complements necessary to study cycles more complex than the three simple examples discussed previously.

It also provided the opportunity to present the powerful methods developed over the past few decades, such as the pinch method and the establishment of exergy balances.

Here too, our approach contrasts with those that are usually used.

The use of productive structures makes it possible to very largely automate the construction of exergy balances, thereby considerably limiting the risks of error.

It was then possible in the third part to show the learners how they can use these powerful tools to improve the performance of simple systems. The main classical cycles were successively studied there, always illustrated by numerous models solved with Thermoptim and about twenty guided explorations.

These constitute a suite allowing learners to initiate themselves into modeling by studying systems of increasing complexity, thus gradually discovering the existing difficulties: combustion calculations, parameterization of heat exchangers, dividers and phase separators...

Finally, the fourth part provided learners with the opportunity to look at many innovative cycles with low environmental impact, usable with a wide variety of energy sources and for multiple applications.

We hope that these examples will enable readers to realize the range of cycles that can be envisioned and will foster their creativity.

Future developments will result from knowledge of both technological constraints and possible thermodynamic cycles.

The advantage of having a tool like Thermoptim is in particular being able to carry out extensive modeling of such new cycles.

The whole book covers a very wide range of fields: 130 realistic cycle models are presented... Many of them correspond to installations that have been the subject of publications, and comparison is made with the published results.

Avenues for further study

Our bias for simplicity has led us not to tackle some very interesting subjects such as for example the calculation of the properties of fluids, the simulation of solar installations, the off-design operation of systems...

You will find in the Thermoptim-UNIT portal (www.thermoptim.org) many additional explanations on these subjects and digital resources to download.

The portal has six main sections: thermodynamics fundamentals, methodological guides, technologies, global issues, education and software.

Do not hesitate to consult it to expand your knowledge of all these subjects.

As indicated in the introduction, we did not deal in this book with the off-design simulation of energy systems, which had been the subject of the whole of part 5 of the first edition (about hundred pages). The two guided explorations presented below will introduce you to these questions if they interest you.

BOX 19.1
Guided educational exploration

Sizing of heat exchangers and exergy balances of a heat network (DTNN-2)

The objective of this exploration is to show you, in a simple example, how to determine the exchange surfaces of the heating network of exploration OPT-1, by using the technological sizing tools of Thermoptim.

Finally, you will study how to obtain the exergy balances of the initial network and the optimized one using their productive structures.

BOX 19.2
Guided educational exploration

Sizing of a displacement air compressor (exploration DTNN-3)

In this guided exploration on positive displacement compressors, we detail the calculation and sizing of a piston compressor used to supply a compressed air storage tank.

You will also find in the Thermoptim-UNIT portal pages indicating where to download Thermoptim, the software documentation, pages on pedagogy...

If you want to get in touch, you can send an e-mail at this address: info@thermoptim.org.

Index